CHEMICAL BIOLOGY

Edited by **Deniz Ekinci**

Chemical Biology
Edited by Deniz Ekinci

Published by InTech
Janeza Trdine 9, 51000 Rijeka, Croatia

Publishing Process Manager Masa Vidovic

Technical Editor Teodora Smiljanic

Cover Designer InTech Design Team

Additional hard copies can be obtained from orders@intechopen.com

Chemical Biology, Edited by Deniz Ekinci
p. cm.
ISBN 978-953-51-0049-2

For EU product safety concerns:
IN TECH d.o.o., Prolaz Marije Krucifikse Kozulić 3, 51000 Rijeka, Croatia, info@intechopen.com or visit our website at intechopen.com.

Meet the editor

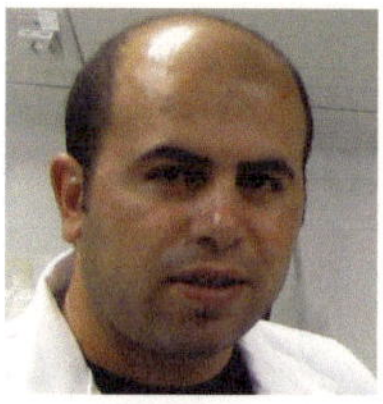

Dr Deniz Ekinci obtained a BSc in Chemistry in 2004, an MSC in Biochemistry in 2006 and a PhD in Biochemistry in 2009 from Atatürk University, Turkey. He studied at Stetson University, USA, in 2007-2008 and at the Max Planck Institute of Molecular Cell Biology and Genetics, Germany, in 2009-2010. Dr Ekinci currently works as an Assistant Professor of Biochemistry in the Faculty of Agriculture, and is the Head of the Enzyme and Microbial Biotechnology Division, Ondokuz Mayıs University, Turkey. He is a member of Turkish Biochemical Society, American Chemical Society and German Genetics society. Dr Ekinci published over fifty scientific papers, reviews and book chapters and presented several conferences to scientists. He has received numerous publication awards from several scientific councils. Dr Ekinci serves as the Editor in Chief of three international books and is involved in the Editorial Board of several international journals.

Contents

Preface

Chemical biology, a scientific discipline connecting the fields of chemistry and biology, refers to the application of chemical techniques and tools, synthetic chemical compounds, to the study and manipulation of biological systems. Chemical biology utilizes chemical principles to modulate systems to either investigate the underlying biology or create new function. The research area of chemical biologists is very close to that of cell biology and pharmacology.

This book aims to provide a brief overview on chemistry and biology of the biomolecules, while focusing on connection of them in chemical biology. The first section opens up this book with an overview of protein purification and enzymology by introducing some important concepts. The second section addresses properties and effects of vitamins and antioxidants. The third section deals with biotransformation and gene delivery, quite popular areas. The last two sections combine signaling, regulation and organization.

The collection of contributions from outstanding scientists and experts from different countries presents detailed information about different topics and provides a general overview about the current issues of chemical biology. The book covers a wide range of current issues, reflecting on actual problems and demonstrating the complexity of the chemical-biological interactions. I would like to tell my gratefulness to the authors around the world for their excellent contributions to the book and hope that the book will enhance the knowledge of scientists in the complexities of chemical and biological approaches; it will stimulate both professionals and students to dedicate part of their future research in understanding relevant mechanisms and applications of chemical biology.

Dr. Deniz Ekinci
Assistant Professor of Biochemistry
Ondokuz Mayıs University
Turkey

Part 1

Protein Purification & Enzymology

1

Protein Purification

Nison Sattayasai
Khon Kaen University
Thailand

1. Introduction

There are many thousand kinds of proteins with different properties and functions in a cell. To study a protein, one has to obtain the highly purified intact form of the protein. A major protein is not so difficult to be purified while a minor one may need many purification steps and high skills on the techniques. This chapter describes the principles underlying techniques for separation and purification of proteins extracted from cells and tissues. The strategy involves extraction and purification.

2. Extraction of protein

Generally both extraction and purification processes should be done under a cold condition, mostly at 0-4°C, except for some proteins. An ice box or another cooling system is normally needed for sample cooling. The extraction procedure varies according to the type of sample and physicochemical properties of the protein. The first step is to disrupt cells or tissues. The more gentle procedure is used the more intact protein is obtained. Most of animal cells and tissues are soft and easy to break, so gentle methods can be applied. For bacteria, fungi and most plant cells which have tough cell envelope, more vigorous procedures are required. Extraction buffer or solvent is also important. Buffer with suitable concentration, ionic strength, and pH is used for extraction of water-soluble protein. In some case, mild detergent or other appropriate dissociating substance is added to the extraction buffer. Protease inhibitors are sometimes necessary to prevent destruction of the extracted proteins by proteases. From the fact that the protein extract obtained from subcellular component has contaminants less than that obtained from whole cell and will be easier to be purified. So separation of subcellular components may be done before the protein extraction. Disruption of cells should be adjusted to get intact sub-cellular components. The required cell component is obtained by centrifugation under appropriate condition.

2.1 Extraction of water-soluble protein from animal cells and tissues

The proteins which are components of fragile unicellular tissues such as animal blood cells can be extracted by using hypotonic buffer solution. If the sample contains different types of cells, separation of the cell types before the extraction will make the purification easier. This osmosis-based method is also used for animal cells grown in culture media. In some case a mild surfactant is included in the extraction buffer. Sonication or freeze/thaw cycle or a mild mechanical agitation may be used to help disruption of the cells and dissociation of the cellular components. For multicellular soft tissue, a hand homogenizer (Fig. 1) or an electrical one with optimal size is a good choice. The tiny pestle using with a micro-centrifuge tube is also commercially available for the sample with a small volume. Equipments with stronger breaking power such as a blade blender are needed for extraction of proteins from tougher tissue such as animal muscle. Cell disruption by nitrogen decompression is another method for animal cells and some plant cells. In this method, large quantities of nitrogen are dissolved in the cells under high pressure in a vessel. When the pressure is suddenly released, the dissolved nitrogen becomes bubbles. The expanding bubbles cause rupture of the cell. Intact organelles are also obtained by this method.

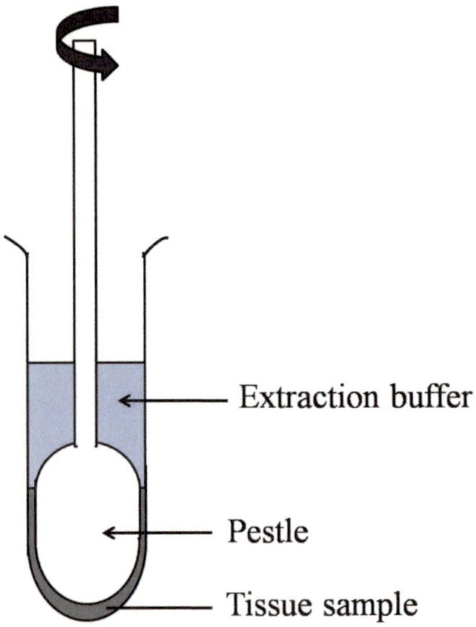

Fig. 1. Homogenization of tissue by a hand homogenizer.

2.2 Extraction of water-soluble protein from unicellular organisms

This group of organisms includes bacteria, yeast, fungi and some algae. Their cell envelopes are tougher than those of animal cells. The stronger methods are needed for disruption of the cells. A simple method is to shake the suspension of cells, as well as spores, with small glass beads or the other kind of beads. If the collision with the beads is done in a blender, the method is called a bead mill. Sonication is another way to break the cells. The solution used to suspend the cells can be only an appropriate buffer, or with the addition of some enzyme and/or mild detergent. The examples of the enzymes are lysozyme, cellulase and chitinase. Nonionic and zwitterionic detergents, especially Triton X-100 and CHAPS (3-[(3-cholamidopropyl) dimethylammonio]-1-propanesulfonate), are commonly used since they are mild and have little effects on protein denaturation. To disrupt the unicellular organisms with very strong envelope, a high-pressure press apparatus may be needed such as French press, Microfluidizer processors. In these systems, the cell suspension in a suitable chamber is pressurized (up to 30,000 psi) by using a hydraulic pump. Shearing force is generated when the pressurized suspension is squeezed pass a very narrow outlet into the atmospheric pressure (Walker, 2005).

2.3 Extraction of water-soluble protein from plant tissue

Grinding with or without sands, in the presence of an appropriate buffer, by using a pestle and mortar actually works well with plant sample. If the plant tissue is too strong to be ground, rapid freeze with liquid nitrogen will make the plant more fragile. The frozen plant tissue is ground before shaking in the buffer (Fido et al., 2004). Some plant cells can be disrupted by the mean of nitrogen decompression.

2.4 Extraction of lipid-soluble protein

Most of lipid-soluble proteins are membrane proteins. These proteins may be called proteolipids which are extracted from the samples by using organic solvents such as a mixture of chloroform and methanol (Velours et al., 1984). An aqueous solution containing mild detergents such as Triton X-100 and CHAPS is an alternative way to dissolve the proteins from cells or organelles. Strong detergents can be used but they may cause permanent denaturation of the proteins.

2.5 Extraction of aggregated protein

High expression of recombinant proteins in bacterial host cell often form insoluble aggregates which is known as inclusion bodies. This form of proteins is very difficult to be solubilized. To extract any protein from bacterial inclusion bodies, strong denaturants such as 6M guanidine-HCl or 6 to 8M urea are included in the solubilizing solution. This solution efficiently extracts the aggregated protein. However, the extracted protein is also denatured and sometimes can not be renatured (Rudolph & Lilie, 1996). The other procedure is to use a mixture of mild detergents such as a combination of Triton X-100, CHAPS and sarkosyl. This method is less efficiency but more native forms of the protein can be recovered (Tao et al., 2010).

3. Purification of protein

To purify any protein, various separation techniques are used depending on physical and chemical properties of the protein. The purification process can be concluded as follows.

Step 1. Crude extract of protein

⇩

Step 2. Detection of the desired protein — Detection by size, activity or property

⇩

Step 3. Separation into fractions
- Separation based on solubility
- Separation based on molecular size and density

⇩

Step 4. Detection of the desired protein — Detection by size, activity or property

⇩

Step 5. Separation for the purified protein
- Chromatographic techniques
- Other separation techniques

⇩

Step 6. Detection of the desired protein — Detection by size, activity or property

⇩

Step 7. Determination of the purity — Electrophoresis

It is not necessary to follow all steps or all techniques in the process for every protein. Some step or some technique is not necessary for some protein. For example, myoglobin is easily observed by its red color or detected by using UV-visible spectrometric method, no need to use any sophisticated procedure. Principles of separation techniques are as followings.

3.1 Separation based on solubility

Solubility of hydrophilic protein depends on its charge and hydrogen bonding with water molecule. Net charge of a protein is zero at its isoelectric pH (pI); so the protein molecule is easy to aggregate and then precipitate. The method is called isoelectric precipitation. If the pI is not known, high concentration of some salt can precipitate protein by destroying hydrogen bond between protein and water molecule. The method is called salting out or salt precipitation. Ammonium sulphate is normally used since it has high ionic strength. Proteins in crude extract can be separated into a number of fractions by gradually addition of the salt. The precipitated proteins are redissolved and used for further steps. Combination of the two methods is used for precipitation of some protein.

3.2 Separation based on size and density

The techniques include centrifugation, dialysis and molecular filtration. A high speed centrifugation (10,000-20,000g) is normally used to remove cell debris and large organelles from crude protein extract if the desired protein is a water-soluble molecule. The supernatant is then used for further purification steps. To know the location of the desired protein in cell, the cell disruption should be a soft procedure to get intact organelles and the other subcellular structures. The subcellular structures including organelles are then fractionated by using differential centrifugation and/or density gradient centrifugation. In differential centrifugation, the subcellular particles are separated according to their sizes into fractions by the stepwise increase of the centrifugal force. Density gradient centrifugation is a procedure to separate particles including protein aggregates by their densities. A liquid density gradient may be pre-formed before use for the separation. Various materials can be used to make the gradients such as sucrose, Ficoll. In addition the liquid density gradient can be formed during the separation. By this way, the sample is layer on top of a suitable concentration of cesium chloride. The strong centrifugation force of ultracentrifuge causes migration of cesium chloride into the bottom of the tube. The liquid density gradient is then formed by the gradient concentration of cesium chloride (Ohlendieck, 2005). The density gradient is also spontaneously generated during centrifugation by using Percoll (polymer-coated silica particle).

Dialysis is a widely used method. The method is simple, but time consuming because the separation depends on diffusion. The sample is placed inside a dialysis bag prepared from a tube made of semipermeable membrane. Rely on commercially available dialysis tube, only small molecules whose sizes less than 10 kDa is removed from the sample to the surrounding medium. So this technique is normally used to remove salts from the solution of protein. Dialysis is also used to concentrate protein solution. Water molecules are removed from the inside of dialysis bag by using a hydrophilic polymer such as polyethylene glycol.

Molecular filtration (or ultrafiltration) is similar to dialysis. A membrane with specific pore size is used to fractionate proteins. By using pressure or centrifugation force, only the molecules smaller than the pore pass through the membrane (Fig 2). There are various sizes of the membrane pore; the largest one can cut off the large molecule whose mass is almost 100 kDa. So proteins in the sample can be separated into various ranges of molecular masses by using the membranes with different pore sizes. The protein solution is also concentrated by this technique.

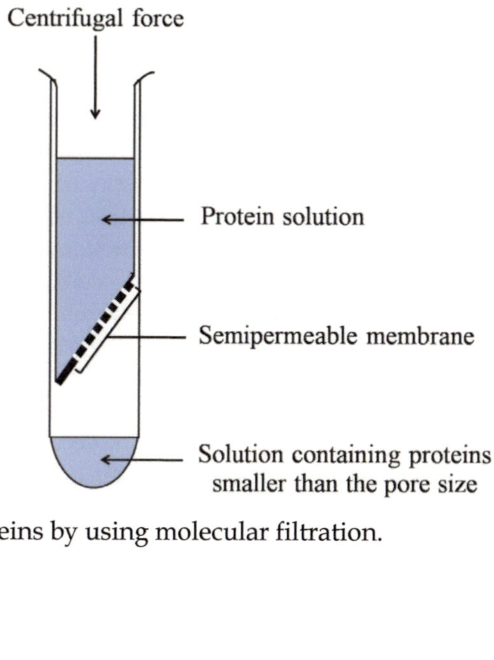

Fig. 2. Separation of proteins by using molecular filtration.

3.3 Chromatographic technique

Chromatography is the powerful method for detection and purification of biological substances. The principle of chromatographic separation is distribution or partition of separated molecules between two immiscible phases called mobile phase and stationary phase. Chromatographic methods are classified by different criteria including physical shape of stationary phase, nature of mobile phase and/or stationary phase, mechanism of separation or the other properties of the chromatographic systems. So the methods are called depending on their popularity. For examples, paper chromatography is called according to the material used as the stationary phase; thin layer chromatography (TLC) and column chromatography are named by the physical shapes of the stationary phases. Gas-liquid chromatography has gas mobile phase and liquid stationary phase. By using mechanisms of separation, chromatography is classified as adsorption chromatography, partition chromatography, size-exclusion chromatography, ion-exchange chromatography and affinity chromatography. Column chromatography is the most popular method for purification of proteins.

The conventional column chromatography is performed under low pressure. The mobile phase flows through the stationary phase in the column by the gravity or by low-pressure pump(s). So it can be called low pressure liquid chromatography (LPLC). High performance liquid chromatography (HPLC) is an advanced version of column chromatography. High pressure pumps, high-quality materials used as stationary phase and detectors with high sensitivity are used to improve the separation and to reduce the operating time. HPLC is now very popular for detection and purification of biological molecules. The other chromatographic methods are fast protein liquid chromatography (FPLC), capillary chromatography, reversed-phase liquid chromatography etc. FPLC is similar to HPLC but its operating pressure is lower than that of HPLC. LPLC, HPLC and FPLC are widely used for purification of proteins. In these methods, proteins are normally separated according to their size, charge or binding affinity. The separated proteins are eluted from the column as peaks which can be seen by various means such as absorption spectrometry, spectrofluorimetry (Wilson, 2005).

3.3.1 Size-exclusion chromatography

The other names of this method are gel filtration chromatography and gel permeation chromatography. This chromatographic technique also separates proteins by molecular sizes. Size-exclusion chromatography is the best to conserve native structure and function of the purified protein since wide varieties of buffers can be selected to obtain the suitable condition for the protein. The stationary phase of size-exclusion chromatography normally contains porous hydrophilic gel beads (Fig. 3). The gel beads used for LPLC are made of agarose, dextran, polyacrylamide, and chemical derivatives of these substances. The beads used for HPLC and FPLC are made of stronger materials such as cross-linked dextran and polystyrene. The cross-linked polystyrene can be used for separation of hydrophobic substances in the presence of organic solvents. The principle of the technique is the diffusion of molecules into the porous cavities of the beads. The molecules larger than the pores can not enter inside the beads where as the smaller ones can do. Since the pores have many sizes, the molecules including proteins are separated according to their molecular masses. The larger molecules pass the column faster than the smaller molecules (Fig. 4). Size-exclusion chromatography can also be used for estimation of molecular mass of protein. A set of suitable proteins with known molecular masses are separated by the same condition as the purified protein. The fraction numbers or elution volumes are then plotted against log molecular masses of the standard proteins. The standard curve is used to estimate molecular mass of the purified protein. To obtain the good result, one have to choose the right beads since different types of beads are suitable for different molecular masses of the proteins.

Fig. 3. A model demonstrates the pores in the hydrophilic gel bead used in size-exclusion chromatography.

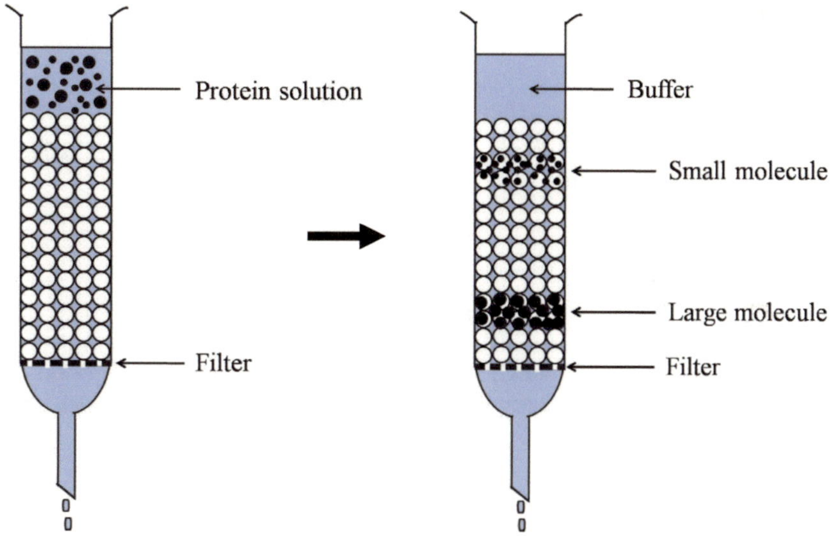

Fig. 4. Separation of two proteins by using size-exclusion chromatography. The protein mixture is loaded on the top of the gel. Then the large molecules pass the column faster than the small molecules.

3.3.2 Ion-exchange chromatography

This chromatographic technique conducts the separation according to magnitude of net electric charge of the proteins. There are two types: anion-exchange chromatography and cation-exchange chromatography. The material packed in the column is called ion exchanger which also have two types, anion and cation exchangers. Anion exchangers possess positively charged groups while cation exchangers have negatively charged groups.

Anion-exchange chromatography consists of anion exchanger and used for separation of proteins containing net negative charges. In contrast cation-exchange chromatography consists of cation exchanger and used for separation of proteins containing net positive charges. There are various varieties of ion exchangers depending on the matrix materials and ionic groups. Cellulose, agarose, dextran and polystyrene are used for LPLC while polystyrene, polyethers and silica are strong enough for HPLC. The matrix is chemically modified to contain ionic groups which are weakly acidic, strong acidic, weakly basic, or strong basic. The matrix with weakly acidic group (cation exchanger) as well as that with weakly basic group (anion exchanger) is suitable for separation and purification of most proteins. Choice of the exchanger depends on the pH range in which the protein is stable and the pI value of the exchanger. Generally the pH of the system should make opposite charges between the stationary phase and the protein (Fig. 5). Elution of proteins from the column can be isocratic or gradient system. The pH and ionic strength of elution buffer is important for getting well separation of proteins.

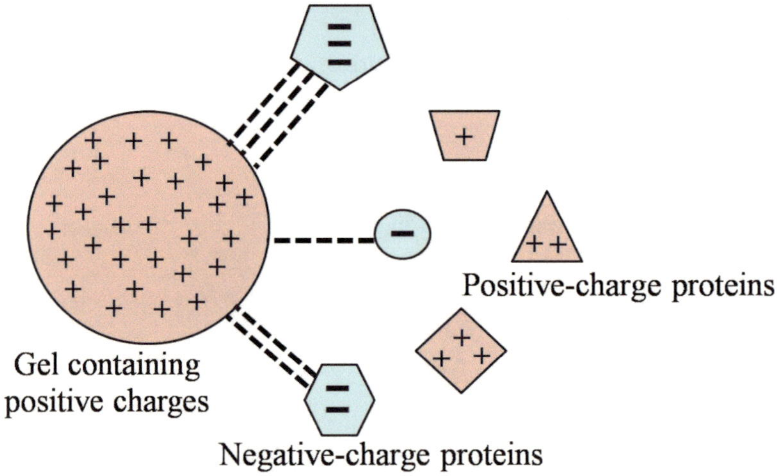

Fig. 5. Principle of ion-exchange chromatography. The anion-exchanger binds to negative-charge proteins but positive-charge proteins. The binding strength varies according to the charges of the proteins.

3.3.3 Affinity chromatography

This is an efficient technique but the material used as stationary phase is costly. The separation of molecules is based on binding affinity between macromolecules and macromolecules, or between macromolecules and low molecular mass ligands. The binding is specific for a certain molecule or a group of molecules. For examples, monoclonal antibody binds to its antigen only while avidin can bind to any biotin-coupled proteins. In this technique, one of the partner molecules is immobilized on a matrix by using a chemical reaction (Fig. 6). The matrix can be polyacrylamide, polystyrene, cross-linked dextrans and agarose etc. There are a number of chemical substances which react to different functional groups of the immobilized molecule. The chemical reaction must not destroy the binding activity. So it is necessary to use the right reaction for the immobilization. However, there are a lot of commercially available affinity matrices for different purposes such as protein A

Sepharose for purification of immunoglobulins, mannan-agarose for purification of mannose-binding lectins. (Fig. 7). After washing off the impurities, the matrix-bound protein is eluted by change of the pH of elution buffer, or by a competing substance. Affinity chromatography is always operated in a simple way by packing the matrix in a small column. Sophisticated system is not necessary for this technique.

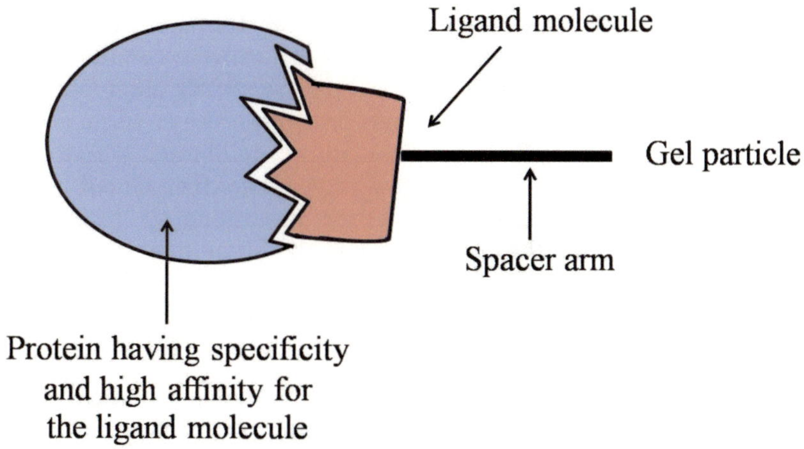

Fig. 6. Principle of affinity chromatography. The ligand is immobilized on a matrix and used for purification of its partner molecule.

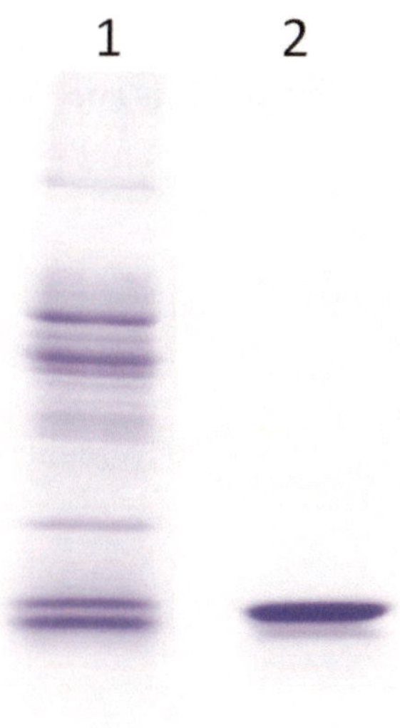

Fig. 7. One-step purification of mannose-binding lectin from *Dendrobium chrysotosum*. The affinity chromatography was performed by using mannan-agarose column. The crude extract (lane 1) and the purified lectin (lane 2) were separated on SDS-PAGE and then stained by Coomassie Brilliant Blue R250.

3.3.4 Hydrophobic interaction chromatography

This chromatographic method separates proteins on the basis of hydrophobicity, the same as reversed-phase liquid chromatography (RPLC). The stationary phases of both methods are hydrophobic ligands attached to a matrix. However, hydrophobicity and number of the ligands in hydrophobic interaction chromatography (HIC) is less than those in RPLC (Fig. 8). So HIC may be referred to as a milder form of RPLC. RPLC is mainly used for separation of peptides and low molecular mass proteins that are stable in aqueous-organic solvents. HIC is suitable for purification of proteins since it uses less extreme condition for elution of the adsorbed proteins. The hydrophobic ligands of HIC are mainly alkyl (ethyl to octyl) or phenyl or polyamide groups; the matrices are cross-linked agarose or silica. For the general procedure of HIC, the sample is applied onto the column equilibrated with a mobile phase of relatively high salt concentration. The adsorbed proteins are then eluted by a solvent of decreasing salt concentration. HIC is suitable for separation of proteins after salt precipitation and/or ion-exchange chromatography since the proteins already dissolves in the solution of high salt. Although HIC gives only moderate resolution, it opens new possibilities for purifying a number of biomolecules such as receptor proteins, membrane proteins (Wilson, 2005).

Which chromatographic procedure should be used for purification of a particular protein? Ion-exchange chromatography is usually the first chromatographic technique to be done for removal of unwanted proteins since the matrix is relatively cheaper and has higher binding capacity. Size-exclusion chromatography is usually operated after that. However the operation sequence can be reversed. One-step purification of protein from crude extract can be successful by using affinity chromatography (Fig. 7). Gel filtration chromatography or ion exchange chromatography is sometimes operated in accompany with affinity chromatography. Hydrophobic interaction chromatography (HIC) is another chromatographic techniques used for purification of some proteins.

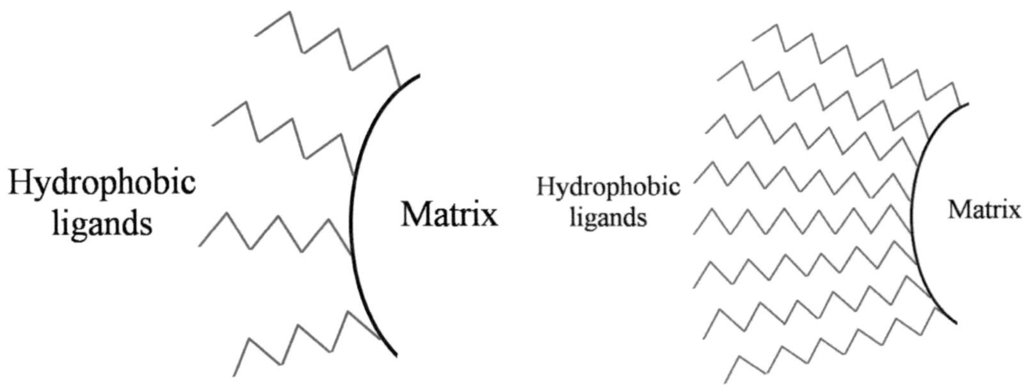

Fig. 8. Comparing the structures of materials used as stationary phases in hydrophobic interaction chromatography (A) and reversed-phase liquid chromatography (B).

3.4 Other separation techniques

There are some minor methods used for purification of protein such as electroelution of the protein separated by SDS-PAGE. The stained protein band is cut out of the gel. The gel slices with a suitable buffer are packed in a dialysis bag. The bag is placed in the buffer between two electrodes. The protein is then electrically eluted from the gel slices. One can make a simple electroelution cell or can take any commercially available apparatus. SDS is removed from the protein by dialysis in the presence of a nonionic detergent. Only a small amount of protein is obtained by electroelution and it is possibly denatured.

Most proteins have been normally purified by the above techniques. To obtained intact hydrophilic protein, the purification process mainly includes salting out and chromatographic technique(s). However some proteins can be purified by simply specific procedures. For an example, molecular filtration in combination with sodium chloride solution was successfully used for purification of glutamine synthetase (Fig. 9) (Arunchaipong et al., 2009). So the simply specific procedure may be made for purification of some protein if the protein has the specific property.

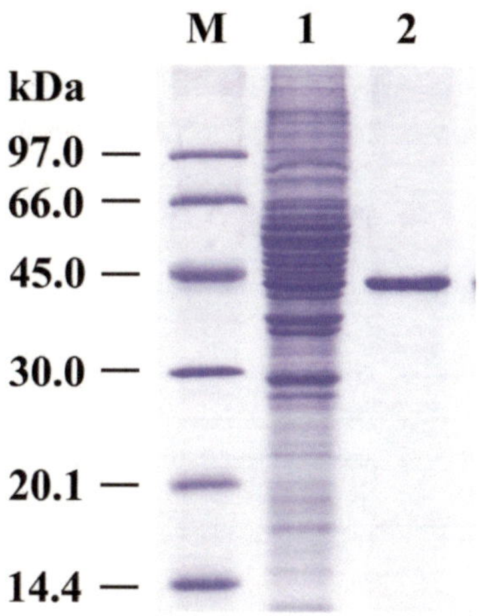

Fig. 9. Purification of glutamine synthetase from chick retina by using molecular filtration in the presence of 0.7M sodium chloride. The crude extract and the purified protein were separated on SDS-PAGE and then dye-stained. Lane M, molecular mass marker; lane 1, crude protein extract; lane2, purified enzyme.

4. Detection of the desired protein

Actually, purification process includes many steps and is laborious. Therefore it is necessary to be certain that the desired protein is present in the crude protein extract and all purification steps. For the protein with known molecular mass or size, the most versatile method is gel electrophoresis, especially SDS-polyacrylamide gel electrophoreis. Determination of enzymatic activities or other properties such as absorption spectra are also widely used. In some case the sophisticated technique, LC-MS/MS, is performed for identification of the protein after separation by electrophoresis.

4.1 Electrophoresis

It is the most widely used technique for separation, detection and determination of the purity of protein. It is sometimes used for protein purification. The technique is based on migration of charged proteins in an electric field. Electrophoresis of proteins is generally conducted in a gel medium, mostly polyacrylamide. The widely used polyacrylamide gel electrophoreses (PAGE) are sodium dodecyl sulphate-polyacrylamide gel electrophoresis (SDS-PAGE) and isoelectric focusing (IEF).

4.1.1 Sodium dodecyl sulphate-polyacrylamide gel electrophoresis

SDS-PAGE is normally employed for estimation of purity and molecular mass of protein. This versatile method is reproducible and low-cost. For the principle, sodium dodecyl sulfate (SDS) in the system binds to most proteins in amount approximately proportional to the molecular mass of the proteins. So each protein has similar charge-to-mass ratio and migrates proportional to its molecular mass. The widely used SDS-PAGE is the Laemmli protocol, as well as its modified protocols, which consists of the discontinuous buffer system. In this buffer system, there are two parts of polyacrylamide gel, stacking gel and separating gel. There are differences in composition, concentration and pH among electrode buffer, stacking buffer and separating buffer (Fig. 10). The usefulness of the buffer stem is stacking effect which occurs in the stacking gel. The effect increases concentration of proteins in the sample which helps well

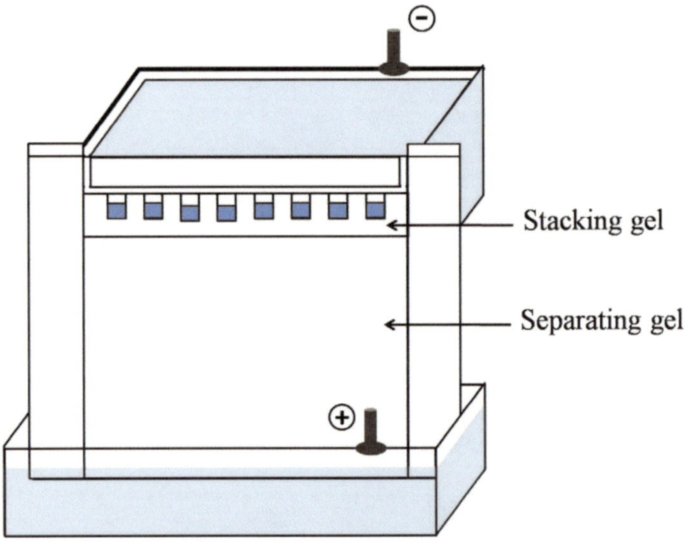

Fig. 10. Polyacrylamide gel electrophoresis. The electrophoresis set composes of the upper and lower chambers which are filled with an electrode buffer. The gel is polymerized in the space between two glass plates and then connected between the two chambers. Protein samples are load in wells at the top of the gel. Protein molecules will migrate into the gel under an electric field. For SDS-PAGE, the proteins migrate from cathode to anode.

separation of the proteins in the separating gel. SDS-PAGE can be used to estimate molecular mass of the protein by using a calibration curve between relative mobility and log molecular mass of standard protein. There are the other protocols of SDS-PAGE including urea-SDS-PAGE, Tricin-SDS-PAGE etc. The other types of polyacrylamide gel electrophoresis are also available such as acid-urea-PAGE, native-PAGE. Although SDS-PAGE is commonly used, some proteins may need specific PAGE for the detection. The proteins separated on polyacrylamide gel can be visualized by many types of staining. The staining with dye especially Coomassie Brilliant Blue R250 is most widely used since it is easy, low cost and effective. Staining with some other dyes and silver staining are also available for specific purposes. Examples are Sudan Black B for staining of proteolipids, Thymol for staining of glycoproteins (Hames, 1981; Holtzhauer, 2006).

4.1.2 Isoelectric focusing

IEF is used for determination of isoelectric pH (pI) of protein. The technique is also useful for separation, detection and determination of the purity of protein. Proteins are separated according to their pI in a medium containing pH gradient. For analytical slap gel, the medium commonly used is polyacrylamide gel or agarose gel. The pH gradient is established by a mixture of low molecular mass organic acids and bases which is called ampholytes (Berg et al., 2002). The proteins in the mixture migrate in an electrical field depending on their charges and then standstill in the gel at the pH equal to their pI (Fig. 11). The separated proteins can be visualized by an activity staining or a dye staining. The activity stain depends on the protein property. There is a problem concerning with the dye staining of IEF gel. The ampholytes can be stained with some dyes especially Coomassie Brilliant Blue R250. So it must be removed by appropriate solutions before staining.

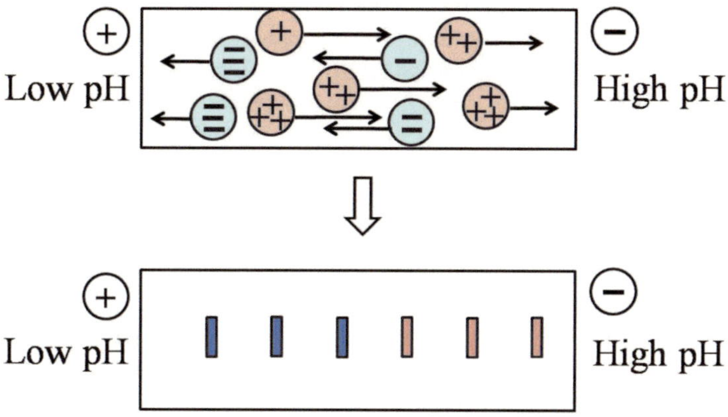

Fig. 11. Principle of isoelectric focusing. A pH gradient is established between two electrodes by using an electric field, a mixture of ampholytes and suitable electrode buffers. Each protein migrates in the gel and then stop moving at the pH equal to its pI.

For some work, two-dimensional polyacrylamide gel electrophoresis (2D-PAGE) is performed to obtain a very high resolution. The technique is a sequential separation of proteins. The protein mixture is separated by using IEF in the first dimension and then SDS-PAGE in the second dimension. High skill is needed to perform the conventional procedure. Fortunately, the commercially available apparatus are designed to make it much easier and more efficient. Since the cost of the technique is rather high, it is done only when necessary.

After dye staining or silver staining, the protein separated by SDS-PAGE or 2D-PAGE may be identified by using Liquid Chromatography-Mass spectrometry (LC-MS). The protein band is cut from the gel and then in-gel digestion with some protease which is mostly trypsin. The peptide fragments are separated by HPLC and injected to a series of Mass Spectrometers. This mean of mass spectrometric analysis is called tandem mass spectrometry. The MS spectra of the peptides are used to search in the data bases through the internet. The amino acid sequences of some peptides are obtained as well as the protein(s) which has them as parts of molecule (Berg et al., 2002).

4.2 Enzyme activity or the other property of protein

Apart from electrophoresis, there are some other methods for detection of the desired protein during purification process. Various specific techniques are used depending on the property of the protein. Many proteins can be stained in gel by using their enzyme activity. Since the native conformation is important for the activity, IEF and native-PAGE are compatible with activity stains. SDS-PAGE is not suitable for activity staining of protein, excepted that the proteins can be re-natured by soaking the gel with non-ionic detergent. The enzyme activity is also determined in solution. Its activity is measured by absorption spectrometry and the activity unit can be determined. Specific activity calculated from enzyme activity and the amount of protein in milligram is normally used for determination of the purity fold during the purification process. Some enzyme activity may be detected by other techniques such as paper chromatography (Arunchaipong et al., 2009).

Many proteins are not enzymes but they may have other activities or properties. Some activities or properties are used for detection of the proteins. For examples, Anti-microbial activities can be detected by microbiological methods. Glycoproteins are pursued during purification process by using lectins. In this case, the binding activity has to be done on a membrane surface since the protein molecules can not freely move in the gel matrix and the specific binding will be interfered. The membrane method is called Western blot analysis. In this method the proteins are separated using SDS-PAGE. The separated proteins are then electrically transferred onto a membrane sheet made of nitrocellulose or polyvinylidene fluoride (PVDF) (Fig. 12). The protein bands are adsorbed on the membrane surface. The membrane is then used for binding with a lectin-linked enzyme. Only glycoprotein band(s) binds to the lectin-linked enzyme. The band(s) can be vistualized by addition of a suitable substrate. Western blot analysis is also suitable for detection of protein by specific antibody.

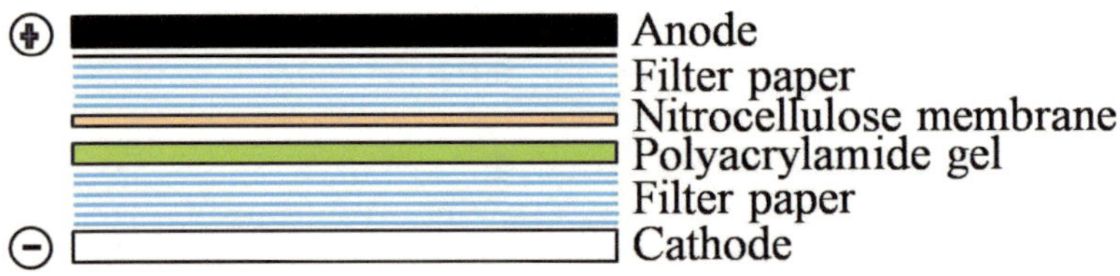

Fig. 12. Diagrammatic representation of electrical transfer of proteins. The transfer of protein from polyacrylamide gel onto a membrane is also called western transfer. The western transfer shown in the picture is the semi-dry procedure. After separation of proteins by using polyacrylamide gel electrophoresis, the gel and the membrane sheet are sandwiched between filter papers and electrodes. The membrane (such as nitrocellulose) and the papers are wet with a transfer buffer. The proteins in the gel are transferred to the membrane by application of an electric field.

5. Determination of protein concentration

Determination of protein concentration is normally required during the purification process. The protein concentration is useful for determination of specific activity. There are a number of protocols to determine protein concentration such as ultraviolet absorption, Bradford method and Lowry method. Ultraviolet absorption (at 280 nm) of protein solution is the simplest method. This method is relatively sensitive and does not destroy protein; so it is used for detection of proteins eluted from chromatographic column. However, the absorption at 280 nm can be interfered by other substances such as nucleic acids. Bradford method relies on the binding between protein and the dye Coomassie Brilliant Blue G. The protein-bound dye has maximal absorption at 595 nm. In Lowry method, the protein solution is mixed with a copper sulphate solution and the Folin reagent (a mixture of sodium tungstate, molybdate and phosphate) to produces blue purple color. The color solution is quantified by the absorbance at 660 nm (Walker, 2005). However all methods used do not give true concentration of protein since it has to used some protein to make a standard calibration curve. Bovine serum albumin (BSA) is mostly used because it is high purity and relatively low cost.

6. Conclusion

Protein purification involves extraction and purification. In extraction process, there are different procedures to disrupt cells or tissues as well as different extraction solvents, depending on the nature of the cells or tissues. Animal cells and tissues can be disrupted by gentle methods while unicellular organisms and plant tissues require the more vigorous procedures to break the cell envelopes. Normally the extraction solvents for most hydrophilic proteins are suitable buffers except that some detergents are included in some cases. Lipid-soluble proteins are extracted by some organic solvents, or buffers containing detergents. A number of procedures are involved in the purification process such as separation based on solubility, separation based on size and density, chromatography. Chromatographic techniques are widely used and can be operated as LPLC, HPLC or FPLC. Many steps are normally required for purification of most proteins but some proteins can be purified by only simple procedures. Detection of the desired protein should be done during the purification process and electrophoreses are widely operated. Detections by activities or other properties are also used for some proteins.

7. Acknowledgment

Thanks go to the Ph.D. students, Miss Suganya Kumla and Miss Patthraporn Siripipatthana for their technical assistance.

8. References

Arunchaipong, K.; Sattayasai, N.; Sattayasai, J.; Svasti, J. & Rimlumduan, T. (2009). A biotin-coupled bifunctional enzyme exhibiting both glutamine synthetase activity and glutamate decarboxylase activity. *Current Eye Research*, Vol.34, No.10, pp.809-818, ISSN 0271-3683

Berg, J. M.; Tymoczko, J. L. & Stryer, L. (2002). *Biochemistry* (5th ed.), pp. 79-91, ISBN 0-7167-4684-0, W.H. Freeman and Company, New York

Fido, R. J.; Mills, E. N. C.; Rigby, N. M. & Shewry, P. R. (2004). Protein extraction from plant tissues. In Protein Purification Protocols, *Methods in Molecular Biology*, Vol.244, pp. 21-27

Hames, B. D. (1981). An introduction to polyacrylamide gel electrophoresis, In: *Gel Electrophoresis of Proteins, A Practical Approach*, Hames, B. D. and Rickwood, D. (Eds.) pp. 1-86. IRL Press, Oxford

Holtzhauer, M. (2006). *Basic Methods for the biochemical Lab*, Springer, Berlin

Ohlendieck, K. (2005). Centrifugation, In: *Principles and Techniques of Biochemistry and Molecular Biology* (6th ed.), Wilson, K. and Walker, J. (Eds.), pp. 103-130, ISBN 0-521-53581-6, Cambridge University Press, Cambridge

Rudolph, R. & Lilie, H. (1996). *In-vitro* folding of inclusion body proteins. *The FASEB Journal*, Vol.10, pp. 49-56

Tao, H.; Liu, W.; Simmons, B. N.; Harris, H. K.; Cox, C. C. & Massiah, M. A. (2010). Purifying natively folded proteins from inclusion bodies using sarkosyl, Triton X-100, and CHAPS. *BioTechniques*, Vol.48, No.1, pp.61-64

Velours, J.; Esparza, M.; Hoppe, J.; Sebald, W. & Guerin, B. (1984). Amino acid sequence of a new mitochondrially synthesized proteo-lipid of the ATP synthase of *Saccharomyces cerevisiae*. *The EMBO Journal*, Vol.3, No.1, pp. 207-212

Walker, J. M. (2005). Protein structure, purification, characterization and function analysis, In: *Principles and Techniques of Biochemistry and Molecular Biology* (6th ed.), Wilson, K. and Walker, J. (Eds.), pp. 349-404, ISBN 0-521-53581-6, Cambridge University Press, Cambridge

Wilson, K. (2005). Chromatographic techniques, In: *Principles and Techniques of Biochemistry and Molecular Biology* (6th ed.), Wilson, K. and Walker, J. (Eds.), pp. 485-550, ISBN 0-521-53581-6, Cambridge University Press, Cambridge

2

The Problems Associated with Enzyme Purification

Etienne Dako[1], Anne-Marie Bernier[2],
Adjéhi Thomas Dadie[3] and Christopher K. Jankowski[1]
[1]*University of Moncton, Moncton, NB*
[2]*University of Saint-Boniface, Winnipeg, MB*
[3]*University of Abobo-Adjamé, Abidjan*
[1,2]*Canada*
[3]*Côte d'Ivoire*

1. Introduction

This chapter aims to highlight the difficulties encountered during the purification of native cellular and membrane-bound enzymes from whole cell extracts. There are many reasons for wanting to purify enzymes, such as to fully characterize them or to mass produce them for commercial purposes. Regardless of the reason for wanting to purify an enzyme, the general extraction and purification procedures are essentially the same. However, depending on the properties of the enzyme, certain modifications of the methods must be considered regarding specific problems that can be encountered throughout the process, such as enzyme insolubility and loss of enzyme activity. Although several classic and more modern methods are available to solve these kinds of challenges, the enzyme purification step remains a major challenge for any method of extraction used.

We will present the general aspects of enzyme purification, followed by a specific example of the problems we encountered and attempted to resolve in the purification of the autolysin PA49.5 from *Lactococcus lactis* subsp. *cremoris*. Autolysins are bacteriolytic enzymes that digest the peptidoglycans of bacterial cell walls (Dako et al, 2008). In several studies it was found that autolysins are linked to lipids or other membranous components during the extraction process and especially during the purification. Since they are very closely linked to the cell walls, their purification presents major challenges associated with their low concentration, their insolubility and the retention of their biological activity (Marshak et al., 1996; Brown et al., 1970). Among these difficulties, their insolubility during purification remains by far the most serious and examples of how we tried to resolve this problem will be presented.

We can look at the process of enzyme purification as a series of steps, each one sequentially enhancing the level of purification from a crude extract. It is essential to minimize protein losses throughout the process; therefore the use of fewer steps during the purification is important since losses can occur in every step due, for example, to linking to separation matrices, insolubilities, or losses into the fringe fractions during separation procedures. We will look at the general steps in the extraction of proteins from the cells, the clarification of the homogenate, the concentration or enrichment of the extract, and finally various steps in the final purification of the enzyme. A purification strategy can be made more efficient if certain physical or chemical properties such as molecular weight, optimal pH, isoelectric point, of the enzyme are known.

To follow the efficacy of the purification of the target enzyme from a whole cell extract there must be a method of measuring its presence throughout the process to ensure the maximum yield. These assays can be an assessment of biological activity or assays based on reactions with antibodies. The antibody assays however do not necessarily indicate if the enzyme is in an active conformation. The assays based on catalytic activity do however require that the enzyme retain its active conformation, which can be problematic since many components of the extractions buffers contain compounds such as detergents that denature the tridimensional structure of the enzyme. Compounds compatible with enzyme assays must therefore be used to minimize losses of activity. The isolation of certain types of enzymes may present problems specific to those enzymes. For example, some problems most often associated with the purification of membrane proteins are that they require organic solvents or detergents for their solubilization. Also, insoluble enzymes may require the use of stronger chaotropic compounds such as lithium chloride or guanidine HCL that can in turn inactivate the enzyme. In order to be able to fully characterize the biochemical properties of a purified enzyme, it must be completely free of contaminants such as cell wall or cell membrane components, nucleic acids or other proteins. The complete removal of contaminants is, therefore, the primary objective of the purification. These contaminants can interfere with subsequent purification steps or with enzyme activity assays. The purity of an enzyme extraction can be confirmed, for example, by the presence of a single band on an electrophoretic gel (SDS-PAGE or isoelectric focusing), a single peak in an HPLC analysis or by N-terminal amino acid sequencing.

The initial step in all protein purification protocols is the release of the enzyme from the cell or tissue material, or cell extract preparation. This is a critical step in the process as it affects protein yield and biological activity of the target enzyme. This naturally requires either a mechanical or chemical lysis of the cells. Methods for homogenizing cells or tissues vary from gentle methods such as osmotic shock, detergent lysis or enzymatic digestion to more vigorous methods such as homogenization in a blender, grinding with an abrasive substance or ultrasonication. The lysis procedure selected must not damage the target enzyme and therefore the conditions must be optimized for each cell type and target enzyme in terms of the pH of the extraction buffer, the temperature and the concentration of certain components of the buffer such as detergents, salts or reducing agents for example. Reducing agents such a dithiothreitol or 2-mercaptoethanol can be used to prevent the oxidation of sulfhydryl groups that could damage the active conformation of the enzyme resulting in a loss of activity (Ward & Swiatek, 2009; Chen et al., 2010). Detergents or organic solvents, generally used to isolate hydrophobic or membrane bound enzymes, can be problematic in subsequent purification steps as well as interfere with the native conformation of the enzyme. It is essential also to consider subsequent purification steps when choosing a detergent. Certain downstream applications such as optical spectroscopy, mass spectrometry and crystallization, can be negatively affected by the presence of some detergents (Linke, 2009). There are various detergents available and these will be discussed further in the section of enzyme purification. The most important criteria for the selection of a detergent for the solubilization of a protein is its ability to not only solubilize but to maintain biological activity. In general, the detergents with higher critical micelle concentrations (CMC) can more easily denature proteins than those with lower CMC (Linke, 2009; Lever et al. 1994; Garavito, 1991; Hjelmeland & Chrambach, 1984), and it must be used at concentrations above its CMC in order to effectively solubilize the proteins (Privé, 2007).

Once the cell or tissue homogenate is produced, the cell debris and other insoluble compounds must be removed by centrifugation. The optimal situation is that following a high speed centrifugation, the solubilized proteins will be in the supernatant along with other soluble compounds, and that all insoluble material will be in precipitate. Sequential or differential precipitations can be performed to progressively remove cell debris in order to rehomogenize it to enhance protein yield or to isolate specific cell organelles if the starting material is eukaryotic. Following this clarification, the cell extract is then ready to be purified in an enrichment or concentration procedure.

A process that can be considered after the clarification is the concentration of the enzyme preparation. Dialysis can be performed prior to the concentration procedure in order to remove salts from the cell extract. The clarified cell extract can therefore be dialyzed against a large volume of buffer to remove salts and the dialysate can then be lyophilized to concentrate the extract. Lyophilisation, or freeze-drying, has the advantage of protecting the protein from the chemical reactions that can damage the proteins in an aqueous solution, such as aggregation, precipitation and protease degradation (O'Fagain 2004). The protein extract can then be stored in the lyophilized state and aliquots resuspended as needed.

The most common concentration procedures are ammonium sulfate precipitation, ultrafiltration and ion exchange chromatography (Ward & Swiatek, 2009).

There are two applications of ammonium sulfate precipitation. The first one is a complete precipitation in which all proteins are precipitated by the addition of the salt. The proteins are recovered by centrifugation and resuspended in a minimal volume of water and are therefore concentrated. The residual salt must be removed by dialysis as it can interfere with protein activity or subsequent purification steps. One of the problems encountered with the dialysis of a protein extract is that the protein can precipitate onto the walls of the dialysis tubing thereby impairing the passage of particles across the membrane. This would be particularly problematic in the dialysis of a crude cell extract (Doonan, 2004a). The salts cause the precipitation of proteins by removing water from the hydrophobic patches on the protein's surface, thus rendering them insoluble (Doonan, 2004b). Another approach to the concentration of a cell extract is a fractional precipitation using a salt. Again ammonium sulfate is the most frequently used salt, because of its high solubility and the fact that the density of saturated solutions is less that of the protein. This allows the collection of proteins by centrifugation (Doonan, 2004b). The fractional precipitation is done by incrementally adding amounts of ammonium sulfate and collecting the fractions precipitated by various salt concentrations by centrifugation. Fractions can, for example, be collected at each increment of 20% concentration of salt. Each fraction can be analyzed for the enzyme in question. This will therefore concentrate the cell extract and also begin the process of purification. Depending on the solubility of the protein it may however be found in more than one fraction. The combination of these fractions will increase the yield of the enzyme however this will also reduce the level of purification.

Ultrafiltration is based on the principal of separation through a semi permeable membrane filter. Many filter type units are available commercially, either applying a centrifugal or stirred cell type of separation (Bonner, 2007). The principal for all ultra centrifugal methods is the filtration through a membrane that has a selective molecular weight cut off point, therefore it will separate the extract based on the size of particles, not the charge.

Ion exchange chromatography is a non affinity absorption method that purifies proteins on the function of molecular charge, not the molecular weight. The resin in the chromatography column is either negatively or positively charged. Proteins also have a charge that will be dependent on the pH of the extract or solution. The principal of ion exchange chromatography is to pass the protein extract through the column at a pH where the protein will adhere to the oppositely charged matrix. The pH is then gradually changed during the chromatography and the proteins will be eluted when their charge is neutralized by the pH of the elution buffer. Fractions are collected from the column and analyzed for enzyme activity to identify the fraction containing the enzyme in question (Bonner, 2007).

There are no set protocols or sequences of procedures that will ensure the purification of an enzyme from a cell extract. These will vary according to the source of the enzyme and the specific properties of the enzyme.

To better understand the difficulties and techniques of extraction and purification, we decided to present a specific example, that of the lytic membrane enzyme PA49.5 so named for its molecular mass of 49.5 kDa. This enzyme was isolated from *Lactococcus lactis* subsp. *cremoris* ATCC 9596 denoted MC5. The objective of our study was to find a non-denaturing method to solubilize the autolysin in order to ensure its purification to a point that would allow its biochemical characterization. The extraction of PA49.5 from a crude cell extract supernatant, from cell wall debris and whole cells was attempted using different approaches in order to attempt to maximize solubilization of the enzyme. Extracts were further solubilized in various detergents and organic compounds and subsequently further purified by HPLC, ion exchange chromatography or ammonium sulfate precipitation.

In view of their richness in ripening enzymes (Dako et al. 2008; Dako et al. 1995), lactococci play an essential role in the production of cheese. These microorganisms also contain lytic enzymes known as autolysins because of their capacity to degrade the membranes of the bacteria that produce them. A few examples of the autolysins are: *N*-acetylmuramidases, N-acetylglucosaminidases, N-acetylmuramyl-L-alanine amidase and peptidases (Dako et al., 2003; Valence et Lortal., 1995; Shockman et Höltje, 1994; Kawagishi et al., 1980). These autolysins have an effect on specific areas in the ß-1,4 bonds between the N-acetylglucosamine and the N-acetylmuramic acid (N-acetylglucosaminidases, N-acetylmuramyl-L-alanine amidases) or between N-acetylmuramic and N-acetylglucosamine (N-acetylmuramidases). Since they are very closely linked to the cell walls of these bacteria, the isolation of autolysins presents major difficulties associated with their low concentration, their insolubility and the retention of their activity (Marshak et al., 1996; Brown et al., 1970). Among these difficulties, the insolubility of autolysins during purification remains by far the most serious problem. This is due to the presence of teichoic (Lortal et al., 1997) and lipoteichoic (Fisher et al., 1980; Höltje & Tomasz, 1975) acids.

According to Brown (1972), the problem of the solubility of enzymes during purification is linked to the presence of material other than autolysins, which absorb strongly at 260 to 280 nm. According to Shockman *et al.* (1967), the treatment of the cell wall by sodium dodecyl sulfate (SDS) would resolve the problem of the insolubility of the autolysins of lactic bacteria during purification. SDS is the most widely used detergent to solubilize most otherwise insoluble proteins such as autolysins (Dolinger et al., 1989), its presence may affect the tertiary structure of proteins because it is one of the most strongly denaturing detergents. It is also one of the most difficult to remove with dialysis. The renaturing of SDS extracted proteins after electrophoresis in denaturing conditions does not totally eliminate the SDS, even if it allows the activity of lytic proteins (Lortal et al., 1997) to be measured. Thus, we propose that the use of other non ionic (Triton X-100) or zwitterion (CHAPS) non-denaturing detergents will have the advantage of resolving the problem of the insolubility of autolysins whilst protecting their structure. In general, the detergents such as CHAPS and those that possess a cholanoic group such as cholate all produce very small micelles and have a critical value of micellar concentration (CMC) below 2mM. The CMC is the value above which there is the formation of micelles (Hjelmeland et Chrambach, 1984; Helenius et al., 1979). These detergents can be eliminated by dialysis without affecting the protein activity (Hjelmeland et Chrambach, 1984). The main aim of this study was to find a non-denaturing method to solubilize the main autolysin (PA49.5) of *L. lactis* subsp. *cremoris* ATCC 9596 (Mc5) so as to ensure its purification from the supernatant of cellular remnants and whole cells.

In several studies it was found that autolysins are linked to lipids or other membranous components during the extraction process and especially during the purification. They are therefore considered to be hydrophobic enzymes; however, it is often unclear whether or not the enzyme of interest is in fact hydrophobic. Although several classic and more modern methods are available to solve these kinds of challenges, there is still much uncertainty, and further studies are still needed to clearly resolve the purification of cell membrane hydrophobic enzymes.

An important observation, reported by Brown et al (1970), is that autolysins are associated to teichoic acid in the cell wall of *Bacillus subtilis*. These authors propose that this acid can be eliminated by ethanol precipitation and by gel filtration on an agarose column. Their finding did not confirm this hypothesis, in fact, they found that only 5% of the techoic acid was eliminated using this method. This acid also persisted after multiple washes with ethanol or ether, and it persisted after separation in a polyacrylamide gradient density gel electrophoresis, ion exchange chromatography and gel filtration using 7M urea.

After many attempts, certain autolysins were purified by electrophoresis using SDS as the detergent (*N*-acétylglucosaminidase, ac-*N*-acetylmuramidase and ac-*N*-acetylmuramyl-L-alanine amidase) (Motoyuki et al. 1995). Brown (1972) suggests that the purification of autolysines from lactic bacteria is only possible in the presence of SDS. While SDS is an excellent approach to the purification of this type of enzyme, we propose that another method must be found in order to eliminate the compounds that bind to the autolysin thereby rendering it insoluble. CHAPS, a zwiterrionic detergent, used at high temperatures was evaluated as a potentially more efficient detergent in the purification of this enzyme.

We will present our attempts to purify the enzymes using these various strategies and the problems encountered during certain steps. Some approaches were more problematic that others while some hold more promise in the purification of the enzyme in question.

Before looking at the specific details of how we attempted to purify this autolysin, we will present the assay we used throughout these experiments to verify the presence and purity of our enzyme. For all extractions, a protein profile was examined by SDS-PAGE stained with Coomassie blue. Molecular weight markers were included in all gels. In order to identify autolysins in the preparations, a sample was analyzed by a denaturing SDS-PAGE containing intact cell walls from *Lactococcus lactis* sp *cremoris* MC5 as the substrate for the autolysin according to the method of Potvin et al. (1988). The gel was renatured in buffer containing Triton X100, and enzyme activity was noted by the presence of a clear zone surrounding the enzyme band after staining with Coomassie blue. Proteins were quantified using the method of Lowry. It is also possible to test for general autolysin activity by agarose gel diffusion in a Petri plate. The gel contains cell wall fragments and the fractions containing the autolysin are introduced into wells perforated into the agarose. The gel is incubated and the presence of autolysins is then observed by the presence of a clear zone surrounding the well.

Three different methods of preparing cell extracts were tried in this work, either using autolysis, cell debris or whole cells. We initially started with a crude cell extract prepared by autolysis and the mechanical breakage of the cell. Washed and precipitated cells were suspended in a citrate lysis buffer containing 2% NaCl (w/v) and 0,3M LiCl. The cells were incubated at 30°C for 48 hours to ensure maximum autolysis. The cell debris precipitated by centrifugation was ground with alumina and the supernatant was combined to the first supernatant. These combined supernatants were dialyzed to eliminate the salts, the dialysate was filter sterilized and lyophilized and the extract was then concentrated 28X. This crude extract (CE) was then separated by preparative HPLC which separates the proteins by molecular weight. High molecular weight proteins are found in the first fractions, in the mobile phase. The fractions collected were dialyzed and lyophilized. Autolytic activity of the HPLC fractions was determined by SDS-PAGE with the cell wall substrate.

Four peaks were observed in the HPLC separation profile, the first and largest peak had a retention time of 6 minutes and contained proteins with molecular weights greater than 30 kDa and represented 57,45% of the total estimated area. The autolysin PA49,5 was found in the first fraction as shown in figure 1. C5 represents the results from *Lactococcus lactis* subsp. *cremoris* ATCC 9596 denoted MC5. A visible clear zone can be seen surrounding the wells. Figure 2 presents the results of a denaturing polyacrylamide gel containing cell walls as substrate to identify specific autolysins separated by molecular weight during the electrophoretic process. We can see on figure 2 that there is one band of activity for a protein with the molecular weight of 49.5.

In comparing the protein profile of the crude extract with the proteins in fraction 1, we saw that fraction 1 had approximately 40% less protein than the crude extract indicating that a certain level of purification had been obtained. While the starting crude extract was concentrated 28x, we required 30 µg of fraction 1 proteins to observe autolytic activity. This confirms the limitations of molecular sieving as a method of purifying autolysins.

The autolysins identified in the peak 1 were evaluated in order to determine whether they are acidic or basic proteins. The Davis gel electrophoresis, which allows the identification of acidic proteins, revealed autolytic activity of a band migrating to molecular weight 49.5. The Reisfeld gel that allows the identification of basic proteins revealed no autolytic activity. We can therefore conclude that autolysin P49.5 is an acidic or neutral protein. Since the protein was in the upper area of the Davis gel it is more probable that the protein tends to be more of a neutral pH.

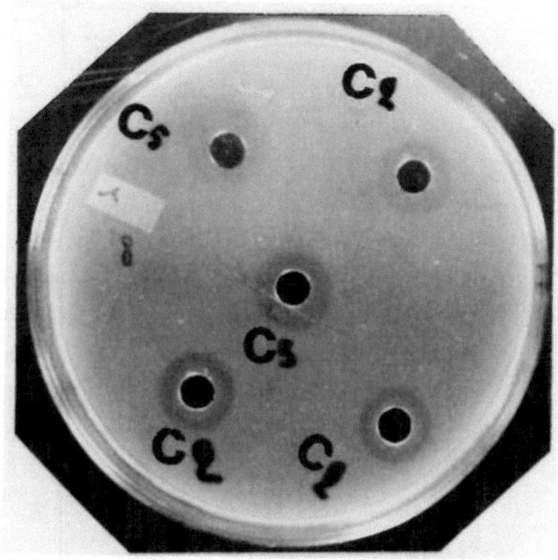

Fig. 1. Gel diffusion assay of HPLC peak 1 on agarose containing cell walls as autolysin substrate.

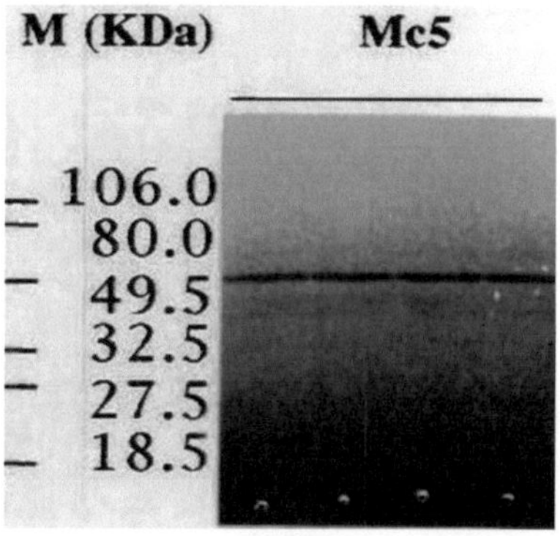

Fig. 2. Autolytic activity of PA49.5 in fraction 1 obtained by HPLC after 48 hours of renaturation.

We determined that the concentration of protein present in 575 mg of fraction 1 extract diluted in 250 ml of buffer was 0.17 mg/ml. Using the following formula, we calculated that proteins constitute only 7.4% of the fraction 1 extract. This again demonstrates that a better method is required if a large quantity of the enzyme in question is to be purified.

$$\% \, Protein = \frac{[protein \; concentration \; x \; dilution \; volume] \; x \; 100}{Amount \; of \; crude \; extract \; (mg)}$$

$$\% \, Protein = \frac{\left[0.17\frac{mg}{ml} x \; 250 \; ml\right] x \; 100}{575 \; mg}$$

$$\% \, Protein = 7.4$$

Large concentrations of salts are generally used to precipitate proteins from cell extract. Ammonium sulfate is the salt of choice as it can be used in lower concentration than other salts and it preserves the biological activity of proteins. In a solid state, the salt can be added gradually to a mixture with constant mixing and fractions collected at various levels of saturation. Ammonium sulfate reaches its saturation point at 760 g/l at 4°C.

Peak 1 from the HPLC preparation was lyophilized and then the lyophilisate was resuspended in buffer, centrifuged to remove insoluble material and the supernatant was precipitated using ammonium sulfate. In order to ensure an efficient precipitation, a minimal concentration of 1.0 mg/ml of protein is required. We therefore measured the protein content of our supernatant or crude extract before proceeding. Four fractions were collected by sequential precipitations (0-20%, 20-40%, 40-60%, 60-80%). For example the first precipitation was obtained adding enough ammonium sulfate to obtain 20% saturation. The mixture was centrifuged, the precipitate was conserved and the supernatant was further precipitated with another addition of ammonium sulfate up to 40% saturation. This process was repeated with 20% increments of ammonium sulfate saturations, to 60% and 80%. The precipitates obtained were dissolved in deionized water, dialyzed and then concentrated by ultrafiltration with a Microcon-10 filtration system. The quantity of protein was measured in all precipitates and supernatants collected.

Our results indicate that we recovered only 26.3% of the initial protein during the sequential precipitations. One of the important steps following the precipitations is the ultrafiltation using the Microcon-10 whose membrane allows the exclusion of proteins based on molecular weight, or size exclusion. This step could explain the loss of protein during the process. This ultrafiltation does however offer the advantage of rapid salt removal and a rapid concentration of proteins. The membrane used in our experiment removed high molecular weight proteins.

The diffusion autolytic test in an agarose plate containing cell walls as the enzyme substrate indicated that all ammonium sulfate fractions except the 0-20% fraction had some autolytic activity as indicated by a clear zone surrounding the well. The control, which was the crude non precipitated extract had no autolytic activity, however this is certainly related to the low concentration of autolysins in this extract. The precipitated fractions were then analyzed for specific PA49.5 activity using the gel electrophoresis in denaturing conditions. The results are presented in figure 3. Lane 1 is the control which was the 0% ammonium sulfate. Lanes 2 to 5 are the precipitated fractions (0-20%, 20-40%, 40-60% and 60-80%). We can see that the 40-60% fraction contained the most autolysin 49.5.

The complete protein profile of peak 1 as determined by SDS-PAGE is shown in Figure 4. Lane T5 is the control unprecipitated extract, lanes 1-4 are 1 mg/ml of protein, lanes 5 to 8 are 1.5 mg/ml protein for each of the fractions 0-20%, 20-40%, 40-60% and 60-80%. The complete protein profile shows us that there are still many contaminating proteins in our extracts, and that there is little difference in the profile between the different concentrations. The crude extract is quite poor in proteins primarily due to a dilution factor. The protein content of fraction 0-20% is also quite low and the concentration of proteins increases with each fraction collected.

The whole cell extract was also precipitated using ammonium sulfate to compare the autolysin profile with that obtained from the HPLC peak 1. The results are presented in figure 5 for each of the four collected fractions at two different protein concentrations. Autolysin PA49.5 is the primary autolysin extracted and it is present in all fractions except the 40-60% fraction (lanes 3 and 7). The intensity of the band in fraction 60-80% (lanes 4 and 8) is approximately half that of the fractions 0-20% and 20-40% and 40-60%. Once again the control does not contain the autolysin due to a low concentration of autolysin in the initial extract. This also demonstrates that ammonium sulfate precipitation not only separates the proteins but also concentrates them. The fact that we found the autolysin in many fractions does however indicate the limitation of this method of its purification and the problem encountered by the difficulty in solubilizing this protein. The ideal situation would be to find the autolysin in one fraction only.

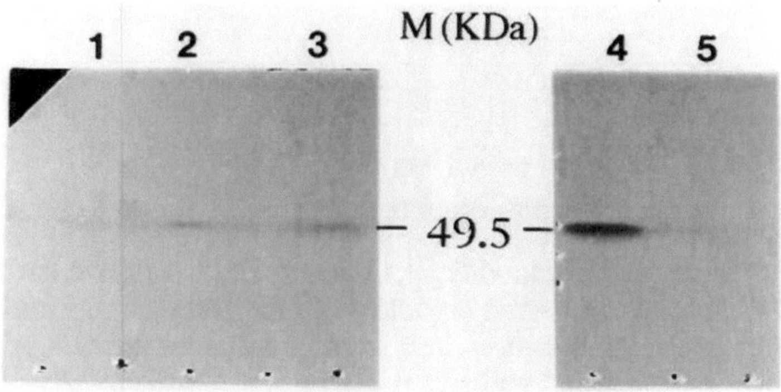

Fig. 3. Autolysin activity of ammonium sulfate precipitated HPLC peak 1.

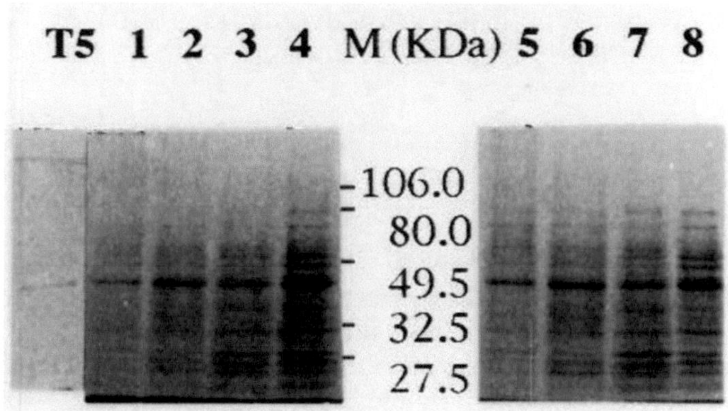

Fig. 4. The complete protein profile of peak 1 of from the HPLC analysis.

Each ammonium sulfate fraction from the peak 1 precipitation was then separated by ion exchange chromatography (DEAE Sephadex A-50). There were important losses of protein content from each fraction after this separation. For example, fraction 20-40% contained 0.510 mg of protein at the start of the chromatography. When the protein content of all the chromatography fractions eluted with NaCl are combined, we recovered 0.246 mg of protein, or 48.24%. The lowest protein recovery was from fraction 0-20% with 26.6% of the protein recovered. When these fractions were assayed for autolysin activity by SDS-PAGE with cell wall incorporated as the substrate, the only fractions that had autolytic activity were from 0.0, 0.25 et 0.5 M NaCl with the the 0-20% ammonium sulfate fraction and the 0.25 M NaCl fraction from the 20-40% et 40-60% ammonium sulfate fractions. Once again we see that the autolysin is present in various fractions. This poses a serious problem regarding the insolubility of PA49.5 which limits the use of ammonium sulfate precipitation and ion exchange chromatography in its purification.

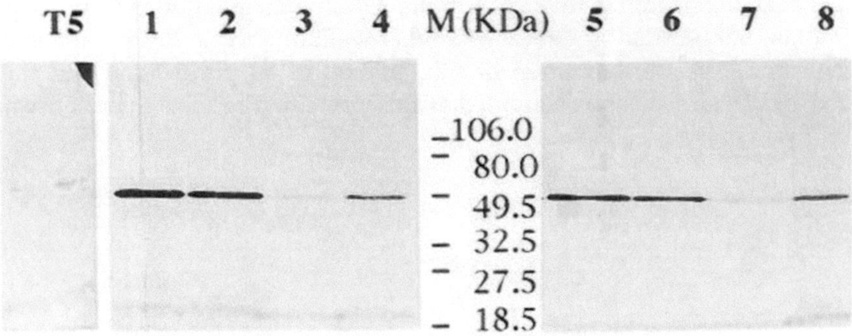

Fig. 5. Autolysin activity of ammonium sulfate precipitated whole cell extract.

In an attempt to enhance the solubility of this enzyme, we proceeded to evaluate the effects of various salts, detergents and organic compounds on the extraction process. Cells were therefore treated as described earlier, however after centrifugation to remove cell debris, the cell debris was treated with either 5M LiCl, 5M NaCl, 60% (v/v) acetonitrile or 20% (w/v) SDS. These cell debris were extracted for one hour, centrifuged and the supernatant was lyophylized to concentrate the product. The use of LiCl and SDS did enhance protein recovery and the extraction of the autolysin. SDS recovered 80.99% of the proteins while LiCl, NaCL and acetonitrile recovered 13.14%, 3.46% and 2.42% protein respectively.

Figure 6 presents the protein profiles (6a) and autolytic activity (6b) for these four different extractions of cell debris. Lanes (1 and 2), (3 and 4), (5 and 6) and (7 and 8) correspond respectively to 1.5 and 15 µg of proteins from treatments LiCl (5M), NaCl (5M), acetonitrile (60%) (v/v) and SDS (20%) (w/v).

We can see from these results that SDS induces a better extraction of protein, followed by LiCl and NaCl. The protein profiles from these three extracts strongly ressemble each other. However, when looking at the specific extraction of autolysin PA49.5 (Fig. 6b) we can see that we observe autolytic activity only when the cell debris is extracted with LiCl or SDS. The SDS extract in fact shows the main autolysin PA49.5 and two minor autolytic bands. The extraction with LiCl shows the main autolysin and one minor autolytic band at the higher protein concentration. Two faint bands are visible in the high protein concentration lane of the NaCl extract and no autolysins were extracted with acetonitrile. These results confirm that SDS is in fact a detergent of choice in the extraction and solubilization of autolysins (Leclerc & Asselin, 1989; Lemee et al., 1994, 1995; Lortal et al.,

1997). One of the limitations of SDS, as previously mentioned, is the fact that it affects the tri-dimensional structure of the protein thereby reducing or eliminating its biological activity. However, we have also shown that LiCl could be efficacious in the extraction process as well, with the added advantage of having less effect on the biological activity of the enzyme. A challenge in the use of LiCl is the amount of protein and enzyme recovered. Our results also indicate that the autolysin is present in the cell wall debris. This confirms the data presented in the literature that autolysins are linked to bacterial cell walls (Brown, 1972; Brown, 1973; Lemee et al., 1995)

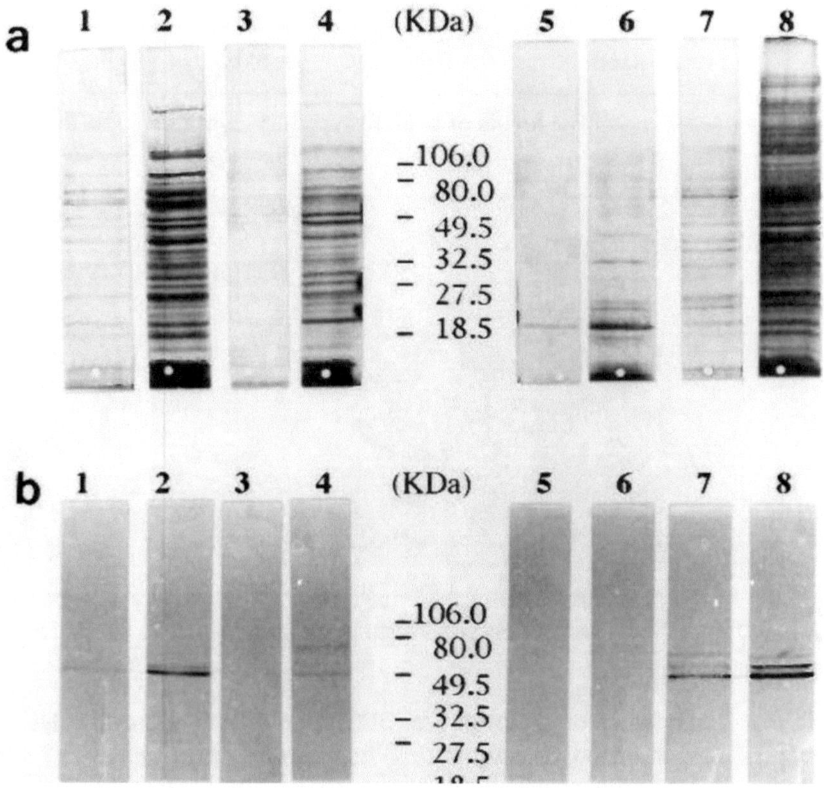

Fig. 6. The protein profile (a) and autolytic activity (b) of cell debris extracted with LiCl, NaCl, acetonitrile and SDS.

In order to more thoroughly investigate the use of LiCl and SDS in the extraction process and to try to increase protein yield, extraction was performed on whole cells rather than the cell debris. Three types of treatments were compared in this study, the LiCl extract from cell debris (A), the LiCl extract from whole cells (Awc) and the SDS extract from whole cells (Dwc).

From these results we can see once again that SDS does allow the best protein extraction and the highest protein recovery. In using it on whole cells the quantity of protein extracted was increased 2-fold compared to that extracted from cell debris (data not shown). The extraction using LiCl on whole cells increased the amount of protein extracted by 3 fold when compared to LiCl used on cell debris. Therefore, the use of whole cells as a starting material certainly yields more interesting results in terms or protein recovery. Figure 7 presents the results of the autolysin assay for the whole cell extracts.

Treatment	Optical density 750 nm	[protein] mg/ml	Volume of the suspension (ml)	Protein (mg)	% protein recovery/1g
A- LiCl cell debris	0.31	1.06	3.10	3.30	0.33
Awc - LiCl on whole cells	0.55	2.35	4.10	9.63	0.96
Dwc - SDS on whole cells	0.68	3.04	13.00	39.56	3.96

Table 1. presents the results of the levels of protein recovery using these different extraction methods.

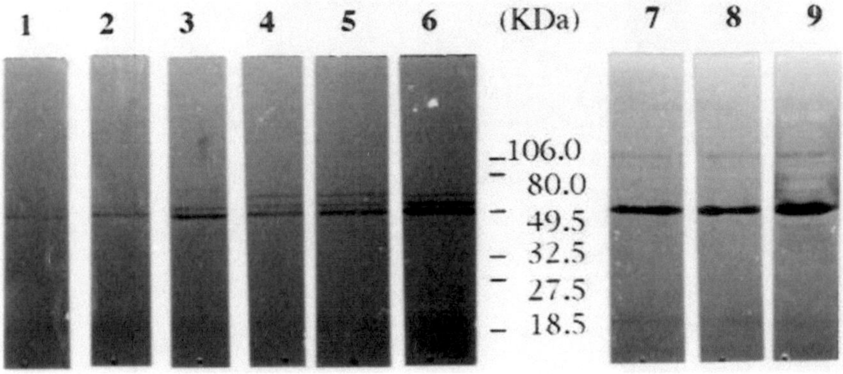

Fig. 7. Autolysin activity for three dilutions of protein (1/10, 1/5 and 1) extracted from cell debris with LiCl (lanes 1-3), from whole cells with SDS (lanes 4-6) and from whole cells with LiCl (lanes 7-9).

We noticed that when used on whole cells, SDS showed the same autolysin extraction pattern (3 bands) as when used on cell debris (Fig. 6b lanes 7 and 8). Lanes 7 to 9 illustrate that the action of LiCl on whole cells allows an increase in autolysin concentration. With this treatment on whole cells, the 1/10 dilution (3 µg/µl) is sufficient to triple the activity when compared to that of the SDS whole cell extract. According to Brown et al (1970), the main approach to the extraction of autolysins, whether soluble or not, requires the use of LiCl. Our results concure with this proposal.

The LiCl treated whole cell extract was centrifuged and the precipitate (Pwc) was conserved and resuspended while the supernatant (Swc) was precipitated with ammonium sulfate as previously described in order to further purify the autolysin in quesiton, PA49.5. The following saturations fractions were obtained: 0-20, 20-40, 40-100%. Protein concentrations were determined for all fractions as well as the initial precipitate (Pwc) and are presented in Table 2. The results demonstrate that protein is still found in the precipitate, representing approximately 19% of the protein. Precipitation with ammonium sulfate recovers 81.81% of proteins from the supernatant. This is a relatively good recovery rate and indicates that ammonium sulfate precipitation is efficient in precipitation proteins. An interesting observation is made regarding protein quantities in the initial whole cell extract (Awc) from Table 1 compared to the amount of protein recovered in precipitated supernatant. We had 9.63 mg of protein in the initial extract (Awc Table 1), however only 1.348 mg of protein were detected in total from all fractions and the precipitate added together (Table 2). This 1.348 g includes the 0.255 mg of proteins found in the initial precipitate (Pwc) that should in fact be considered insoluble proteins. Thus, only 14% of the protein is in fact recovered overall after ammonium sulfate precipitation. We noticed a white precipitate on the walls of the Microcon-10 ultrafiltration column during these tests. This white precipitate became less prevalent as the concentration of ammonium sulfate increased. This could therefore be the source of our lost proteins during this process.

Treatment	Sample type		[protein] mg/ml	Volume of suspension (ml)	Protein (mg)	% protein recovered
Precipitate of whole cell extract	Pwc		0.850	0.300	0.255	18.92
Supernatant of whole cell extract	Swc	0-20%	1.500	0.300	0.450	33.38
		20-40%	0.860	0.645	0.555	41.17
		40-100%	0.250	0.250	0.088	6.53

Table 2. Protein assays of ammonium sulfate precipitation fractions of the whole cell extract (Awc) supernatant (Swc) and precipitate (Pwc). Total protein recovered is 1.348 mg.

Since the protein amount was so high in the first whole cell precipitate (Pwc) we decided to analyse it further to obtain its protein profile and evaluate the presence of autolysin PA49.5 in this precipitate. The results of this analysis proved to be very interesting. PA49.5 was found to be present and active in the precipitate and in all the ammonium sulfate fractions. Our evaluation of the intensity of the band in the precipitate lead us to believe that approximately 80% of the autolysin from the original whole cell extract treated with LiCl (Awc) is found in the precipitate (Pwc). This confirms our initial tests that indicate that the autolysin is indeed insoluble in an aqueous solution.

There are three possible explanations to the insolubility observed for our enzyme. Firstly, we may not be working at the appropriate pH to ensure enzyme activity. A study on the effects of different pHs would allow us to discover the ideal pH for the maximum solubilization of the enzyme in question. Secondly, it is possible that the concentrations of the enzyme during the extraction cause the aggregation of the enzyme in micelles whereby the hydrophobic regions would be oriented towards the outside or inside, or that there is still a lot of cell wall debris left that contain the lytic enzymes. In these cases, the use of organic solvant or detergents or even a natural autolysis would facilitate solubilisation. Thirdly, it is possible that the autolysin in question is linked to a fatty acid or to teichoic acide or lipoteichoic acid. This would render the enzyme insoluble if the complexe is not disrupted.

Three hypotheses and experimental approaches to resolving the problem were then proposed.

Hypothesis 1: A study of the effects of different pH on the Awc extract will allow us to find the ideal pH for the solubilization of the protein. This supposes that after centrifugation, the autolysin with then be found in the supernatant, not the precipitate as it is now.

Hypothesis 2: If the autolysins are still linked to the cell wall debris, a natural hydrolysis will allow the breakage of this bond and therefore enhance solubilization.

Hypothesis 3: If the autolysins of the whole cell extract (Awc) are linked to hydrophobic compounds such as fatty acids or teichoic acide, the use of detergents, either non ionic or zwitterionic, will break these bonds and solubilize the enzyme.

Hypothesis 1: The effect of pH on the solubility of PA49.5

Whole cells were extracted as previously described in 5M LiCl. The lyophilisat was suspended in deionised water and the protein concentration was determined. The solubilization of the proteins was performed in different buffers at pHs ranging from 4 to 9 for a final concentration of 1 mg/ml of protein in a final volume of 100 µl. These were incubated overnight and then centrifuged. The supernatant (Swc) was removed and the precipitate was resuspended in the same volume of water. The amount of protein was measured in all precipitates and supernatants and autolytic activity was evaluated as previously presented in a renatured SDS gel containing cell walls as the substrate. The main results of this experiment was that the amount of protein in the precipitate was found to be approximately half of that found in the supernatant (Swc) for all the pH values tried except for that of pH 4.0. If this protein found in the precipitate is the autolysin in question, this would mean that the pH has no effect on solubilisation of the protein. Figure 8 presents the results of the autolysis activity assay in a denaturing gel containing the cell walls as substrates. Gel 1 (Lanes 1 to 9) contain the precipitates (Pwc) and Gel 2 (lanes 1′ to 9′)contain the supernatants (Swc) at the following pHs: 4, 5, 6, 6.5, 7, 7.5, 8, 8.5, 9), Lane 10 is the uncentrifuged whole cell extract Awc as the control.

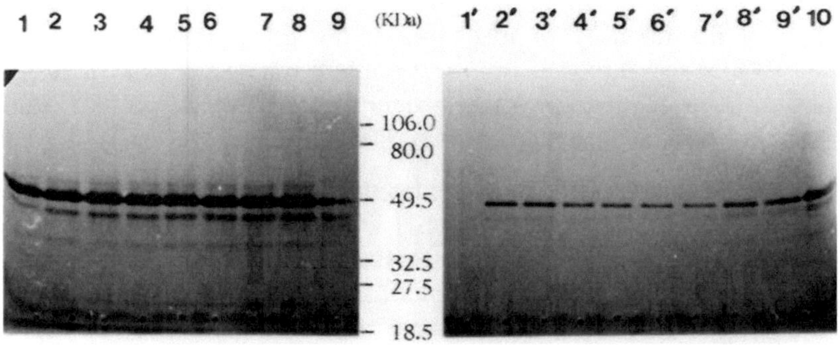

Fig. 8. The effect of pH on the solubility of PA49.5.

We can conclude from the results presented in figure 8 that the precipitate contains approximately 4x more autolysin than does the supernatant and that the pH has little effect on this results. We can therefore refute our first hypothesis that the working pH was reducing the solubility of the enzyme.

Hypothesis 2: The effect of a natural hydrolysis over time on the solubility of PA49.5

The whole cell extract (Awc) is incubated at 37°C for 8 h, and samples removed at hourly intervals. Autolytic activity was evaluated in an acrylamide gel containing cell walls as the substrate. Figure 9 presents the results of the autolysis analysis of the aliquots. Lanes 1 to 8 and 1′ to 8′ represents respectively the precipitates of the extract (Pwc) and the supernatants of the extract (Swc) for each hourly aliquot. Lanes 9 and 9′ are the precipitate and supernatant of the control at time 0. We can see the main autolysin with the molecular weight of 49.5 and a smaller minor autolysin. The incubation time did not have an effect on the extraction of the enzyme. All precipitates are identical to each other. The activity in the supernatants was much lower than that in the precipitates and did vary somewhat with time, the most activity being obtained after 6 hours of incubation. It is, however, difficult to assess if this difference is significant.

The use of an acetate buffer did not improve natural hydrolysis. A simple dilution in water indicates that the actual dilution is the primary reason for the presence of PA49.5 in the supernatant and not the pH nor the time of incubation. We already know it is insoluble in water, however, a dilution of the Awc in water could explain why the PA49.5 is in an aggregate form thereby inhibiting its solubilization.

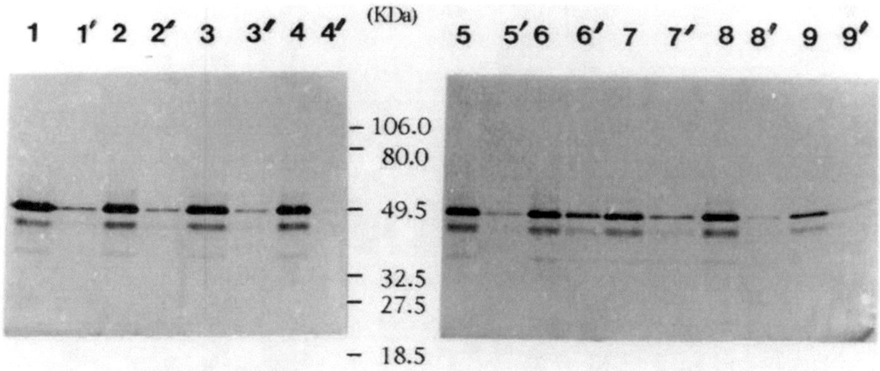

Fig. 9. The effect of natural autolysis of the whole cell extract (Awc) over time.

In an assay to study the effects of autolysis *in vitro* on enzyme solubility, the cell wall substrate was added to the whole cell extract (Awc) and incubated for 16 h. Autolysis was determined by measuring the reduction of optical density over time. Autolysis induced *in vitro* by the Awc autolysins on the cell wall substrate increased with time. Figure 10 shows that after 16 h of incubation autolysis reached 52%. At the 8 h point on Figure 9 only 35.55% autolysis has actually been achieved.

The presence of lytic activity only after extraction in LiCl shows that most of the autolysines of the Awc are soluble, contrary to the results obtained with the precipitate and the supernatant (results not shown). Analysed after treatment in 3 M and 5 M LiCl for 5 hours, the enzymes in the precipitate and the lyophilisate of the supernatant showed no lytic activity in the acrylamide gel containing the cell wall substrate, except for the untreated control. This suggests that the free autolysines were unstable in the LiCl and can induce the inhibitory effects of these free enzymes. It has been previously reported that LiCL can be inhibitory when the extracted product is not immediately suspended in water before dialysis (Brown, 1973; Shockman et Höltje, 1994; Lawrence et Glaser, 1972; Herbold et Glaser, 1975b).

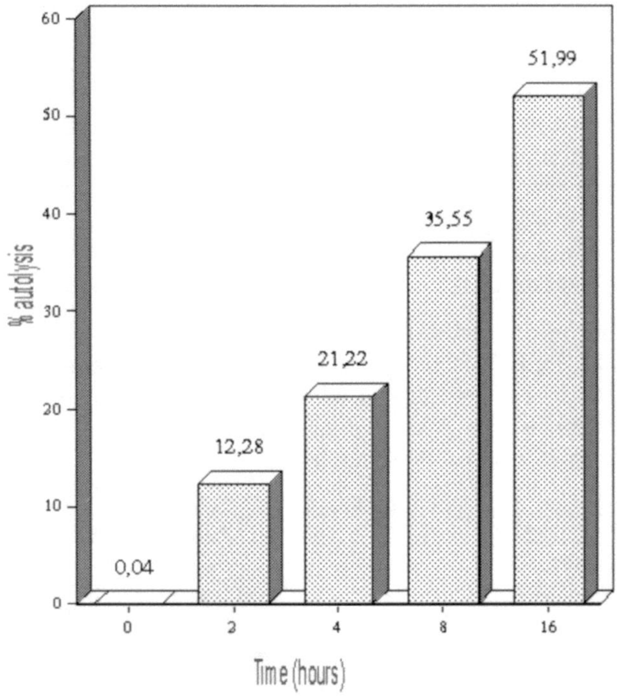

Fig. 10. The effects of autolysis by whole cell extract (Awc) *in vitro* on suspended cell walls

Hypothesis 3: The action of detergents on the solubility of PA49.5

Various concentrations of CHAPS (Zwitterionic; 0.2, 2.0 and 6.0%), Triton X-100 (non ionic; 0.2, 2.0 and 6.0%) and LiCl (3 M, 5 M and 7 M) were separately added to the whole cell extract (Awc). After incubation, the mixtures were centrifuged, the precipitate was resuspended in water and both precipitate and supernatant were assayed for autolytic activity as previously described. The results show that the precipitate and supernatants for the CHAPS and Triton-X treatments are very similar to each other. There does appear to be more solubilization of the autolysin with these detergents as the activity in the supernatant is now approximately half of that found in the precipitate. At the highest CHAPS concentration there does appear to be some inhibition of solubility as the activity in the supernatant is reduced with this treatment. In fact, for both these detergents, the enzyme activity decreases as the concentration of the detergent increases. The difference between these two detergents is that CHAPS is easier to eliminate by dialysis, while Triton X-100 is very difficult to eliminate from the extract. A concentration of between 0.2 and 2% CHAPS would provide excellent results.

After the treatment with LiCl and the subsequent dialysis, most of the autolysin was found in the precipitate. This would suggest that the compound that renders the enzyme insoluble is also found in the supernatant due to the effects of LiCl. It is only after dialysis that the insoluble complex is reconstituted. A dilution of 50% shows that a part of this complex is solubilized. The concentration of the initial whole cell extract (Awc) can explain this phenomenon; the lower the concentration of the extract, the greater the solubilization. Dilution of the extract by 50% is not however sufficient to resolve the insolubility problem.

We also evaluated the effect of incubation time (2, 4, 8 and 24 hours) in the presence of CHAPS, lysozyme and LiCl. CHAPS had a positive effect on the solubilization of the enzyme, however the incubation time did not affect the extraction of the enzyme. For the samples treated with lysozyme we found that the intensities of the activity in the supernatants tested was identical to that of the control. Higher concentrations of lysozyme did however negatively affect the amount of autolysin in the supernatant. We did notice the presence of a new autolysin in the extracts treated with lysozyme. This was a very important discovery as we can conclude that there must be trace amounts of cell wall still present in the extract on which the lysozyme is working. Lysozyme is involved in the degradation of the cell wall, in particular the cleavage of bonds between specific peptidoglycans (N-acetylmuramic acid and N-acetlyglucosamine) (Cottagnoud & Tomawsz, 1993). The fact that we did not observe more PA49.5 in the supernatant of extracts treated with lysozyme indicates that the compounds involved in binding the PA49.5 into an insoluble complex are not in fact cell wall debris. The time course study of the LiCl treatment indicates once again that most of the enzyme is present in the precipitate. This indicates that the dialysis performed during the extraction must be the determining step in the production of this insoluble complex.

When the original cell extract (Awc) produced in 5M LiCl is diluted and incubated for 2 or 4 hours, we observed interesting results. The incubation time did have an effect on enzyme activity as compared to the control, and we found that the more dilute extract had more enzyme activity. At a dilution of 1% the activity in the supernatant was approximately equal to that of the precipitate. At 10% concentration, the activity was found primarily in the precipitate. Since there is an effect of extract concentration on the solubility, we treated several dilutions of cell extract (Acw; 1, 3.5 and 7%) with different concentrations of LiCl (1, 2.5 and 5M) to solubilize the lipid-PA49.5 complexes. These extracts were then incubated and dialyzed after centrifugation as previously described. The combination of 0.25M LiCl with the dilution of 7% Awc was the only one to have intense activity in the supernatant, however most of the activity still remains in the precipitate.

The combined effects of detergents and heat were then evaluated on a dilute sample of the whole cell extract. Various concentrations of LiCl, SDS (0,1M, 0.5M and 1M), and CHAPS (0.04, 0.08, 0.1, 0.2% and 0.4%, w/v) were added to a dilution of 3% Awc treated and incubated at 100°C. After incubation with the various treatments, the extracts were centrifuged and the precipitation was resuspended in water. The supernatant was once again dialysed then concentrated using a Speed-Vac and the lyophilisate was resuspended in a denaturing buffer. Activity was assayed as previously described.

The results show that the combination of temperature and detergents does influence the solubility of the enzyme. We found that the treatments with LiCl did allow the solubilization of the enzyme, however some activity still remained in the precipitate. The LiCl was however less efficient than SDS which completely solubilized the enzyme, with all the activity being present in the supernatant. We observed that this intensity of the activity in the supernatant decreased as the concentration of SDS increased. When analyzed in a non denaturing native gel electrophoresis, the controls and the SDS treated extracts did not have any activity, however the LiCl treated extracts did have some weak activity. This suggests that the presence of LiCl, at least in the short term, does protect the solubilized protein against effects of the temperature. Also, the absence of activity in the SDS treated extracts confirms that this detergent in fact inhibits the solubilization of the enzyme.

Contrary to the results obtained with the detergent SDS, the detergent CHAPS completely solubilized the enzyme under these conditions. A concentration of 0.1% CHAPS is sufficient for the solubilization. It has been suggested that the presence of CHAPS in a solubilized protein can interfere with enzyme activity assays (Hjelmeland & Chrambach, 1984). In this case the conditions of the assay may have to be adjusted. The solubilizing effect of CHAPS does require a heat treatment. Unheated samples have enzyme activity only in the precipitate, not the supernatant, illustrating the importance of the heat treatment with CHAPS.

In an attempt to eliminate contaminating proteins from the cell extract, we performed sequential precipitations in the presence of CHAPS. In this process, the initial extract (Awc) was centrifuged and the supernatant (S1) removed. The precipitate (P1) was resuspended in water and centrifuged. The supernatant was removed and the precipitate (P2) was resuspended in water and the process was repeated 3 more times to produce ultimately S1 to S5 and P5. The final precipitate (P5) was resuspended in 1% CHAPS and heated for 2 minutes at 100°C. After cooling, the suspension was centrifuged and the supernatant was dialysed and lyophilized giving the CHAPS supernatant (Sc). The precipitate was dissolved in water to give the CHAPS precipitate (Pc). The CHAPS supernatant was further processed by centrifugation to produce the final supernatant (Sf) and the final precipitate (Pf). The following samples were then evaluated for PA49.5 activity in a polyacrylamide gel containing the cell wall as the substrate: S1-S5 combined, Sc, Sf, P1, P5, Pc and Pf. The results of this experiment are presented in figure 10. We notice that there is more enzyme activity in the last precipitate P5 as compared to the first precipitate (P1). The absence of activity in the CHAPS precipitate (Pc) indicate the complete solubilization of the enzyme, as confirmed by the strong level of activity in the CHAPS supernatant (Sc).The presence of the enzyme in the first five supernatants (S1-S5) indicate that there is a loss of enzyme throughout the process, representing approximately one half of the enzyme. We can also observe on figure 10 that when the CHAPS supernatant (Sc) is dialysed, the PA49.5 once again is unsolubilized and is found in the final precipitate (Pf). No enzyme activity was found in the final supernatant (Sf). Therefore, we can conclude that the elimination of the CHAPS detergent by dialysis stimulates the production of the insoluble PA49.5 complex. We also conclude that the lipophylic compound that binds the PA49.5 is present in the solubilized extract and that neither dialysis nor centrifugation can remove it completely. The detergent CHAPS is the only compound capable of being dialysed that does not compromise the activity of the autolysin (Hjelmeland & Chrambach, 1984), however, it must not be complexed.

All these results bring us back to our starting point, the difficulty in purifying this autolysin. We propose that during dialysis three types of insoluble complexes can be formed: PA49.5-insoluble compound complex, PA49.5-CHAPS complex, or insoluble compound-CHAPS complex. In the first case, we have the situation where the insoluble complex is present in the actual extraction conditions used. However, in the second case, the PA49.5-CHAPS complex, there is the high probability that this will be eliminated by dialysis. This bond is weaker than that between the PA49.5 and the insoluble compound that is probably a teichoic or fatty acid (Brown et al., 1970). Thus, after dialysis, it would be possible to identify specific activity if this second complex was formed. The third complex is more difficult to explain. It could in fact intervene as a contaminant in the second type of complex proposed. The presence of the insoluble compound-CHAPS complex would greatly impact the purification of PA49.5. It would therefore be advantageous to find an alternative detergent, perhaps non-ionic, that could more easily be dialyzed and does not denature the enzyme in question or facilitate the production of unwanted bonds. CHAPs is a zwitterionic detergent, having both a positive and negative charge that cancel each other making it an uncharged detergent. PA49.5 is an acidic autolysin, meaning it has a negative charge. It could therefore form a complex with the positively charged end of the CHAPS detergent. The insoluble compound could also be charged or neutral. If it is negatively or positively charged, it could bind to the autolysine and/or the CHAPS. If this is the case, the problem of solubility would present itself as we experienced it in our work.

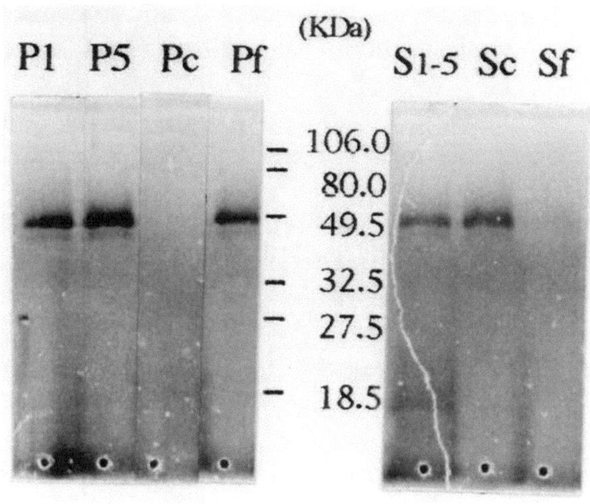

Fig. 11. The effects of sequential precipitations combined with CHAPS on the solubility of the enzyme PA49.5

Our work allowed us to comprehend more clearly the challenges encountered in the purification of an autolysin such as PA49.5 from *Lactococcus lactis* sp *cremoris*. The extraction from whole cells rather than cell wall debris was much more effective. Our results from the ammonium sulfate precipitation indicate the limitations of this procedure for the isolation of an insoluble protein. Our results were quite interesting regarding the use of LiCl during the extraction of the enzyme, however, the use of non-ionic detergents should be investigated. Our study will also allow other researchers to avoid some of the difficulties that might be encountered during the purification of cytoplasmic or membrane bound enzymes. The results of our study clearly show that we must absolutely take into account the nature of the various detergents, the extractions conditions and the specific type of enzyme when deciding on a purification strategy

2. References

Bonner, P.L.R. (2007). Protein purification. Taylor and Francis Group, New York, NY.

Brown, W.C. (1972). Binding and release from cell walls: a unique approach to the purification of autolysins, *Biochemical and Biophysical Research Communications* Vol 47: 993–996.

Brown, W.C., Fraser, D.K. & Young, F.E. (1970). Problems in purification of a *Bacillus subtilis* autolytic enzyme caused by association with teichoic acid, *Biochemica Biophysica Acta*, Vol.198: 308-315.

Brown, W.C. (1973). Rapid method for extracting autolysins from Bacillus subtilis. Applied Microbiology 25, 295–300.

Chen W., Druhan L.J., Chen C.-A., Hemann C., Chen Y.-R., Berka V., Tsai A.-L. & Zweier J.L. (2010). Peroxynitrite Induces Destruction of the Tetrahydrobiopterin and Heme in Endothelial Nitric Oxide Synthase: Transition from Reversible to Irreversible Enzyme Inhibition, *Biochemistry* Vol.49: 3129–3137.

Cottagnoud, p. & Tomasz, A. (1993). Triggering of pneumococcal autolysis by lysozyme, *The Journal of infectious diseases* Vol.167: 684-690.

Dako, E., Asselin, A., & Simard, E.R. (2003b). Partial purification and characterization of the main autolysins from Lactococcus lactis subsp, cremoris atcc 9596 *Journal of Food Technology* Vol.1 (2): 63–74.

Dako, E., El Soda, M., Vuillemard, J.C. & Simard, R.E. (1995). Autolytic properties and aminopeptidase activities of lactic acid bacteria, *Food Research International* Vol.28: 503-509.

Dako, E., Jankowski, C.J. Bernier, A.M., Asselin, A., & Simard, R.E. (2008). A new approach for the purification and characterisation of PA49.5, the main prebiotic of Lactococcus lactis subsp. Cremoris, *International Journal of Food Microbiology* Vol.126: 186-194.

Dolinger, D.L., Daneo-Moore, L. & Shockman, G.D. (1989). The second peptidoglycan hydrolase of Streptococcus faecium ATCC 9790 covalently binds penicillin, *Journal of Bacteriology* Vol.171(8): 4355-4361.

Doonan, S. (2004a). Concentration of extracts, *in* Paul Cutler, Protein Purification Protocoles: second edition, Humana Press Inc., Totowa, NJ. pp. 85-90.

Doonan, S. (2004b). Bulk purification by fractional precipitation, *in* Paul Cutler, Protein Purification Protocoles: second edition, Humana Press Inc., Totowa, NJ. pp. 117-124

Fisher, W., Koch, H.U. & Rösel, P. (1980). Alanine ester-containing native lipoteichoic acids do not act as lipoteichoic acid carrier. Isolation, structural and functional Characterization, *Journal of Biological Chemistry* Vol.255: 4555–4562.

Garavito, R.M. (1991). In *Crystallisation membrane* (Michel. Ed). CRC Press, Boca Raton, FL, P. 89.

Helenius, A., McCaslin, D.F.R. & Tanford, C. (1979). Properties of detergents, *Methods in Enzymology* Vol.56: 734–749.

Herbold, D.R. & Glaser, L. (1975). Interaction of N-acetylmuramic acid L-alanine amidase with cell wall polymers. Journal of Biological Chemistry 250, 7231–7238.

Hjelmeland, L.M. & Chrambach, A. (1984). Solubilization of functional membrane proteins, *Methods in Enzymology* Vol.104: 305-318.

Höltje, J.V. & Tomasz, A. (1975). Specific recognition of choline residues in the cell wall teichoic acid by the *N*-acetylmuramyl-L-alanine amidase of *Pneumococcus, Journal of Biological Chemistry* Vol.250: 6072-6076.

Kawagishi, S., Araki, Y. & Ito, E. (1980). Separation and characterization of an autolytic endo-ß-glucosaminidase from *Bacillus cereus, European Journal of Biochemistry* Vol.112: 273-281.

Lawrence, C. & Glaser, L. (1972). Purification of *N*-acetylmuramic acid-L-alanine amidase from *Bacillus magneterium, Journal of biological Chemistry* Vol.247: 5391-5397.

Leclerc, D. & Asselin, A. (1989). Detection of bacterial cell wall hydrolases after denaturing polyacrylamide gel electrophoresis. Canadian Journal of Microbiology 35, 749–753.

Lemee, R., Lortal, S., Cesselin, B. & Heijenoort, J.V. (1994). Involvement of an N-Acetylglucosaminidase in Autolysis of *Propionibacterium freudenreichii* CNRZ 725, *Applied and Environmental Microbiology* Vol.60(2): 4351-4358.

Lemee, R., Lortal, S. & Heijenoort, J.V. (1995). Autolysis of dairy propionibacteria: isolation and renaturing gel electrophoresis of the autolysins of Propionibacterium freudenreichii CNRZ 725. Lait 75, 345–365.

Lever T.M., Cogdell R.J. & Lindsay J.G. (1994). In Membrane Protein Expression System - A Guide (G.W. Gould, ed). Portland Press, London, pp. 1.

Linke, D. (2009). Detergents: an overview, *Methods of Enzymology* Vol.463: 603-617.

Lortal, S., Lemée, R. & Valence, F. (1997). Autolysis of thermophilic lactobacilli and dairy propionibacteria: a review, *Lait* Vol.77: 133–150.

Marshak, D.R., Kadonaga, J.T., Burgess, R.R., Knuth, M.W., Brennan,W.A. & Lin, S.-H. (1996). Strategies for Protein Purification and Characterization. *Cold Spring Harbor Laboratory Press*, pp. 1-396.

Motoyuki S., Komatsuzawa H., Akiyama T., Hong Y.-M., Oshida T., Miyake Y., Yamaguchi T. & Suginaka H. (1995). Identification of Endo-b-*N*-Acetylglucosaminidase and *N*-Acetylmuramyl-L-Alanine Amidase as Cluster-Dispersing Enzymes in *Staphylococcus aureus, Journal of Bacteriology*, pp. 1491–1496

O'Fagain, C. (2004). Lyophylisation of proteins, *in* Paul Cutler, Protein Purification Protocoles: second edition, Humana Press Inc., Totowa, NJ. pp. 309-322.

Potvin, C., Leclerc, D., Tremblay, G., Asselin, A. & Bellmare, G. (1988). Cloning, sequencingand expression of a Bacillus bacteriolytic enzyme in E, coli, *Molecular and General Genetics* Vol.214: 241–248.

Privé, J.J. (2007). Detergents for the stabilization and crystallization of membrane proteins. *Methods* Vol.41: 388-397.

Shockman, G.D. & Höltje, J.V. (1994). Microbial peptidoglycan (murein) hydrolases. Comprehensive biochemistry. I. In: Ghuysen, J.M., Hakenbeck, R. (Eds.), *Bacterial Wall Elsevier of Science B.V.*, London, pp. 133–166.

Shockman, G.D., Pooley, H.M., & Thompson, J.S. (1967). Autolytic enzyme system of Streptococcus faecalis. III. Localization of the autolysin at the sites of cell wall synthesis, *Journal of Bacteriology* Vol.94 (5): 1525–1530.

Valence, F. & Lortal, S. (1995). Zymogram and preliminary characterization of Lactobacillus helveticus autolysins, *Applied and Environmental Microbiology* Vol.61: 3391–3399.

Ward, W. W. & Swiatek, G. (2009). Protein purification. *Current Analytical Chemistry* Vol.5: 85-105.

3

Nicotinamide Phosphoribosyltransferase Inhibitors

Dan Wu[1], Dilyara Cheranova[1], Daniel P. Heruth[1], Li Qin Zhang[1] and Shui Qing Ye[1,2]
[1]*Department of Pediatrics, University of Missouri School of Medicine, Kansas City*
[2]*Department of Biomedical and Health Informatics, Children's Mercy Hospitals and Clinics, University of Missouri School of Medicine, Kansas City*
USA

1. Introduction

Nicotinamide phosphoribosyltransferase (NAMPT, EC 2.4.2.12) catalyzes the condensation of nicotinamide with 5-phosphoribosyl-1-pyrophosphate (PRPP) to yield nicotinamide mononucleotide (NMN), a rate limiting enzyme in a mammalian salvage pathway of nicotinamide adenine dinucleotide (NAD) synthesis. Human NAMPT consists of 491 amino acids with a molecular weight of 52 kDa **(Samal et al., 1994)**. NAMPT was initially named pre-B-cell colony-enhancing factor (PBEF) for its growth factor like function on promoting pre-B-cell colony formation in the presence of stem cell factor plus interleukin 7 **(Samal et al., 1994)**. Martin et al. **(2001)** found that the gene encoding the bacterial *Haemophilus ducreyi* nicotinamide phosphoribosyltransferase (nadV) had a significant homology to the mammalian PBEF gene. Since then, Rongvaux et al.**(2002)**, Revollo et al. **(2004)** and others **(van der Veer et al., 2005)** have characterized the enzymological features of mammalian NAMPT. In 2005, NAMPT/PBEF was named visfatin, a "new visceral fat-derived hormone", which is an adipocyte-derived adipokine that induces insulin mimetic effects **(Fukuhara et al., 2005)**. To avoid the confusion, the name NAMPT will be used throughout this chapter since NAMPT was approved as the official name of this gene by the Human Genome Organization Gene Nomenclature Committee.

Because of NAMPT's pleiotropic functions in a variety of physiological processes, the dysregulation of NAMPT activity has been implicated in the pathogenesis of a number of human diseases or conditions such as acute lung injury, aging, atherosclerosis, cancer, diabetes, obesity related disease, rheumatoid arthritis and sepsis **(Borradaile & Pickering, 2009; Galli et al., 2010; Moschen et al., 2010)**. Therefore, targeted inhibition of NAMPT has become an attractive therapeutic strategy for these related diseases. Inhibition of NAMPT has been actively pursued as a potential new therapeutic modality to treat patients with cancer in clinical trials, to inhibit rheumatoid arthritis and to attenuate acute lung injury. The list of 'inhibition of NAMPT based therapy' is expanding rapidly. The initially tested inhibitors include FK866 (now called APO866, a small chemical molecule), antisense oligo, siRNA or shRNA and antibody.

This chapter will review the latest findings on NAMPT inhibitors in human clinical trials, animal studies and cell cultural experiments from published literature and our research findings. The first part will briefly cover the current understanding of NAMPT physiology. The second part will describe the pathological roles of NAMPT in various human diseases. The third and major component of this chapter will focus on the development, action mechanisms, and applications of various inhibitors of NAMPT. The biochemical and molecular characterization of various inhibitors to NAMPT looms large in this part. Perspective remarks at the end will provide some food for thought for future directions on the development of new and improved NAMPT inhibitors.

2. Physiology of NAMPT

This section deals with the three major functions of NAMPT: growth factor, cytokine and nicotinamide phosphoribosyltransferase. Accumulating evidence suggests that NAMPT can function as a growth factor or a cytokine though the underlying molecular mechanisms remain to be established. It is beyond dispute that NAMPT can function as a nicotinamide phosphoribosyltransferase.

2.1 Growth factor

Growth factor generally refers to a naturally occurring protein capable of stimulating cellular growth, proliferation and differentiation. Growth factors are important for regulating a variety of cellular processes. Several studies indicate that NAMPT may function as a growth factor. Samal et al. **(1994)** first found that NAMPT can enhance significantly the number of pre-B-cell colonies derived from normal human or mouse bone marrow by at least 70% in the presence of both IL-7 and stem cell factor. Thus, the authors first named this protein as pre-B-cell colony enhancing factor. Van der Veer et al. **(2005)** reported that NAMPT can promote vascular smooth muscle cell maturation. Human smooth muscle cells transduced with the NAMPT gene had enhanced survival. Fukuhara and co-workers **(2005)** proposed NAMPT as a visfatin, an adipokine produced by visceral fat that can engage and activate the insulin receptor (IR). Although this original publication was retracted **(Fukuhara et al., 2007)** because of questions regarding the reproducibility of the NAMPT/IR interaction from different preparations of recombinant NAMPT protein, Xie et al. found that NAMPT exerts an insulin-like activity as a growth factor for osteoblasts. They noticed that the effects of NAMPT, such as glucose uptake, proliferation, and type I collagen enhancement in cultured human osteoblast-like cells, bore a close resemblance to those of insulin and were inhibited by hydroxy-2-naphthalenylmethylphosphonic acid tris-acetoxymethyl ester (HNMPA-[AM]3), a specific inhibitor of IR tyrosine kinase activity **(Xie et al., 2007)**.

2.2 Cytokine

Cytokine is sometimes used interchangeably among scientists with the term growth factor. The term cytokine encompasses a large and diverse family of polypeptide regulators that are produced widely throughout the body by cells of diverse embryological origin. Their actions may be grouped as autocrine, paracrine and endocrine. NAMPT may be added to the list of cytokines. The first NAMPT cDNA was screened out using a degenerate oligonucleotide probe designed on the basis of the similarity in the coding sequences of five different cytokines (GM-CSF, IL-2, IL-1β, IL-6 and IL-13), at the signal peptidase processing site, though the DNA or protein sequence of NAMPT bears no homology to other known cytokines **(Samal et al., 1994)**. Ognjanovic and colleagues reported that recombinant human NAMPT (rhNAMPT) treatment of WISH cells and fetal membrane explants significantly increased IL-6 and IL-8 gene expression **(Ognjanovic et al., 2001; Ognjanovic & Bryant-Greenwood, 2002)**. We also found that an overexpression of NAMPT significantly augmented IL-8 secretion and mRNA expression in A549 cells, a human pulmonary carcinoma type II epithelial cell line, and in human pulmonary artery endothelial cells **(Li, H. et al., 2008; Liu, P. et al., 2009)**. It also significantly augmented IL-1β-mediated cell permeability. The opposite results were obtained with the knockdown of NAMPT expression. NAMPT expression also affected the expression of two other inflammatory cytokines (IL-16 and CCR3) **(Li, H. et al., 2008; Liu, P. et al., 2009)**. Hong et al. **(2008)** demonstrated that rhNAMPT functions as a direct rat neutrophil chemotactic factor in in vitro studies. They also detected a marked increase in bronchoalveolar lavage leukocytes after the intratracheal injection of rhNAMPT into C57BL/6J mice. Thus, NAMPT behaves like a chemokine.

2.3 Nicotinamide phosphoribosyltransferase

The clue that PBEF could be a nicotinamide phosphoribosyl transferase was first obtained by the work of Martin et al. **(2001)**. They found that the sequence of nadV gene is homologous to that of human NAMPT, suggesting that mammalian PBEF may also function as a NAMPT. Rongvaux et al. **(2002)** verified that similarly to its microbial counterpart, mouse PBEF is a NAMPT, catalyzing the condensation of nicotinamide with PRPP to yield NMN. Revollo et al. **(2004)** demonstrated further that NAMPT catalyzes a rate-limiting step in a salvage pathway of the mammalian NAD biosynthesis. Van der Veer et al. **(2005)** established that enhanced NAMPT activity is linked directly to the lengthening of the cellular lifespan of both human smooth muscle cells and fibroblasts. Recent work by Revollo et al. **(2007)** revealed that NAMPT regulates insulin secretion in beta cells as a systemic NAD biosynthetic enzyme. Because the salvage pathway of NAD synthesis has a faster rate and is more efficient than that of de novo NAD synthesis, it is conceivable that NAMPT plays an important role in a variety of physiological processes via the regulation of NAD synthesis.

3. Pathophysiology of NAMPT

This section presents potential roles of NAMPT in human diseases. The dysregulation of the NAMPT gene has been implicated in the susceptibility and pathogenesis of a number of human diseases and conditions because of its pleiotropic physiological functions. There is strong supporting evidence that both tissue and circulating NAMPT levels change in acute respiratory distress syndrome, aging, atherosclerosis, cancer, diabetes, rheumatoid arthritis, and sepsis.

3.1 Acute respiratory distress syndrome (ARDS)

ARDS is the severe form of acute lung injury (ALI). ALI is characterized by pulmonary inflammation, non-cardiogenic edema, and severe systemic hypoxemia **(Ware & Matthay, 2000; Wheeler & Bernard, 2007)**. One of the earliest manifestations of ALI is a diffuse, intense inflammatory process and damage to both endothelial and epithelial cell barriers, resulting in marked extravasation of vascular fluid into the alveolar airspace **(Matthay et al.,**

2003). A number of inflammatory cytokines including tumor necrosis factor-alpha (TNFα) and interleukin 8 (IL-8) can induce or aggravate the inflammation of endothelial and epithelial cells, leading to these barrier dysfunctions **(Frank et al., 2006)** and pathogenesis of ALI. The mortality and morbidity of ALI/ARDS remain high since the etiology and molecular pathogenesis are still not understood completely.

To identify novel candidate ALI genes, our lab employed a high-throughput functional genomics approach and found that NAMPT was a highly expressed gene in canine, murine and human ALI **(Ye et al., 2005a)**. These results suggest that NAMPT may be a potential biomarker in ALI. Analysis of single nucleotide polymorphisms (SNPs) in the NAMPT gene proximal promoter region indicated that a GC haplotype had a higher risk (nearly 8 fold) of ALI, while a TT haplotype had a lower risk of ALI **(Ye et al., 2005a)**. Our findings were confirmed and extended by Bajwa et al. **(2007)**. These results support that NAMPT is a genetic marker for ALI. To investigate further the role and molecular mechanism underlying NAMPT in the pathogenesis of ALI, we showed that heterozygous NAMPT (+/-) mice were protected significantly from severe ventilator associated lung injury (VALI) **(Hong et al., 2008)**. We also found that the NAMPT-specific siRNA would attenuate thrombin-induced decreases in human lung endothelial cell barrier function, increased cytoskeletal rearrangement, and secretion of the proinflammatory cytokine IL-8 **(Ye et al., 2005b)**. Overexpression of NAMPT significantly augmented IL-8 secretion and IL-8 mRNA expression in A549 cells and human pulmonary artery endothelial cells (HPAEC), respectively. NAMPT expression also affected the expression of two other inflammatory cytokines (IL-16 and CCR3) **(Li, H. et al., 2008; Liu, P. et al., 2009)**. These results reveal that NAMPT overexpression may adversely affect pulmonary cell barrier function, the deregulation of which is the well-recognized feature in the pathogenesis of ALI.

3.2 Aging

Aging is the accumulation of changes in an organism over time. Several evidences suggest that NAMPT may be an important regulator in aging. Axonal degeneration occurs in many neurodegenerative diseases. Sasaki et al. **(2006)** found that NAMPT can delay axon degeneration in the presence of nicotinamide in an in vitro Wallerian degeneration assay. These results suggest that increased activity of the NAD biosynthetic pathway stemming from nicotinamide promotes axonal protection. Van der Veer et al. **(2007)** reported that NAMPT can extend the lifespan of human smooth muscle cells. They found that replicative senescence of smooth muscle cells was preceded by a marked decline in the expression and activity of NAMPT. Furthermore, reducing NAMPT activity with the antagonist FK866 induced premature senescence in smooth muscle cells. NAMPT overexpression also reduced the fraction of p53 that was acetylated on lysine 382, a target of SIRT1, suppressed an age-related increase in p53 expression, and increased the rate of p53 degradation. Moreover, add-back of p53 with recombinant adenovirus blocked the anti-aging effects of NAMPT **(van der Veer et al., 2007)**. These data indicate that NAMPT is a longevity protein that can add stress-resistant life to human smooth muscle cells by optimizing SIRT1-mediated p53 degradation. Recently, Benigi et al. **(2009)** noticed that the longevity phenotype in angiotension II type 1 receptor knockout mice was associated with an increased number of mitochondria and up-regulation of the prosurvival genes NAMPT and sirtuin 3 (Sirt3) in the kidney. They postulated that disruption of angiotension II type 1 receptor promotes longevity in mice, possibly through the attenuation of oxidative stress and overexpression of prosurvival genes such as NAMPT and Sirt 3.

3.3 Atherosclerosis

Atherosclerosis is a disease affecting arterial blood vessels. In a microarray experiment, Dahl et al. **(2007)** identified that NAMPT expression was markedly enhanced in carotid plaques from symptomatic individuals compared with plaques from asymptomatic individuals. Zhong et al. **(2008)** also reported that serum NAMPT was increased in patients with carotid plaques. Cheng et al. **(2008)** noticed that NAMPT levels in epicardial and abdominal adipose tissues were significantly higher in coronary artery disease (CAD) patients relative to control subjects. In addition, significantly higher tissue NAMPT levels from abdominal fat depots were found compared to those from epicardial fat in CAD patients. These findings suggest that abdominal adiposity may play a more significant role than epicardial fat in the pathogenesis of coronary atherosclerosis. More studies are warranted to firmly establish the relationship between NAMPT and atherosclerosis and to elucidate the role(s) and molecular mechanisms of NAMPT in the pathogenesis of atherosclerosis.

3.4 Cancer

Molecular screening, epidemiological survey and pharmacological studies have indicated that NAMPT may be an attractive diagnostic and drug target for cancer therapy. Hufton et al. **(1999)** first noticed that NAMPT expression was increased 6 fold in primary colorectal cancer over the normal control using the suppression subtractive cDNA hybridization technique. This result was confirmed at the protein and tissue levels by both western blotting and immunohistochemical analyses **(Van Beijnum et al., 2002)**. Using cDNA microarray based expression profiling of different grades of astrocytomas, Reddy et al. **(2008)** identified several fold increased levels of NAMPT transcripts and protein in glioblastoma samples, suggesting that NAMPT could be a potential malignant astrocytoma serum marker and prognostic indicator in glioblastoma.

Through a chemical screen to find new antitumor drugs, Hasmann and Schemainda **(2003)** identified the first low molecular weight compound, designated FK866 {the chemical name: (E)-N-[4-(1-benzoylpiperidin-4-yl) butyl]-3-(pyridin-3-yl) acrylamide}, which induced apoptosis by highly specific and potent inhibition of nicotinamide phosphoribosyltransferase in HepG2 human liver carcinoma cells. Using ^{1}H-decoupled phosphorus (^{31}P) magnetic resonance spectroscopy, Muruganandham et al. **(2005)** observed that FK866 (also known as APO866) increased apoptosis and subsequent radiation sensitivity in the mammary carcinoma. FK866 has been shown to have anti-tumor, anti-metastatic and antiangiogenic activities in a murine renal cell carcinoma model **(Drevs et al., 2003)**. The three dimensional structural analysis of the NAMPT–FK866 complex by three groups revealed that the FK866 compound binds at the nicotinamide-binding site of NAMPT to competitively compete directly with the nicotinamide substrate to inhibit its activity **(Khan et al., 2006; Kim, M.K. et al., 2006; Wang et al., 2006)**. These structural analyses provided a molecular basis for the inhibition of FK866 on NAMPT and a starting point for the development of new anticancer agents.

Accumulated evidence indicates that at least three molecular mechanisms may implicate NAMPT in the pathogenesis of cancer. First, NAMPT inhibits apoptosis of tumor cells via its role as a key enzyme in NAD biosynthetic salvage pathway. Second, Li Y. et al. **(2008)** reported that NAMPT could activate an IL-6/STAT3 survival signaling pathway via a non-enzymatic mechanism. Third, increased NAMPT activity has been associated with angiogenesis and neovascularization. Kim et al. **(2008)** found that NAMPT potently stimulates in vivo neovascularization in chick chorioallantoic membranes and in implanted mouse Matrigel plugs. Furthermore, Bae et al. **(2009)** reported that NAMPT-induced angiogenesis is mediated by endothelial fibroblast growth factor-2 (FGF-2). Therefore, inhibition of NAMPT activity provides a important anticancer target, since neovascularization by angiogenesis is the prerequisite for tumor growth and expansion **(Chamberlain, 2008)**.

3.5 Diabetes mellitus

Diabetes mellitus, simply referred to as diabetes, is a syndrome of disordered metabolism with hyperglycemia as a hallmark phenotype. Its long and extensively-sought hereditary and environmental causes are not fully known. Initial attention brought to the relationship between NAMPT and Type 2 diabetes was due to the work by Fukuhara et al. **(2005)**. They found that NAMPT is a secreted factor produced abundantly by visceral fat. Plasma NAMPT levels in NAMPT gene heterozygous knockout mice were only 2/3 of those in wild type mice but their plasma glucose level were significantly higher. This suggests that like insulin, NAMPT may have a physiological role in lowering plasma glucose levels. The authors found that NAMPT bound to the insulin receptor (IR) but not to the same site to which insulin binds. Similar to insulin, NAMPT could stimulate insulin signaling, such as inducing tyrosine phosphorylation of the IR, insulin receptor substrate-1 (IRS-1), and IRS-2 in the liver. Taken together, the authors considered NAMPT as an insulin-mimetic and dubbed it as visfatin. Although this paper was withdrawn from *Science* due to the variation of different batches of recombinant NAMPT for their adipogenic and insulin-mimetic activities **(Fukuhara et al., 2007)**, this report has immediately and continuously drawn increased interest among biomedical researchers to the roles and mechanisms underpinning NAMPT in the pathogenesis of diabetes, obesity, insulin resistance, and metabolic syndrome.

The epidemiological survey of the relationship between NAMPT and Type 2 diabetes by Fukuhara et al. **(2005)** has been supported by a number of subsequent studies which reported that the level of plasma NAMPT and/or the amount of visceral fat NAMPT mRNA were increased significantly in or positively associated with Type 2 diabetes or obesity or insulin resistance or metabolic syndrome in patients under the baseline without any intervention of medication, surgery, and other reagents **(Berndt et al., 2005; Chen et al., 2006; Krzyzanowska et al., 2006; Dogru et al., 2007; Fernandez-Real et al., 2007; Lewandowski et al., 2007; Sandeep et al., 2007; Alghasham & Barakat, 2008; Botella-Carretero et al., 2008; Liang et al., 2008; Mazaki-Tovi et al., 2008; Retnakaran et al., 2008; Ziegelmeier et al., 2008; Hallschmid et al., 2009)**. Patients with Type 1 diabetes also had higher NAMPT concentrations than controls **(Haider et al., 2006; Lopez-Bermejo et al., 2006)**. However, other groups have obtained opposite findings **(Jian et al., 2006; Akturk et al., 2008; Kato et al., 2009)** or no changes of plasma NAMPT level in Type 2 diabetes **(Chan et al., 2006; Pagano et al., 2006; Takebayashi et al., 2007; Tsiotra et al., 2007; Palin et al., 2008)**. These conflicting findings on the correlation of plasma NAMPT level with diabetes may stem from four major reasons: small sample sizes, variable phenotyping criteria, different populations and ethnicities and assay variation. The biological role and molecular mechanism of NAMPT related to diabetes remains to be fully elucidated.

3.6 Rheumatoid arthritis

Rheumatoid arthritis [RA] is a chronic, systemic autoimmune disorder that most commonly causes inflammation and tissue damage in joints. Despite that the first recognized description of rheumatoid arthritis was made in 1800 by Dr. Augustin Jacob Landré-Beauvais in Paris **(Landre-Beauvais, 2001),** its pathogenesis remains incompletely understood. Otero et al. **(2006)** first reported that patients with rheumatoid arthritis showed higher plasma levels of NAMPT, leptin, and adiponectin than healthy controls. These findings were confirmed by Nowell et al. **(2006)**, Bretano et al. **(2007)** and Matsui et al. **(2008)**. When compared with osteoarthritis (OA) patients, Nowell and colleagues detected elevated levels of NAMPT in synovial fluid from RA patients **(Nowell et al., 2006)**. In a large survey of 167 RA patients and 91 control subjects, Rho et al. **(2009)** found that elevated levels of NAMPT correlated with both radiological joint destruction and mediators of inflammation.

NAMPT is a pleiotropic protein which can induce the expression of a number of genes (e.g. CCR2, CCR3, Cox-2, IL-6, IL-8, IL-16, ICAM1, MCP-1, MMP-2, MMP-9, VCAM1, VEGF) **(Ognjanovic et al., 2001; Kim, S.R. et al., 2007; Adya et al., 2008; Adya et al., 2008; Kim, S.R. et al., 2008; Li, H. et al., 2008; Adya et al., 2009; Liu, P. et al., 2009; Liu, S.W. et al., 2009)**. It is suggested that NAMPT affects the innate immune system's inflammatory response in the pathogenesis of rheumatoid arthritis via regulating expression of those inflammatory cytokines or activators. However, the mechanism by which NAMPT mediates the cytokine signaling cascade has not yet been determined fully.

3.7 Sepsis

Sepsis, a life-threatening disorder characterized by a whole-body inflammatory state caused by infection, is a frequent cause of ALI/ARDS. Jia et al. **(2004)** reported that NAMPT functions as a novel inflammatory cytokine that plays a requisite role in the delayed neutrophil apoptosis of clinical and experimental sepsis. They found that transcription of the NAMPT gene is increased in neutrophils from septic patients, while the prevention of NAMPT translation through the use of an antisense oligonucleotide largely restored the normal kinetics of apoptosis. Moreover, the incubation of quiescent neutrophils from healthy volunteers with recombinant NAMPT results in dose-dependent inhibition of apoptosis, while antisense NAMPT prevents the inhibition of apoptosis that results from exposure to lipopolysaccharide (LPS) or to a variety of host-derived inflammatory cytokines **(Jia et al., 2004)**. The authors postulate that the prolonged survival of activated neutrophils may be linked to sustained inflammation and the organ injury of sepsis.

Because of its multiple functional roles in physiology, the list of NAMPT involvement in various human diseases is expected to grow.

4. NAMPT inhibitors

As summarized in the first part of this chapter, NAMPT is considered as a rate-limiting enzyme of a mammalian synthetic pathway of NAD synthesis. Thus, NAMPT may exert its physiological and pathological roles by regulating the synthesis of NAD. NAD plays a major role in the regulation of several essential cellular processes. It's an essential coenzyme in metabolic pathways, and important in several biological processes including signal transduction **(Corda & Di Girolamo, 2003)**, DNA repair **(Menissier de Murcia et al., 2003)**, calcium homeostasis **(Lee, H.C., 2001)**, gene regulation **(Blander & Guarente, 2004)**, longevity **(Lin et al., 2000)**, genomic integrity **(Schreiber et al., 2006)** and apoptosis **(Wright et al., 1996)**. Additionally, as reviewed above, NAMPT is a pleiotropic protein which also functions via non-enzymatic mechanisms. Thus, NAMPT has become an attractive target in the treatment of many diseases. This section will focus on the development and progress of various NAMPT inhibitors and their applications.

4.1 Chemical inhibitors of NAMPT

4.1.1 FK866

The first NAMPT inhibitor, FK866 (also called APO866 or WK175, (E)-N-[4-(1-benzoylpiperidin-4-yl) butyl] acrylamide-3-(pyridin-3-yl), was reported by Hasmann and Schemainda, who found that FK866 induced apoptosis by a highly specific and potent inhibition of nicotinamide phosphoribosyltransferase in HepG2 human liver carcinoma cells **(Hasmann & Schemainda, 2003)**. FK866 has no primary effect on cellular energy metabolism and thus has no direct and immediate cytotoxicity, but rather gradually depletes the cells of a vital factor, NAD, by inhibiting NAMPT, which eventually triggers apoptosis. The authors proposed that FK866 may be used for treatment of diseases involving deregulated apoptosis, such as cancer, or as a sensitizer for genotoxic agents. Furthermore, FK866 may provide an important tool for investigation of the molecular triggers of the mitochondrial pathway leading to apoptosis through enabling temporal separation of decreased NAD levels from ATP breakdown and apoptosis **(Hasmann & Schemainda, 2003; Pittelli et al., 2010)**.

Depletion of cellular NAD levels leads to lowered ATP levels and the inhibition of poly (ADP-ribose) polymerases (PARPs) **(Khan et al., 2007)**. Cancer cells have a high demand of both PARP and ATP, and they also display higher energy requirements **(Hufton et al., 1999)**. Thus, cancer cells would be expected to be more sensitive than normal cells to the inhibition of NAD synthesis **(Hasmann & Schemainda, 2003; Billington et al., 2008; Nahimana et al., 2009)**. Therefore, since NAMPT acts as a key enzyme of NAD synthesis, its inhibitors could provide an effective cancer therapy. Additional reports have demonstrated that FK866 elicited massive cell death in primary leukemia cells and in numerous leukemia/lymphoma cell lines **(Nahimana et al., 2009; Zoppoli et al., 2010)**. FK866 has been shown to have anti-tumor, anti-metastatic and antiangiogenic activities in a murine renal cell carcinoma model. Nahimana et al. **(2009)** investigated the cytotoxic effects of FK866 in both in vitro and in vivo assays, using a cell panel of human hematologic malignancies and early and established human hematologic cancers, respectively. They observed in in vitro assays that FK866-induced apoptosis involved an initial decrease in intracellular NAD levels that were subsequently accompanied by decreases in intracellular ATP levels. In animal models of human acute myeloid leukemia, lymphoblastic lymphoma, and leukemia, FK866 as a single agent prevented and abrogated tumor growth without significant toxicity to the animals. These findings demonstrated that FK866 displayed strong anticancer activity in hematologic malignancies both in vitro and in vivo.

FK866 was also found to improve the sensitivity of other anticancer agents. The main objectives of combination chemotherapy are an increased response rate against the tumor and minimization of adverse effects of drugs without compromising efficacy of treatment. Using ^{1}H-decoupled phosphorus (^{31}P) magnetic resonance spectroscopy, Muruganandham et al. **(2005)** observed that FK866 increased cell death (apoptosis) and subsequent radiation sensitivity in the mammary carcinoma. Pogrebniak et al. **(2006)** treated THP-1 and K562 cells with FK866 and various cytotoxic agents: the antimetabolite Ara-C, the DNA-intercalating agent daunorubicin and the alkylating compounds 1-methyl-3-nitro-1-nitrosoguanidinium (MNNG) and melphalan. The results showed the anticancer activity of FK866 was particularly obvious in combination with substances like MNNG that cause NAD depletion chemo-sensitizing. Additionally, Yang et al. **(2010)** combined FK866 and the indoleamine 2,3-dioxygenase inhibitor L-1-methyl-tryptophan (L-1MT) in the treatment of mouse tumor models. The combination of FK866 and L-1MT had a better therapeutic effect than did either L-1MT or FK866 alone.

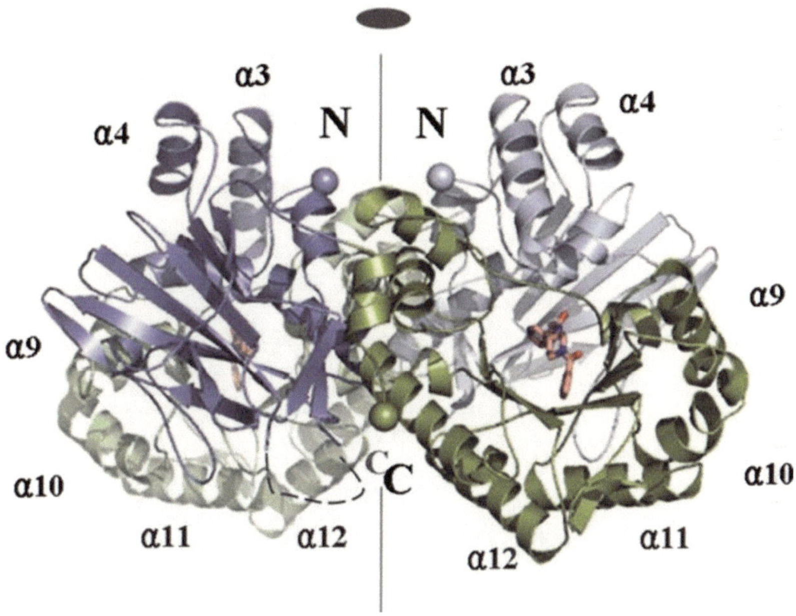

Fig. 1. Representative ribbon diagram of two FK-866 molecules binding to the NAMPT dimer. The two FK-866 molecules are shown in red. The two monomers are shown in slate and green (subunit A) and pale blue and green (subunit B), respectively. This figure is copied from Kim M.K. et al (2006) with permission from Elsevier.

The mechanism of FK866 inhibition of NAMPT has been well characterized. Three dimensional structural analyses of the NAMPT-FK866 complex revealed that FK866 binds at the nicotinamide-binding site of NAMPT, thus functioning as a competitive inhibitor of NAMPT enzymatic activity **(Khan et al., 2006; Kim, M.K. et al., 2006; Wang et al., 2006)**. A representative crystal structure of NAMPT-FK866 binding is shown in Figure 1 **(Kim, M.K. et al., 2006)**. FK-866 binds to the active site of NAMPT with higher affinity than either the substrate or the product does. The benzoylpiperidin group of FK-866 plays a key role in binding to NAMPT in hydrophobic interactions **(Kim, M.K. et al., 2006)**. These structural analyses provided a molecular basis for the inhibition of FK866 on NAMPT and a starting point for the development of new anticancer agents.

The first human study of FK866 in the treatment of cancer was reported by Holen et al. **(2008)**. They collected serial plasma and blood samples from 24 patients, with advanced solid tumors, treated with increasing doses of FK866. The patient age ranged from 34-78 and the dose of FK866 ranged from 0.144 mg/m2/h to 0.018mg/m2/h. The recommended dose for phase II clinical trial was 0.126mg/m2/h, which was given as a continuous 96 h infusion every 28 days. Thrombocytopenia was the main toxicity of FK866. Plasma analysis showed that FK866 did not affect VEGF concentration. FK866 has been used in phase II and I/II clinical trials against cancer, including advanced melanoma, cutaneous T-cell lymphoma and B-chronic lymphocytic leukemia **(Holen et al., 2008; Olesen et al., 2010a, 2010b)**.

FK866 has also been applied to treat a mouse model of collagen induced arthritis (CIA). Busso et al. (2008) carried out a CIA curative experiment using the optimal dose of FK866. Twenty mice with CIA were treated twice daily with 10 mg/kg/ip of FK866 from the first day onward of appearance of clinical arthritis for 14 consecutive days. They found that FK866 effectively reduced arthritis severity with comparable activity to etanercept, and decreased proinflammatory cytokine secretion in affected joints. Paws from FK866-treated mice showed minimal signs of inflammation after 2 weeks of treatment whereas paws from placebo-treated mice were still inflamed (Figure 2). This study indicates that the inhibition of NAMPT might have therapeutic efficacy in immune-mediated inflammatory disorders.

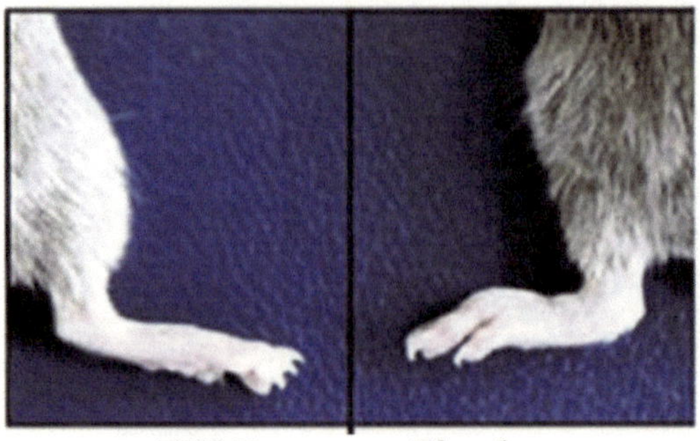

FK866 **Placebo**

Fig. 2. FK866 treatment alleviated inflammation in a CIA mouse model. Collagen induced arthritis was initiated in male DBA/1 mice. Mice were treated for 14 consecutive days with two injections of 10 mg/kg/IP/day of APO866 from the first day onward of appearance of clinical arthritis (clinical score >1, mild swelling and/or erythema). Placebo mice received vehicle only. One of parameters to evaluate the FK866 therapeutic effect on arthritis is presented here with paws of FK866-treated (left) showing diminished swelling and inflammation compared to those in placebo-treated (right) arthritic mouse. This figure was copied from Busso et al. (2008).

4.1.2 CHS 828

CHS 828 (N-(6-chlorophenoxyhexyl)-N′-cyano-N″-4-pyridylguanidine, GMX1778, active form of GMN1777), a cyanoguanidine compound, which displayed promising preclinical anticancer activity, is another important NAMPT inhibitor currently in phase II clinical trials in oncology **(Schoua C., 1997; Hjarnaa et al., 1999; Olesen et al., 2010b; von Heideman et al., 2010)**. Hjarnaa et al. **(1999)** first demonstrated the anticancer properties of CHS 828 both in vitro and in vivo using established cancer cell lines and in rodent models of tumor growth, respectively. The mechanism of this small molecular inhibitor has been hypothesized to function through NF-κB inhibition **(Schoua C., 1997; Hassan et al., 2006)**, but the exact mechanism was undefined. Olesen et al. **(2004)** originally found no common resistance mechanism between CHS 828 resistant human small cell lung carcinoma NYH cells (NYH/CHS) and NF-κB inhibitors resistant NYH cells. For FK866, it was observed that its cross-resistance was comparable to CHS 828 in NYH/CHS cells. CHS 828 was subsequently defined as an inhibitor of NAD synthesis as increasing cellular levels of NAD were able to completely block cytotoxicity of CHS 828, which is similar to FK866. Furthermore, crystal structure and in vitro biochemistry results showed that FK866 and CHS 828 shared a binding site in the active site of NAMPT **(Kim, M.K. et al., 2006; Olesen et al., 2008, 2010a)**. Thus, CHS 828 is conclusively identified as a competitive inhibitor of NAMPT.

4.1.3 Other chemical inhibitors

Kang et al. **(2009)** designed and synthesized IS001, in which a ribose ring was added to the FK866 pyridyl ring. The ribose ring of IS001 did not improve its solubility and binding interactions. The structures of the NAMPT-IS001 and NAMPT-FK866 complexes are nearly identical. You et al. **(2011)** replaced the pyridine ring of FK866 with various heteroaromatic rings for combination with a simple ribose-mimicking moiety containing a diol group. One analogue, compound 7, showed superior anti-cancer activity to FK866 (IC50 > 20μM) in human leukemia cells (K562) with an IC50 value of 1.4μM, but not in other cancer cell lines. Crystal structure of NAMPT-compound 7 demonstrated the binding site of compound 7 was nearly identical to that of FK866 and NMN.

TP201565, a potent analogue of CHS 828, displayed inhibition activity in xenograft models **(Hjarnaa et al., 1999)**. TP201565 shows more than 10 fold increased activity in sensitive cancer cell lines compared to FK866 and CHS 828. Additionally, computer modeling analysis predicted that TP201565 inhibits NAMPT by binding to the same site as FK866 and CHS 828. Myriad Pharmaceuticals (http://www.myrexis.com) is developing a series of NAMPT inhibitors, such as MPC-9528, CB30865, MPI0479883. These inhibitors, which are designed analogues of FK866 and CHS 828, potently inhibited NAMPT activity and cancer cell growth with lower IC_{50} values than previously reported inhibitors.

4.2 NAMPT siRNA/shRNA

RNA interference (RNAi) is a natural biological mechanism where gene expression is silenced in a highly specific manner through the addition of double stranded RNA (dsRNA). Once dsRNA enters the cell, it is cleaved by an RNase III -like enzyme, Dicer, into double stranded small interfering RNAs (siRNA). The resulting siRNA with 21-23 nucleotides in length, containing 2 nucleotide overhangs on the 3′ ends, integrate into a multi-subunit protein complex, RNAi induced silencing complex (RISC). The RISC-bound antisense strand then serves as a guide for targeting the activated complex to complementary mRNA sequences, resulting in subsequent mRNA cleavage and degradation. Small hairpin RNA or short hairpin RNA (shRNA) is a sequence of RNA that makes a tight hairpin turn that can be used for RNA interference. shRNA uses a vector introduced into cells and utilizes the U6 or H1 promoter to ensure that the shRNA is constitutively expressed. This vector is usually passed on to daughter cells, allowing the gene silencing to be inherited. The shRNA hairpin structure is cleaved into siRNA, which is then bound to the RISC. This complex binds to and cleaves target mRNAs which match the siRNA that is bound to it. Knock down of gene expression by siRNA and shRNA technology has become a popular and effective tool to dissect gene function in the current functional genomics era.

Ye et al. **(2005b)** employed NAMPT stealth siRNAs to explore the pathophysiological relevance of the altered NAMPT expression to the lung endothelial barrier dysregulation induced by thrombin and the pathogenesis of ALI. Acute lung injury (ALI) is characterized by a diffuse intense inflammatory process and by damage to both endothelial and epithelial cell barriers. The NAMPT gene was highly expressed in ALI and our group reported that NAMPT was a biomarker in ALI **(Ye et al., 2005a)**. We studied whether downregulation of NAMPT protein expression by the NAMPT-specific siRNA would affect thrombin-induced decreases in human lung endothelial cell barrier function **(Ye et al., 2005b)**. The results revealed that NAMPT expression was decreased significantly and the thrombin effect on cell barrier function was attenuated by NAMPT siRNA treatment in cultured human pulmonary artery endothelial cells (HPAEC), while siRNA had no effect on the protein expression level of β-actin, a house-keeping gene. Similar results were also obtained in A549 cells, a lung alveolar type II epithelial cell line.

NAMPT levels in serum and synovial fluid are elevated in RA patients. RA synovial fibroblasts (RASFs) were major NAMPT expressing cells. Brentano et al. **(2007)** used siRNA to silence NAMPT expression in RASF, which significantly inhibited basal and TLR ligand-induced production of IL-6, IL-8, MMP-1, and MMP-3. Van der Veer et al **(2005)** knocked down endogenous NAMPT in smooth muscle cells (SMC) to explore if the conversion of the SMC phenotype from a proliferative state to a nonproliferative state was accompanied by up-regulation of NAMPT. This shift of SMC phenotype is essential for conferring vasomotor function to developing and remodeling blood vessels. SMC, transfected with NAMPT siRNA, exhibited significant decrease in NAMPT mRNA and protein. Furthermore, they found that NAMPT siRNA treatment increased SMC apoptosis and reduced the capacity of synthetic SMCs to mature to a contractile state. NAMPT expression also increased during macrophage differentiation. To examine the role of NAMPT in macrophages differentiation, NAMPT siRNA was transfected into macrophages to silence NAMPT. Dahl et al. **(2007)** found that silencing of NAMPT increased lipid accumulation in THP-1 macrophages, increased ADRP (adipose differentiation-related protein) and cholesterol levels in oxidized LDL stimulated macrophages, and enhanced the binding of acetylated LDL in these cells. Cumulatively, these results indicate that NAMPT siRNAs could effectively and specifically inhibit the NAMPT expression and its function, which may be exploited as a promising strategy for NAMPT expression based therapy in relevant human diseases.

Sigma Company (http://www.sigmaaldrich.com/) now markets MISSION® NAMPT esiRNA. MISSION esiRNAs are endoribonuclease-prepared siRNA pools comprised of a heterogeneous mixture of siRNAs that all target the same mRNA sequence. These multiple silencing triggers lead to highly specific and effective gene silencing with lower off-target effects than single or pooled siRNAs. Although no experimental report on the true effectiveness of MISSION® NAMPT esiRNA to silence the NAMPT expression appears in the literature yet, MISSION® NAMPT esiRNA may gain steam over other NAMPT siRNA tools soon.

4.3 Antisense oligonucleotide to NAMPT

An antisense oligonucleotide is a synthesized strand of nucleic acid which binds to and inactivates its corresponding mRNA molecule effectively making the target gene silent. Generally, they are relatively short (13–25 nucleotides) and hybridize (at least in theory) to a unique sequence of target present in cells.

Jia et al. **(2004)** used a phosphorothioate-modified antisense oligonucleotide to block NAMPT mRNA expression in neutrophils. Neutrophils play an important role in sepsis. Activated neutrophils have been implicated in the increased micro vascular permeability of systemic inflammation **(Gautam et al., 2001)** and in the pathogenesis of inflammatory injury to the lung **(Lee, W.L. & Downey, 2001)**, liver **(Ho et al., 1996)**, gastrointestinal tract **(Kubes et al., 1992)**, and kidney **(Lauriat & Linas, 1998)**. The number and activity of neutrophils are tightly linked to infection and inflammatory injury, and regulated by apoptotic program. LPS and other inflammatory cytokines can inhibit neutrophil apoptosis. The research by Jia et al. demonstrated that NAMPT is synthesized and released by neutrophils in response to inflammatory stimuli and that it plays a requisite role in the inhibition of apoptosis in neutrophils **(Jia et al., 2004)**. The NAMPT antisense oligonucleotide prevented neutrophil apoptosis induced by either LPS or other inflammatory cytokines. In neutrophils of patients with sepsis, addition of NAMPT antisense oligonucleotide resulted in a greater than two fold increase in rates of apoptosis. This study demonstrated that the antisense oligonucleotide to NAMPT could effectively inhibit NAMPT expression, which could prove useful as a new therapeutic tool.

4.4 NAMPT miRNA

MicroRNAs (miRNAs), short ribonucleic acid (RNA) molecules, are post-transcriptional regulators, about 22 nucleotides long. miRNAs could bind to 3′ UTR of messenger RNA transcripts (mRNAs) resulting in translational repression and gene silencing. Gene silencing may occur either via mRNA degradation or preventing mRNA from being translated. It has been demonstrated that if there is complete complementation between the miRNA and target mRNA sequence, mRNA can be cleaved and degraded.

Elangovan et al. **(2011)** reported at the American Thoracic Society Annual Conference, 2011 that two miRNA hsa-miR-374a and hsa-miR-568 potentially bound to the 3′ UTR of NAMPT mRNA and thereby inhibited its expression. In human pulmonary artery endothelia cell (HPAEC) transfected with the luc-Nampt-3′UTR reporter construct and miRNAs, they found that hsa-miR-374a and hsa-miR-568 effectively reduced LPS- and cyclic stretch-stimulated NAMPT expression in cultured HPAEC cell by both dual luciferase assay and immunoblotting. This study indicates that synthetic NAMPT miRNAs have potential to be a novel class of therapeutic molecules.

Currently, several companies, such as Sigma (http://www.sigmaaldrich.com/) and Dharmacon (http://www.dharmacon.com/), offer the products of NAMPT miRNA mimics. MicroRNA mimics are double-stranded RNA oligonucleotides and are chemically modified. They effectively mimic endogenous mature miRNA functions with a superior performance in comparison to native double-stranded miRNA. Although there is yet a report about the application of NAMPT miRNA mimics, NAMPT miRNA mimics may gain an advantage over natural NAMPT miRNA as more effective inhibitors of NAMPT for the reasons as described above.

4.5 Antibody to NAMPT

An antibody, also known as an immunoglobulin, is produced by the immune system to identify and neutralize a foreign object by recognizing its antigen, a unique part of the foreign target. The current progress in biotechnology and genetic engineering has kindled a growing interest in antibody based therapy, which has quickly become incorporated into the therapeutic armamentarium for various human diseases. It will be no exception for the development of NAMPT antibody based therapy.

Our group has evaluated whether NAMPT antibody can block the function of NAMPT and thus attenuate ventilator induced lung injury (VILI) in a mouse model since NAMPT overexpression seemed to play a significant role in the pathogenesis of acute lung injury **(Hong et al., 2008)**. We found that simultaneous instillation of rhNAMPT (20 mg/mouse) and NAMPT neutralizing antibody produced dramatic reductions in rhNAMPT-induced PMN recruitment. The intratracheal delivery of NAMPT neutralizing antibody (30 min before mechanical ventilation) abolished VILI-induced increases in total BAL cell counts and significantly decreased PMN influx into the alveolar space as well as VILI-mediated increases in lung tissue albumin. This result indicates that NAMPT antibody based strategy may be a viable therapeutic modality to acute lung injury. Although this was only one case report **(Hong et al., 2008)**, it could ignite an increasing interest at NAMPT antibody based therapy to other diseases where a dysregulated overexpression of NAMPT gene is one of the key features.

4.6 NAMPT gene knockout

Gene knockout (KO), namely the inactivation of a specific gene within an organism, is a potential important gene therapy technique. In mouse experiments, typically, a gene target vector contains a selective marker in the center with both 5'- and 3'- arms, whose sequences are homologous to the targeted gene. This gene target vector is introduced into the mouse embryonic stem cell by electroporation. As the embryonic stem cell divides, the original gene segment is replaced through homologous recombination in a small number of the divided cells. After the target vector is integrated in the cognate position, the stem cells are then introduced into a female mouse. Subsequently, a knock-out mouse line would be engendered after further characterization.

To explore the effect of NAMPT to ventilator-induced lung injury (VILI) in vivo, Hong et al. **(2008)** generated a heterozygous NAMPT+/- mouse line and subsequently exposed the mouse to a model of severe VILI. The results were opposite between the rhNAMPT and NAMPT+/- mice when exposed to a model of severe VILI. In rhNAMPT mice, it was observed that dramatic increases in bronchoalveolar lavage (BAL) leukocytes, BAL protein, and cytokine levels (IL-6, TNF-a, KC). In contrast, NAMPT+/- mice exhibited significantly decreased expression of VILI associated genes, inflammatory lung injury and lower peak aspiratory pressures compared with control mice. These results suggest that NAMPT is a critical effecter in the development of ventilator induced lung pathobiology and it shows a promise that reducing NAMPT gene expression could protect lungs from VILI in human patients down the road.

5. Perspective

NAMPT has drawn an ever-increasing attention in biomedical fields because of its presumed pleiotropic physiological functions and its dysregulation implicated in a number of human diseases and conditions. As of Aug 15, 2011, 836 publications have appeared in PubMed when searched with key words: NAMPT or PBEF or visfatin.

A growing interest has been to develop inhibitors of NAMPT as a potential new therapeutic strategy to several human diseases where the dysregulated overexpression of NAMPT is implicated in their pathogeneses. Although there have been several reported methods to effectively inhibit NAMPT activity, such as small molecular chemicals, siRNA, miRNA, antisense oligonucleotide, and gene knock out, further improvement of these inhibitors are warranted. These include improving the efficacy, reducing side effects and enhancing the specificity. For FK866, or other synthetic chemical inhibitors, the main obstacle to be overcome is toxicity. Enhancing the potency, lowering the effective dosage and reducing the side effects are necessary for the improvement of this NAMPT inhibitor class. Although RNA interference technology by siRNA and miRNA has become a standard research tool for genetic studies and a new class of drugs designed to silence disease-causing genes, it is still in the development stage. RNA in general is more unstable than DNA and tends to be degraded easily. siRNA or miRNA may also cause an immune response which causes the body to reject the foreign RNA. Transfection inefficiency is a problem both in treatment and in study of RNA interference. Off target effects are a major concern. The stability issue of antisense oligonucleotides in blood and tissue is still not satisfactorily resolved. Although our group demonstrated that antibody blocking of NAMPT could attenuate acute lung injury in a mouse model, antibody based inhibition of NAMPT is far from fully developed as a effective therapy to other diseases. Gene therapy still has a long way to go before clinical fruition. NAMPT aptamers or other forms of NAMPT inhibitors have yet come into the world. Nevertheless, it is anticipated that improved and new NAMPT inhibitors will be developed within next few years. NAMPT based strategy holds an immense promise in management of various human conditions.

6. Acknowledgements

Some cited studies in this chapter were in part supported by NIH grants (R01HL080042 and RO1 HL080042-S1, SQY) as well as the start-up funds of Children's Mercy Hospitals and Clinics, University of Missouri (SQY).

7. References

Adya, R., Tan, B.K., Chen, J. and Randeva, H.S. (2008). Nuclear factor-kappaB induction by visfatin in human vascular endothelial cells: its role in MMP-2/9 production and activation. Diabetes Care 31(4): 758-760.

Adya, R., Tan, B.K., Chen, J. and Randeva, H.S. (2009). Pre-B cell colony enhancing factor (PBEF)/visfatin induces secretion of MCP-1 in human endothelial cells: role in visfatin-induced angiogenesis. Atherosclerosis 205(1): 113-119.

Adya, R., Tan, B.K., Punn, A., Chen, J. and Randeva, H.S. (2008). Visfatin induces human endothelial VEGF and MMP-2/9 production via MAPK and PI3K/Akt signalling pathways: novel insights into visfatin-induced angiogenesis. Cardiovasc Res 78(2): 356-365.

Akturk, M., Altinova, A.E., Mert, I., Buyukkagnici, U., Sargin, A., et al. (2008). Visfatin concentration is decreased in women with gestational diabetes mellitus in the third trimester. J Endocrinol Invest 31(7): 610-613.

Alghasham, A.A. and Barakat, Y.A. (2008). Serum visfatin and its relation to insulin resistance and inflammation in type 2 diabetic patients with and without macroangiopathy. Saudi Med J 29(2): 185-192.

Bae, Y.H., Bae, M.K., Kim, S.R., Lee, J.H., Wee, H.J., et al. (2009). Upregulation of fibroblast growth factor-2 by visfatin that promotes endothelial angiogenesis. Biochem Biophys Res Commun 379(2): 206-211.

Bajwa, E.K., Yu, C.L., Gong, M.N., Thompson, B.T. and Christiani, D.C. (2007). Pre-B-cell colony-enhancing factor gene polymorphisms and risk of acute respiratory distress syndrome. Crit Care Med 35(5): 1290-1295.

Benigni, A., Corna, D., Zoja, C., Sonzogni, A., Latini, R., et al. (2009). Disruption of the Ang II type 1 receptor promotes longevity in mice. J Clin Invest 119(3): 524-530.

Berndt, J., Kloting, N., Kralisch, S., Kovacs, P., Fasshauer, M., et al. (2005). Plasma visfatin concentrations and fat depot-specific mRNA expression in humans. Diabetes 54(10): 2911-2916.

Billington, R.A., Travelli, C., Ercolano, E., Galli, U., Roman, C.B., et al. (2008). Characterization of NAD uptake in mammalian cells. J Biol Chem 283(10): 6367-6374.

Blander, G. and Guarente, L. (2004). The Sir2 family of protein deacetylases. Annu Rev Biochem 73: 417-435.

Borradaile, N.M. and Pickering, J.G. (2009). NAD(+), sirtuins, and cardiovascular disease. Curr Pharm Des 15(1): 110-117.

Botella-Carretero, J.I., Luque-Ramirez, M., Alvarez-Blasco, F., Peromingo, R., San Millan, J.L., et al. (2008). The increase in serum visfatin after bariatric surgery in morbidly obese women is modulated by weight loss, waist circumference, and presence or absence of diabetes before surgery. Obes Surg 18(8): 1000-1006.

Brentano, F., Schorr, O., Ospelt, C., Stanczyk, J., Gay, R.E., et al. (2007). Pre-B cell colony-enhancing factor/visfatin, a new marker of inflammation in rheumatoid arthritis with proinflammatory and matrix-degrading activities. Arthritis Rheum 56(9): 2829-2839.

Busso, N., Karababa, M., Nobile, M., Rolaz, A., Van Gool, F., et al. (2008). Pharmacological inhibition of nicotinamide phosphoribosyltransferase/visfatin enzymatic activity identifies a new inflammatory pathway linked to NAD. PLoS One 3(5): e2267.

Chamberlain, M.C. (2008). Antiangiogenesis: biology and utility in the treatment of gliomas. Expert Rev Neurother 8(10): 1419-1423.

Chan, T.F., Chen, Y.L., Lee, C.H., Chou, F.H., Wu, L.C., et al. (2006). Decreased plasma visfatin concentrations in women with gestational diabetes mellitus. J Soc Gynecol Investig 13(5): 364-367.

Chen, M.P., Chung, F.M., Chang, D.M., Tsai, J.C., Huang, H.F., et al. (2006). Elevated plasma level of visfatin/pre-B cell colony-enhancing factor in patients with type 2 diabetes mellitus. J Clin Endocrinol Metab 91(1): 295-299.

Cheng, K.H., Chu, C.S., Lee, K.T., Lin, T.H., Hsieh, C.C., et al. (2008). Adipocytokines and proinflammatory mediators from abdominal and epicardial adipose tissue in patients with coronary artery disease. Int J Obes (Lond) 32(2): 268-274.

Corda, D. and Di Girolamo, M. (2003). Functional aspects of protein mono-ADP-ribosylation. EMBO J 22(9): 1953-1958.

Dahl, T.B., Yndestad, A., Skjelland, M., Oie, E., Dahl, A., et al. (2007). Increased expression of visfatin in macrophages of human unstable carotid and coronary atherosclerosis: possible role in inflammation and plaque destabilization. Circulation 115(8): 972-980.

Dogru, T., Sonmez, A., Tasci, I., Bozoglu, E., Yilmaz, M.I., et al. (2007). Plasma visfatin levels in patients with newly diagnosed and untreated type 2 diabetes mellitus and impaired glucose tolerance. Diabetes Res Clin Pract 76(1): 24-29.

Drevs, J., Loser, R., Rattel, B. and Esser, N. (2003). Antiangiogenic potency of FK866/K22.175, a new inhibitor of intracellular NAD biosynthesis, in murine renal cell carcinoma. Anticancer Res 23(6C): 4853-4858.

Elangovan, V.R., et al. (2011). Hsa-Mir-374a And Hsa-Mir-568 Attenuate LPS- And Cyclic Stretch-Induced Pbef Gene Expression In Human Lung Endothelium. American Journal of Respirator and Critical Care Medicine 183: A3523.

Fernandez-Real, J.M., Moreno, J.M., Chico, B., Lopez-Bermejo, A. and Ricart, W. (2007). Circulating visfatin is associated with parameters of iron metabolism in subjects with altered glucose tolerance. Diabetes Care 30(3): 616-621.

Frank, J.A., Parsons, P.E. and Matthay, M.A. (2006). Pathogenetic significance of biological markers of ventilator-associated lung injury in experimental and clinical studies. Chest 130(6): 1906-1914.

Fukuhara, A., Matsuda, M., Nishizawa, M., Segawa, K., Tanaka, M., et al. (2005). Visfatin: a protein secreted by visceral fat that mimics the effects of insulin. Science 307(5708): 426-430.

Fukuhara, A., Matsuda, M., Nishizawa, M., Segawa, K., Tanaka, M., et al. (2007). Retraction. Science 318(5850): 565.

Galli, M., Van Gool, F., Rongvaux, A., Andris, F. and Leo, O. (2010). The nicotinamide phosphoribosyltransferase: a molecular link between metabolism, inflammation, and cancer. Cancer Res 70(1): 8-11.

Gautam, N., Olofsson, A.M., Herwald, H., Iversen, L.F., Lundgren-Akerlund, E., et al. (2001). Heparin-binding protein (HBP/CAP37): a missing link in neutrophil-evoked alteration of vascular permeability. Nat Med 7(10): 1123-1127.

Haider, D.G., Pleiner, J., Francesconi, M., Wiesinger, G.F., Muller, M., et al. (2006). Exercise training lowers plasma visfatin concentrations in patients with type 1 diabetes. J Clin Endocrinol Metab 91(11): 4702-4704.

Hallschmid, M., Randeva, H., Tan, B.K., Kern, W. and Lehnert, H. (2009). Relationship between cerebrospinal fluid visfatin (PBEF/Nampt) levels and adiposity in humans. Diabetes 58(3): 637-640.

Hasmann, M. and Schemainda, I. (2003). FK866, a highly specific noncompetitive inhibitor of nicotinamide phosphoribosyltransferase, represents a novel mechanism for induction of tumor cell apoptosis. Cancer Res 63(21): 7436-7442.

Hassan, S.B., Lovborg, H., Lindhagen, E., Karlsson, M.O. and Larsson, R. (2006). CHS 828 kill tumour cells by inhibiting the nuclear factor-kappaB translocation but unlikely through down-regulation of proteasome. Anticancer Res 26(6B): 4431-4436.

Hjarnaa, P.J., Jonsson, E., Latini, S., Dhar, S., Larsson, R., et al. (1999). CHS 828, a novel pyridyl cyanoguanidine with potent antitumor activity in vitro and in vivo. Cancer Res 59(22): 5751-5757.

Ho, J.S., Buchweitz, J.P., Roth, R.A. and Ganey, P.E. (1996). Identification of factors from rat neutrophils responsible for cytotoxicity to isolated hepatocytes. J Leukoc Biol 59(5): 716-724.

Holen, K., Saltz, L.B., Hollywood, E., Burk, K. and Hanauske, A.R. (2008). The pharmacokinetics, toxicities, and biologic effects of FK866, a nicotinamide adenine dinucleotide biosynthesis inhibitor. Invest New Drugs 26(1): 45-51.

Hong, S.B., Huang, Y., Moreno-Vinasco, L., Sammani, S., Moitra, J., et al. (2008). Essential role of pre-B-cell colony enhancing factor in ventilator-induced lung injury. Am J Respir Crit Care Med 178(6): 605-617.

Hufton, S.E., Moerkerk, P.T., Brandwijk, R., de Bruine, A.P., Arends, J.W., et al. (1999). A profile of differentially expressed genes in primary colorectal cancer using suppression subtractive hybridization. FEBS Lett 463(1-2): 77-82.

Jia, S.H., Li, Y., Parodo, J., Kapus, A., Fan, L., et al. (2004). Pre-B cell colony-enhancing factor inhibits neutrophil apoptosis in experimental inflammation and clinical sepsis. J Clin Invest 113(9): 1318-1327.

Jian, W.X., Luo, T.H., Gu, Y.Y., Zhang, H.L., Zheng, S., et al. (2006). The visfatin gene is associated with glucose and lipid metabolism in a Chinese population. Diabet Med 23(9): 967-973.

Kang, G.B., Bae, M.H., Kim, M.K., Im, I., Kim, Y.C., et al. (2009). Crystal structure of Rattus norvegicus Visfatin/PBEF/Nampt in complex with an FK866-based inhibitor. Mol Cells 27(6): 667-671.

Kato, A., Odamaki, M., Ishida, J. and Hishida, A. (2009). Relationship between serum pre-B cell colony-enhancing factor/visfatin and atherosclerotic parameters in chronic hemodialysis patients. Am J Nephrol 29(1): 31-35.

Khan, J.A., Forouhar, F., Tao, X. and Tong, L. (2007). Nicotinamide adenine dinucleotide metabolism as an attractive target for drug discovery. Expert Opin Ther Targets 11(5): 695-705.

Khan, J.A., Tao, X. and Tong, L. (2006). Molecular basis for the inhibition of human NMPRTase, a novel target for anticancer agents. Nat Struct Mol Biol 13(7): 582-588.

Kim, M.K., Lee, J.H., Kim, H., Park, S.J., Kim, S.H., et al. (2006). Crystal structure of visfatin/pre-B cell colony-enhancing factor 1/nicotinamide phosphoribosyltransferase, free and in complex with the anti-cancer agent FK-866. J Mol Biol 362(1): 66-77.

Kim, S.R., Bae, S.K., Choi, K.S., Park, S.Y., Jun, H.O., et al. (2007). Visfatin promotes angiogenesis by activation of extracellular signal-regulated kinase 1/2. Biochem Biophys Res Commun 357(1): 150-156.

Kim, S.R., Bae, Y.H., Bae, S.K., Choi, K.S., Yoon, K.H., et al. (2008). Visfatin enhances ICAM-1 and VCAM-1 expression through ROS-dependent NF-kappaB activation in endothelial cells. Biochim Biophys Acta 1783(5): 886-895.

Krzyzanowska, K., Krugluger, W., Mittermayer, F., Rahman, R., Haider, D., et al. (2006). Increased visfatin concentrations in women with gestational diabetes mellitus. Clin Sci (Lond) 110(5): 605-609.

Kubes, P., Hunter, J. and Granger, D.N. (1992). Ischemia/reperfusion-induced feline intestinal dysfunction: importance of granulocyte recruitment. Gastroenterology 103(3): 807-812.

Landre-Beauvais, A.J. (2001). The first description of rheumatoid arthritis. Unabridged text of the doctoral dissertation presented in 1800. Joint Bone Spine 68(2): 130-143.

Lauriat, S. and Linas, S.L. (1998). The role of neutrophils in acute renal failure. Semin Nephrol 18(5): 498-504.

Lee, H.C. (2001). Physiological functions of cyclic ADP-ribose and NAADP as calcium messengers. Annu Rev Pharmacol Toxicol 41: 317-345.

Lee, W.L. and Downey, G.P. (2001). Leukocyte elastase: physiological functions and role in acute lung injury. Am J Respir Crit Care Med 164(5): 896-904.

Lewandowski, K.C., Stojanovic, N., Press, M., Tuck, S.M., Szosland, K., et al. (2007). Elevated serum levels of visfatin in gestational diabetes: a comparative study across various degrees of glucose tolerance. Diabetologia 50(5): 1033-1037.

Li, H., Liu, P., Cepeda, J., Fang, D., Easley, R.B., et al. (2008). Augmentation of Pulmonary Epithelial Cell IL-8 Expression and Permeability by Pre-B-cell Colony Enhancing Factor. J Inflamm (Lond) 5: 15.

Li, Y., Zhang, Y., Dorweiler, B., Cui, D., Wang, T., et al. (2008). Extracellular Nampt promotes macrophage survival via a nonenzymatic interleukin-6/STAT3 signaling mechanism. J Biol Chem 283(50): 34833-34843.

Liang, Y., Xu, X.M., Wang, H.S. and Wang, P.W. (2008). [Correlation between the expression of gastrocolic omentum visfatin mRNA and gestational diabetes mellitus]. Zhonghua Fu Chan Ke Za Zhi 43(11): 824-827.

Lin, S.J., Defossez, P.A. and Guarente, L. (2000). Requirement of NAD and SIR2 for life-span extension by calorie restriction in Saccharomyces cerevisiae. Science 289(5487): 2126-2128.

Liu, P., Li, H., Cepeda, J., Zhang, L.Q., Cui, X., et al. (2009). Critical role of PBEF expression in pulmonary cell inflammation and permeability. Cell Biol Int 33(1): 19-30.

Liu, S.W., Qiao, S.B., Yuan, J.S. and Liu, D.Q. (2009). Visfatin stimulates production of monocyte chemotactic protein-1 and interleukin-6 in human vein umbilical endothelial cells. Horm Metab Res 41(4): 281-286.

Lopez-Bermejo, A., Chico-Julia, B., Fernandez-Balsells, M., Recasens, M., Esteve, E., et al. (2006). Serum visfatin increases with progressive beta-cell deterioration. Diabetes 55(10): 2871-2875.

Martin, P.R., Shea, R.J. and Mulks, M.H. (2001). Identification of a plasmid-encoded gene from Haemophilus ducreyi which confers NAD independence. J Bacteriol 183(4): 1168-1174.

Matsui, H., Tsutsumi, A., Sugihara, M., Suzuki, T., Iwanami, K., et al. (2008). Visfatin (pre-B cell colony-enhancing factor) gene expression in patients with rheumatoid arthritis. Ann Rheum Dis 67(4): 571-572.

Matthay, M.A., Zimmerman, G.A., Esmon, C., Bhattacharya, J., Coller, B., et al. (2003). Future research directions in acute lung injury: summary of a National Heart, Lung, and Blood Institute working group. Am J Respir Crit Care Med 167(7): 1027-1035.

Mazaki-Tovi, S., Romero, R., Kusanovic, J.P., Erez, O., Gotsch, F., et al. (2008). Visfatin/Pre-B cell colony-enhancing factor in amniotic fluid in normal pregnancy, spontaneous labor at term, preterm labor and prelabor rupture of membranes: an association with subclinical intrauterine infection in preterm parturition. J Perinat Med 36(6): 485-496.

Menissier de Murcia, J., Ricoul, M., Tartier, L., Niedergang, C., Huber, A., et al. (2003). Functional interaction between PARP-1 and PARP-2 in chromosome stability and embryonic development in mouse. EMBO J 22(9): 2255-2263.

Moschen, A.R., Gerner, R.R. and Tilg, H. (2010). Pre-B cell colony enhancing factor/NAMPT/visfatin in inflammation and obesity-related disorders. Curr Pharm Des 16(17): 1913-1920.

Muruganandham, M., Alfieri, A.A., Matei, C., Chen, Y., Sukenick, G., et al. (2005). Metabolic signatures associated with a NAD synthesis inhibitor-induced tumor apoptosis identified by 1H-decoupled-31P magnetic resonance spectroscopy. Clin Cancer Res 11(9): 3503-3513.

Nahimana, A., Attinger, A., Aubry, D., Greaney, P., Ireson, C., et al. (2009). The NAD biosynthesis inhibitor APO866 has potent antitumor activity against hematologic malignancies. Blood 113(14): 3276-3286.

Nowell, M.A., Richards, P.J., Fielding, C.A., Ognjanovic, S., Topley, N., et al. (2006). Regulation of pre-B cell colony-enhancing factor by STAT-3-dependent interleukin-6 trans-signaling: implications in the pathogenesis of rheumatoid arthritis. Arthritis Rheum 54(7): 2084-2095.

Ognjanovic, S., Bao, S., Yamamoto, S.Y., Garibay-Tupas, J., Samal, B., et al. (2001). Genomic organization of the gene coding for human pre-B-cell colony enhancing factor and expression in human fetal membranes. J Mol Endocrinol 26(2): 107-117.

Ognjanovic, S. and Bryant-Greenwood, G.D. (2002). Pre-B-cell colony-enhancing factor, a novel cytokine of human fetal membranes. Am J Obstet Gynecol 187(4): 1051-1058.

Olesen, U.H., Christensen, M.K., Bjorkling, F., Jaattela, M., Jensen, P.B., et al. (2008). Anticancer agent CHS-828 inhibits cellular synthesis of NAD. Biochem Biophys Res Commun 367(4): 799-804.

Olesen, U.H., Petersen, J.G., Garten, A., Kiess, W., Yoshino, J., et al. (2010a). Target enzyme mutations are the molecular basis for resistance towards pharmacological inhibition of nicotinamide phosphoribosyltransferase. BMC Cancer 10: 677.

Olesen, U.H., Thougaard, A.V., Jensen, P.B. and Sehested, M. (2010b). A preclinical study on the rescue of normal tissue by nicotinic acid in high-dose treatment with APO866, a specific nicotinamide phosphoribosyltransferase inhibitor. Mol Cancer Ther 9(6): 1609-1617.

Olsen, L.S., Hjarnaa, P.J., Latini, S., Holm, P.K., Larsson, R., et al. (2004). Anticancer agent CHS 828 suppresses nuclear factor-kappa B activity in cancer cells through downregulation of IKK activity. Int J Cancer 111(2): 198-205.

Otero, M., Lago, R., Gomez, R., Lago, F., Dieguez, C., et al. (2006). Changes in plasma levels of fat-derived hormones adiponectin, leptin, resistin and visfatin in patients with rheumatoid arthritis. Ann Rheum Dis 65(9): 1198-1201.

Pagano, C., Pilon, C., Olivieri, M., Mason, P., Fabris, R., et al. (2006). Reduced plasma visfatin/pre-B cell colony-enhancing factor in obesity is not related to insulin resistance in humans. J Clin Endocrinol Metab 91(8): 3165-3170.

Palin, M.F., Labrecque, B., Beaudry, D., Mayhue, M., Bordignon, V., et al. (2008). Visfatin expression is not associated with adipose tissue abundance in the porcine model. Domest Anim Endocrinol 35(1): 58-73.

Pittelli, M., Formentini, L., Faraco, G., Lapucci, A., Rapizzi, E., et al. (2010). Inhibition of nicotinamide phosphoribosyltransferase: cellular bioenergetics reveals a mitochondrial insensitive NAD pool. J Biol Chem 285(44): 34106-34114.

Pogrebniak, A., Schemainda, I., Azzam, K., Pelka-Fleischer, R., Nussler, V., et al. (2006). Chemopotentiating effects of a novel NAD biosynthesis inhibitor, FK866, in combination with antineoplastic agents. Eur J Med Res 11(8): 313-321.

Reddy, P.S., Umesh, S., Thota, B., Tandon, A., Pandey, P., et al. (2008). PBEF1/NAmPRTase/Visfatin: a potential malignant astrocytoma/glioblastoma serum marker with prognostic value. Cancer Biol Ther 7(5): 663-668.

Retnakaran, R., Youn, B.S., Liu, Y., Hanley, A.J., Lee, N.S., et al. (2008). Correlation of circulating full-length visfatin (PBEF/NAMPT) with metabolic parameters in subjects with and without diabetes: a cross-sectional study. Clin Endocrinol (Oxf) 69(6): 885-893.

Revollo, J.R., Grimm, A.A. and Imai, S. (2004). The NAD biosynthesis pathway mediated by nicotinamide phosphoribosyltransferase regulates Sir2 activity in mammalian cells. J Biol Chem 279(49): 50754-50763.

Revollo, J.R., Korner, A., Mills, K.F., Satoh, A., Wang, T., et al. (2007). Nampt/PBEF/Visfatin regulates insulin secretion in beta cells as a systemic NAD biosynthetic enzyme. Cell Metab 6(5): 363-375.

Rho, Y.H., Solus, J., Sokka, T., Oeser, A., Chung, C.P., et al. (2009). Adipocytokines are associated with radiographic joint damage in rheumatoid arthritis. Arthritis Rheum 60(7): 1906-1914.

Rongvaux, A., Shea, R.J., Mulks, M.H., Gigot, D., Urbain, J., et al. (2002). Pre-B-cell colony-enhancing factor, whose expression is up-regulated in activated lymphocytes, is a nicotinamide phosphoribosyltransferase, a cytosolic enzyme involved in NAD biosynthesis. Eur J Immunol 32(11): 3225-3234.

Samal, B., Sun, Y., Stearns, G., Xie, C., Suggs, S., et al. (1994). Cloning and characterization of the cDNA encoding a novel human pre-B-cell colony-enhancing factor. Mol Cell Biol 14(2): 1431-1437.

Sandeep, S., Velmurugan, K., Deepa, R. and Mohan, V. (2007). Serum visfatin in relation to visceral fat, obesity, and type 2 diabetes mellitus in Asian Indians. Metabolism 56(4): 565-570.

Sasaki, Y., Araki, T. and Milbrandt, J. (2006). Stimulation of nicotinamide adenine dinucleotide biosynthetic pathways delays axonal degeneration after axotomy. J Neurosci 26(33): 8484-8491.

Schoua C., O.E.R., Petersena H.J., Björkling F., * Latinib S., Hjarnaac P.V., Brammc E. and Binderup L. (1997). Novel cyanoguanidines with potent oral antitumour activity Bioorganic & Medicinal Chemistry Letters 7(24): 3095-3100.

Schreiber, V., Dantzer, F., Ame, J.C. and de Murcia, G. (2006). Poly(ADP-ribose): novel functions for an old molecule. Nat Rev Mol Cell Biol 7(7): 517-528.

Takebayashi, K., Suetsugu, M., Wakabayashi, S., Aso, Y. and Inukai, T. (2007). Association between plasma visfatin and vascular endothelial function in patients with type 2 diabetes mellitus. Metabolism 56(4): 451-458.

Tsiotra, P.C., Tsigos, C., Yfanti, E., Anastasiou, E., Vikentiou, M., et al. (2007). Visfatin, TNF-alpha and IL-6 mRNA expression is increased in mononuclear cells from type 2 diabetic women. Horm Metab Res 39(10): 758-763.

Van Beijnum, J.R., Moerkerk, P.T., Gerbers, A.J., De Bruine, A.P., Arends, J.W., et al. (2002). Target validation for genomics using peptide-specific phage antibodies: a study of five gene products overexpressed in colorectal cancer. Int J Cancer 101(2): 118-127.

van der Veer, E., Ho, C., O'Neil, C., Barbosa, N., Scott, R., et al. (2007). Extension of human cell lifespan by nicotinamide phosphoribosyltransferase. J Biol Chem 282(15): 10841-10845.

van der Veer, E., Nong, Z., O'Neil, C., Urquhart, B., Freeman, D., et al. (2005). Pre-B-cell colony-enhancing factor regulates NAD+-dependent protein deacetylase activity and promotes vascular smooth muscle cell maturation. Circ Res 97(1): 25-34.

von Heideman, A., Berglund, A., Larsson, R. and Nygren, P. (2010). Safety and efficacy of NAD depleting cancer drugs: results of a phase I clinical trial of CHS 828 and overview of published data. Cancer Chemother Pharmacol 65(6): 1165-1172.

Wang, T., Zhang, X., Bheda, P., Revollo, J.R., Imai, S., et al. (2006). Structure of Nampt/PBEF/visfatin, a mammalian NAD+ biosynthetic enzyme. Nat Struct Mol Biol 13(7): 661-662.

Ware, L.B. and Matthay, M.A. (2000). The acute respiratory distress syndrome. N Engl J Med 342(18): 1334-1349.

Wheeler, A.P. and Bernard, G.R. (2007). Acute lung injury and the acute respiratory distress syndrome: a clinical review. Lancet 369(9572): 1553-1564.

Wright, S.C., Wei, Q.S., Kinder, D.H. and Larrick, J.W. (1996). Biochemical pathways of apoptosis: nicotinamide adenine dinucleotide-deficient cells are resistant to tumor necrosis factor or ultraviolet light activation of the 24-kD apoptotic protease and DNA fragmentation. J Exp Med 183(2): 463-471.

Xie, H., Tang, S.Y., Luo, X.H., Huang, J., Cui, R.R., et al. (2007). Insulin-like effects of visfatin on human osteoblasts. Calcif Tissue Int 80(3): 201-210.

Yang, H.J., Yen, M.C., Lin, C.C., Lin, C.M., Chen, Y.L., et al. (2010). A combination of the metabolic enzyme inhibitor APO866 and the immune adjuvant L-1-methyl tryptophan induces additive antitumor activity. Exp Biol Med (Maywood) 235(7): 869-876.

Ye, S.Q., Simon, B.A., Maloney, J.P., Zambelli-Weiner, A., Gao, L., et al. (2005a). Pre-B-cell colony-enhancing factor as a potential novel biomarker in acute lung injury. Am J Respir Crit Care Med 171(4): 361-370.

Ye, S.Q., Zhang, L.Q., Adyshev, D., Usatyuk, P.V., Garcia, A.N., et al. (2005b). Pre-B-cell-colony-enhancing factor is critically involved in thrombin-induced lung endothelial cell barrier dysregulation. Microvasc Res 70(3): 142-151.

You, H., Youn, H.S., Im, I., Bae, M.H., Lee, S.K., et al. (2011). Design, synthesis and X-ray crystallographic study of NAmPRTase inhibitors as anti-cancer agents. Eur J Med Chem 46(4): 1153-1164.

Zhong, M., Tan, H.W., Gong, H.P., Wang, S.F., Zhang, Y., et al. (2008). Increased serum visfatin in patients with metabolic syndrome and carotid atherosclerosis. Clin Endocrinol (Oxf) 69(6): 878-884.

Ziegelmeier, M., Bachmann, A., Seeger, J., Lossner, U., Kratzsch, J., et al. (2008). Adipokines influencing metabolic and cardiovascular disease are differentially regulated in maintenance hemodialysis. Metabolism 57(10): 1414-1421.

Zoppoli, G., Cea, M., Soncini, D., Fruscione, F., Rudner, J., et al. (2010). Potent synergistic interaction between the Nampt inhibitor APO866 and the apoptosis activator TRAIL in human leukemia cells. Exp Hematol 38(11): 979-988.

4

Cofactor Recycling Using a Thermostable NADH Oxidase

Jun-ichiro Hirano, Hiromichi Ohta, Shosuke Yoshida and Kenji Miyamoto*
Department of Biosciences and Informatics
Keio University, Hiyoshi, Yokohama
Japan

1. Introduction

From the standpoint of enzymatic organic synthesis, NADH oxidase (NOX) will be a key enzyme that plays an essential role in the cofactor regeneration of NAD^+ dependent enzymatic reactions. For example, enzymatic enantioselective oxidations of racemic secondary alcohols (Geueke et al., 2003; Riebel et al., 2003; Hummel & Riebel, 1996) and amino acids (Hummel et al., 2003a) have been reported as utilizing this enzyme (Fig. 1). The kinetic resolutions of secondary alcohols are important processes in cases where preparation of the corresponding ketones is difficult. In these reported reactions, NAD^+ dependent enantioselective alcohol dehydrogenase (or amino acid dehydrogenase) was used with NOX for regeneration of the oxidized form of cofactor NAD^+.

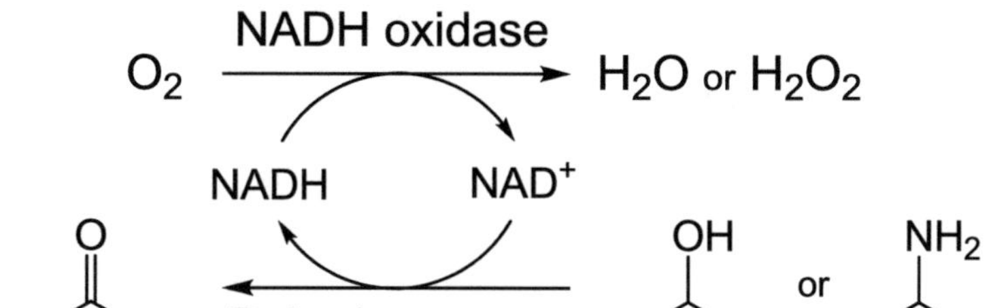

Fig. 1. Enzymatic oxidation of alcohols using a cofactor regeneration system

NADH oxidase catalyzes the oxidation of NADH to NAD^+ using molecular oxygen as the electron acceptor. The NADH oxidase family of enzymes is divided into two major types corresponding to the mode of oxygen reduction. One catalyzes a two-electron reduction of oxygen to give hydrogen peroxide, and the other catalyzes a four electron reduction of oxygen to give water with NADH oxidation. So far, NOXs have been isolated from anaerobic bacteria such as *Streptococcus* (Matsumoto et al., 1996; Higuchi et al., 1993, 1994; Ross and Claiborne, 1992; Schmidt et al., 1986), *Thermotoga* (Yang & Ma, 2005, 2007), *Clostridium* (Kawasaki et al., 2004; Maeda et al., 1992), *Eubacterium* (Herles et al., 2002), and *Lactobacillus* (Hummel et al., 2003b; Riebel et al., 2002, 2003), and from archaea such as

* Corresponding author

Sulfolobus (Masullo et al., 1996), *Thermus* (Toomey & Mayhew, 1998; Park et al., 1992), and *Archaeoglobus* (Kengen et al., 2003). In these anaerobic organisms, NOX plays an important role as an oxygen scavenger under oxidative stress (Miyoshi et al., 2003). In aerobic microorganisms, NOX activity results from the electron transfer reaction from NADH to O_2 through cytochromes by membrane-bound enzymes. In fact, NADH oxidase of *Corynebacterium* (Matsushita et al., 2001; Ginson et al., 2000) has been isolated from the membrane. Characterization of NOX from aerobic bacteria has rarely been reported because of the difficulty in purifying the membrane proteins.

Generally, the activity of alcohol dehydrogenase in oxidation reactions is higher under basic conditions, because this reaction produces a proton. Therefore, highly active and stable NOX under basic conditions is desirable for enzymatic oxidation. We have recently isolated *Brevibacterium* sp. KU1309 from soil. This Gram-positive aerobic bacterium can grow in medium containing 2-phenylethanol as the sole source of carbon and has soluble NOX in spite of the fact that the NOXs of aerobic bacteria are generally membrane-bound enzymes. Utilizing this strain, various alcohols have been oxidized to the corresponding carboxylic acids and ketones (Miyamoto et al., 2004). Herein, we describe the isolation of a hydrogen peroxide-producing NOX from *Brevibacterium* sp. KU1309, which exhibits high thermal stability in a broad range of pH values.

2. Biocatalytic oxidation of various alcohols

The development of effective catalytic oxidation of alcohols using environmentally benign and inexpensive oxidants, such as O_2, is an important challenge (Sheldon et al., 2000). Acetic acid bacteria have been employed for the biotransformation of various alcohols (Ohta & Tetsukawa, 1979, 1981), including the enantioselective oxidation of racemic primary alcohols (Gandolfi et al., 2001; Romano et al., 2002). Whole cells are preferable as they have regeneration systems for the cofactors. A wide variety of $NAD(P)^+$-dependent dehydrogenases (EC 1.1.1.-), which catalyze the asymmetric reduction of ketones to optically active alcohols, have been reported and are used to synthesize industrially important chemicals (Nakamura et al., 2003). The application of $NAD(P)^+$-dependent enzymes to the oxidative direction is limited because few efficient methods for $NAD(P)^+$ regeneration are applicable at the preparative scale (Hummel et al., 2003a; Riebel et al., 2003). Flavin-dependent alcohol oxidases (EC 1.1.3.-) are able to oxidize alcohols to carbonyl compounds with simultaneous reduction of O_2 to H_2O_2 (Burton, 2003). Several oxidases (oxidases of methanol/ethanol, glucose, etc.) are used as analytical tools (Karube & Nomura, 2000). However, only a limited amount of information exists on the enzymatic synthesis of carbonyl compounds using alcohol oxidases. Previously, we reported the enantioselective oxidation of mandelic acid using mandelate oxidase of *Alcaligenes bronchisepticus* KU1201 (Miyamoto & Ohta, 1992; Tsuchiya et al., 1992). However, the oxidizing enzyme system was limited to mandelate derivatives. Thus, we tried to screen an oxidation system that has broad substrate specificity and high activity. We herein describe a novel method for the synthesis of carbonyl compounds via oxidation by a 2-phenylethanol-degrading microorganism.

2.1 Screening of a 2-phenylethanol-degrading microorganism

We selected 2-phenylethanol (PE, **1**) as a screening compound. Because of the toxicity of this alcohol, we expected PE-degrading microorganisms to have strong metabolic enzymes for it.

The enrichment culture technique (Asano, 2002) was used to screen for various PE-degrading microorganisms. Of the various microorganisms isolated after growing on PE, one strain was assigned to the pathway in Fig. 2. If the microorganism responsible for this oxidation reaction also acted on 2-phenylpropanol (**3**), then 2-phenylpropanoate (**4**) would be expected to be obtained because further metabolism would be impossible. Strain KU 1309 was found to be the most active in the oxidation of propanol (**3**) to propanoate (**4**). The bacterium was Gram-positive, formed rods, non-motile, non-spore-forming, and without flagella. No acid or gas was produced from glucose. The 16S rDNA sequence showed that 2-phenylethanol-utilizing strain KU 1309 is closely related to *Brevibacterium iodium* (98.2%) and *Brevibacterium epidermidis* (97.8%). Based on these results, strain KU 1309 was identified as *Brevibacterium* sp.

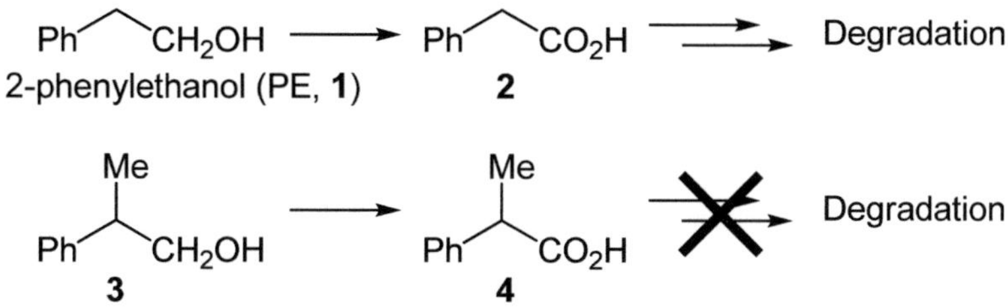

Fig. 2. 2-Phenylethanol-degrading pathway of *Brevibacterium* sp. KU1309

Although the enzyme system of the bacterium KU 1309 was demonstrated to have broad substrate specificity and high activity, it showed no enantioselectivity toward the substrate (**3**). Recently, we succeded in isolating microorganisms that oxidize 2-phenyl-1-propanol (**3**) to enantiomerically enriched (*S*)-2-phenylpropanoic acid (**4**) (Miyamoto et al., 2009). The resulting (*S*)-2-arylpropanoic acids are important compounds known as non-steroidal anti-inflammatory drugs (NSAIDs) (Kourist et al., 2011).

2.1.1 Optimization of microbial oxidation conditions by *Brevibacterium* sp. KU1309

The cultivation conditions suitable for inducing potent oxidative activity in intact cells were examined. The oxidation level was constant in the cells grown without PE, although the addition of alcohol increased activity. Consistent PE oxidation was obtained by using a nutrient medium containing PE. The optimum concentration of PE and the adequate cultivation period were 0.4% (w/v) and 3 d, respectively. Cell growth was inhibited by a higher PE concentration (>0.5% (w/v)). Next, the conditions for the production of acid (**4**) from alcohol (**3**) were examined. The strain exhibited highest productivity at around pH 10. When the concentration of PE was 0.1–0.4% (w/v), oxidation proceeded smoothly and the corresponding carboxylic acid was obtained in high yield (>80%). Oxidation heavily decreased when the alcohol concentration exceeded 0.6% (w/v). Therefore, the enzymatic reaction was performed at alcohol concentrations of under 0.4% (w/v).

2.1.2 Substrate specificity of microbial oxidation system

Under optimal conditions, microbial oxidation was extended to other alcohols (Table 1). First, the effect of variations in the aromatic part (R^1) was examined (Entries 1-3). These substrates were readily oxidized, and the yields of the corresponding carboxylic acids were high. The oxidizing enzyme was not enantioselective, giving racemic products. Next, the R^2 group was changed to hydrogen (Entries 4-6). Relatively high yields were realized by quenching the reaction after 6 h. When the R^2 group was an isopropyl (Entry 7), oxidation did not proceed and the starting material was recovered intact. It is considered that the isopropyl group was too bulky for the enzyme to bind with the substrate. To investigate the cofactor requirement of the oxidizing enzyme system, we studied the effect of the reaction atmosphere. When the reaction was performed under argon, the product yield decreased drastically; thus, the oxidizing system was O_2 dependent. In addition, we tried to detect enzyme activity for the reduction of 2,6-dichloroindophenol (DCIP) at 600 nm in the presence of *N*-methylphenazonium methosulphate (PMS), or the formation of NAD(P)H at 340 nm. We detected 2-phenylethanol oxidase activity in the wall-membrane fraction and NAD^+-dependent dehydrogenase in the cytoplasm.

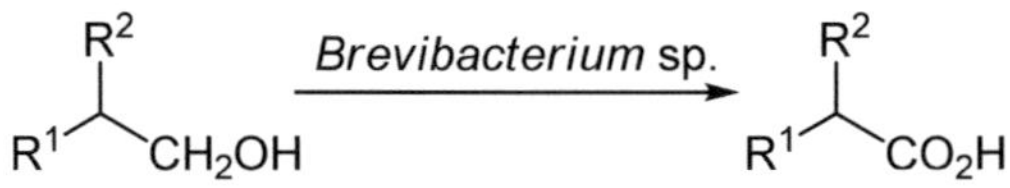

Entry	Substrate R¹	Substrate R²	Reaction time (h)	Yield (%)
1	Ph	Me	24	86
2	PhO	Me	24	87
3	2-Naphthyl	Me	96	80
4	Ph	H	6	70
5	$PhCH_2$	H	6	71
6	$PhCH_2CH_2$	H	6	77
7	4-Chlorophenyl	^iPr	96	0

Table 1. Substrate specificity of primary alcohols

2.2 Purification and characterization of oxidation enzymes

We have shown that soil bacterium *Brevibacterium* sp. KU 1309 can be efficiently applied to the oxidation of various alcohols. The strain assimilated PE, and exhibited high oxidative activity towards various alcohols (Miyamoto et al., 2004). Endogenous 2-phenylethanol was found in the metabolic pathway of styrene in *Xantobacter* strain 124X (Hartmans et al., 1989), 2-phenylethylamine in *Escherichia coli* (Parrott et al., 1987), and phenylalanine in *Saccharomyces cerevisiae* (Dickinson et al., 2003). However, there was not sufficient information concerning the enzymes that participate in the metabolism of 2-phenylethanol, although 2-phenylethanol dehydrogenase (PEDH) was estimated to be the key enzyme of this metabolic pathway. Thus we tried to purify PEDH from *Brevibacterium* sp.

2.2.1 Purification and characterization of alcohol dehydrogenase with broad substrate specificity

The summary of the purification procedures for PEDH from *Brevibacterium* sp. is shown in Table 2 (Hirano et al., 2005). PEDH was purified about 1400-fold from the cell-free extract by sequential column chromatography.

Step	Total protein (mg)	Total activity (unit)	Specific activity (Unit/mg)	Yield (%)	Purification (fold)
Cell-free extract	1381	18.8	0.0136	100	1
$(NH_4)_2SO_4$	412	12.2	0.0296	65	2
Ether-Toyopearl	68.9	16.1	0.234	85	17
DEAE-Toyopearl	3.28	13.3	4.06	71	299
DEAE-Sepharose	1.65	7.00	4.24	37	312
Phenyl-Toyopearl	0.360	6.38	17.7	34	1302
Butyl-Toyopearl	0.198	3.80	19.2	20	1412

Table 2. Purification of PEDH from *Brevibacterium* sp. KU1309

The enzyme did not bind to Toyopearl AF-Blue and AF-Red in spite of its NAD^+ dependent oxidoreductase activity. The specific activity of PEDH increased to 19.2 U/mg of protein with a 20% yield from the cell-free extract. The purity of the enzyme was checked by SDS-PAGE, and analytical HPLC with Superose 12 gel filtration column chromatography. These analyses showed that the enzyme sample was homogenous. The molecular mass of PEDH was estimated to be 29 kDa by analytical HPLC on gel filtration column chromatography. The molecular weight under denaturing conditions was determined to be 39 kDa by SDS-PAGE. These results suggested that the purified enzyme is a monomeric protein. The N-terminal amino acid sequence of PEDH was determined to be MKASLATAIGGEFTVHD. A database search of the sequence revealed no similar proteins. Thus, the family this enzyme belongs to is very interesting. The activity of PEDH was measured at various pH values. The maximum activity of 2-phenylethanol oxidation was observed at pH 10.4. In contrast, the enzyme exhibited the maximum activity of phenylacetaldehyde reduction at pH 6.0. No significant loss of enzyme activity was observed after heat treatment for 30 min at 35°C, while about 50% of the initial activity was lost after heat treatment at 50°C. The optimal reaction temperature for the enzyme was estimated to be around 60°C. Further heat treatment above 60°C caused a rapid decrease in enzyme activity. The kinetic properties and substrate specificity fitting to Michaelis–Menten equation are shown in Tables 3–5. The kcat/Km value of NADH was about 10-fold greater than that of NAD^+ (Table 3). As a result, the reduction of aldehyde proceeded faster than the oxidation of alcohol. This enzyme did not oxidize alcohols with $NADP^+$ as cofactor: the relative reaction rate with $NADP^+$ was less than 0.1% of that with NAD^+.

cofactor	Relative activity (%)	kcat/Km (s^{-1}/mM)
NAD^+	100	22.4
$NADP^+$	<0.1	-
NADH	100	210.3
NADPH	<0.1	-

Table 3. Effect of cofactors on the relative activities

The substrate specificity of this enzyme is shown in Table 4. This enzyme oxidized various primary alcohols with aromatic rings (2-phenylethanol, 2-phenylpropanol, benzyl alcohol, 3-phenylpropanol) and primary aliphatic alcohols (ethanol, 1-butanol, 1-octanol, and 1-decanol). On the other hand, the enzyme showed lower activity toward secondary alcohols, such as 1-phenylethanol and 2-propanol. Furthermore, this enzyme preferred (*S*)-2-phenylpropanol to the (*R*)-enantiomer as a substrate.

Substrate	Relative activity (%)	Km (mM)
2-Phenylethanol	100	0.025
(*S*)-2-Phenylpropanol	156	0.157
(*R*)-2-Phenylpropanol	63	0.020
Benzyl alcohol	199	0.012
3-Phenylpropanol	135	0.033
Ethanol	76	-
1-Butanol	111	-
1-Octanol	101	-
1-Dodecanol	68	-
1-Phenylethanol	46	-
2-Propanol	54	-

Table 4. Relative activities in the oxidation of various alcohols

Substrate	Relative activity (%)	Km (mM)
Phenylacetaldehyde	100	0.261
2-phenylpropionaldehyde	188	0.864
1-Octylaldehyde	87	-
Acetophenone	0	-

Table 5. Relative activities in the reduction of various carbonyl compounds

This enzyme reduced aldehydes with an aromatic ring (phenylacetaldehyde, 2-phenylpropionaldehyde) and an aliphatic aldehyde (1-octylaldehyde) in the reverse reaction, that is, the reduction of aldehyde. However, similar to the case of the oxidation of secondary alcohols, the enzyme could not reduce ketones (acetophenone) (Table 5). Various compounds were investigated for their effects on enzyme activity. The enzyme activity was completely inhibited by 1 mM of $CuCl_2$ (0%), $NiCl_2$ (7%), $BaCl_2$ (5%), and $HgCl_2$ (0%). Other inorganic compounds, such as $MgCl_2$, $CaCl_2$, $MnCl_2$, and $ZnCl_2$ had no influence on activity. While the enzyme was completely inhibited by *p*-chloromercuribenzoate, other thiol reagents such as iodoacetate and N-ethylmaleimide did not have inhibitory effects. Metal chelating reagents, such as 8-quinolinol (98%) and EDTA (99%), carbonyl reagents such as NaN_3 (94%), and serine inhibitors, for example, phenylsulfonyl fluoride (85%), had no significant effect on the enzyme. We isolated PEDH from 2-phenylethanol-assimilating soil bacterium *Brevibacterium* sp. which was grown in medium containing 2-phenylethanol as the sole source of carbon. PEDH oxidized 2-phenylethanol to phenylacetaldehyde, and utilized NAD^+ but not $NADP^+$ as the cofactor. Previously, we detected 2-phenylethanol oxidase activity in the wall-membrane fraction (Miyamoto et al., 2004). Phenylacetaldehyde dehydrogenase (PADH) activity was also recognized in the cell-free extracts of the strain. The present results suggest that the oxidative degradation of 2-phenylethanol (**1**) via phenylacetaldehyde (**5**) to phenylacetate (**2**) is one of the metabolic pathways of 2-phenylethanol by *Brevibacterium* sp. PEDH plays a role, in part, in the oxidation of 2-phenylethanol into phenylacetaldehyde in this strain (Fig. 3). The substrate specificity of PEDH is different from that of NAD^+-dependent aromatic alcohol dehydrogenase. Aryl alcohol dehydrogenase (EC 1.1.1.90) from *Pseudomonas putida* could catalyze the oxidation of benzyl alcohol but not 2-phenylethanol and alkanol (Shaw & Harayama, 1990). Benzyl alcohol dehydrogenase from benzyl alcohol-assimilating *Thauera* sp. similarly could not oxidize alkanol (Biegert et al., 1995). Although phenylacetaldehyde reductase from styrene-assimilating *Corynebacterium sp.* strain ST-10 reduced various aldehydes, it did not catalyze the reverse reaction (Itoh et al., 1996). The dehydrogenase of *P. putida* appears as a homodimer composed of a 42 kDa subunit. Alcohol dehydrogenases (ADHs) from *Thauera* sp. and *Corynebacterium* sp. display similar subunit sizes (40–42 kDa) and were found to be homotetramers. The active form of PEDH is a monomer, while the size (39 kDa) of PEDH is similar to those of the ADHs listed above. The alcohol dehydrogenase superfamily can be divided into three groups: group I is the zinc-dependent long chain ADHs (approximately 350 residues per subunit); group II is the short chain zinc-independent ADHs (approximately 250 residues per subunit); and group III is the iron-activated ADHs (approximately 385 residues per subunit) (Reid & Fewson, 1994). From the results of inhibition studies, PEDH is a zinc-independent alcohol dehydrogenase containing approximately 350 residues per subunit. Thus, PEDH does not fit into any groups. Furthermore, the N-terminal amino acid sequence of PEDH showed no similarity to other reported dehydrogenases. Based on these results, it can be said that PEDH is a novel enzyme in terms of these characteristic properties and its broad substrate specificity. Thus, PEDH from *Brevibacterium* sp., in combination with an NAD^+ regeneration system, would be a new entry of a clean and versatile alcohol oxidizing system.

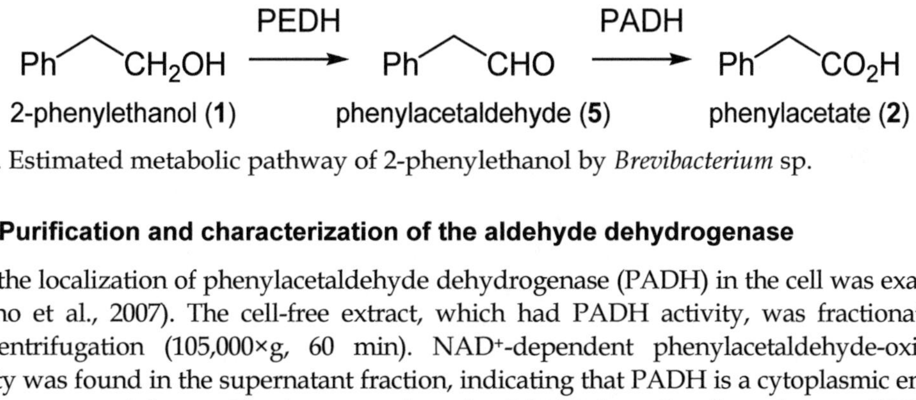

Fig. 3. Estimated metabolic pathway of 2-phenylethanol by *Brevibacterium* sp.

2.2.2 Purification and characterization of the aldehyde dehydrogenase

First, the localization of phenylacetaldehyde dehydrogenase (PADH) in the cell was examined (Hirano et al., 2007). The cell-free extract, which had PADH activity, was fractionated by ultracentrifugation (105,000×g, 60 min). NAD^+-dependent phenylacetaldehyde-oxidizing activity was found in the supernatant fraction, indicating that PADH is a cytoplasmic enzyme. The summary of the purification procedure for PADH from *Brevibacterium* sp. KU1309 is shown in Table 6. The specific activity of the enzyme increased by 16-fold to 4.16 U/mg protein with a 0.21% yield from the cell-free extract. The low yield of purified PADH through this procedure is considered to be due to the low stability of the enzyme. The molecular weight of the denatured protein was estimated to be approximately 61 kDa based on a comparison with the mobility of the marker proteins. The molecular mass of native PADH was estimated to be 219 kDa by gel filtration on high-performance liquid chromatography. These results indicate that the enzyme is a homotetrameric protein. Because the reported PadA from *E. coli* was a homodimer, PADH is obviously different from this enzyme.

Step	Total protein (mg)	Total activity (unit)	Specific activity (Unit/mg)	Yield (%)	Purification (fold)
Cell-free extract	480	125	0.261	100	1
Butyl-Toyopearl	6.15	2.01	0.327	1.6	1.25
DEAE-Toyopearl	0.620	0.864	1.39	0.69	5.34
DEAE-Sepharose	0.132	0.250	1.89	0.20	7.26
Ether-Toyopearl	0.063	0.262	4.16	0.21	15.9

Table 6. Purification of PADH from *Brevibacterium* sp. KU1309

The N-terminal sequence of PADH was found to be TTTVESPARSP. A database search for protein sequences resembling this protein was conducted using BLAST or FASTA systems. However, no protein with similar aldehyde dehydrogenase activity was found. The optimum pH and pH stability of PADH were determined. PADH exhibited oxidizing activity at pH 7-10, with the maximum activity observed at pH 9. The pH stability was examined by measuring activity at pH 9 after incubation for 30 min at various levels of pH. This enzyme was relatively stable between pH 6 and 8. At pH 8.0, PADH retained 60% of its initial activity. PADH showed highest activity at 35°C. At this temperature, residual activity was 40% after 30 min. The effects of various compounds on the activity of PADH were determined. PADH was activated upon incubation in the presence of several divalent cations such as Mg^{2+} (142%), Ca^{2+} (153%), and Mn^{2+} (117%). A similar effect was reported for ALDH1 from *Saccharomyces cerevisiae* (Wang et al., 1998). Enzyme activity was inhibited by sulfhydryl agents such as Hg^{2+} (23%), *p*-CMBA (0%), iodoacetamide (0%), and N-methylmaleimide (0%), which indicates that a thiol group plays an important role in oxidation. PADH was not stimulated by potassium ions, unlike ALDH2 from *S. cerevisiae* and BADH from *Pseudomonas putida* (McLeish et al., 2003). PADH activity was inhibited by a high concentration (0.1 mM) of phenylacetaldehyde. This property is the same as that for PadA from *E. coli*. The kinetic parameters of PADH were determined by Lineweaver–Burk plotting (Table 7). The Km of PADH for phenylacetaldehyde was low (1.24 µM), and substrate inhibition was observed even at low substrate concentrations. The low Km value is on the same order as the Km of PAD from *E. coli* K12 (7 µM; Hanlon et al., 1997). This result suggested that phenylacetaldehyde is probably the natural substrate for PADH. On the other hand, the Km of PADH for NAD^+ (116 µM) is greater than that of phenylacetaldehyde dehydrogenase from *Achromobacter eurydice* (70 µM). The substrate specificity of PADH was investigated using a variety of aldehydes (Table 8). PADH oxidized a wide range of aldehydes, such as aromatic aldehydes (2-phenylethanol, benzylalcohol, 3-phenylpropanol, and 1-naphthaldehyde) and aliphatic aldehydes (hexanal, octanal, and decanal). The enzyme showed highest activity toward phenylacetaldehyde, and the activity for octylaldehyde was highest. This enzyme prefers NAD^+ as the cofactor rather than $NADP^+$ (6%).

Substrate	Km (µM)	Vmax (µmol min^{-1} mg^{-1})
Phenylacetaldehyde	1.24	4.63
NAD^+	116	4.40

Table 7. Kinetic parameters for PADH

Substrate	Relative activity (%)
Phenylacetaldehyde	100
Benzaldehyde	54
3-Phenylpropionaldehyde	34
1-Naphthaldehyde	24
Hexanal	21
Octanal	31
Decanal	17
NAD^+	100
$NADP^+$	6

Table 8. Substrate specificity of PADH

We have isolated phenylacetaldehyde dehydrogenase from *Brevibacterium* sp. KU1309 as the key enzyme in the 2-phenylethanol metabolic pathway. The anticipated catabolic pathway of 2-phenylethanol is shown in Fig. 4; *Brevibacterium* sp. KU1309 oxidizes 2-phenylethanol to phenylacetaldehyde by NAD^+-dependent PEDH. As PEDH could not oxidize aldehyde (data not shown), the resulting phenylacetaldehyde is oxidized to phenylacetate by NAD^+-dependent PADH. Phenylacetate should be further degraded via the general pathway of aromatic compounds. To the best of our knowledge, the metabolic pathway of 2-phenylethanol via phenylacetate from 2-phenylethanol-assimilating bacteria has not been reported.

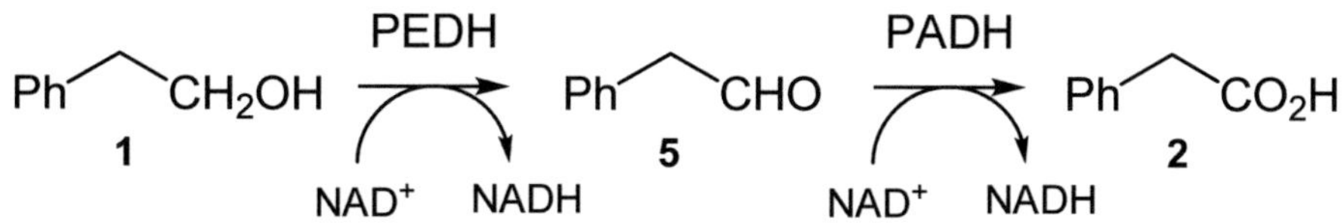

Fig. 4. The metabolic pathway of 2-phenylethanol in *Brevibacterium* sp.

As mentioned earlier, the intermediate of the metabolic pathway of phenylalanine, styrene, and 2-phenylethylamine is also phenylacetaldehyde. However, the N-terminal amino acid sequence of purified PADH in the present work was not similar to the sequences of the other phenylacetaldehyde dehydrogenases. Thus, a genetic comparison of the gene encoding PADH with those of the corresponding aldehyde dehydrogenases of the other three pathways is of great interest and the target of future studies. The property of PADH is also different from other aldehyde dehydrogenases relating to phenylacetaldehyde. For example, PadA, which originates from *E. coli* w3110 and is a homodimeric enzyme, could not oxidize benzaldehyde. MdlD from *Pseoudomonas putida,* which has a broad substrate range of aryl or aliphatic aldehydes, is a K^+-activated enzyme. However, PADH is a homotetrameric enzyme and activated by divalent metal ions. Moreover, PADH also oxidizes benzaldehyde and aliphatic aldehydes. PADH has broad substrate specificity and would be extremely useful as an aldehyde-oxidizing biocatalyst such as commercially available ALDH originating from *S. serevisiae.* We isolated and characterized PADH from *Brevibacterium* sp. KU1309. The enzyme has broad substrate specificity and a different N-terminal sequence from those of known aldehyde dehydrogenases

2.2.3 Purification and characterization of thermostable H_2O_2-forming NADH oxidase

A summary of the purification of NADH oxidase (BreNOX) from *Brevibacterium* sp. is shown in Table 9 (Hirano et al., 2008). About one third of NOX activity was detected in the cytoplasmic fraction, and the remaining 60% was detected in the membrane fraction after ultracentrifugation (105,000 ×g, 60 min). We purified BreNOX from the cytoplasmic fraction. BreNOX was purified to homogeneity by ammonium sulfate fractionation and three steps of column chromatography.

Step	Total protein (mg)	Total activity (unit)	Specific activity (Unit/mg)	Yield (%)	Purification (fold)
Cell-free extract	800.9	69.1	0.086	100	1
$(NH_4)_2SO_4$	466.7	33.3	0.071	48	0.83
Phenyl-Toyopearl	16.55	15.7	0.94	23	10.9
DEAE-Toyopearl	2.31	9.73	4.2	14	48.8
Butyl-Toyopearl	1.35	7.04	5.2	10	60.5

Table 9. Purification of BreNOX

Purified BreNOX showed a single band on SDS/PAGE stained with Coomassie brilliant blue. The molecular weight was found to be about 57 kDa. The molecular weight of native BreNOX was estimated to be about 102 kDa by gel filtration. These results show that the native enzyme has a homodimeric structure. The enzyme shows a yellow color, which is probably because of the binding of FAD as the cofactor. NADH oxidases can be divided into two major categories: one group produces H_2O_2 from O_2 and the other produces H_2O from O_2. The reduction of molecular oxygen was examined to clarify the category in which the present enzyme belongs. Oxidation of NADH catalyzed by this enzyme resulted in the stoichiometric production of H_2O_2, which was confirmed by the spectrophotometrical method (Kengen et al., 2003), i.e., detection of a dye compound formed by electron transfer from H_2O_2 using peroxidase. Thus, it was found that BreNOX belongs to the H_2O_2-producing type. The effects of various salts on the activity of BreNOX were examined (Table 10). The activity of the enzyme increased about ten-fold with the addition of ammonium sulfate and ammonium chloride (Fig. 5). On the other hand, sodium chloride and potassium chloride had no influence on NOX activity. Accordingly, the activation of this enzyme upon addition of the two salts would be due to the ammonium ion. Although the precise mechanism is not clear, it is supposed that the ammonium cation may bind to some anionic part of the enzyme resulting in a change of conformation.

Salts	Relative activity (%)
None	100
$(NH_4)_2SO_4$	1000
NH_4Cl	940
NaCl	80
KCl	100

Table 10. Effect of salts on NOX activity

The optimal reaction temperature for the enzyme was determined to be 70°C, and the activity at 30°C was about 40% of the maximal. As shown in Fig. 6, no significant loss of enzyme activity was observed after heat treatment at 70°C for 60 min, while about 30% of the initial activity was lost after incubation at 80°C for 60 min. Thus, it can be said that BreNOX is a thermostable enzyme. To the best of our knowledge, such a thermally stable enzyme has not been reported except for the NADH oxidase from hyperthermophilic archaeon *Archaeoglobus fulgidus*. It is very interesting that the enzyme from *Brevibacterium* is stable at high temperature regardless of the fact that this strain grows at normal temperatures under aerobic conditions.

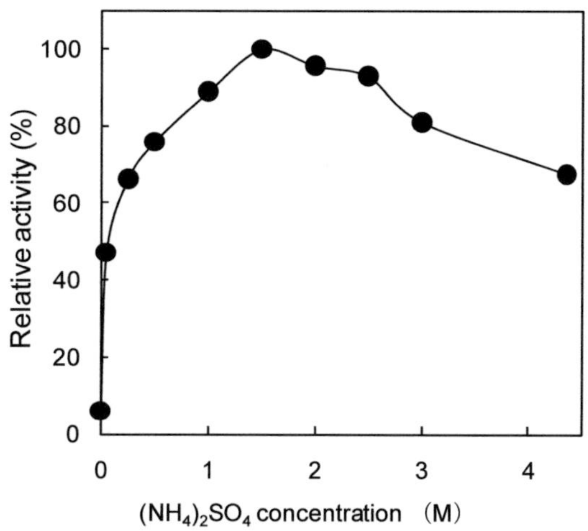

Fig. 5. The effect of ammonium sulphate concentration

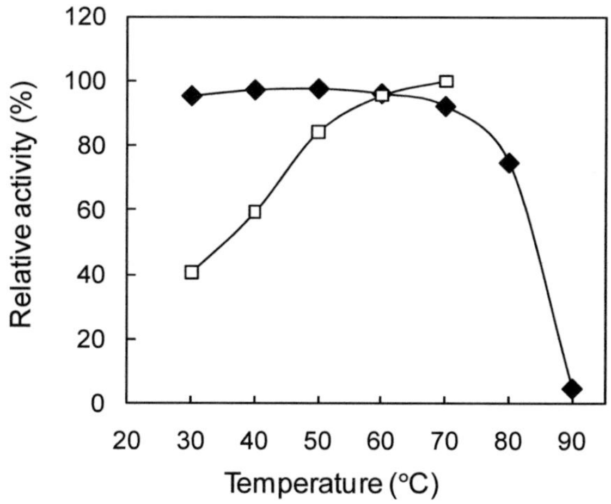

Fig. 6. The effect of temperature on activity and stability. Remaining activity after incubation for 1 h at various temperatures (closed diamond). Relative activity at various temperatures (open square)

The effect of pH on the activity and stability of BreNOX was examined. The enzyme was active and stable in the pH range 6–10 (Figs. 7 and 8), while maximal activity was observed at pH 9. These results indicated that the enzyme is an alkalophilic enzyme.

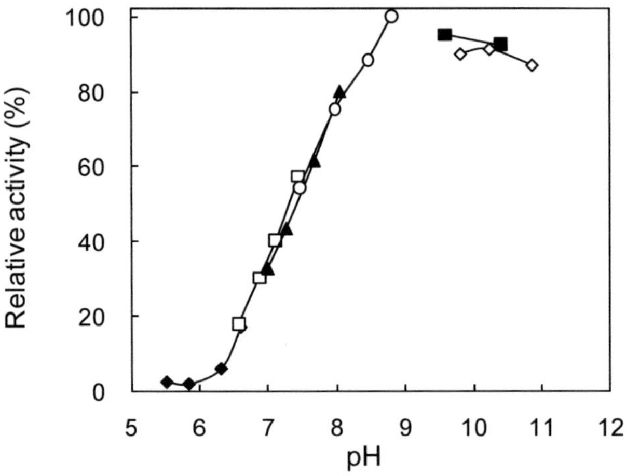

Fig. 7. The effect of pH on NOX activity. The activity of BreNOX was measured at various pH levels. The buffers used were MES (pH 5.5-6.6; closed diamond), MOPS (pH 6.6-7.4; open square), HEPES (pH 7.0-8.1; closed triangle), Tris (pH 7.5-8.8; open circle), Glycine (pH 9.6-10.4; closed square), and CAPS (pH 9.8-10.9; open diamond)

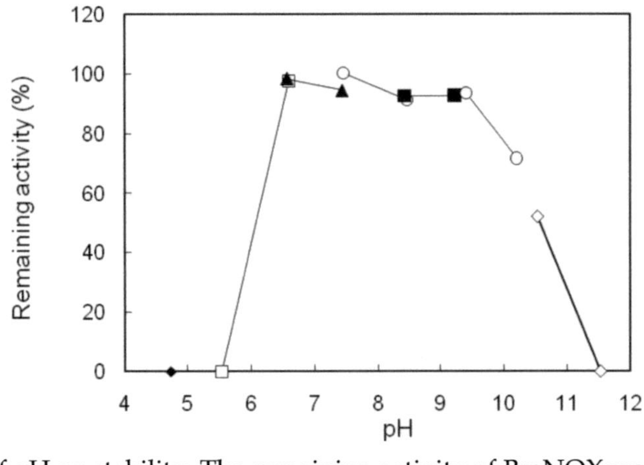

Fig. 8. The effect of pH on stability. The remaining activity of BreNOX was measured after incubation for 1 h at various pH levels. The buffers were citric acid (pH 4.7; closed diamond), MES (pH 5.5-6.6; open square), MOPS (pH 6.6-7.4; closed triangle), Tris (pH 7.5-8.5; open circle), TAPS (pH 8.4-9.2; closed square), CAPS (pH 9.4–10.2; open circle), and Na_2HPO_4 (pH 10.5–11.5; open diamond)

The effects of various compounds on enzyme activity were investigated. NOX activity was inhibited by the presence of 1 mM of some metal ions, such as Zn^{2+} (39%), Cu^{2+} (43%), Hg^{2+} (71%), and Ag^{+} (37%). Curiously, enzyme activity was nearly doubled by adding *p*-CMBA.

Cofactors such as FAD and FMN had no effect on activity. The Km and Vmax values of NOX were found to be 0.022 mM and 8.86 U/mg, respectively, in the presence of 500 mM ammonium sulfate. The Km value of the enzyme was similar to that of other types of NOX. In addition, the enzyme showed little activity toward the oxidation of NADPH. The N-terminal amino acid sequence was found to be XDELTYDLVVLGGGTGG. A FASTA-based search of a protein database indicated that this sequence exhibited high identity with those of the pyridine nucleotide–disulfide oxidoreductase family (dihydrolipoamide dehydrogenase and glutathione reductase). Because the N-terminal amino acid residue could not be determined by Edman degradation, it is estimated that the amino group of this amino acid residue is protected.

2.3 Coupling reaction with dehydrogenases and BreNOX

Recycling of the cofactor is the key process for effective oxidative biotransformation. If the newly discovered BreNOX can be applied to cofactor regeneration of oxidative reactions that require NAD^+, the whole system will be extremely useful for the green process. Thus, we tried the coupled reaction as demonstrated below (Fig. 9). First, BreNOX was coupled with mandelate dehydrogenase (MDH) for cofactor regeneration (eq. 1). MDH from *E. faecalis* (Tamura et al., 2002) oxidized (*R*)-mandelic acid (**6**) using NAD^+ as the electron acceptor. The enantioselective oxidation of racemic mandelic acid using MDH and NOX proceeded smoothly. The (*R*)-mandelic acid was oxidized to benzoylformate (**7**), and the recovered mandelic acid was found to be *S* configuration with over 99% e.e. The result showed that (*R*)-mandelate was oxidized completely in the presence of only 0.05 mol equivalent of NAD^+. L-phenylalanine (**8**) was completely oxidized to phenylpyruvate (**9**) in the pure D-phenylalanine (99% e.e.) in a coupled reaction with L-phenylalanine dehydrogenase (LPADH) (eq. 2) using racemic phenylalanine.

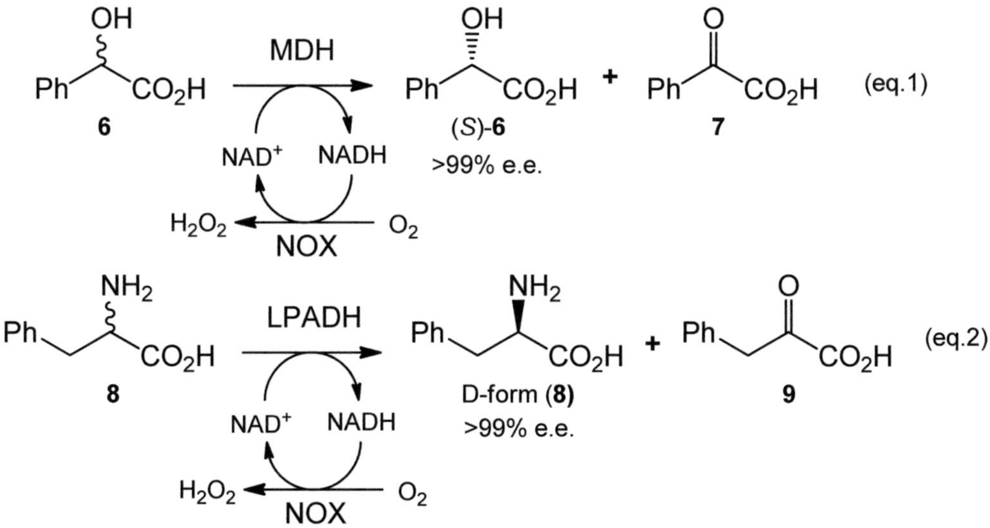

Fig. 9. Coupling reaction with dehydrogenase and NOX

Next, BreNOX was coupled with 2-phenylethanol dehydrogenase (PEDH) and phenylacetaldehyde dehydrogenase (PADH) from *Brevibacterium* sp. (Table 11). The system for the coupled reaction was composed of 5 mM of various alcohols, 1 mM of NAD^+, and Tris-HCl buffer (100 mM, pH 8.8). The mixture was incubated overnight at 25°C. Hydrochloric acid (2 M) was added to the reaction mixture, and the mixture was extracted with diethyl ether. After esterification by $TMSCH_2N_2$, the esters were analyzed by GLC to determine the yield. The oxidation products were obtained quantitatively. As demonstrated by the above two examples, BreNOX can be successfully coupled with various NAD^+-dependent enzymes, because of high activity under a wide range of reaction conditions.

$$\underset{5\ \text{mM}}{R{-}CH_2OH} \xrightarrow[\substack{NAD^+\ (1\ \text{mM}),\ \text{Buffer (pH9)} \\ 25\ °C,\ \text{Overnight}}]{\substack{\text{PEDH, PADH \& BreNOX} \\ \text{from } \textit{Brevibacterium} \text{ sp.}}} R{-}CO_2H$$

Entry	Substrate	Carboxylic acid (%)	Alcohol (%)
1	$PhCH_2OH$	74	2
2	$PhCH_2CH_2OH$	86	3
3	$PhCH_2CH_2CH_2OH$	87	0
4	$CH_3(CH_2)_4CH_2OH$	60	5
5	$CH_3(CH_2)_6CH_2OH$	57	3

Table 11. Oxidation of primary alcohols using PEDH, PADH and BreNOX

3. Conclusion

We have succeeded in isolating a hydrogen peroxide producing NOX from the aerobic bacterium *Brevibacterium* sp. KU1309, which showed activity and stability in a broad range of pH. The alignment of N-terminal amino acid sequence is shown in Fig. 10.

```
Brevibacterium sp. KU1309                        -------XDE LTYDLVVLGG GTGG
Bacillus sp. NRRL B-14911                        --------MA QEYDLVILGG GTGG
Limnobacterium sp. MED105                        -----MSTPN APYDLVVIGG GSGG
Mycobacterium smegmatis                          MS-HPGATAS DRHKVVIIGS GFGG
Psedomonas fluorescens                           -------MT- --HRIVIVGG GAGG
Corynebacterium glutamicum ATCC13032             MSVNPTRPEG GRHHVVVIGS GFGG
Escherichia coli                                 -------MTT PLKKIVIVGG GAGG
```

Fig. 10. Alignment of the N-terminal amino acid sequence of various flavoproteins from bacteria. *Brevibacterium* sp. KU1309; Dihydrolipoamide dehydrogenase (AC: Q2B5N0) from *Bacillus* sp. NRRL B-14911; Glutathione reductase (AC: A6GLK6) from *Limnobacter* sp. MED105; NADH dehydrogenase (AC: A0QYD6) from *Mycobacterium smegmatis*; NADH dehydrogenase (AC: Q9KGX0) from *Pseudomonas fluorescens*; NADH dehydrogenase (AC: Q79VG1) from *Corynebacterium glutamicum* ATCC13032; and NADH dehydrogenase (AC: Q0TIW5) from *Escherichia coli*. Italic letters represent the consensus hydrophobic region. Bold letters represent the binding sites for FAD or NAD.

The sequence of dihydrolipoamide dehydrogenase from *Bacillus* sp. shows the highest similarity to the proteins originating from this aerobic bacterium. The enzyme from *Limnobacterium* sp. was a glutathione reductase. The other four enzymes were NADH dehydrogenases. These enzymes have characteristic consensus sequences in the N-terminal region, which is the hydrophobic and glycine rich region, such as GXGXXG for binding NADH or FAD. Accordingly, the NOX obtained in this study can be regarded as belonging to the FAD-dependent pyridine nucleotide reductase family. However, while the enzyme has NADH-oxidizing activity, the sequence of the N-terminus is more like that of dihydrolipoamide dehydrogenases than that of NADH oxidases. Therefore, BreNOX may have other catalytic activities because homologous enzymes of this family exhibit various activities, such as those related to peroxidases and disulfide reductases. The cloning and analysis of the sequence of BreNOX is an interesting future subject. It was found that BreNOX had properties characteristic of the dihydrolipoamide dehydrogenase family. Although a NOX has been isolated from the membrane fraction of an aerobic bacterium, such as NOX from *Corynebacterium glutamicum* (Matsushita et al., 2001), the NOX in this study was found in the cytoplasm. Moreover, to the best of our knowledge, this is the first observation of the activation of NOX by ammonium ions. It is estimated that a change in the secondary or tertiary structure of the enzyme due to high ionic strength, as well as the effect of pH, might bring about an increase in the activity of the enzyme. In addition, the pH profile of this enzyme characteristically shows maximum activity under alkaline conditions, whereas other members of the NOX family are most active under neutral or acidic pH conditions (Table 12).

Organisms	M.W. (kDa)	Km (μM)	Specific activity (U/mg)	Product of O_2 reduction	Optimal pH
Aerobic bacteria					
Brevibacterium sp.	57, α2	22	8.86 (25°C)	H_2O_2	9
C. glutamicum	55	n.d.	46.5 (25°C)	n.d.	6.5
Obligate anaerobic bacteria					
T. hypogea	50, α2	7.5	37 (80°C)	H_2O_2	7
C. aminovalericum	45, α2	19.2	130 (37°C)	H_2O	7
Thermophilic anaerobic archaea					
A.flugidus NOXA-1	48	130	8.7 (80°C)	H_2O_2	8
A.flugidus NOXB-1	61	11	1.5 (70°C)	H_2O_2	6.5

Table 12. Comparison of the properties of NADH oxidases

Moreover, this enzyme exhibits high activity in a broader range of temperatures and high thermal stability compared with NOX enzymes from thermophilic bacteria and archaea. Although NOX enzymes obtained from thermophilic archaea have high thermal stability, they show low or no activity at room temperature. On the other hand, because BreNOX exhibits high activity even at lower temperatures, in addition to thermal stability, it would be more applicable to coupled reactions with various dehydrogenases. In conclusion, we have purified and characterized NADH oxidase (BreNOX) from *Brevibacterium* sp. KU1309.

Although the physiological roles of the enzyme have not been fully disclosed, enzymatic properties such as stability and responses to salt and pH have been characterized and are considered to be useful for organic synthesis. Problems in cloning and over-expression of the gene will need to be overcome to raise the efficiency of the enzyme.

4. References

Asano, Y. (2002). Overview of screening for new microbial catalysts and their uses in organic synthesis - selection and optimization of biocatalysts. *Journal of Biotechnology*, Vol.94, No.1, pp. 65-72, ISSN 0168-1656

Biegert, T.; Altenschmidt, U.; Eckerskorn, C. & Fuchs, G. (1995). Purification and properties of benzyl alcohol dehydrogenase from a denitrifying *Thauera* sp. *Archives of Microbiology*, Vol.163, No.6, pp. 418-423, ISSN 0302-8933

Burton, S.G. (2003). Oxidizing enzymes as biocatalysts. *Trends in Biotechnology*, Vol.21, No.12, pp. 543-549, ISSN 0167-7799

Dickinson, J. R.; Salgado, L. E. J. & Hewlins, M. J. E. (2003). The catabolism of amino acids to long chain and complex alcohols in *Saccharomyces cerevisiae*. *Journal of Biological Chemistry*, Vol.278, No.10, pp. 8028-8034, ISSN 0021-9258

Gandolfi, R.; Ferrara, N. & Molinari, F. (2001). An easy and efficient method for the production of carboxylic acids and aldehydes by microbial oxidation of primary alcohols. *Tetrahedron Letters*, Vol.42, No.3, pp. 513-514, ISSN 0040-4039

Geueke, B.; Riebel, B. & Hummel, W. (2003). NADH oxidase from *Lactobacillus brevis*: a new catalyst for the regeneration of NAD. *Enzyme and Microbial Technology*, Vol.32, No.2, pp. 205-211, ISSN 0141-0229

Ginson, C.M.; Mallett, T.C.; Claiborne, A.I. & Caparon, M.G. (2000). Contribution of NADH oxidase to aerobic metabolism of *Streptococcus pyogenes*. *Journal of Bacteriology*, Vol.182, No.2, pp. 448-455, ISSN 0021-9193

Hanlon, S.P.; Hill, T.K.; Flavell, M.A.; Joseph, M.; Stringfellow, J.M. & Cooper, R.A. (1997). 2-Phenylethylamine catabolism by *Escherichia coli* K-12: gene organization and expression. *Microbiology*, Vol.143, No.2, pp. 513-518, ISSN 1350-0872

Hartmans, S.; Smits, J. P.; Van der Werf, M. J.; Volkering, F. & De Bont, J. A.M. (1989). Metabolism of styrene oxide and 2-phenylethanol in the styrene-degrading *Xanthobacter* strain 124X. *Applied and Environmental Microbiology*, Vol.55, No.11, pp. 2850-2855, ISSN 0099-2240

Herles, C.; Braune, A. & Blaut, M. (2002). Purification and characterization of an NADH oxidase from *Eubacterium ramulus*. *Archives of Microbiology*, Vol.178, No.1, pp. 71-74, ISSN 0302-8933

Higuchi, M.; Shimada, M.; Matsumoto, J.; Yamamoto, Y.; Rhaman, A. & Kamio, Y. (1994). Molecular cloning and sequence analysis of the gene encoding the H_2O_2-forming NADH oxidase from *Streptococcus* mutans. *Bioscience, Biotechnology, and Biochemistry*, Vol. 58, No.9, pp. 1603-1607, ISSN 0916-8451

Higuchi, M.; Shimada, M.; Yamamoto, Y.; Hayashi, T.; Koga, T. & Kamio, Y. (1993). Identification of two distinct NADH oxidases corresponding to H_2O_2-forming oxidase and H_2O-forming oxidase induced in *Streptococcus* mutans. *Journal of General Microbiology*, Vol.139, No.10, pp. 2343-2351, ISSN 0022-1287

Hirano, J.; Miyamoto, K. & Ohta, H. (2005). Purification and characterization of the alcohol dehydrogenase with a broad substrate specificity originated from 2-phenylethanol-assimilating *Brevibacterium* sp. KU 1309. *Journal of Bioscience and Bioengineering*, Vol.100, No.3, pp. 318-322, ISSN 1389-1723

Hirano, J.; Miyamoto, K. & Ohta, H. (2007). Purification and characterization of aldehyde dehydrogenase with a broad substrate specificity originated from 2-phenylethanol-assimilating *Brevibacterium* sp. KU1309. *Applied Microbiology and Biotechnology*, Vol.76, No.2, pp. 357-363, ISSN 0175-7598

Hirano, J.; Miyamoto, K. & Ohta, H. (2008). Purification and characterization of thermostable H_2O_2-forming NADH oxidase from 2-phenylethanol-assimilating *Brevibacterium* sp. KU1309. *Applied Microbiology and Biotechnology*, Vol.80, No.1, pp. 71-78, ISSN 0175-7598

Hummel, W. & Riebel, B. (1996). Chiral alcohols by enantioselective enzymatic oxidation. *Annals of the New York Academy of Sciences*, Vol.799 (Enzyme Engineering XIII), pp. 713-716, ISSN 0077-8923

Hummel, W.; Kuzu, M. & Geueke, B. (2003a). An efficient and selective enzymatic oxidation system for the synthesis of enantiomerically pure D-tert-leucine. *Organic Letters*, Vol.5, No.20, pp. 3649-3650, ISSN 1523-7060

Hummel, W. & Riebel, B. (2003b). Isolation and biochemical characterization of a new NADH oxidase from *Lactobacillus brevis*. *Biotechnology Letters*, Vol.25, No.1, pp. 51-54, ISSN 0141-5492

Itoh, N.; Yoshida, K. & Okada, K. (1996). Isolation and identification of styrene-degrading *Corynebacterium* strains, and their styrene metabolism. *Bioscience, Biotechnology, and Biochemistry*, Vol.60, No.11, pp. 1826-1830, ISSN 0916-8451

Karube, I. & Nomura, Y. (2000). Enzyme sensors for environmental analysis. *Journal of Molecular Catalysis B: Enzymatic*, Vol.10, No.1-3, pp. 177-181, ISSN 1381-1177

Kawasaki, S.; Ishikura, J.; Chiba, D.; Nishino, T. & Niimura, Y. (2004). Purification and characterization of an H_2O-forming NADH oxidase from *Clostridium aminovalericum*: existence of an oxygen-detoxifying enzyme in an obligate anaerobic bacteria. *Archives of Microbiology*, Vol.181, No.4, pp. 324-330, ISSN 0302-8933

Kengen, S.W.; van der Oost, J. & de Vos, W.M. (2003). Molecular characterization of H_2O_2-forming NADH oxidases from *Archaeoglobus fulgidus*. *European Journal of Biochemistry*, Vol.270, No.13, pp. 2885-2894, ISSN 0014-2956

Kourist, R.; Domínguez de María, P. & Miyamoto, K. (2011). Biocatalytic strategies for the asymmetric synthesis of profens - recent trends and developments. *Green Chemistry*, Vol.13, No.10, pp. 2607-2618, ISSN 1463-9262

Maeda, K.; Truscott, K.; Liu, X.L. & Scopes, R.K. (1992). A thermostable NADH oxidase from anaerobic extreme thermophiles. *Biochemical Journal*, Vol.284, No.2, pp. 551-555, ISSN 0306-3275

Masullo, M.; Raimo, G.; Dello Russo, A.; Bocchini, V. & Bannister, J.V. (1996). Purification and characterization of NADH oxidase from the archaea *Sulfolobus acidocaldarius* and *Sulfolobus solfataricus*. *Biotechnology and Applied Biochemistry*, Vol.23, No.1, pp. 47-54, ISSN 0885-4513

Matsumoto, J.; Higuchi, M.; Shimada, M.; Yamamoto, Y. & Kamio, Y. (1996). Molecular cloning and sequence analysis of the gene encoding the H_2O-forming NADH oxidase from *Streptococcus* mutans. *Bioscience, Biotechnology, and Biochemistry*, Vol.60, No.1, pp. 39-43, ISSN 0916-8451

Matsushita, K.; Otofuji, A.; Iwahashi, M.; Toyama, H. & Adachi, O. (2001). NADH dehydrogenase of *Corynebacterium glutamicum*. Purification of an NADH dehydrogenase II homolog able to oxidize NADPH. *FEMS Microbiology Letters*, Vol.204, No.2, pp. 271-276, ISSN 0378-1097

McLeish, M.J.; Kneen, M.M.; Gopalakrishna, K.N.; Koo, C.W.; Babbitt, P.C.; Gerlt, J.A. & Kenyon, G.L. (2003). Identification and characterization of a mandelamide hydrolase and an $NAD(P)^+$-dependent benzaldehyde dehydrogenase from *Pseudomonas putida* ATCC 12633. *Journal of Bacteriology*, Vol.185, No.8, pp. 2451-2456, ISSN 0021-9193

Miyamoto, K. & Ohta, H. (1992). Enantioselective oxidation of mandelic acid using a phenylmalonate metabolizing pathway of a soil bacterium *Alcaligenes bronchisepticus* KU 1201. *Biotechnology Letters*, Vol.14, No.5, pp. 363-366, ISSN 0141-5492

Miyamoto, K.; Hirano, J. & Ohta, H. (2004). Efficient oxidation of alcohols by a 2-phenylethanol-degrading *Brevibacterium* sp. *Biotechnology Letters*, Vol.26, No.17, pp. 1385-1388, ISSN 0141-5492

Miyamoto, K.; Fujimori, K.; Hirano, J. & Ohta, H., (2009). Microbial kinetic resolution of 2-substituted-1-propanol, *Biocatalysis and Biotransformation*, Vol.27, No.1, pp. 66-70, ISSN 1024-2422

Miyoshi, A.; Rochat, T.; Gratadoux, J.J.; Le Loir, Y.; Oliveira, S.C.; Langella, P.& Azevedo, V. (2003). Oxidative stress in *Lactococcus lactis*. *Genetics and Molecular Research*, Vol.2, No.4, pp. 348-359, ISSN 1676-5680

Nakamura, K.; Yamanaka, R.; Matsuda, T. & Harada, T. (2003). Recent development in asymmetric reduction of ketones with biocatalysts. *Tetrahedron: Asymmetry*, Vol.14, No.18, pp. 2659-2681, ISSN 0957-4166

Ohta, H. & Tetsukawa, H. (1979). Enantiotopically selective oxidation of 1,3-diols with microorganism. *Chemistry Letters*, No.11, pp. 1379-1380, ISSN 0366-7022

Ohta, H. & Tetsukawa, H. (1981). Enantiotopically selective oxidation of 1,5-diols with *Gluconobacters* preparation of (*S*)-mevalonolactone. *Agricultural and Biological Chemistry*, Vol.45, No.8, pp. 1895-1896, ISSN 0002-1369

Park, H.J.; Reiser, C.O.; Kondruweit, S.; Erdmann, H.; Schmid, R.D. & Sprinzl, M. (1992). Purification and characterization of a NADH oxidase from the thermophile *Thermus thermophilus* HB8. *European Journal of Biochemistry*, Vol.205, No.3, pp. 881-885, ISSN 0014-2956

Parrott, S.; Jones, S. & Cooper, R.A. (1987). 2-Phenylethylamine catabolism by *Escherichia coli* K12. *Journal of General Microbiology*, Vol.133, No.2, pp. 347-351, ISSN 0022-1287

Reid, M. F. & Fewson, C. A. (1994). Molecular characterization of microbial alcohol dehydrogenases. *Critical Reviews in Microbiology*, Vol.20, No.1, pp. 13-56 ISSN 1040-841X

Riebel, B.R.; Gibbs, P.R.; Wellborn, W.B. & Bommarius, A.S. (2002). Cofactor regeneration of NAD^+ from NADH: novel water-forming NADH oxidase. *Advanced Synthesis and Catalysis*, Vol.344, No.10, pp. 1156-1168, ISSN 1615-4150

Riebel, B.R.; Gibbs, P.R.; Wellborn, W.B. & Bommarius, A.S. (2003). Cofactor regeneration of both NAD^+ from NADH and $NADP^+$ from NADPH:NADH oxidase from *Lactobacillus sanfranciscensis. Advanced Synthesis and Catalysis*, Vol.345, No.6+7, pp. 707-712, ISSN 1615-4150

Romano, A.; Gandolfi, R.; Nitti, P.; Rollini, M. & Molonari, F. (2002). Acetic acid bacteria as enantioselective biocatalysts. *Journal of Molecular Catalysis B: Enzymatic*, Vol.17, No.6, pp. 235-240, ISSN 1381-1177

Ross, R.P. & Claiborne, A. (1992). Molecular cloning and analysis of the gene encoding the NADH oxidase from *Streptococcus faecalis* 10C1. Comparison with NADH peroxidase and the flavoprotein disulfide reductases. *Journal of Molecular Biology*, Vol.227, No.3, pp. 658-671, ISSN 0022-2836

Schmidt, H.L.; Stocklein, W.; Danzer, J.; Kirch, P. & Limbach, B. (1986). Isolation and properties of an H_2O-forming NADH oxidase from *Streptococcus faecalis. European Journal of Biochemistry*, Vol.156, No.1, pp. 149-155, ISSN 0014-2956

Shaw, J. P. & Harayama, S. (1990). Purification and characterization of TOL plasmid-encoded benzyl alcohol dehydrogenase and benzaldehyde dehydrogenase of *Pseudomonas putida. European Journal of Biochemistry*, Vol.191, No.3, pp. 705-714, ISSN 0014-2956

Sheldon, R.A.; Arends, I.W.C.E. & Dijksman, A. (2000). New developments in catalytic alcohol oxidations for fine chemicals synthesis. *Catalysis Today*, Vol.57, No.1-2, pp. 157-166, ISSN 0920-5861

Tamura, Y.; Ohkubo, A.; Iwai, S.; Wada, Y.; Shinoda, T.; Arai, K.; Mineki, S.; Iida, M. & Taguchi, H. (2002). Two forms of NAD-dependent D-mandelate dehydrogenase in *Enterococcus faecalis* IAM 10071. *Applied and Environmental Microbiology*, Vol.68, No.2, pp. 947-951, ISSN 0099-2240

Toomey, D. &Mayhew, S.G. (1998). Purification and characterisation of NADH oxidase from *Thermus aquaticus* YT-1 and evidence that it functions in a peroxide-reduction system. *European Journal of Biochemistry*, Vol.251, No.3, pp. 935-945, ISSN 0014-2956

Tsuchiya, S.; Miyamoto, K. & Ohta, H. (1992). Highly efficient conversion of (±)-madelic acid to its (*R*)-enantiomer by combination of enzyme-mediated oxidation and reduction. *Biotechnology Letters*, Vol.14, No.12, pp. 1137-1142, ISSN 0141-5492

Wang, X.; Mann, C.J.; Bai, Y.; Ni, L. & Weiner, H. (1998). Molecular cloning, characterization, and potential roles of cytosolic and mitochondrial aldehyde dehydrogenases in ethanol metabolism in *Saccharomyces cerevisiae. Journal of Bacteriology*, Vol.180, No.4, pp. 822-830, ISSN 0021-9193

Yang, X. & Ma, K. (2005). Purification and characterization of an NADH oxidase from extremely thermophilic anaerobic bacterium *Thermotoga hypogea. Archives of Microbiology*, Vol.183, No.5, pp. 331-337, ISSN 0302-8933

Yang, X. & Ma, K. (2007). Characterization of an exceedingly active NADH oxidase from the anaerobic hyperthermophilic bacterium *Thermotoga maritima*. *Journal of Bacteriology*, Vol.189, No.8, pp. 3312-3317, ISSN 0021-9193

5

Analyses of Sequences of (β/α) Barrel Proteins Based on the Inter-Residue Average Distance Statistics to Elucidate Folding Processes

Masanari Matsuoka, Michirou Kabata, Yosuke Kawai and Takeshi Kikuchi
Department of Bioinformatics, College of Life Sciences, Ritsumeikan University
Japan

1. Introduction

It is well-known that many proteins fold into unique 3D structure (Anfinsen & Scheraga, 1975; Pain, 2000). The mechanisms by which an amino acid chain forms a complicated tertiary structure have been studied extensively (Sato et al,, 2006; Sato & Fersht, 2007; Sosnick & Barrick, 2011; Schaeffer & Daggett, 2011) but are still not understood in a comprehensive way (Bowman et al., 2011). It is well-recognized that all information on the 3D structure of a protein is coded in its amino acid sequence (Anfinsen & Scheraga, 1975; Pain, 2000). Hence we should be able to extract the information from the sequence. However, this is a significant, long-standing and unsolved problem in structural bioinformatics. Gross features of protein 3D structures or protein folds are characterized by some combination of secondary structural elements, and the ways of combination of secondary units are full of variety (Lesk, 2010). This situation does not allow us to construct a simple picture of the protein folding mechanisms. Among such a variety of protein folds, frequently appearing common folds, so-called superfolds (Orengo et al., 1994), are attractive targets for studying their folding mechanisms. The crucial point is that in general the fold of proteins tends to be more conservative than their sequences, i.e., sometimes proteins sharing the same fold show low sequence homology (Orengo et al., 1994; Jennings & Wright, 1993, Cavagnero et al., 1999; Nishimura et al., 2000). This fact implies the difficulty in approaching this problem using standard bioinformatics techniques such as multiple alignment techniques and so on.

We have been investigating this problem for several proteins using inter-residue average distance statistics of proteins. The tool we introduced is a kind of predicted contact map constructed from the sequence of a protein disregarding the knowledge of its 3D structure. We call this map the Average Distance Map (ADM) (Kikuchi et al., 1988; Kikuchi, 2002; Kikuchi 2011). The ADMs have seen used to analyze the folding problems of proteins in the fatty acid binding protein family (Ichimaru & Kikuchi, 2003; Kikuchi, 2011), the globin family (Ichimaru & Kikuchi, 2003; Nakajima et. al, 2005; Kikuchi, 2011), the c-type lysozyme family (Nakajima & Kikuchi, 2007; Kikuchi, 2010), IgG binding domains (Kikuchi, 2008; Kikuchi, 2011) and β-sandwich proteins (Ishizuka & Kikuchi, 2011). In this chapter, a new application of this technique to (β/α) barrel protein is presented. We further discuss the evolutionary variation of predicted folding units in a (β/α) barrel protein by analyzing sequences of homologues in a family.

The folding scenario of a protein may be like the following. One or several portions in a sequence form (a) partial hydrophobic collapse(s). Each portion may grow to a larger assembly with a native-like configuration or two or more portions may merge into a larger block to form a native-like configuration. Such regions assemble to form a folding transition state. Then, the final native structure emerges very quickly. The ultimate goal is to learn how a protein folds into a complicated fold via above scenario. Thus we would like to know which parts of the sequence form the native-like 3D configurations in the early stage of folding. These parts may correspond to a 3D structural formation portion at the folding transition state. In the present study, we try to predict a position of the initial hydrophobic collapse.

In our experiences so far, a protein in the T4-phage lysozyme family consists of two well-structured domains, and our method predicts two distinct folding units (Kawai et al., 2011). One clear folding unit and a short relatively weak one are predicted for a protein in the globin family (Kawai et al., 2011). The sequence of a β sandwich protein is predicted to contain two folding units, and this is interpreted to mean that these two portions merge into a larger block in the protein and form a native-like structure (Ishizuka & Kikuchi, 2011). Thus, it is interesting to see how our method predicts for the sequence of a (β/α) barrel protein and how its folding process can be interpreted from the predictions. It is also observed that location of such predicted regions is robust among evolutionally related proteins (Kawai & Kikuchi, unpublished). This suggests a robustness of the location of folding units during evolution of life on Earth. We seek to reveal this point for (β/α) barrel proteins.

2. (β/α) barrel protein

A (β/α) barrel protein shows a remarkable feature of the 3D scaffold which is constituted eight cyclically arranged successive (β/α) units with high symmetry. This fold is called the "TIM barrel" because the first discovered protein with this scaffold was triose phosphate isomerase (see Fig. 1). Since then, a huge number of proteins with the TIM barrel fold have

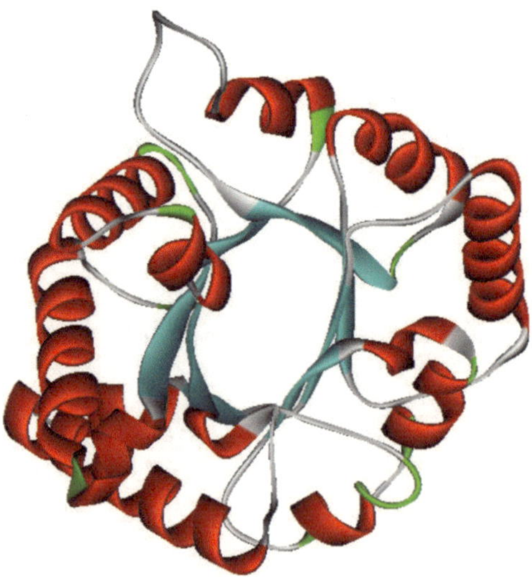

Fig. 1. An example of the 3D structure of a (β/α) barrel protein, triose phosphate isomerase (PDB: 1TIM). This is a typical (β/α) barrel protein consisting of eight (βα) units.

been found, and they show a variety of functions. Thus, elucidation of the folding mechanism and evolution of the (β/α) barrel scaffold is a quite interesting and challenging problem. There are many studies on folding mechanisms of several proteins with TIM barrel folds (Akanuma & Yamagishi, 2008; Gu et al., 2007; Silverman & Harbury, 2002; Seitz at al., 2007).

Among them, the recent studies on 3D structures and sequences of imidazole glycerol phosphate synthase (HisF) from Thermotoga maritime and N'-[(5'-phosphoribosyl) formimino]-5-aminoimidazole-4-carboxamide ribonucleotide isomerase (HisA) (Lang at al., 2000; Höcker et al., 2001) are remarkable. The 3D structures of HisF and HisA are presented in Fig. 2. According to these studies, 3D structural similarities are observed among the N terminal and C terminal halves of HisA and HisF (Lang et al., 2000). The sequence similarities among the sequences of the N terminal and C terminal halves of HisA and HisF are not high, but some conserved residues with similar properties are observed (Lang et al., 2000). These findings suggest that that the $(\beta/\alpha)_4$ half-barrel may fuse to yield a $(\beta/\alpha)_8$ complete barrel protein. This hypothesis that a half barrel of each protein can fold independently was confirmed by Höcker et al. (2001), and a $(\beta/\alpha)_8$ barrel protein designed as a fused identical half barrels can also form a stable folded structure (Seitz et al., 2007).

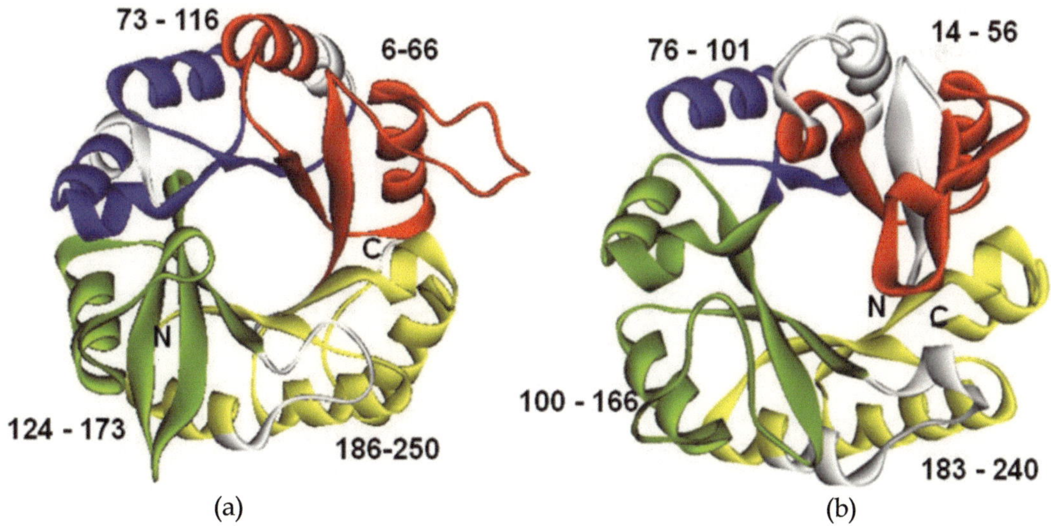

Fig. 2. 3D structures of 1THF(a) and 1QO2(b). Each region colored in red, blue, green or yellow denotes, respectively, the first, second, third or fourth folding unit predicted by each ADM (see text).

The two-fold symmetry may be further broken down into a four-fold symmetry, that is, a $(\beta\alpha)_2$ quarter-barrel is a symmetry unit. For example, the conserved GXD/GXG motif is repeatedly observed within α1-β2, α3- β4, α5-β5 and α7- β8 loops in HisF. Such symmetry is also observed for other $(\alpha\beta)_8$ barrel proteins based on the location of functional sites in them (Nagano at al., 2002). The phosphate binding sites in HisF corresponds to the four-fold symmetric active sites (see Fig. 3). Furthermore, Richter et al. (2010) designed a $(\alpha\beta)_8$-barrel protein composed of four identical $(\beta\alpha)_2$ quarter-barrel units that form a stable 3D structure by the introduction of disulfide bridges. Their results suggests that HisF evolved from an ancestral $(\beta\alpha)_2$ quarter-barrel via a $(\beta\alpha)_4$ half-barrel into the $(\alpha\beta)_8$-barrel (Richter et al., 2010).

On the other hand, the NMR study by Setiyaputra et al. (2011) demonstrated that the truncated phosphoribosylanthranilate isomerase (trPRAI), which is three-quarter-barrel-sized fragment of a $(\beta\alpha)_8$ barrel, forms the distinct 3D structure (Setiyaputra et al., 2011). This observation may suggest that the $(\beta\alpha)_2$ quarter-barrel is a kind of structural module in a $(\beta/\alpha)_8$ barrel protein but all 4 modules are not always indispensable.

(a) **HisF**

(b) **HisA**

Fig. 3. Sequences of (a) HisF and (b) HisA. A residue in a β-strand is labeled by "b" and that in an α-helix by "a". A region with residues written by white letters with black background denotes the predicted region by ADM (see text). The symbol "PPPP" denotes an active site where phosphate binds in HisF and "(PPPP)" denotes a site in HisA that corresponds to a phosphate binding site in HisF. An arrow with a double line points to a boundary between two $(\beta\alpha)_2$ units.

Thus, HisF and HisA have attracted a lot of interest from researchers studying folding mechanisms and evolution. Hence we take HisF and HisA as examples of $(\beta/\alpha)_8$ barrel proteins in the present study.

3. Techniques used in the present work

3.1 Average distance map (ADM) method

The average distance map (ADM) method is a technique to predict structure-forming or compact portions in the amino acid sequence of a protein, and details of the method are described in (Kikuchi et al., 1988; Kikuchi, 2011). We have been confirming that ADMs contain variety of information on 3D structures and folding of proteins in spite of the simplicity of the method. The regions predicted by the ADM of a protein sequence correspond to; (1) nuclei of structural domains, (2) a nucleus of a structural domain and a portion forming a structure by interaction with the nucleus of the domain, or (3) two regions that form a stabilized structure by merging (Kawai, Matsuoka & Kikuchi, 2011). The essence of the procedure is as follows.

1. Calculation of inter-residue average distances in proteins

Inter-residue average distances were calculated using proteins with known 3D structures taking separation of two residues along the amino acid sequence of a protein into consideration. That is, the average distances were calculated for a residue pair within each group, i.e., 1≦k≦8, 9≦k≦20, 21≦k≦30, 31≦k≦40 and so on where k=|i-j| and i and j mean the i-th and j-th residues of the sequence. Each group of separation is referred to as a range, and each range is defined as; M=1 for 1≦k≦8, M=2 for 9≦k≦20, M=3 for 21≦k≦30 and so on (Refer to Kikuchi et al. (1988) for proteins used in the calculations of average distances). Here, an inter-residue distance means the distance between Cα atoms in Cartesian space.

2. Construction of a predicted contact map

Our ADM analysis entailes making some plots showing when the average distance between a pair of residues in the sequence of a protein is less than some threshold value. A threshold value is defined for each range to reproduce density of plots on a contact map constructed from the spatial 3D structure of a protein under consideration. The prediction of density of plot on a contact map is made according to the way described by Kikuchi et al. (1988). Examples of ADMs are presented in Fig. 4. These ADMs were constructed based on the sequences of HisF and HisA.

3. Prediction of a compact region or a folding unit in a given sequence with ADM

When two segments in a protein form many contacts, the ADM should show be a region with a high density of contacts corresponding to plots on the contact map of a protein. Such a region shows a sudden change of the density of plots at its boundary on a map. Suppose that a map is divided into two parts by a line parallel to the ordinate of the map, thereby creating triangular and trapezoid parts. The difference in the density of plots between these two parts should be minimum or maximum at the boundary of the region with high density of plots. The same thing is also true when a map is divided into two parts by a line parallel to the abscissa. Let $\Delta\rho_i^v \left(\Delta\rho_j^h\right)$ be the differences of plot densities between two parts defined by lines parallel to the ordinate (abscissa) of a map. Then, the boundary of a region with high density contacts on a map can be detected by maxima and minima of the values of and as depicted in Fig. 5(a). In the example of Fig. 5(a), there are valleys at A and D , and peaks at B and C, and thus the interactions between the segments

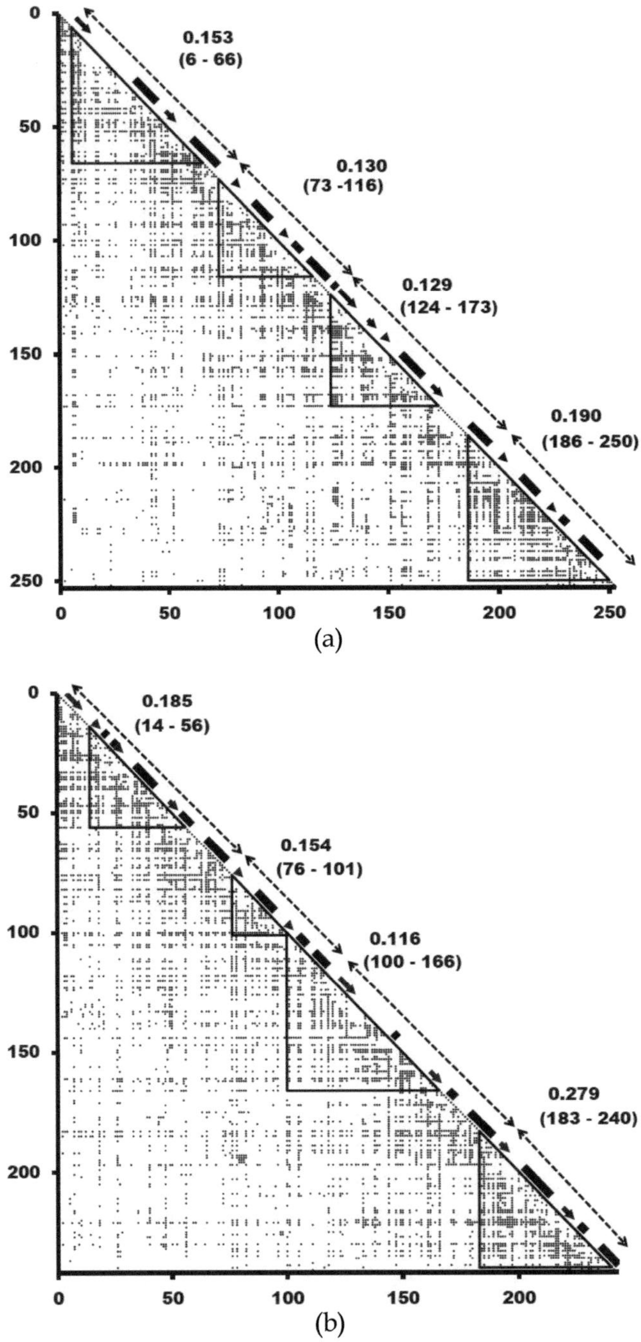

Fig. 4. ADMs for (a) HisF and (b) HisA. A bar along the diagonal denotes a position of a β-strand and an arrow does that of an α-helix. A region of a predicted folding unit is enclosed by solid lines in a map with a parenthesis and a numeral denotes the η value corresponding to the corresponding predicted region. A broken double arrow means the portion of a corresponding $(\beta\alpha)_2$ unit.

A-B and C-D are observed as a high density region of contacts on a map. Sometimes, a compact region in a protein can be observed as a high density region of contacts near the diagonal of a map. Such a region can be detected by peaks in the values in and as shown in Fig. 5(b). Peaks at E and F are observed in Fig. 5(b), and this predicts the compact region E-F in the sequence. It is convenient to define η value $= \Delta\rho_E^v + \Delta\rho_F^h$ in Fig. 5(b) as a measure of compactness of the region E-F. For several proteins, we have confirmed that such a compact region on ADM can also be regarded as a folding unit in the sequence of a protein (Ichimaru & Kikuchi, 2003; Nakajima et. al, 2005; Nakajima & Kikuchi, 2007; Kikuchi, 2011).

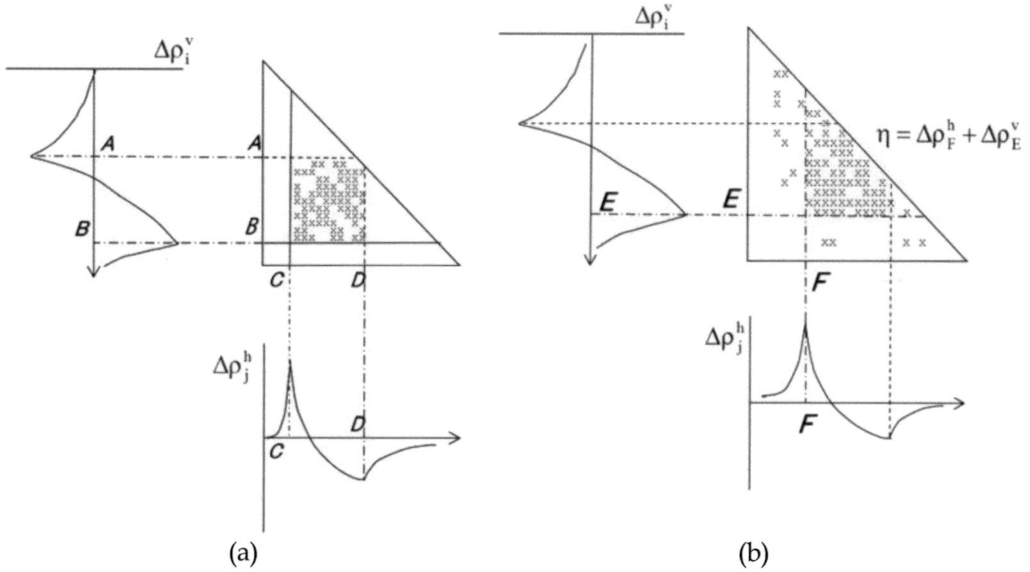

Fig. 5. (a) When a map is divided into two parts by a line parallel to the ordinate of the map creating triangular and trapezoid parts, the difference of the density of plots between these two parts should be minimum or maximum at the boundary of the region with high density of plots. The same thing is also true when a map is divided into two parts by a line parallel to the abscissa. $\Delta\rho_i^v\left(\Delta\rho_j^h\right)$ denote the differences of plot densities between two parts (the triangular and trapezoid parts) defined by lines parallel to the ordinate (abscissa) of a map. A peak and a valley appear at the boundaries of a high dense region of a plot of or. This hypothetical map suggests the interaction between the segments A-B and C-D. (b) A compact region or a domain in a given protein can be observed as a highly dense region of plots along the diagonal of a map. This figure shows a schematic drawing of a compact region at F-E. We define η as a measure of the compactness of the region, i.e. $\eta = \Delta\rho_r^h + \Delta\rho_q^v$.

In Fig. 4, we show the predicted folding units with η values for (a) HisF and (b) HisA in respective ADMs. Such a region corresponds to that enclosed by triangle in each map.

3.2 Multiple alignment analyses with homologues of HisF and HisA

It is interesting to see how predicted folding units in the sequences of the homologues of HisA and HisF appear in the sequences, i.e., whether predicted folding units are common among homologues. If commonality of folding units were observed, folding units of a protein could be considered to be robust during evolution, and thus the folding process should be conserved during evolution. In the present study, evolutionary analyses with multiple alignments for homologues are also tried in the following procedure.

1. Homologous sequences of HisF and HisA were searched within the Uniprot and Swiss-prot databases. We used BLAST (Altschul et al, 1990) as a search algorithm and collected homologues less than 0.01 of their e value.
2. The multiple alignments of the collected sequences were made with ClustalW.
3. Phylogenetic trees of the collected sequences of homologues of HisF and HisA obtained in the way above were made based on the multiple alignments using the Neighbor-Joint method (Saitou & Nei, 1987) with 100 times bootstrapping.
4. For all sequences, the predictions of the positions of folding units by ADMs were made with AutoADM (Kawai et al., unpublished).

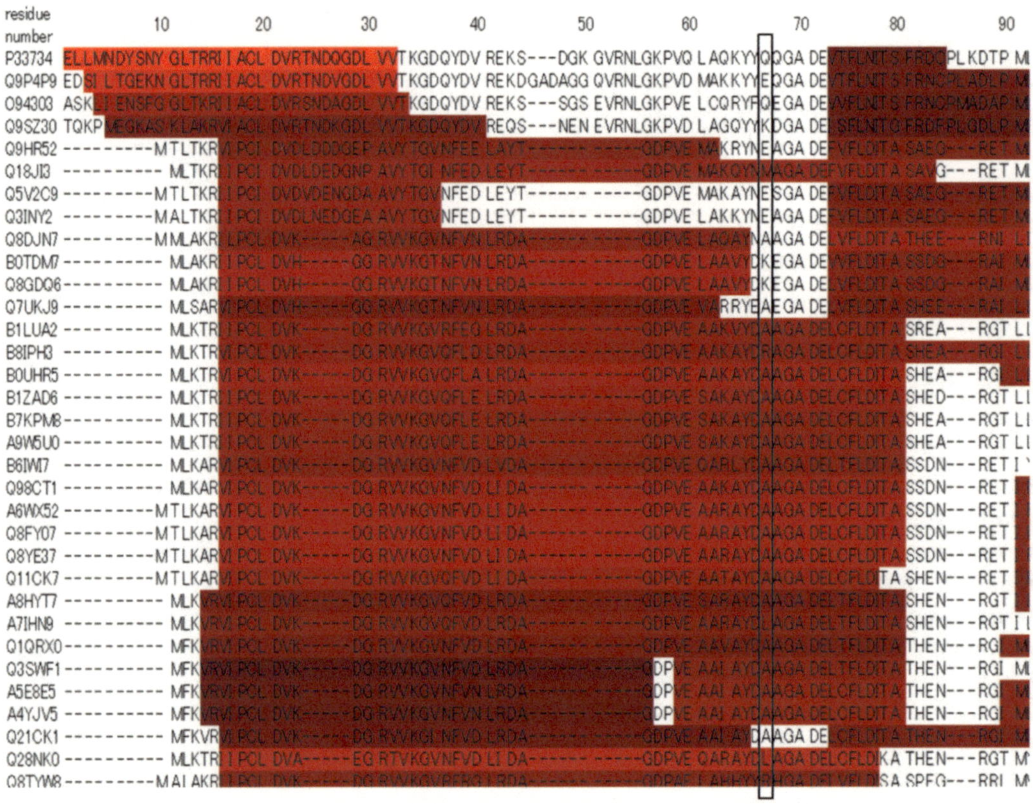

Fig. 6. A part of aligned sequences of aligned sequences in a multiple alignment. A brighter red region indicates a larger η value region. The number of residues within predicted folding units is counted at each site in a multiple alignment, e.g., the number of residues in red zones in the region enclosed by a black rectangle is counted and a histogram of such numbers is constructed for all sites of the multiple alignment.

A part of aligned sequences is presented in Fig. 6. In this figure, a region colored by red in each sequence denotes a predicted folding unit in the protein. A brighter red region means a larger η value region.

5. The number of residues within predicted folding units is counted at each position of a residue in a multiple alignment. That is, a histogram of the number of residues within the predicted region for an each position in a multiple alignment is made.

4. Analyses for HisA and HisF

4.1 Predicted folding units in HisF and HisA by ADMs

The ADM for HisF presented in Fig. 4(a) predicts four folding regions; 6-66, 73-116, 124-173 and 186-250 (regions enclosed by solid lines) as illustrated in this figure. Each region contains a phosphate binding site as shown in Fig. 3 (and also in Fig. 7). That is, this ADM predicts that HisF contains four folding units. The η values of these regions are 0.153, 0.130, 0.129 and 0.190, respectively, and thus the η values of these portions are relatively similar suggesting that each of these regions constitutes a folding unit with equal significance. From Figs. 3(a) and 4(a) it is easily confirmed that these predicted parts correspond to β1-α1- β2-α2, β3- α3-β4- α4, β5- α5- β6 and α6- β7- α7- β8- α8 in the 3D structure of HisF, and each of these regions corresponds well to the positions of each $(\beta\alpha)_2$ unit. The structures of these parts are colored in red, blue, green and yellow in Fig. 2(a).

On the other hand, the folding units of HisA are predicted as 14-56, 76-101, 100-166 and 183-240 according to its ADM as shown in Fig. 4(b). In Fig. 3 (and also in Fig. 7), the corresponding phosphate binding sites in HisF are presented, and the second, third and fourth predicted folding units contain such regions. Thus, HisA can be regarded as consisting of four folding units. The η values of these regions are 0.185, 0.154, 0.116 and 0.279, respectively. The η values of these regions are relatively similar except the larger value of the fourth region. This result may suggest the stronger significance of the fourth region and other three parts would contribute equally to the scaffold of the protein. These parts correspond to β1- α1- β2- α2, β3-α3- β4, α4- β5- α5- β6 and α6- β7- α7- β8- α8 in the 3D structure of HisA, and thus again each of these regions corresponds well to the positions of each of $(\beta\alpha)_2$ unit as confirmed in Figs. 3(b) and 4(b). The 3D structures of these parts are depicted in Fig. 2(b).

4.2 Multiple sequence alignment analyses of homologues of HisF and HisA and the location of folding units

Collected in Fig. 7 are 184 and 232 homologues of HisF and HisA. For these we made ADM predictions and multiple alignments for these sequences. For a detailed explanation of Fig. 7, see the caption. A colored bar below the alignment shows a possible common folding unit in this group of proteins. Each of these regions was defined as a region where the majority of sequences in the multiple alignment shows the positions of folding units, i.e., the values in the histogram are relatively large. The histogram of the number of residues within predicted folding units is shown at the bottom of the figure.

Taking the four predicted folding units in HisF by the ADM into account, careful observation of the multiple alignment in Fig. 7 reveals conservation or robustness of the region of a predicted folding unit. For example, we can observe in Fig. 7 that the N terminal part of each sequence always shows a predicted folding unit, i.e. red zone in each sequence in Fig. 7. This conservation or robustness does not mean conservation or robustness of residues in the sequence but rather that of the properties of residues in the sequence. From the histogram, a common folding unit can be assigned in the N terminal part as a region with high values in the histogram as symbolized by the red bar below of the alignment. The boundaries of the red bar were defined by the height of the histogram around this region and visual inspection of the location of the first folding units in the alignment, i.e., the red zones. We call a region defined in this way a common folding unit. Thus, some robustness of the first folding unit can be observed and this region corresponds well to the first $(\beta\alpha)_2$ units as seen in Fig. 7. The fourth region symbolized by the yellow bar also shows relatively large robustness and corresponds to the fourth $(\beta\alpha)_2$ unit. This region shows another common folding unit. We confirm that such analyses with the multiple alignment in this manner reveal clearly the location of folding units and these regions correspond well to the portion of $(\beta\alpha)_2$ units.

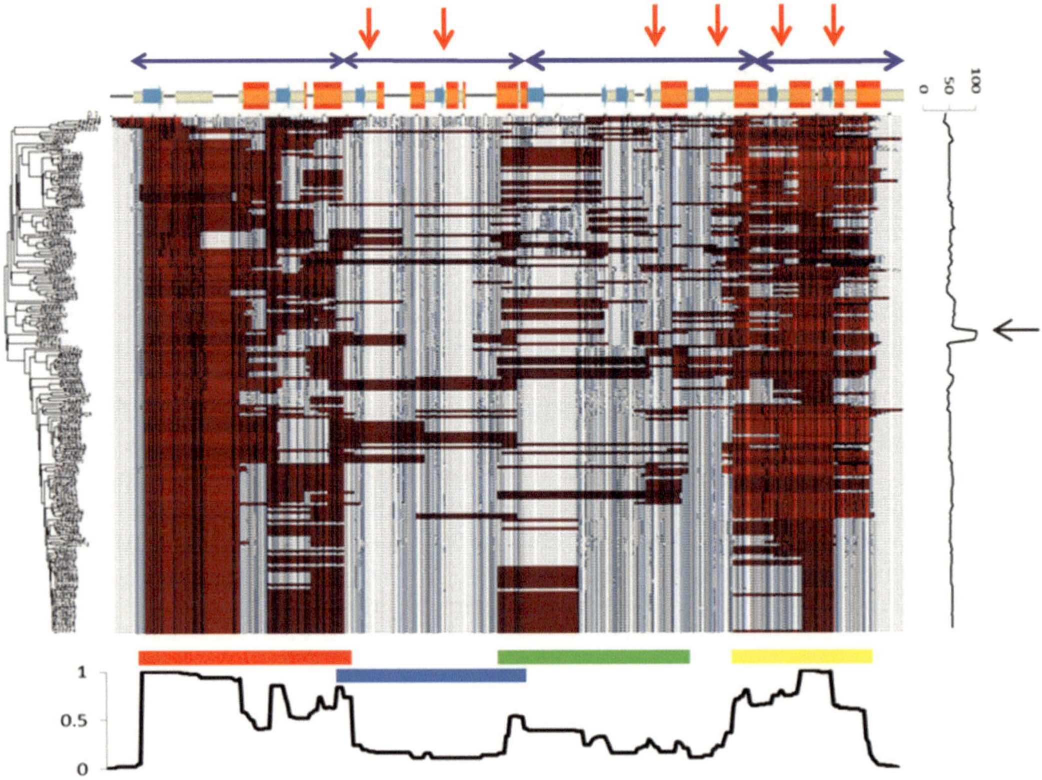

Fig. 7. Whole multiple alignment for HIsF and its homologues. The red arrows at the top of the figure denote a position of phosphate binding sites. A dark blue two headed arrows just below the red arrows indicate a position of $(\beta\alpha)_2$ barrel unit. The bars just below these blue arrows show secondary structures, i.e., a blue bar denotes a β strand and red bar an α helix. The phylogenetic tree is presented on the left side of the figure. A colored bar at the bottom of the alignment means a possible common folding unit in this group of proteins. Each of these regions was defined as a region at where the majority of sequences in the multiple alignment shows the positions of folding units, i.e., the values in the histogram are relatively large. The boundaries of a region can be defined by peaks of the histogram at the bottom. The histogram graphs the number of residues within predicted folding units. A graph on the right side of the figure denotes a plot of sequence homology (%) to that of HisF. The average homology of homologue sequences to that of HisF is about 50%. The arrow at the peak means the 100% homology with HisF sequence.

The third region of Fig. 7 seems to have varied frequently during the evolution from the ancestral HisF, and only a modestly high region can be observed in the histogram. That is symbolized by the green bar. Furthermore, we cannot observe the second folding unit in the histogram, and we just see predicted folding units, i.e., red zones, in some sequences of homologues by visual inspection in this region. Temporarily, we assign two common folding units at the regions symbolized by the blue and green bars.

In Fig. 9(a), we show the 3D structures of the common folding units in HisF. Each color of the region in the structure corresponds to the color of each bar in Fig. 7. The reason why the second region implies the low robustness of sequence properties is not clear. It may relate to the folding property or functional property.

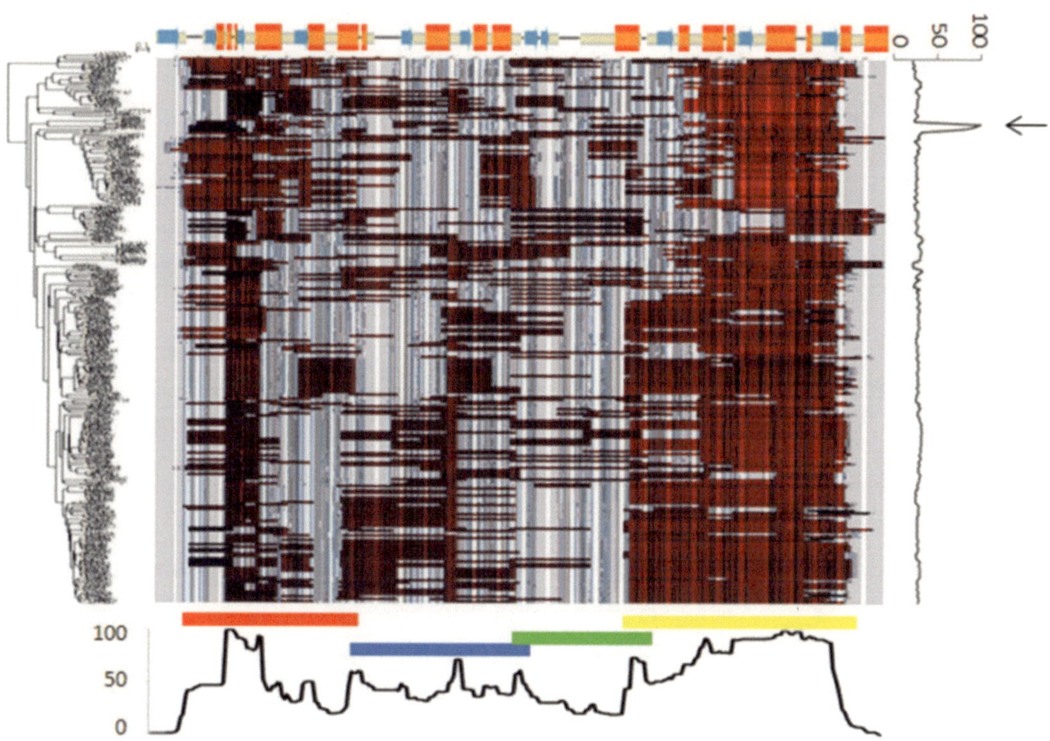

Fig. 8. Whole multiple alignment for HisA and its homologues. See the caption of Fig. 7 for detailes. A graph on the right side of the figure denotes a plot of sequence homology (%) with that of HisF. The arrow at the peak means the 100% homology with HisA sequence.

The whole multiple alignment for HisA and its homologues is presented in Fig. 8. The meaning of each symbol is the same as in Fig. 7. In HisA, a strong robustness is observed in the C terminal part of the sequences, and the region is symbolized by the yellow bar just below the multiple alignment in Fig. 8. This region is longer than the fourth $(\beta\alpha)_2$ unit suggesting that this part is the main region of the folding of HisA, and we define this region as a common folding unit. The first region also shows modest robustness (strong robustness of the shorter part in this region) indicating the existence of a common folding unit but not as strong as in HisF. This observation suggests some significance of the C terminal region for folding mechanism of a protein in this group. Compared with the case of HisF, the histogram for HisA in Fig. 8 shows the modestly high region corresponding to the second predicted folding unit by ADM, and this region is symbolized by the blue below the alignment in Fig. 8. The corresponding third region cannot be assigned by the histogram, but by visual inspection predicted folding regions (red zones) appear in several sequences of homologues. So we make an additional assignment of the third common folding unit symbolized by the green bar. The 3D structures of the common folding units in HisA are presented in Fig. 9(b).

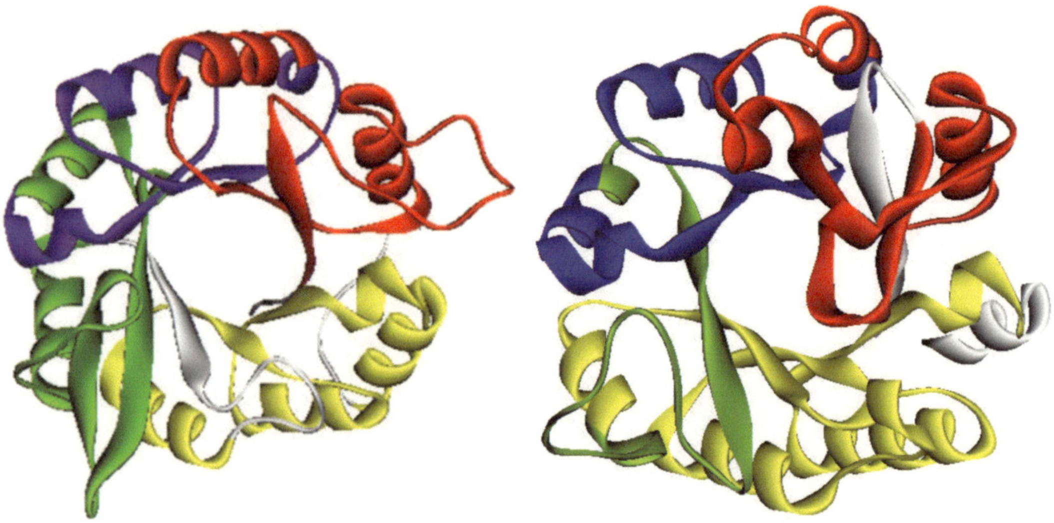

(a) (b)

Fig. 9. Region assigned as common folding units in (a) HisF and (b) HisA from Figs. 7 and 8, respectively.

Our ADM analyses predict the existence of four folding units in HisF and HisA, which correspond to $(\beta\alpha)_2$ units proposed by Richiter et al., 2010, only from their sequences. For HisA, the significance of the C terminal unit predicted by the ADM for its 3D structure formation can be speculated based on the η value of this region. The combination of the multiple alignment analyses with the ADM analyses reveals that the N and C terminal common folding units are robust and would be significant for the folding process of these proteins commonly. On the other hand, the second and third regions are easily varied and the participation of these units to folding mechanisms would not be common among the homologues.

Thus, we can predict the four folding units from properties of sequences of $(\beta\alpha)_8$ proteins through ADMs. It is rather difficult to obtain such information by standard techniques of sequence analyses (Richter et al., 2010). Furthermore, our method specifies a kind of plasticity of each region from the present multiple alignment analysis. We think that this information may useful for predicting the folding properties of a protein, this is discussed below.

5. Concluding remarks

As we see above, the ADM method predicts positions of possible folding units in a protein. Regions predicted by the ADMs in homologues of a protein reveal the conservation or robustness of a predicted region during evolution. Proteins treated in this chapter show high (8-fold) symmetry. The folding process of such proteins with high symmetry may be rather different from other folds. The present analyses along with the preceding investigations (Lang et al., 2000; Höcker et al., 2001; Richter et al., 2010) imply that $(\beta\alpha)_8$ proteins such as HisF and HisA consist of four independent $(\beta\alpha)_2$ units and these proteins start to fold at each $(\beta\alpha)_2$ unit. However, huge number of proteins with the $(\beta\alpha)_8$ fold exist, so the folding mechanisms of $(\beta\alpha)_8$ proteins may show wide variations. Actually, various folding mechanisms of $(\beta\alpha)_8$ proteins have been reported (Akanuma & Yamagishi, 2008; Gu et al., 2007; Silverman & Harbury, 2002; Seitz at al., 2007; Setiyaputra et al. (2011)). Such variation of folding may be related to the nonrobustness of the second and third common folding units in HisF and HisA. We will perform ADM analyses on additional $(\beta\alpha)_8$ proteins.

As mentioned in the Introduction, the folding scenario of proteins may be put in the following categories:

1. A one domain protein

A portion in the sequence of a protein starts to form a hydrophobic collapse, and this part grows to a whole sequence to form the native structure. Protein G, Protein A and so on may belong to this category. In this case, the ADM for a protein tends to predict the location of one folding core in the sequence (Kikuchi et al, 1988; Kikuchi, 2008).

2. A protein composed by two (or more) distinct domains

Two (or more) portions in the sequence of a protein start simultaneously to form hydrophobic collapses, and these parts grow independently to domains in the native structure. T4-lysozyme, papain and so on belong to this category (Kikuchi et al, 1988; Kawai, Matsuoka and Kikuchi, 2011). In this case, the ADM for a protein tends to predict the location of distinct folding cores corresponding to the domains in the sequence.

3. A protein formed by a main part with an interacting short fragment

A main portion in the sequence of a protein starts to form hydrophobic collapse, and this part grows to the partial native structure, and the rest (relatively short fragment compared with the main part) interacts with the main portion followed by the formation of the final native structure. According to our investigations with the ADM method, proteins with the Globin fold, fatty-acid binding protein and so on may belong to this category (Ichimaru & Kikuchi, 2003; Nakajima et. al, 2005; Kawai, Matsuoka & Kikuchi, 2011). In this case, the ADM for a protein tends to predict the location of the main part in the sequence.

4. A protein containing long range interactions along its sequence

Two (or more) portions in the sequence of a protein starts to form partial (or imperfect) hydrophobic collapses followed by aggregation of these two (or more) portions to form a more perfect hydrophobic core. This part grows to the final native structure. β-Sandwich proteins such as titin, azurin (Ishizuka & Kikuchi, 2011) and so on, proteins in the c-type lysozyme fold (Nakajima & Kikuchi, 2007), αβ sandwich proteins (Matsuoka & Kikuchi, unpublished) such as ribosomal protein S6, and so on may belong to this category. In this case, the ADM for a protein tends to predict the location of folding cores which should aggregate into the larger hydrophobic core.

5. A protein with a highly symmetrical structure

Three (or more) portions in the sequence in the sequence of a protein start to form hydrophobic collapses independently, and each portion forms the native structure. The formation of the native structure may occur cooperatively, i.e., once a symmetrical unit starts to form the partial native structure, the rest of the protein tends to form the native structure. This case corresponds to the present work.

The scenarios of protein folding mechanisms are proposed based on the investigations of many researchers (Fersht, 1997; Sato et al,, 2006; Sato & Fersht, 2007; Sosnick & Barrick, 2011; Schaeffer & Daggett, 2011) and the results from our ADM analyses. A portion of a protein sequence contains a folding unit predicted by ADM, and the folding process proceeds by one of the scenarios above, we consider that with ADM analysis it is possible to predict the folding process of a related protein from only a sequence. Of course, there many unclear properties in ADM analyses for a protein, and we are continuing to elucidate the meanings of outcomes from the ADM analyses.

As presented in this chapter, ADM analyses combined with multiple alignment analyses provide fruitful results on evolutionary changes in the folding process of proteins in a family.

Finally, it would be quite nice if our techniques were able to contribute to ab initio protein 3D structure prediction. Unfortunately, the present stage is still far from this goal, but we are trying to do it by analyzing the relationships between the 3D structure of the folding unit of a protein and common properties of residues in this portion among homologues of a protein. We hope that we will be close to this goal in the near future.

6. References

Akanuma, S. & Yamagishi, A. (2008). Experimental evidence for the existence of a stable half-barrel subdomain in the $(\beta\alpha)_8$-barrel fold. J. Mol. Biol., 382, 458–466.

Altschul, SF., Gish, G., Miller, W., Myers, EW. & Lipman, DJ. (1990) Basic local alignment search tool. *J. Mol. Biol.*, 215, 403-410.

Anfinsen, CB & Scheraga, HA. (1975). Experimental and theoretical aspects of protein folding. *Adv. Protein Chem.*, 29, 205-300.

Bowman, GR., Voelz, VA. & Pande, VS. (2011). Taming the complexity of protein folding. *Curr Opin. Str. Biol.*, 21, 4-11.

Cavagnero, S., Dyson, HJ & Wright, PE. (1999). Effect of H helix destabilizing mutations on the kinetic and equilibrium folding of apomyoglobin. *J. Mol. Biol.*, 285, 269-282.

Fersht, AR. (1997). Nucleation mechanisms in protein folding. *Curr. Opin. Struct. Biol.*, 7, 3–9.

Gu, Z., Rao, MK., Forsyth WR., Finke, JM. & Mathews CR. (2007) Structural analysis of kinetic folding intermediates for a tim barrel protein, indole-3-glycerol phosphate synthase, by hydrogen exchange mass spectrometry and Gō model simulation. J. Mol. Biol., 374, 528–546.

Höcker, B., Beismann-Driemeyer, S., Hettwer, S., Lustig, A. & Sterner R. (2001). Dissection of a (βα)8-barrel enzyme into two folded halves. Nature Str. Biol., 8, 32-36.

Ichimaru, T. & Kikuchi, T. (2003). Analysis of the differences in the folding kinetics of structurally homologous proteins based on predictions of the gross features of residue contacts. Proteins, 51, 515-530.

Ishizuka, Y. & Kikuchi, T. (2011). Analysis of the local sequences of folding sites in β sandwich proteins with the interresidue average distance statistics. The Open Bioinf. J., *5*, 59-68.

Jennings, PA. & Wright, PE. (1993). Formation of a molten globule intermediate early in the kinetic folding pathway of apomyoglobin. *Science*, 262, 892-896.

Kawai, Y., Matsuoka, M. & Kikuchi, T. (2011). Analyses of protein sequences using inter-residue average distance statistics to study folding processes and the significance of their partial sequences. Protein & Peptide Let. 18, 979-990.

Kikuchi, T., Némethy, G. Scheraga, HA. (1988). Prediction of the location of structural domains in globular proteins. *J. Protein Chem.*, 7, 427-471.

Kikuchi, T. (2002). Contact maps derived from the statistics of average distances between residues in proteins. Application to the prediction of structures and active sites of proteins and peptides, In: *Recent Research Developments in Protein Engineering.* Pandalai SG. (ed.), 1-48, Research Signpost, Kerala, India, ISBN 81-7736-147-3.

Kikuchi, T. (2008). Analysis of 3D structural differences in the IgG binding domains based on the interresidue average distance statistics. *Amino Acids*, 35, 541-549.

Kikuchi T. (2011). Decoding amino acid sequences of proteins using inter-residue average distance statistics to extract information on protein folding mechanisms, In: *Protein Folding*. Walters EC. (ed.), 465-487, Nova Science Publishers. Inc., NY, USA, ISBN 978-1-61728-990-32011.

Lang, D., Thoma, R., Henn-Sax, M., Sterner R. & Wilmanns, M. (2000). Structural Evidence for evolution of the β/α barrel scaffold by gene duplication and fusion. Science, 289, 1546-1550.

Lesk, AM. (2010). Introduction to protein science, architecture, function, and genetics (2nd edition), Oxford University Press, ISBN 978-0-19-954130-0, Oxford, UK.

Nagano, N., Orengo, OA & Thornton, JM. (2002). One fold with many functions: the evolutionary relationships between tim barrel families based on their sequences, structures and functions. J. Mol. Biol., 321, 741–765.

Nakajima, S., Álvarez-Salgado, E., Kikuchi, T., Arredondo-Peter, R. (2005). Prediction of folding pathway and kinetics among plant hemoglobins by using an average distance map method. *Proteins*, 61, 500-506.

Nakajima, S. & Kikuchi, T. (2007). Analysis of the Differences in the folding mechanisms of c-type lysozymes based on contact maps constructed with interresidue average distances. *J. Mol. Model.*, 13, 587-594.

Orengo, CA., Jones, DT. & Thornton, JM. (1994). Protein superfamilies and domain superfolds. *Nature*, 372, 631-634.

Nishimura, C., Prytulla, S., Dyson, HJ. & Wright, PE. (2000). Conservation of folding pathways in evolutionarily distant globin sequences. *Nature Struct. Biol.*, 7, 679-686.

Pain, RH. (ed.), (2000). *Mechanisms of Protein Folding* (2nd Edition), Oxford University Press, ISBN 0-19-963788-1, Oxford, UK.

Richter, M., Bosnali, M., Carstensen, L., Seitz, T., Durchschlag, H., Blanquart, S., Rainer Merk, R. & Sterner R. (2010). Computational and experimental evidence for the evolution of a (βα)8-barrel protein from an ancestral quarter-barrel stabilised by disulfide bonds. *J. Mol. Biol.*, 398, 763–773.

Saitou, N., & Nei, N. (1987). A neighbor-joining method: a new method for constructing phylogenetic tree. *Mol. Biol. Evol.*, 44, 406-425.

Sato, S., Religa, TL. & Fersht, AR. (2006). Φ-Analysis of the folding of the B domain of protein a using multiple optical probes. *J. Mol. Biol.*, 360, 850-864.

Sato S. & Fersht, AR. (2007). Searching for multiple folding pathways of a nearly symmetrical protein: Temperature dependent Φ-value analysis of the b domain of protein A. *J. Mol. Biol.*, 372, 254-267.

Schaeffer, RD. & Daggett, V. (2011). Protein folds and protein folding. *Protein Eng. Des. Sel.*, 24, 11–19.

Seitz, T., Bocola, M., Claren, J. & Sterner, R. (2007). Stabilisation of a (βα)8-barrel protein designed from identical half barrels. *J. Mol. Biol.*, 372, 114–129.

Silverman, JA. & Harbury PB. (2002). The equilibrium unfolding pathway of a (β/α) 8 barrel. *J. Mol. Biol.*, 324, 1031–1040.

Setiyaputra, S., Mackay, JP. & Patrick, WM. (2011). The structure of a truncated phosphoribosylanthranilate isomerase suggests a unified model for evolution of the $(\beta\alpha)_8$ barrel fold. *J. Mol. Biol.*, 408, 291–303.

Sosnick, TR. & Barrick, D. (2011). The folding of single domain proteins—Have we reached a consensus? *Curr Opin. Str. Biol.*, 21, 12–24.

6

Functional Implication Guided by Structure-Based Study on *Catalase-Peroxidase* (KatG) from *Haloarcula Marismortui*

Takao Sato
Depertment of Life Science, Graduate School of Bioscience and Biotechnology, Tokyo Institue of Technology, Midori-ku, Yokohama
Japan

1. Introduction

The *catalase–peroxidase* (KatG; EC 1.11.1.7) has a variety of roles in different organisms, which distributes widely among fungi, bacteria and archaea. However, the protein sequences of all procaryotic KatGs belong to a member of the class I *peroxidase* superfamily involving Cytochrome c *peroxidase* (CcP) and Ascorbate *peroxidase* (APX) (Welinder, 1991, 1992). As indicated by the enzyme name, KatG is a bi-functional enzyme exhibiting both *catalase* and *peroxidase* activities to prevent potential damage to cellular components by exogenous hydrogen peroxide (H_2O_2) and its deprotonation product. KatG is also interested in its involvement of the activation of anti-*tuberculotic* pro-drug isonicotinic acid hydrazide (isoniazide, INH) (Bertrand et al., 2004). INH is activated as its *peroxidase* substrate by KatG from *Mycobacterium tuberculosis* (*Mt*). Resulting radical via oxidation (Zhang et al., 1992; Heym et al., 1993) prevents growth of the pathogenic microorganism by inhibiting the synthesis of mycolic acid component of the mycobacterial cell wall. *Mt*KatG shares 55% identity and 69% similarity in its amino acid sequence with KatG from *Haroarcula marismortui* (*Hm)* as homologous protein. Reaction process of *catalase* and *peroxidase* activities in the heme-containing hydroperoxidases including class I, class II (fungus lignin *peroxidase*), class III (classical secretory *peroxidase*, example being horseradish *peroxidase*) enzymes and manmalian liver *catalase* is explained uniformly (Hillar et al., 2000). Ionic (heterolytic) cleavage of the first H_2O_2 molecule on the heme iron leads to compound I formation, an oxyferryl-radical cation intermediate [Fe^{IV}=O–Por(•+) or Fe^{IV}=O–Por–aa(•+), aa: side-chain of amino acid of Tyr218(Y218) or Trp95(W95) (all numbering is for *Hm*KatG unless otherwise stated)], is the first step of hydroperoxidase reaction as reaction 1. KatG forms a Por(•+) as well as Y218 and W95 radicals in the absence of reducing substrate (reaction 2; Ivancich et al., 2003). In *peroxidases*, compound I is reduced in two sequential one-electron transfers, usually from donor (AH, *O*DA: *o*-dianisidine), and involve an intermediate called compound II (reactions 3 and 5). Two resonance structures for compound II could exist (reaction 4).

$$\text{KatG [Por - Fe}^{\text{III}}] + H_2O_2 \rightarrow \text{Compound I [Por}\bullet\text{-Fe}^{\text{IV}} = \text{O]} + H_2O \quad (1)$$

$$\text{Compound I}\left[\text{Por}\left(\bullet^{+}\right) - \text{Fe}^{\text{IV}} = \text{O}\right] \leftrightarrow \text{Compound I}\left[\text{Por Y218}\left(\bullet^{+}\right) - \text{Fe}^{\text{IV}} = \text{O}\right] \leftrightarrow \\ \leftrightarrow \text{Compound I}\left[\text{Por W95}\left(\bullet^{+}\right) - \text{Fe}^{\text{IV}} = \text{O}\right] \tag{2}$$

$$\text{Compound I}\left[\text{Por}\left(\bullet^{+}\right)/\ \text{Y218}\left(\bullet^{+}\right)/\ \text{W95}\left(\bullet^{+}\right) - \text{Fe}^{\text{IV}} = \text{O}\right] + \text{AH} \rightarrow \\ \rightarrow \text{Compound II}\left[\text{Por} - \text{Fe}^{\text{IV}} = \text{O}\right] \tag{3}$$

$$\text{Compound II}\left[\text{Por} - \text{Fe}^{\text{IV}} = \text{O})\right] \leftrightarrow \text{Compound II}\left[\text{Por} - \text{Fe}^{\text{III}} - \text{OH}\right] \tag{4}$$

$$\text{Compound II}\left[\text{Por} - \text{Fe}^{\text{IV}} = \text{O}\right] + \text{AH} \rightarrow \text{KatG}\left[\text{Por} - \text{Fe}^{\text{III}}\right] + \text{H}_2\text{O} + \text{A} \tag{5}$$

$$\text{AH}\left[\text{ODA}\ /\ \text{ODA}\ (\bullet^{+})\right] \rightarrow \text{A}\left[\text{ODA}_{\text{red}}\right] \tag{6}$$

One of the most common causes of INH resistance is the Ser305Thr (S305T) mutation and, while remaining significant *catalase* and *peroxidase* activities, the [S305T] variant enzyme maintains the equivalent affinity for INH (Musser et al., 1996; Haas et al., 1997; Dobner et al., 1997; Marttila et al., 1998; Wengenack et al., 1997; Heym et al., 1995). The *Mt*KatG [S305N] variant completely lost the ability to convert INH into the InhA inhibitor (Wei et al., 2003). A role for the radicals in activation of the antibiotic is known to proceed by one-electron oxidation producing drug-based radical intermediates (Slayden & Barry III et al., 2000). It indicates that the direct reduction of compound I KatG by INH with pyridine ring is arguing against an activation mechanism requiring amino-acid based radicals. Structural and functional information is available for the crystallographic, kinetics and site-directed mutagenesis studies for *Hm*KatG. The crystal structure of wild-type (WT) *Hm*KatG, [S305T] or [S305A], [R409L], and [M244A] variants is reported otherwise (Sato et al., to be submitted; 2011a; 2011b; 2011c; 2011d). Both *catalase* and *peroxidase* activities of *Hm*KatG [S305T] have the equal to or lower than those of WT (Sato et al., 2011a; 2011b. [S305A] variants exhibit the equal to or lower *catalase* and the equal to or higher *peroxidase* activities, respectively, than those of WT (Sato et al., 2011c). [S305A] would yield the prospective complexes with aromatic inhibitor for the higher affinity than that of [S305T]. [R409L] variant, in Arg substitutions of Arg409 to Leu, reveals the higher *peroxidase* property. These structures in combination with the biochemical characterization of variants lead to identify a few of KatG-specific residues, including the cross-linkage covalent adduct among W95, Y218 and M244, unique to KatGs, G99, Y101, D125, and E194, which is conserved across all KatGs. The mechanistic aspects of KatG catalysis and kinetic studies on site-directed mutagenesis from *Hm* of equivalent variants from *Mt* have allowed a fully quantitative rationalization of the *catalase* reactivity of [S305T] and [R409L] for *Hm*KatG (Sato at al., 2011a). The structure-based computational techniques have provided a mechanistic framework that the binding affinity and rate of turnover for H_2O_2 of these enzymes may be controlled by bonding (or nonbonding) frontier-orbital due to the presence (or absence) of π-orbital overlap between W95 and Heme. Based on the structures of *Hm*KatG [R409L] variant complexes with (electrophilic reagent) substrate for *o*-dianisidine (*O*DA) and (nucleophilic) inhibitor for cyanide anion (CN^-), the elevated *peroxidase* reactivity is gained in binding *O*DA from reorientation of the side-chain of specific residues in D125, W95 and heme (Sato et al., 2011b). The inhibition of both H_2O_2 and *O*DA is observed by such as CN^-

and salicylhydroxamic acid (2-hydroxybenzohydroxamic acid; SHA) with phenolic ring that are classical heme enzyme inhibitor, respectively, and an inhibitor of compound II KatG with respect to substrate aromatic donor as *O*DA with dimeric phenolic rings, leading to insight to INH with pyridine ring. SHA and exogenous heme iron ligands compete in binding to the resting enzyme, while phenolic substrates (phenol, p-cresol, and resorcinol) do not competitively inhibit liganding to the heme iron (Modi et al., 1989; Ikeda-Saito et al., 1991). The aromatic substrate binding site on heme iron has been identified by determining the structures of [S305A] and its complexes with SHA. These provide a complete description of substrate binding in KatG (Sato et al., 2011b; 2011c). It is implied for SHA inhibition reaction that the proposal model of INH- *Mt*KatG complex may share structural similarity with SHA in binding site of *Hm*KatG [S305A] variant. Hence, *Hm*KatG [S305T] or [S305A] and [R409L] variants are expressed with an attempt at rational catalytic redesign, to prepare the inhibitor- or substrate-enzyme complex, because the [S305A] and [R409L] variants are expected to have the higher affinity for SHA and *O*DA than that of the WT, respectively, and exhibits the equal to or higher *peroxidase* activity than that of WT.

On the other hand, in [R409A], [R409K], and [R409L] variants of other KatGs, the *catalase* activity decreased while the *peroxidase* activity did not affected (Carpena et al., 2005; Jakopitsch et al., 2004; Ghiladi et al., 2004). One of the commonest causes of resistance about INH is also the [R409L] mutation in *Mt*KatG and known for *peroxidase* reaction (Ghiladi et al., 2004; Saint-joanis et al., 1999). It is required for [R409L] equivalent variants from *Hm*KatG to be increasing sensitivity to *O*DA and, activating significant *peroxidase* while remaining slightly *catalase* activities (Sato et al., 2011a; 2011b). The single-base mutation between [R (cgg) 409L (ctg)] and [S (agc) 305T (acc)] would be possible from mutational events in natural evolution. It is expected that attempts to generate a "new" function within a laboratory time scale can imitate the entire process used by Nature to evolve and optimize a new function, but these studies illustrate that the acquisition of sequential mutations can produce a continuous pathway for evolution of a "new" function (*peroxidase* here).

Notably, KatG-specific Met244 sulfur coordination to two carbonyl backbone of Tyr101 (Y101) and Gly99 (G99) and including Tyr218, Trp95 on the distal side of the heme, the unique covalent modification among the side chains of three amino acids (M244, Y218 and W95) are present. It was reported that additionally, phenyl oxygen (Oη) of Y218 and amide nitrogen (N) backbone of M244 in the covalent adduct have hydrogen (H)-bonds with the guanisyl N*ε1* and Nε2 of Arg409 (R409), respectively (Yamada et al., 2002). The covalent adduct has been shown to be generated by autocatalytic process (Ghiladi et al., 2005a). Crystallographic analysis of KatG from *Bulkhorderia pseudomallei* (*Bp*) has indicated that the side-chain of R409 (corresponding to *Hm*KatG numbering) shows conformational change depending on both of pH and the physicochemical state of the active site (Carpena et al., 2005). Five amino acid residues, Y101, G99, M244, Y218, and W95 are conserved in all the KatGs but not in the other class I *peroxidases*. [W95A], [W95F], and [M244I] variants in KatG from *Synechocystis* (*Sy*) or [M244A], [M244L] in *Bp*KatG considerably decreased the *catalase* activity, whereas the *peroxidase* activity was even enhanced (Regelsberger et al., 2000; Jakopitsch et al., 2004; Singh et al., 2004). It has a slight difference in [M244A], [M244L] variants in *Bp*KatG and [M244I] variant in *Sy*KatG. Substitution of Tyr218 to Ala [Y218A] or Phe [Y218F] in the M244–Y218–W95 covalent adducts induced complete loss of the *catalase* activity, whereas the *peroxidase* activity was highly enhanced in *Sy* KatG (Jakopitsch et al., 2003a; Jakopitsch,et al., 2003b). M244 in the adduct also has any significant role in the enzyme, whereas the substitution of

Met to Ile considerably attenuated the *catalase* activity but did not lose at all, and enhanced the *peroxidase* activity (Jakopitsch et al., 2004; Singh et al., 2004). Site-specific mutagenesis indicated that the covalent adduct and its relevant was significant to the *catalase* activity, whereas their effect on the activity was complicated and was hardly explained. Substitution for [M244A] affected the *Hm*KatG structure of the access channel and therefore the enzymatic parameters for the *peroxidase* activity (Ten-i et al., 2007). This mutation also induced the significant functional change in the active site that would trigger a loss of *catalase* activity and a high enhancement of *peroxidase* activity (Sato et al., 2011d). It was also investigated whether the absence of the structural feature unique to KatG, namely M244–Y218–W95 adduct in which are covalently linked together through these side chains, can be correlated with exhibiting the *peroxidase* reactivity. Resulting in the [M(atg)244A(gca)] substitution, the designed three-base mutation leads to cleavage the covalent bond between M244 and Y218–W95, as was identified in the adducts. It has been constructed the successive triple-base substitution of Met244 to Ala [M244A] and to cleavage the covalent bond amongst M244–Y218–W95 adduct, though this variant would not be expected in natural evolution. Using rounds of site-specific mutagenesis coupled with the structure based-evolution, the unique access will be provided to structural changes that allow enhancement and eventual optimization of the "new" function.

In this paper, the structure-based computational chemical technique can confirm a fully quantitative rationalization of the reactivity mechanism of electron transfer (ET). Interestingly, it was strongly suggested by this structure-base calculation that this variant did not alter compound I formation but can have capacity to restore compound I from compound II, possibly suggesting a yet-to-be defined mechanism for *peroxidase* that can be allowed to perform in *Hm*KatG.

2. Materials and method

2.1 Structure based molecular orbital and docking analyses

Starting structure models contained 370 atoms for WT *Hm*KatG, 375 atoms for [S305T], 385 to 389 atoms between [S305A] and SHA–[S305A], 368 to 405 atoms among [R409L], CN–[R409L], CN - [R409L] complexes with *O*DA and 364 atoms for [M244A], after hydrogen addition to their crystal structures. The residues were confined to the immediate vicinity (surrounded around 3.6Å) among the heme and ligand coordinate of G99 or Y101, M244-Y218-W95 covalent linkages adduct. Geometries were determined by mechanics optimization using augmented MM3. The semi-empirical molecular orbital (MO) method was performed by a MOPAC2002 program / AM1 wavefunction (Stewart, 2002; Dewar et al., 1985). All the sets of molecular orbitals are generated from the highest occupied molecular orbital (HOMO) to the lowest unoccupied molecular orbital (LUMO), exhibiting the negative value for MOPAC-specific calculation. When ionization energy (HOMO) is greater than -8 eV including with HOMO density, electrophilic frontier orbital density (fr) and electrophilic superdelocalizability (Sr), the residues have the electrophilic reactivity. Electron affinity (the LUMO) for nucleophilic group is less than that of -2eV with LUMO density, nuclophilic fr and nuclophilic Sr, exhibiting the reactive indices, fr and Sr to be most reactive (Fukui et al., 1954; 1957) with the excitation of electron in the energy gap. An π–electron is transferred from HOMO to LUMO, when HOMO and LUMO polarities are homologous and uniformed, the geometries of the HOMO/LOMO states that have the distance within 3.4Å of van der Waals contact, the matching phases (overlapping the bonding orbitals) and the gap energies within 6eV (Pearson, 1986). It is assumed that the π-electron delocalization of adduct might influence ability of the intrinsic properties of KatG to transfer electron in its conjugate system and then to exhibit the *catalase* activity (Sato et al., 2011a). For understanding the chemical reactivity and site, the MO calculations were carried out on a PC using BioMedCAChe ver6.1.12.34 (Fujitsu, Tokyo). The outcomes of the *in vacuo* calculations on coordinate complex, covalent adduct and heme are related to the electronic characteristics of the covalent adducts, SHA, and *O*DA which bound to heme into the hydrophobic cleft of the variants and its complexes. *O*DA as a *peroxidatic* substrate and an electron in the HOMO of the donor molecule as *O*DA (·+) cation radical is transferred to the LUMO in the heme of compound I. The reverse is also taking place for SHA.

In addition, when it was difficult for the value of K_m (substrate affinity) to measure, conveniently it was quantified by binding energy calculated from the docking study. The binding energy for a given ligand (ΔE_{ligand}) can be expressed in (eq.7) as the difference in the energy between complex and components (Fukuzawa et al., 2003).

$$\Delta E_{ligand} = E_{complex} - \left(E_{enzyme} + E_{ligand}\right), \quad (7)$$

where are the heat of formation energy of each of the three systems, i.e., E_{ligand} of H_2O_2, SHA, or *O*DA and, E_{enzyme}, of the KatG enzyme, and $E_{complex}$ of the enzyme complexes with H_2O_2, SHA, or *O*DA. The binding energy can be estimated to subtract the sum of heat of formation energy values of each system from that of pair of an enzyme and a ligand.

It is indicated that H_2O_2 molecule leads to compound I formation on the heme iron. In docking exogenous ligands to heme, the binding energy between ligand (H_2O_2 as an initiator, SHA as an inhibitor of *peroxidase* compound I and *O*DA as *peroxidase* substrate) and enzyme can be approximately estimated from the relative energy difference which subtracts each heat of formation calculated between H_2O_2, SHA, *O*DA and enzyme from that of ligand-enzyme complex.

3. Results and discussion

3.1 Coordinate, covalent-adduct, heme distal side in the active center and substrate access channel

The heme moiety of the [R409L] variant is found to be deeply buried inside the *Hm*KatG and *O*DA as the *peroxidase* substrate can access to the active centre only through the narrow cylindrical channel with ~20Å depth from the entrance to the bottom, as shown in Fig1. Side-chain of Asp125 located at the bottom of the channel and showed the structural change in [R409L] complexes with *O*DA.

As shown in Fig. 2(a), the access channel was composed of LL1 loop, *helices B* and *E*, which involved in the active site of R92, H96, and heme. There is three KatG-specific structural characteristics, which consist of first component (π-complex between W95 and Heme), second component (M244-Y218-W95 covalent adduct) and third component (sulfur cation radical centered coordinate complex). One of the unique structural features (of G99, Y101, M244-Y218-W95 covalent adduct and N126) have been revealed in the structures of WT *Hm*KatG, [S305T], [S305A] and [R409L]. It consists of the octahedral (six) coordinate sulfur radical cations ($>S^{\bullet+}$) center of M244 which is capped at the *N-ter* end of *Helix E* by a positively charged amino acid in order to neutralize this helix dipole, ligands among phenyl carbon of (KatG specific) Y218 -W95, carbonyl oxygens of Y101 (KatG specific) and G99 (KatG-typical and conserved with APX) at the *C-ter* end of the *Helix B* in KatG. SHA—[S305A] complex shows the phenolic group oxygen of SHA to be of 3.2Å within van der Waals contact with the the δ-heme edge. In *O*DA— [R409L] complex, the *δ*-heme edge is also confirmed as the primary site oxidation of *O*DA. The phenol ring part of SHA in the [S305A] overlaps and binds to the equivalent binding site as one aniline ring part of *O*DA in the [R409L] variant (Fig.2(b)).

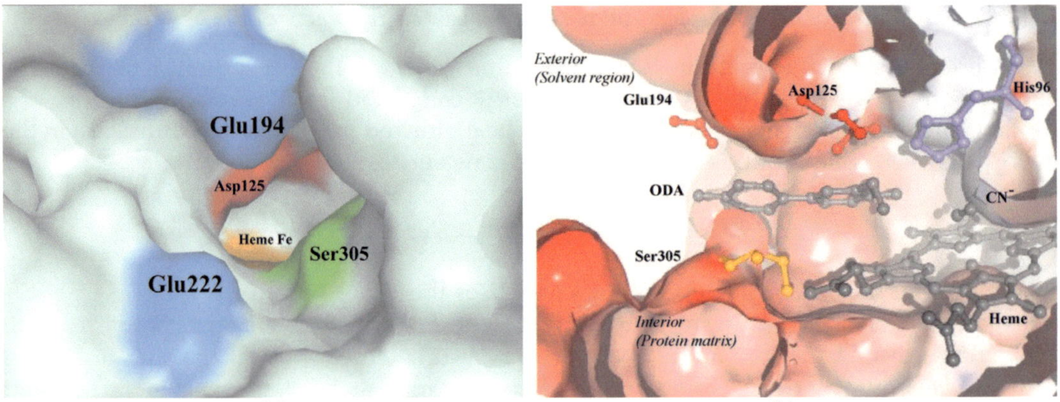

(a) (b)

Fig. 1. Active and *O*DA Binding Sites of *Hm*KatG. (a; left panel) Space-filling representation of the binding site in showing locations of the five critical residues, Glu194 and Glu222 (blue), Ser305 (green), Asp125 (red) and heme iron (orange). (b; right panel) The vertical cross-sectional view of active site of *O*DA and CN (gray), surrounded with the residues His96 (cyan), Asp125 and Glu194 (red), Ser305 (yellow) and heme (gray). These figures were constructed using MolFeat v3.0 (FiatLux Corp., Tokyo, Japan).

As shown in Fig.3 (A), entering the distal side cavity of [M244A] variant through the constricted portion of the channel, the initiator H_2O_2 would come into contact with the active-site residues W95 and H96 on *Helix B* (§3.6 *vide infra* in Table 5). While the WT, [S305T], [S305A], and [R409L] variants have been stabilized LL1 loop by covalent adduct linkage between Y218 on the mobile LL1 and M244 on *Helix E*, the [M244A] exhibits the cleavage of the linkage with Y218. When the displacement of E194 and E222 was not endured by the flexible response of the mobile D125 of upstream residues to the downstream portion of LL1 loop, it allows the mouth of the channel to open and to facilitate adequate uptake of substrate into the heme cavity. Side-chains of E194 and E222 on LL1 loop locates at the entrance of the channel and was also affected by this mutation. In subunit A, the side chain D125 disrupted H-bonding interaction with amino nitrogen of backbone I217 in the LL1 loop and the side oxygen *Oδ1* formed H-bond with its own carbonyl oxygen in backbone of D125, resulted perpendicular rotation of $\chi 2$ of side D125 to face the imidazole of H96 (Fig. 3 (B)) while D125 has been known to be important in the H_2O_2 oxidation to date (Jakopitsch et al., 2003a; Singh et al., 2004). However, there is a displacement of 2.69Å toward 2.80 Å and dramatic reversal of the dihedral angle $\chi 2$ of side D125, which is to bind *peroxidase* substrate as *O*DA or water molecule as deriving from H_2O_2, respectively. Hence, the mobile D125 residue will also be suggestive of utilizing as either initiator H_2O_2 or substrate *O*DA recognition, making it effective in binding substrate, though disruption of π-complexes with heme and W95 known to act on as molecular switch from the *catalase* to the

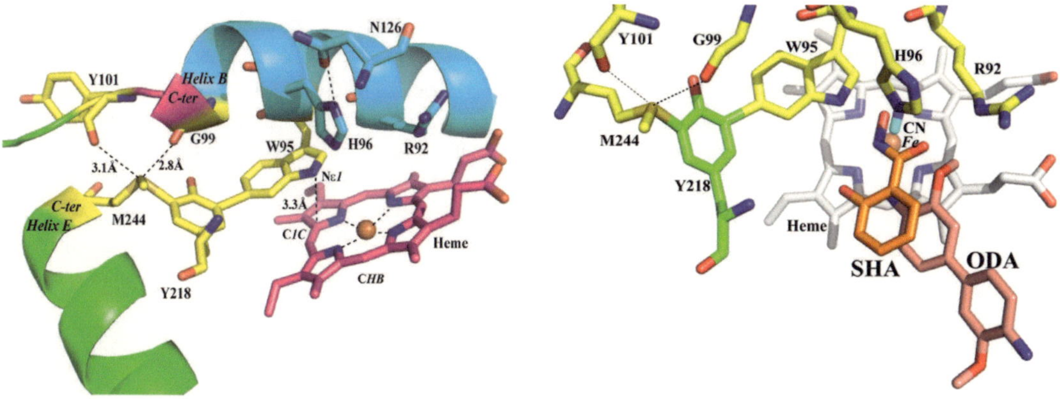

(a) (b)

Fig. 2. (A) Helixes B and E, the 6th Coordinate, the Covalent Adduct and Heme π-complexes in Active Site Structure of WT, and three variants. (B) Structure-based Overlays of the [S305A]-SHA and Cyanide [R409L]-*O*DA complexes. (a; left panel).The distal R92, H96, N126 on Helix B are shown in blue, W95 on Helix B, Y218 and M244 on Helix E residues in yellow. Heme is shown in purple and orenge sphere represent heme iron atom. The π–π* stacking interaction between W95 and Heme is represented by dashed line between N*ε1*, respectively, and *C1C* atoms near α-meso heme edge. The H- and coordinated-bonds for the structures of [WT], [S305T], [S305A] and [R409L] are represented by dashed lines (black). (b; right panel).The ligand structures are coincided with δ-meso heme edge and are avoided overlap considerably between the inhibitor SHA (yellow) and substrate *O*DA (magenta). The heme is shown in gray, and two residues, W95, H96 and R92, which involved in the binding site for the [S305A] and [R409L], directly superimposed in all cases; for clarity, only one residue is indicated. The coordinated-bonds for the S305A-SHA structure are represented by dashed lines (black).

peroxidase (Carpena et al., 2005). The amino N of backbone I217 forms an H-bond to the oxygen O*δ1* of carbonyl group of the side D125. Thus no H-atom can be in the C=O*δ1* group. When O*δ2* of the D125 can be an ionized carboxyl group at optimum pH6 near pKa value of 4.0, it may be a powerful proton donor for binding the *peroxidase* substrate with extremely high affinity (Km^A = 0.974 μM) for *O*DA would be attributed to rotate the dihedral angle χ2 by 61.3° of the mobile side D125 in the subunit A of [M244A] variant. Hence it reorients the sp^2 orbital of the –O*δ2* (-H) which functions as powerful proton acceptor and has a high affinity for the *O*DA molecule, when the *sp* orbital of the —C=O*δ1* which acts as proton donor and a lower affinity for H_2O_2 molecule. The role of D125 may form the binding site for the water molecule which derived from the first H_2O_2 molecule (and then compound I formation), may carry its water molecule out of the active site and may induce and control entry of the second *O*DA molecule into the cavity. In other words, O*δ1* with *syn* lone pairs binds *O*DA molecule. The one (*syn*) direction of lone pairs of the carbonyl oxygen (-C=O*δ1*) of side D125 form the more stable H-bind with cation of *O*DA rather than that of another (*anti*) direction bound with the amide nitrogen (-NH) of backbone I217 respectively. Thus, it is also indicated by *ab initio* quantum chemical studies that *syn* protonation from (-C=Oδ*1*) is more favorable to (C-Oδ2 -H) than *anti* protonation, implying the *syn* lone pairs are more basic and therefore bind *O*DA more readily than do the anti lone pairs (Petersen and Csizmadia, 1979). While the side chain of D125 is rotated around the dihedral angel $\chi 2$, since the oxygen orbital changes between (-Oδ2-H) sp^2 and (=Oδ*1*) *sp* is caused with and without *O*DA respectively, the *O*DA is bound to the δ–meso heme edge by the molecular mechanics and orbital changes of the D125. The D125 has different affinity for *O*DA from either strong subunit A or weak B, respectively, and facilitates the uptake of *O*DA in *peroxidase*, with draining away the water molecule. Therefore, the geometry of side chain of D125 in subunit A allows for higher binding affinity for *O*DA molecule than that of subunit B, which is consistent with the *peroxidase* in the [R409L] (Sato,T., et al., 2011b).

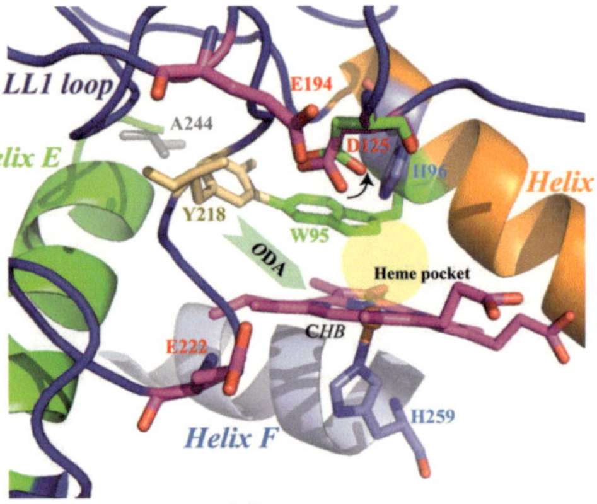

(a)

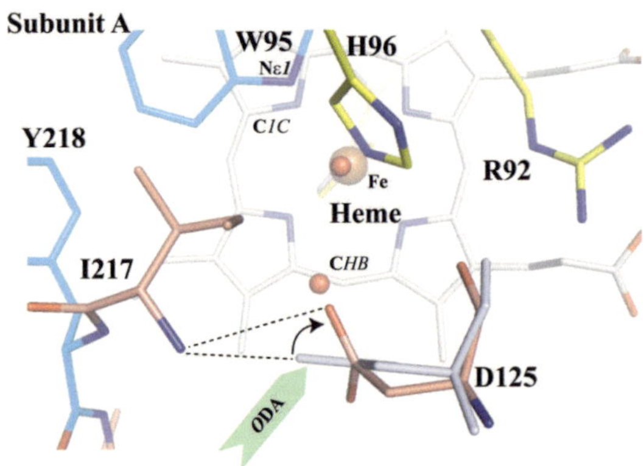

(b)

Fig. 3. (A) Proposed Active Site Structure in [M244A] variant. (B)The mobile D125 in subunit A superposed that of WT. (a; left panel)In Subunit B of [M244A] variant, the distal H96 (on helix B; orange) is shown in blue. The W95 (on helix B), Y218 (on LL1 loop; cyan) and A 244 (on helix E; green) residues are in green, orange, and gray. The heme and its iron atom are represented in magenta sticks and orange sphere. The latent access channel residues of D125, E194 and E222 locate on LL1 loop, showing in red (left panel). (b; right panel) In subunit A of M244A, the Mobile D125 and I217 in pink, distal R92 and H96 are presented in yellow, the Y218 and W95 covalent adduct in cyan. The nitrogen, oxygen and heme iron atoms are colored for dark blue, red, and orange. D125 in subunit B is superimposed in colored blue, white and water molecule is shown as red spheres. The figure is view from the distal side of left panel. The figures were constructed using Pymol (DeLano, 2002).

3.2 Docking and molecular orbital calculations on ODA among WT, S305T and R409L variants

While the difference between [S305T] and [R409L] crystal structures are very subtle changes (with overall root mean square deviation of 0.492 Å) and the structural integrity is highly maintained, the binding energy of such *peroxidatic* substrate as *O*DA appears to affect functioning of the enzyme. Each of WT *Hm*KatG and [S305T] variant structures had not exhibited as a potential active site for *O*DA substrate at all. Since the propionate carbonyl group of heme was H-bonded with either the hydroxyl group of Ser or Thr305, it assumed that it would interfere with the transport of *O*DA and cause a blockage of the channel. A calculated binding energy in the [R409L] variant (subunit A for -27.6 kcal/mol; B for -73.3 kcal/mol, which is negative and can be discussed by the magnitude of its absolute value as the following from here) also reveals to have significantly high affinity for *O*DA (Table 1) and a possible site has been proposed for [R409L] KatG, in a cavity on the distal side of the heme. Consequently, it is suggested that the porphyrin carbon atom (*CHB*) of δ-meso heme edge may serve as a docking site for *O*DA. Neither *catalase* nor *peroxidase* activities may be expected from docking energy (and site) in [S305T]. The promising *peroxidase* despite a lack of *catalase* activity in [R409L] may result from the localized electronic state of the heme, arising from the H-bonding interaction with *O*DA.

	WT		S305T		R409L	
Subunit	A	B	A	B	A	B
Affinity of *O*DA for *Peroxidase*						
E222–D125(kcal/mol)					-27.6	
D125–Heme (kcal/mol)						-73.3

Table 1. *O*DA binding affinity against the docking calculation. It is suggested as binding affinity that the value of ΔE_{ligand} is negative to be predicted by docking calculation with the substrate *O*DA, according to eq.7.)

Only when there are very subtle conformational changes with side-chain of residues and without backbone distortion in the substrate access channel, this structure-based molecular orbital (MO) calculation method appears to provide a useful means of assessing ET pathways and reactivity in heme-containing system. Therefore, an advantage of *semi-empirical* calculation is the ability to calculate the energies of all the MO of the individual active residues in the protein, which may be expected to be practically much better than no data obtained overtime by *ab initio* molecular orbital study combined with the large scale computer system of supercomputer.

By examining each subunits of [R409L] structure, the *CHB* atom is the most reactive site in heme that has electron-withdrawing group, where exhibits as well or greater energy of LUMO, e.g. the eigenvalues of LUMO molecular orbital has -1.43eV and -1.12 eV (Table 2) than those of WT with Subunit A of -1.79 eV and B of -1.96 eV (Table 4). Additionally, such index *peroxidase* reaction as LUMO density (0.15, 0.153), nucleophilic frontier fr (for subunit A of 0.023 and B of 0.004), and nucleophilic Sr (0.891, 0.943) in [R409L] are indicative of either equaling or surpassing the corresponding parameters (0.058, 0.159; 0.001, 0.017; 0.832, 0.912) to WT. The most likely site of binding *peroxidase* substrate would be the *CHB* atom in the δ-meso heme edge where is consistent with *O*DA recognition site suggested by the docking study in the prospective *peroxidase* activity. It may be possible to determine the geometries (phases and energies) of the states that would lead to *peroxidase* activity. It is indicated that the electronic nature of the compound I might be different as could be the mechanism of H_2O_2 oxidation. Reduction of compound I rather than compound I formation by KatG is the intersection point of the *peroxidase* and *catalase* cycle.

		WT		S305T		R409L	
Subunit		A	B	A	B	A	B
	Catalase						
	π–complex						
LUMO (eV)	W95 (Nε*1*)			-1.15	-0.26		
HOMO (eV)	Heme (C*1*C)			-6.71	-5.10		
The π-π* energy gap (eV)				5.56	4.85		
LUMO (eV)	Heme (C*1*C)	-1.05	-1.60	-2.22	-1.49	-1.43	-1.31
HOMO (eV)	W95 (Nε1)	-7.61	-7.77	-8.75	-7.74	-7.50	-7.47
The π-π* energy gap (eV)		6.56	6.17	6.53	6.25	6.07	6.16
	Octahedoral	Coordinate		complex			
LUMO (eV)	G99,Y101,Y218	-2.88	-2.84	-2.22	-2.64	-2.49	-1.93
HOMO (eV)	M244	-7.32	-7.48	-7.17	-7.28	-7.21	-7.47
HOMO-LUMO gap (eV)		4.44	4.64	4.95	4.64	4.72	5.54
	Peroxidase						
LUMO(eV)	Heme C*HB*	-1.73	-1.96	-1.62	-1.48	-1.43	-1.12

Table 2. The frontier molecular orbital of (a) HOMO and (b) LUMO Energies for Active Site Associated with *Catalase* and *Peroxidase* Reactions in WT *Hm*KatG, [S305T] and [R409L] variants.

3.3 Active sites confirmed by structure based on cyanide R409L complex with or without ODA

The *peroxidase* substrate acts as the electron donor in the *peroxidase* reaction. In the *O*DA substrate, the geometry is confined to the biphenyl and binitrogen moieties. The reaction point in *O*DA (•+) cation radical is also confined to one side of the central part of molecule. When one electron goes from neutral *O*DA to the *O*DA (+) positive ion or when changing from the *O*DA (•) neutral radical to the *O*DA (•+) radical-cation states, the *O*DA molecule undergoes a strain reduction of bicyclic geometry occurring in quinonedimine moiety different from what can be expected from the biphenyl, while the similarity between the central part of the *O*DA molecule and quinonedimine is maintained, the group of N-C bonds remains planar. The earlier geometry results were obtained from crystallographic structure of complex with *O*DA. This geometry is consistent with the analysis of the frontier orbitals, confirming that the HOMO level of *O*DA and LUMO levels of heme are mainly localized over two nitrogen atoms of amino group of the central biphenyl core and *CHB* atom of the δ-meso heme edge. The difference in HOMO/LUMO gap energy between 5.7 eV for *O*DA and 4.9 eV for *O*DA quinonedimine is the same electronic configurations at different geometries. The lower value of 0.8 eV integrate the information given by the geometry, showing the substantial difference between the strain reduction processes taking place in *O*DA with respect to binitrogen of quinonedimine and obtained from the results of energy calculations on the basis of AM1 calculations.

The frontier orbital approach based on the structure of cyanide [R409L] complexes with *ODA* also suggests that it is enough to transfer electron from the HOMO ([-7.60eV; -7.72eV]) of *ODA* cation radical with the electrophilic Sr (0.372, 0.352) to the LUMO ([-1.62 eV; -1.42eV]) of the nucleophilic *CHB* in heme with nucleophile Sr (0.863, 1.017) greater than that of WT (0.832, 0.912). The HOMO/LUMO difference between *ODA* cation radical and *CHB* is of 5.98eV, respectively, and 6.3eV between A and B subunits as shown in Table 3. As suggested by the energy gaps, ET may complete and its cation radical can generate from neutral *ODA*.

The *ODA* acts as the cation radical ($\bullet^{+}$) and is poised for attack by the carbon atom (*CHB*) of δ-meso heme edge. The present work also gives evidence for an essential role of *CHB* as the KatG-typical active site on the *peroxidatic* activity. The LUMO energy of *CHB* has the nucleophilic reactivity with and plays a role in binding with *ODA* of which the HOMO has the electrophilic and forms a positive charge or lone electron pair on N atom in amino groups. Therefore, it is suggested that the *peroxidase* activity reflects advantageously on the binding affinity. Involvement of the HOMO of the (electrophilic) cationic amino group of *ODA* in the compound I is oxidized by the *CHB* of the (nucleophilic) anionic heme group, and an increase in the cavity volume through decay of π-complex between W95 and heme which cause a loss of the *catalase* reaction, may account for the increased *peroxidatic* activity and more catalytic efficiency of *kcat/Km* because of easier binding of *ODA* than that of WT. The redox intermediate compound I, which has iron oxidized to Fe^{4+}=O (oxyferryl species), is formed and a second oxidation equivalent as either a porphyrin cation radical (the LUMO energy of *CHB*) or the HOMO energy of *ODA* cation radical, which are phase-matched orbital in respective subunits.

On the other hand, the MO calculation based on structure of cyanide [R409L] is predictive that the unpaired electrons are localized on the iron (IV)-oxo moiety and on W95 by one electron shift of the unpaired electron from W95 to the porphyrin (*C1C*) near α-meso heme edge. In the absence of *ODA*, porphyrin cation radical ($\bullet^{+}$) (LUMO energy for *C1C* heme of -1.56eV; -1.42eV) has the nucleophilic Sr (0.636, 0.612) greater than that of WT (0.613, 0.583) with the initial formation of anion (-) heme group rather than W95 protein radical (HOMO for Nε1 W95 [-7.57eV and -7.54eV]) has formed the electrophilic Sr (0.275, 0.281) less than that of WT (0.285, 0.283). The respective energy gap between HOMO and LUMO has 6.01 eV and 6.13eV, ET can be complete occurring within 6eV (Pearson et al., 1986) of the energy gap for the excitation of the electron. The favourably stable neutral radical formation at W95 with concomitant loss of *catalase* function is followed by due to reduced stabilization of the iron (IV)-oxo containing compound II and increased reversion to compound I when ET in porphyrin ring has been implicated from *C1C* to *CHB*. It that a reduction in the cavity volume is suggested through occurrence of π-π* interaction between W95 and heme, which cause a loss of electron with concomitant of reversion to resting state during the *catalase* reaction, may account for reverting Compound II to Compound I with faster and slower rate of turnover of *ODA*, respectively, than that of WT in spite of increasing *peroxidatic* activity and more catalytic efficiency of *kcat/Km*. Since *ODA* cation radical ($\bullet^{+}$) is more erratic and unstable than W95 protein radical, if *ODA* is able to reach the active site (*CHB*), the *peroxidase* reaction occur in close proximity to the *CHB* after the ferry oxygen abstracts proton from *ODA*. We have identified a variant, R409L, which provides direct evidence for a bifurcated pathway in which *ODA* is able to reach the active site (*CHB*) of heme, and is able to be oxidized by Compound I or Compound II heme intermediates. The presence of substrate is necessary to switch from *catalase* to *peroxidase* activities.

		Cyanide [R409L]			Cyanide [R409L]-*O*DA complex	
Subunit		A	B		A	B
	Electron source					
HOMO(eV)	M244(>S•+)	-7.30	-7.54	M244	-7.78	-7.44
LUMO(eV)	G99Y101Y218	-2.62	-2.11	G99Y101Y218	-2.19	-2.04
HOMO-LUMO gap (eV)		4.68	5.43		5.6	5.4
phase	binding			binding		
Interaction	6th coordinate			Geometry of	Centred M244	
compound I reversion	but not *catalase*					
HOMO(eV)	W95Nε1(•Trp)	-7.57	-7.54			
LUMO(eV)	(Por•)*C1C*	-1.56	-1.42			
The H/L	Energy gap(eV)	6.01	6.12			
phase Interaction	anti-binding π-π*	3.4Å	3.4Å			
				Peroxidase		
HOMO(eV)			*O*DA(•+)	cation radical	-7.60	-7.72
LUMO(eV)			*CHB*	nucleophilic	-1.62	-1.42
HOMO-LUMO gap(eV)					5.98	6.30
phase Interaction	binding π-π*				3.3Å	3.3Å

Table 3. Atomic orbital (a) HOMO and (b) LUMO Compositions of the MO for active site in R409L Cyanide Complexes with or without *O*DA. The geometries, phases and energies of the states would lead to *peroxidase* with the intramolecular interactions (coordinate, π–complexes, covalent and H–bonds) within van der Waals contacts and the limited energy gap.

3.4 MO study based on the structures of S305A complexes with or without SHA

Tripartite architecture is involved in the *catalase* function, which composed of the first π-complex between W95 residue and heme, the second covalent adduct of M244–Y218–W95 and the nearly octahedral coordinate complexes among G99, Y101 and M244. As first component stacks of alternating donor and acceptor residues are found in the *Hm*KatG. The relative orientations within the parallel planes of these residues are determined by the energies and charge distributions of the [HOMO] from which an electron will come, and the [LUMO] to which the electron will go. As presented in Table 4, both energies of [E (HOMO)] and [E (LUMO)] were calculated using the AM1 Hamiltonian, of which PM3, PM5 and PM6 wave functions (Stewart, 1989a; 1989b; 2007) denotes the same tendency. The [HOMO] energy represents the ionization potential of the indole ring in W95, i.e. the [E (HOMO)] of the negative value required to transmit an electron to [E (LUMO)] in heme edge of the porphyrin ring. (In such ET complexes the interplanar spacing of the flat residues is smaller than the normal spacing is of 3.4Å or can be equal to 3.1 Å). It is considered that ET has occurred in such donor-acceptor complexes. The indole ring of W95 in the covalent adduct is almost coplanar with the heme and the "general" ET is likely to occur to the lowest unoccupied heme orbital from the highest occupied π-orbital of W95. When LUMO/HOMO interactions are phase-matched energy gap with small difference between HOMO and LUMO, as well as strong intermolecular interactions, ET can be complete.

3.4.1 Third component (sulfur cation radical formation) in WT *Hm*KatG, [S305A] variants with and without SHA

Since the central S*δ* atom in side M244 of third component makes use of six d^2sp^3 hybrid orbitals, the shape is expected to be octahedral. Octahedral molecule geometries of sulfur coordination complexes have its steric number of 6. When five ligands can be used with a lone electron pair ":" in the octahedral system, an nearly octahedral arrangement complex around the S*δ* is created with the five ligands which contains two C atoms in the ethylmethylsulfide group of M244, two O atoms of G99 and Y101, and C*ε1* atom of the phenol side chain of Y218. The O atom of Y101 and C*ε1* atom of Y218 takes up an axial position of octahedral SO_2C_3":", whereas one O atom of G99, two C atom of ethylmethylsulfide group and a lone pair electron may take up the equatorial position. When the central S*δ* atom contains more than 6 electrons, the extra electrons form a lone electron pair. If a cation, charged 1+, gathers 6 equivalent ligands around it, S^+L_6, then each of these anions contributes a partial negative charge of -1/6 to the area around the cation, so that local electroneutrality results. For example, if each oxygen atom will experience 1/6 of the charge of the S^+ ion (=$1/6^+$) and S ion will experience 1/6 of an electron (=0.16^-) from each carbon and oxygen atoms, since a sulfur ion $S^+(0.96+)$ for WT is surrounded by two carbonyl oxygens and three carbon atoms in G99–C=O (0.47^-) and G101–C=O (0.51^-), Cε2(0.48^-) in Y218, Cγ(0.39^-), and Cε (0.45^-) which each of partial charge was calculated in M244 (SATO et al., 2011a; to be submitted). This value of 1/6 is called a bond valence of surrounding anions, respective atom was added a ground-state orbital slightly larger than the bond valence of 0.16^- or at least 0.2^-, which provides a value for the strength of each interaction between a cation and various anions. The shorter $S^+\cdots O$ interactions would be assumed to be stronger, with a greater partial charge donated by the oxygen atom to the S^+ ion in order to attain electroneutrality and to maintain the $S^+\cdots O$ distances that in S305A

(SATO et al., 2011c) are of [2.86 Å and 2.89 Å for C=O of G99] and [3.14 Å and 3.21Å for C=O of Y101] which about equal to [2.9Å and 2.85Å of G99] and [3.14Å and 3.28Å of Y101] in the subunit A, relatively, and B of WT. But the distances between S in M244 and C=O in G99 have longer M244_S$^+$···O_G99 for SHA-S305A (2.91Å; 2.91Å) than that of S305A (2.86Å; 2.89Å) whereas shorter M244_S$^+$···O_G101 for SHA-S305A (2.98Å; 3.08Å) than that of S305A (3.14Å; 3.21Å). When the SHA bound to [S305A], the electronegativity of O ligand atom of G99 in equatorial position of octahedral coordination was acquired by ET from S$^+$ with maintenance of electropositivity M244. At the same time, ET occurred to the M244 from the O atom of Y101 and C*ε1* atom of Y218 of an axial position. By close to Y101 and extension of Y218, electron from lone paired electron of S$^+$ was divided to W95 at one end of the covalent adducts, the W95 cation radical was caused by electron localization which electron transmitted to W95 from G99 and Y101 through Y218.

Electron must flow from the donor [HOMO] to the acceptor [LUMO], which need to be relatively close (~6eV) in energy (Pearson, 1986). As shown in Table 4, in [S305A] and SHA- [S305A], the sulfur center atom of M244 and the coordinate ligand atoms of carbonyl oxygen atoms of backbone G99 and Y101 plus phenol carbon of side Y218 have [E (HOMO) =-7.15, -7.41; -7.34, -7.88 eV] and [E(LUMO)=-2.52, -2.76; -2.91, -2.87eV], respectively. However, in spite of HOMO, sulfur atoms can be estimated as nucleophilic fr (0.192, 0.379; 0.336; 0.13), nucleophilic Sr (1.263, 1.004; 3.241, 1.141) as well as HOMO density (0.127, 0.217; 0.195, 0.085), and then may fill both roles as HOMO and LUMO by exhibiting the sulfur-ion complexation for MOPAC-specific calculation. In such a coordinate ligand as the carbonyl oxygen atoms of backbone G99 and Y101 plus phenol carbon atom of side Y218, there is also nucleophilic Sr in addition to only HOMO density and electrophilic fr. Thus the third component becomes an excellent source of electron, which an electron was transferred to attractive destination for W95 by way of the second component. Qualitatively, the properties and reactions of indoles have been explained by the resonance theory. π-electrons of an aromatic system is approximated as a linear combination of the functions of suitable contributing LUMO nucleophic phenol coordinate to HOMO electrophilic sulfur center structure and then may produce the electrophilic Sr and positive charge on nitrogen atom in indole group of side W95, reflecting the known electron-donating properties of the indole nitrogen.

3.4.2 First component (π–complex between W95 and heme formation) in S305T, S305A with or without SHA

At first component it was predicted from the MO calculation that a direct ET from W95 to heme and back donation took place during the *catalase* reaction as a result of a strong electronic interaction between W95 and heme. As shown in Table 2, the W95 π-orbital in [S305T] became both HOMO and LUMO while heme is just the contrary, exhibiting *catalase* activity. The W95 π-orbital became the HOMO of the complex system in cyanide [R409L] and [S305A], as shown in Tables 2, 3 and 4. The electron orbital is also best explained by the ET from the HOMO of W95 to the LUMO of *C1C* near α–meso heme edge. Although a transition between W95 orbital and *C1C* atomic orbital near the point of attachment was a strong contributor to the lower-energy level, when the accepting orbital on the δ–meso heme edge had no contribution from W95, including the LUMO of *CHB* as acceptor made a complete preparation for ET from *peroxidase* substrate, therefore no π-π* interaction has favorably shifted from LUMO energy levels of *C1C* to *CHB*, allowing ET to the LUMO of *CHB* (from the HOMO of *ODA* as *peroxidase* substrate to be bound). The *CHB* is poised for nucleophilic attack by the amino nitrogen lone pair of *ODA*.

		WT	WT	S305A	S305A	SHA-S305A	SHA - S305A
	Subunit	A	B	A	B	A	B
LUMO(eV)	SHA					-0.33	-0.51
HOMO(eV)	Heme *CHB*					-7.66	-7.69
HOMO - LUMO gap (eV)						7.33	7.18
Phase						binding	binding
ET						impossible	impossible
LUMO(eV)	Heme *CHB*	-1.73	-1.96	-1.83	-1.75	-1.98	**-2.33**
LUMO(eV)	Heme *C1C*	-1.59	**-2.27**	**-1.34**	**-1.48**	-1.52	-1.47
HOMO(eV)	W95Nε1	-7.92	-8.20	-7.55	-7.69	-7.8	-7.99
The π-π* energy gap (eV)		6.33	5.93	6.21	6.21	6.33	6.52
Phase		bonding	bonding	bonding	bonding	bonding	bonding
ET		possible	possible	possible	possible	possible	possible
LUMO(eV)	G99 Y101 Y218	-2.86	-2.84	-2.52	-2.76	-2.91	-2.87
HOMO(eV)	M244	-7.32	-7.48	-7.15	-7.41	-7.34	-7.88
HOMO - LUMO gap (eV)		4.46	4.64	4.63	4.65	4.43	5.01
Phase		binding	binding	binding	binding	binding	binding
ET		Possible	possible	possible	possible	possible	Possible

Table 4. Computer-calculated parameters representative of the one-electron oxidation between porphyrin ring of Heme and either indole ring of W95 or aromatic inhibitor of SHA.

The calculated energies between HOMO of *ODA* and LUMO of *CHB* show that ET occurs in the [R409L] complex with *ODA* molecule. The transition from the ground to the first excited state and is mainly described by one-electron excitation from the HOMO of *ODA* (•+) cation radical to LUMO of *CHB*. The LUMO ":" of π nature, (i.e. porphyrin ring) is delocalized over the whole C-C bond. The HOMO is located over >C=N (•+) H atom of *ODA* quinonedimine, consequently the HOMO→LUMO transition implies an ET to aromatic part of π-conjugated system from =NH group. Moreover, these orbitals significantly overlap. The HOMO–LUMO energy gap reveals that they are closely bound and both phase matched therefore become lower in energy. The net effect of ET from the HOMO to LUMO must correspond to the π-π* interaction bonds to be broken in the *peroxidase* reaction, instead of the π-bonds between W95 and heme to be made during the *catalase* reaction. In [S305A] variant structure, the potential ET is evident from the energies gap between orbital suggestive of [E (HOMO) =-7.55; -7.68 eV] on the indole nitrogen atom (N*ε1*) of W95 involved with nuclophilic Sr (0.331, 0.329) and [E (LUMO) =-1.34; -1.48 eV] on the carbon atom (*C1C*) electrophilic characteristics associated with nucleophilic superdelocalizability (0.676, 0.675) of the heme in the third component. In contrast, in the SHA-[S305A] complexes it was suggestive of an impossible ET from [E(HOMO)=-7.8; -7.99 eV] on the N*ε1* to [E(LUMO)=-1.47;-1.47eV] on the *C1C* due to deficiency in π–π* interaction formation of first component (Pearson RG. 1986). Despite the success of disrupting the π-complexes between heme and W95, one electron will require to be excited by more than 7eV of energy difference from [E (HOMO) =-7.66;-7.69 eV] on the (C*HB*) carbon of heme edge to [E (LUMO) =-0.33;-0.51 eV] on the SHA without electrophilic Sr, which is of *O*DA in *O*DA-[R409L]. It was guided by MO calculation that the LUMO of SHA and the HOMO of C*HB* of δ-heme edge cannot act as the electron donor, respectively, and the acceptor. It is indicative of MO results that SHA act as inhibitor, when HOMO/LUMO occurs in reverse from heme to SHA against the order from *O*DA to heme.

3.5 [R409L] structure based molecular orbital study with *O*DA and CN

It was predicted by [R409L] structure-based calculation that the new band originated as a result of a strong electronic interaction between *O*DA and *δ*-meso carbon (C*HB*) in δ–heme edge by 5.98–6.3 eV than that of N*ε1* of W95 and C*1C* near α–meso heme edge by 6.01–6.12 eV (Table 3). The former gap energy became more than the latter, increasing the *peroxidase* and reducing the *catalase* activities. If SHA serve as hypothetical substrate despite the original inhibitor, a direct ET from SHA to heme would not able to take place during the *peroxidase* reaction by 7.33–7.18eV (Table 4), since [S305A] cannot have the less gap energy than that of 6.21eV and then exhibits the *catalase* activity, which added a ground-state π-orbital slightly above the valence bond edge. A transition between SHA orbital and C*HB* with strong contributions from Fe (3d) atomic orbitals of heme near the point of attachment may also be too strong contributor to generate the new lower-energy band. SHA acts as inhibitor but not substrate, since it has not electrophile Sr which is essential for electorphilic characteristics of *O*DA to transfer an electron to nucleophilic *CHB* during *peroxidase* reaction process.

When the accepting orbital of C*1C* near the α– meso heme edge, including the LUMO, had no contribution from W95, it can be explained by allowing hole transfer from C*1C* to C*HB* in the δ–heme edge that the C*HB* π-orbital became the LUMO of the combined system and so ET was complete from *O*DA to C*HB* . But the reverse is hardly the case. The interaction has unfavorably changed HOMO and LUMO energy levels, not allowing ET to the LUMO of SHA from the HOMO of the C*HB*. The calculated HOMO of the C*HB* and LUMO of the SHA energies show so high that ET cannot occur in the molecule.

The *peroxidase* reaction has lower activation energy of the phase matched bonding orbital than that of *catalase* in [R409L]. A small HOMO-LUMO energy gap reveals a more reactive compound of *peroxidase* substrate ligand than that of covalent adduct, which antibonding orbital has out -of -phase interaction and then are higher in energy than bonding orbitals that combine to produce binding interactions. A compound of substrate ligand as *O*DA with a small HUMO-LUMO energy gap could be considered as a soft base, since the electrophilc Sr of ODA (0.372, 0.352) has also the greater than that of Nε1 in W95 (0.275, 0.281) whereas the C*HB* (0.863, 1.017) exhibits much greater nucleophilic Sr than that of C*1C* (0.636, 0.612).

This three-part structural feature may generally occur in KatG molecules, over quite long distances from M244 via W95 to C*HB* at the heme edge, where an electron is localized by disrupting the π-conjugated system between Nε*1* of W95 and C*1C* near the heme edge. Then C*HB* become the LUMO orbital to accept an electron from *peroxidase* substrate. The energy gap of C*HB* in the transition- iron-based heme is determined by the π-π* energy difference from property inherent in *peroxidase* substrate, but cannot be tuned regardless of π-π* splitting between W95 and heme. Therefore, the inhibitor ligand as SHA attached to the transition iron is relatively small. The π-conjugated system of SHA can be made smaller by including ferryl oxygen [Fe^{4+}=O] within the conjugated system of heme than that of as *O*DA. The ferryl oxygen lowers the excited-state energy owing to a stronger electron affinity relative to carbon. Small π-chromophores such as SHA are used to inhibit resting (π-ground) and compound II (π*-excited) states of heme *peroxidase*. It involves a direct transition from the π-ground state into the heme, by passing the π*-excited state, which is high in energy.

In the *catalase* reaction, it would be, for instance, an electron in the HOMO of W95 is transferred to the LUMO of the heme by way of the "hot spot" between heme (C*1C*) and W95(Nε*1*). In the *peroxidase* reaction, the HOMO→LUMO transition implies an ET from the HOMO of *O*DA as a donor molecule to the LUMO of the heme (C*HB*) as acceptor. An ET prone part of heme and ODA is common in site, where has the large gap and is of opposite phase in LUMO orbital to SHA as inhibitor and the *O*DA*red* as product. The pathway is suggested by suppression of ET from the HOMO of C*HB* to the LUMO of SHA and the *O*DA_{red}, of course, as the contradiction is untrue. The hydroxamic acid oxygen atom (-NH-OH) of the SHA is thought to be of [-0.33 for subunit A and -0.51 eV for B] as LUMO electron acceptor but in fact acts as HOMO nucleophilic Sr (0.345, 0.358) rather than LUMO electrophilic Sr (0.286, 0.240). The C*HB* in the δ-meso carbon and ferryl oxygen por (•+) cation radical is of [-7.66 – -7.69eV] which is considered to be electron donor as HOMO nucleophilic Sr (0.860, 0.766). It should give the intermolecular space to the possible ET mechanism of *peroxidase* reaction that distance between the -NH-OH oxygen atoms of SHA and the nitrogen atom of cyanide heme Fe is shorter than those the ferryl oxygen atom of Fe^{IV}=O of CCP(Itakura et al.,1997) .

The distance between the hydroxamic acid oxygen atom of aromatic donor and the oxygen atom of Fe^{IV}=O is 1.5–1.6Å in KatG, 3.3Å of BHA and 3.4Å of SHA in ARP. These distances are shorter than those of cytochrome C system or CCP (Poulos et al., 1993). Therefore SHA cause to inhibit the KatG.

3.6 [M244A] structure based docking and frontier orbital studies with H_2O_2, SHA and *O*DA

Both k_{cat} (the rate of turnover) and K_m values (affinity) for H_2O_2 to [M244A] cannot be detected from kinetic study (Ten-i et al., 2007; SATO et al., 2011d). From the docking study, however, H_2O_2 binding affinity can be estimated and the proposal space (§3.1 *vide supra* in Fig.3 (A)) among three target residues can be predicted as W95, H96 and D125 from the docking energies which each subunit A and B is of (-30.3kcal/mol; -23.9kcal/mol) for W95, (-29.9 kcal/mol; -18.2 kcal/mol) for H96 and (-32.1 kcal/mol;-30.3kcal/mol) for D125. It is suggested by the structure-based calculation that [M244A] variant has almost the same as affinity for H_2O_2 between subunit A and B, respectively.

A calculated binding energy of subunit A for -76.5 kcal/mol also reveals to have significantly 2-fold higher affinity for inhibitor SHA than that of B for -30.9 kcal/mol. In subunit A, SHA acts as an inhibitor of compound II with respect to a substrate aromatic donor. It is conceivable that SHA behaves as a substrate of compound II but does not imply the competitive binding with other donor such as *O*DA due to binding at the different sites in reactions with compound II intermediate. In the subunit B, SHA bind to the resting state when an ionisable carbonyl group of D125 with pKa value of 4 is not protonated. It is deduced that SHA also inhibits compound I formation by retarding H_2O_2 binding the iron. A possible inhibitor binding site has been proposed for target residue D125, in a cavity on the distal side of the δ- meso heme edge as shown in Fig.5.

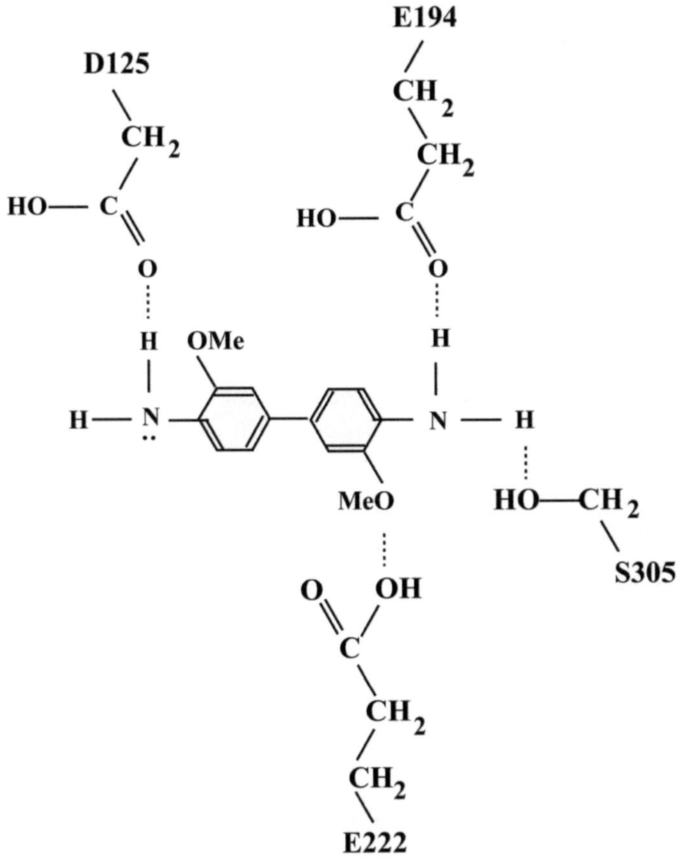

Fig. 4. The Bind Site Predicted from Docking Calculation with *O*DA and Confirmed X-ray Structure of Complex.

As shown in Fig.4, the proxy end N atom of the amino side chain in substrate *ODA* is H-bonded with the Oδ2 atom of D125 in the heme distal pocket. The opposite end N atom forms salt bridge with the side-chain of E194, the carboxyl group of E222 and the carbonyl oxygen atom of backbone S305, with the flexible side chain and without distortion of main chain after binding *ODA*. The *ODA* affinity would be increased by the displacement of the side chain of D125 and E194, two active residues on LL1 loop. Oxidation of *ODA* at nitrogen atom of quinoneimine groups can be also shown by reaction 6 as illustrated in scheme 1.

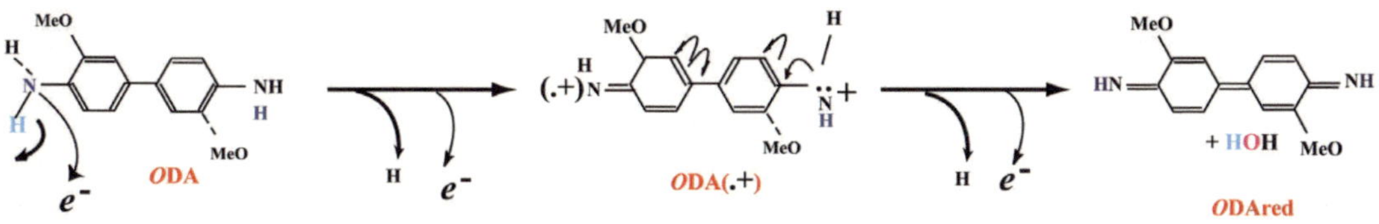

Scheme 1.

Indeed the specific enhancements in *peroxidise* isoenzyme pattern can be influenced by the *ODA* affinity difference between each subunit. The binding site of another *peroxidise* substrate as *ODA* cation radical (*ODA* (·+) had been estimated with -46.6 kcal/mol from docking calculations against D125 for subunit A but expected from repulsive energy for subunit B in Table 5. Thus it can be postulated that the active site of subunit A exhibits the higher affinity for the imino (>C=NH) group of *ODA* (·+) and deprotonate the amino ($-NH_2$) group of *ODA* more efficiently than that of the subunit B. The difference of catalytic efficiency in M244A here is of the 18 fold higher subunit A than that of B (Sato et al., 2011d). It has established that subunit B may be correlated with the slow rate of Compound I formation and the fast rate of reduction of Compound II. Subunit A occur the reduction of compound II and can release the product molecule as ODA_{red} resulting from complete reaction of the *peroxidase*, if subunit A of compound II may adsorb *ODA* (·+). Subunit B cannot has the *catalase* reactivity but regenerate compound I, since its compound II is insensitive to H_2O_2 and desorbed *ODA* (·+).

M244A						
Subunit	A	B	A	B	A	B
	initiator		inhibitor		substrate	
Ligand	H_2O_2		SHA		*ODA*	
Target residue						
W95 (kcal/mol)	-30.3	-23.9				
H96 (kcal/mol)	-29.9	-18.2				
D125 (kcal/mol)	-32.1	-30.3	-76.5	-30.9	-46. 6	repulsive
Heme (kcal/mol)	-26.6	-23.9	-69.3	-26.9		

Table 5. Binding Energies for *Hm*KatG [M244A] Variant Associated with Initiator H_2O_2, Inhibitor SHA and *Peroxidise* Substrate *ODA*.

As shown in Table 6, it is supported by the frontier orbital calculation for M244A that *catalase* activity lost when ET cannot complete between *C1C* carbon and N*ε1* nitrogen atom. Though the *C1C* carbon atoms [for subunit A of -0.96eV and B of -1.27eV] of the heme is of LUMO orbital balanced with existing nucleophilic Sr (0.523,0.771), the *C1C* can shrink neither less than 3.3 Å nor always link the *π-π* interaction to W95, if W95 N*ε1* (·+) cation radical show transient HOMO orbital on the covalent adduct of Y218–W95 [-7.57eV, respectively, and -7.62eV] with electrophilic Sr (0.462, 0.518), since both energy gaps exceed the capacity of ET over 6 eV. C*HB* carbon atoms of heme are of LUMO in subunit A [-1.87eV] and B [-1.39V] and also exhibit nucleophilic Sr (1.113, 0.924). It is possible to recognize as the *peroxidase* substrate with the presence of the *O*DA. However, without the *O*DA in especially subunit B of M244A, Compound I and W95 were elaborated by ET to W95 (•+) cation radical with electrophilc Sr from Por(•+) via п-complex between them. It is suggested that the *C1C* coexist with HOMO density, electrophilc fr and electrophilic Sr. When electron donation to the W95(•+) cation radical is made from (•-)*C1C* atom on Por (•+), compound II reverts to compound I. The working hypothesis of the present study therefore includes the assumption that outcomes of the structure-based calculations can be used as a role to assist reproduction of compound I in *peroxidase* reaction.

			M244A			
	Subunit		A		B	
HOMO(radical)	Y218- W95	Nε1(·+)	-7.57eV		-7.62eV	
LUMO+1	Heme	*C1C*	-0.96eV		-1.27eV	
The energy gap & distance			6.61eV	3.9Å	6.35eV	3.3Å
Phase & ET			bonding	impossible	bonding	possible
LUMO(nucleophilic)	Heme	C*HB*		-1.87eV		-1.39V

Table 6. Each Subunit of MO Energy in the п-complexes with Y218-W95 covalent adduct of *Hm*KatG [M244A] Variant

Geometry, which used for the calculations of its model compounds I and II ,was consist of 15 residues which included in R92, W95, H96, G99, T100, Y101, D125, N126, I217, Y218, A244, H259, S305, R409, and heme. In subunit B, the orbital between *C1C* Heme and N*ε1* W95 become the binding orbital (in green and yellow) and promotes the bonding of п-system of indole rings of Trp and pyrole ring of heme, which may be reclaiming compound I. In subunit A, the two atoms between N*ε1* W95 and *C1C* Heme is distantly-positioned from 3.9Å, which opposes the п-bonding interaction and cannot transfer electrons, even if each orbital (in red and blue) are attractive and may normally overlap between the bonding molecular orbital. Therefore the reduction of compound II would occur when *O*DA (·+) cation radical bound to C*HB*.

Despite a lack of *peroxidase* substrate in [M244A] variant, no *catalase* reactivity against the second H_2O_2 exhibits at all so that the electron cannot transfer from M244 to the covalent adducts W95 via Y218. In spite of a п-п* electron interaction of the heme with W95 in the covalent adduct, the ability to transfer electron between an electrophile of tyrosinate of

Y218 and the nucleophile of sulfur cation was lost by the deletion mutation at the position 244. Therefore KatG is considered the *catalase* function to use a methionine nucleophile intramolecularly- and octahedrally-coordinated complexes with the carbonyl O atoms of Y101 and G99 as well as known to use tyrosinate-indole electrophiles (Carpena et al., 2005). Crystallographic analysis of *Hm*KatG [M244A] variant as indicated that the Y218 could move its side chain closer to the indole group of W95 in subunit B (SATO et al., 2011d). It was suggested that oxygenation of heme (compound I) was activated in ET where heme formed π-complexes with W95 in the covalent adduct, and reduction of ferryl-intermediate by *O*DA was activated in resting (ferric) KatG. This suggestion is supported by MO calculations for [M244A] that compounds I and II would be sufficient to initiate substrate oxidation as *O*DA. As summarized in the scheme shown in scheme 2, a proposed mechanism for twice the one-electron oxidation of *O*DA substrate to the radical product (*O*DA$\cdot^{+}$) by KatG intermediate is presented as follows: one electron could transfer from compound II (*A2*) reduction back to the resting state (*C1*) and another electron could be attribute to converse compound I (Fe^{IV}=O) (*B1*) to the compound II (Fe^{IV}-OH) formation (*A1*) in Subunit A. It is common to both subunit A and B. While exhibiting the isoenzyme pattern of *peroxidase* reaction, it is indicated as the bifurcated pathway that this recoverable form of compound I (*B2*) occurs directly from the activated form of compound II (*A1*) in Subunit B. It has proposed for KatG that M244A mutation effectively follows the *peroxidase* dual pathway.

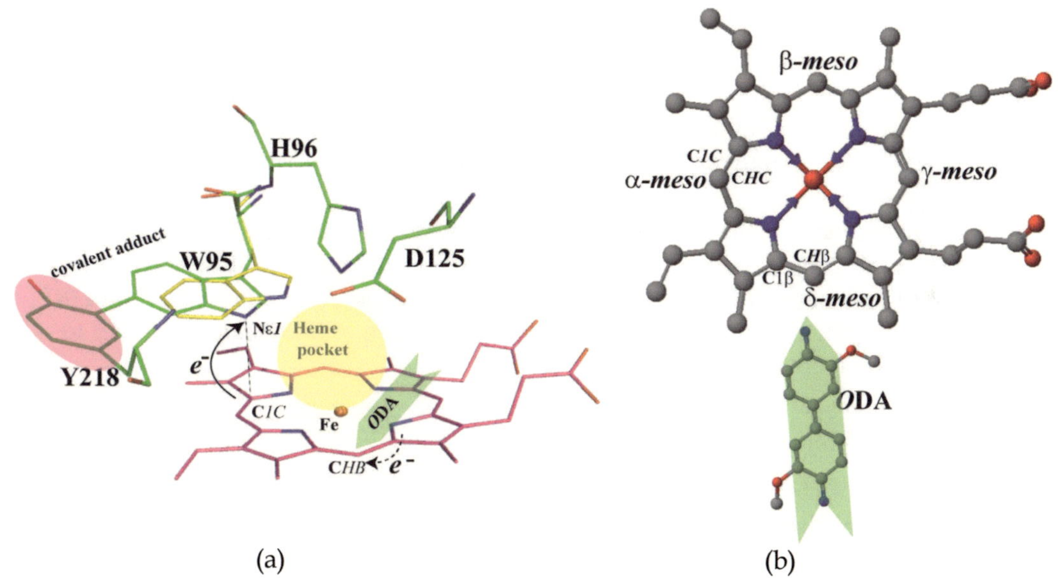

Fig. 5. Electron Transfer in iron Protoporphyrin IX (heme) and the Adduct or *O*DA. View showing the stacking of the indole ring of W95 on 3.25Å above pyrol ring C of heme (left panel). The intramolecular ET from porphyrin to the covalent adduct can be explained by the formation π-complexes with W95 ($^{\cdot+}$) and heme, having reverted to compound I from the compound II with no substrate. If *O*DA is bound to *C*HB of δ-heme edge, it allows electron from *O*DA to the heme to promote oxidation by substrate *O*DA. Structure of heme and *O*DA (right panel). This nomenclature is consistent with using in crystallographic work. The figures were constructed using Pymol (DeLano, 2002).

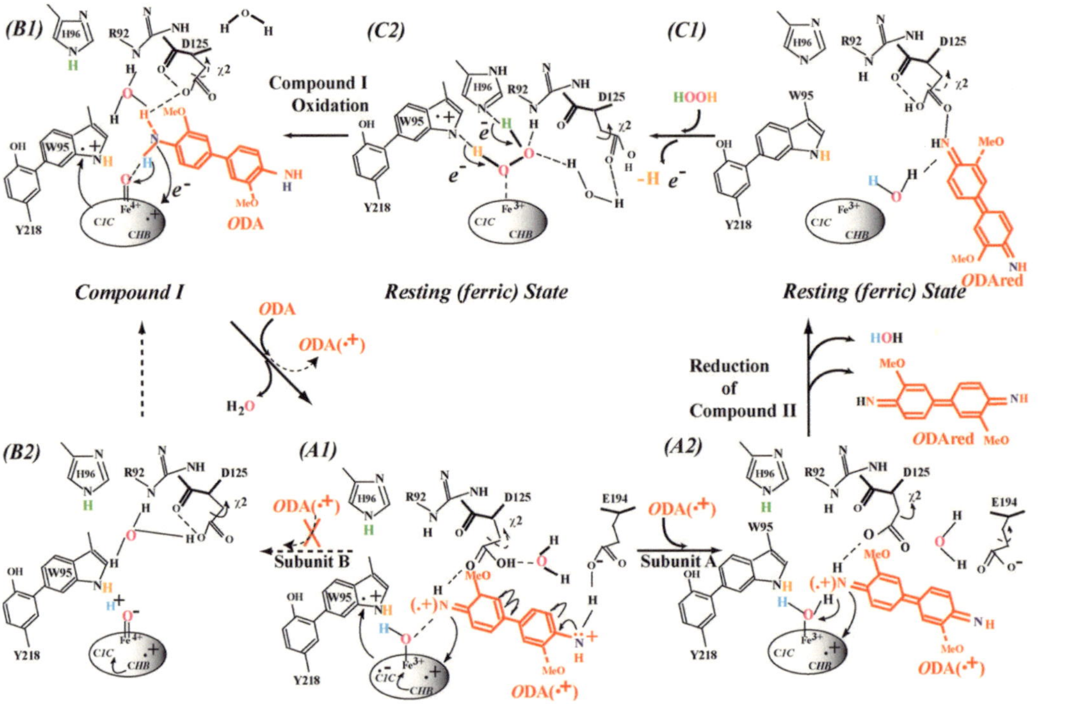

Scheme2. Summary of the Effect of the [M244A] Mutation on the Scheme Combined Compound I Regeneration of Subunit B and the Reduction of Compound II of Subunit A. Showing the involvement of key active site residues of [M244A] variant, compound II can decay into one of two active forms. The Compound II in the active center of subunit B can be converted back to compound I (*B2*) without rebinding the inverted *O*DA(·+) cation radical, proceeding to the left, but the *peroxidase* reaction, proceeding to the right from Compound II (*A1*) in the active center of subunit A with binding inverted *O*DA(·+) results in conversion to resting state (*C1*). The initiator H_2O_2 molecule involved in the Compound I oxidation (*B1*), and the substrate *O*DA (·+) involved in the reduction of compound II (*A2*) are shown in potential H-bonded conformations.

4. Conclusion

Crystal structure of *Hm*KatG has been first determined by 2.0Å resolution and reported the proposal structural characteristics of covalent adduct between Sδ of M244–Cε2 of Y218 and Cε1 of Y218–Cη2 of W95 in maintaining the *catalase* activity of the KatG (Yamada et al., 2002). The residues which make up the active site catalytic triad are highly conserved in all KatGs, identified both through functional studies and by comparison with homologues identified in sequenced genomes. X-ray analysis of other KatGs, including *Mt*, indicates that the corresponding residues are also arranged in a similar catalytic triad. Crystal structure analysis has demonstrated that the Met variant significantly affects W95 in the protein structure on the distal side of the heme since slight changes were observed for the puckering (shrinking and stretching of the interatomic bonds between) heme *C1C* –N*ε1* of indole ring as compared to the WT protein (Heering et al., 2002; Santoni et al., 2004).

However, unlike these Met variants from other sources, in the *Hm*KatG [M244A] variant the *catalase* activity was not detected at all but *peroxidase* activity was even enhanced. All the reported kinetic constants (rate of turnover, k_{cat}, the affinity of substrate, K_m, and catalytic efficiency, k_{cat}/K_m) of otherwise KatG for *catalase* and *peroxidase* activities are "apparent" values as conventional *Catalase* and *Peroxidase* do not follow typical Michaelis-Mentan kinetics. Kinetic parameters for *catalase* and *peroxidase* are determined by fitting the kinetic data to non-linear (mixed) Michaelis-Menten equation and show that isoenzyme pattern of active two catalytic center motifs typical of *catalase* and *peroxidase*. That's the reason why KatG is a functional heterodimer in governing KatG dimeric subunits structure.

Crystal structure of the [M244A] variant of *Hm*KatG was of not identical subunits (SATO et al., 2011d). Subunit A disrupted a possible π-interaction between W95 and heme. Including the differences in active site geometry, it would be sufficiently stronger to facilitate the oxoferryl (Fe (IV) =O) reduction in the *peroxidase* reaction. And back donation of electron from heme edge to W95 would suffice for compound I revitalization in subunit B, suggesting from HOMO/LUMO energy gap and nucleophilic Sr of N*ε1* based on S305T structure (SATO et al., 2011a).

Characterization of the electron path may identify each site of electrophilic W95 N*ε1* and nucleophilic heme *C1C* in *Hm*KatG and influence of heme *CHB* on *peroxidase* substrate reduction. Further functional studied with the use of X-ray crystal structure based calculation has given to better understanding of ET pathways and radical formation sites in KatG. This approach was investigated how frontier molecular orbital theory applies to either the inhibitory of a SHA (SATO et al., 2011c; to be submitted) or the reactivity of *O*DA compounds (SATO et al., 2011b; to be submitted) with, respectively, single and double phenolic ring. For SHA binding, the side-chain of D125 is rotated around the dihedral angle $\chi 1$ by 100.8° whereas *O*DA binding affinity was enhanced by $\chi 2$ of 61.3°of D125 residue. On the contrary, in subunit B, the consequent ET from heme to W95 could explain the enhancement of *peroxidase* and iteratively-generated compound I intermediate. The isoenzyme pattern of *peroxidase* was discussed in terms of its hetero-dimeric character of *peroxidatic* subunit A for reduction of compound II and subunit B for reclaiming compound I. The value of catalytic efficiency ($k_{\mathrm{cat}}/K_{\mathrm{m}}$) for the *peroxidatic* reaction catalyzed by the *Hm*KatG [M244A] variant falls within the expected range for an efficient enzyme (Albery & Knowles, 1976).

The M244–Y218–W95 covalent adduct confirms to be essential for the *catalase* activity in *Hm*KatG as well as *Mt*KatG (Ghiladi et al., 2005b; 2005c). It was also constructed to explore the effect of successive triple-base substitutes for Met244 to Ala and to cleavage the covalent bond amongst the tripeptide. The [M244A] variant that coupled with the structure based-evolution within laboratory time scale is not biochemically associated with INH susceptibility. Despite of INH resistance-conferring variants, this "unnatural" protein engineering for *Hm*KatG, can confirm the inherent *catalase* functional capability of the *C-ter* end of *E- helix* in KatG. Perhaps the most intriguing feature of the *Mt*KatG is its ability to mediate INH susceptibility. In our current working hypothesis for the cause of isoenzyme pattern of *peroxidase* activity, it is suggest that the drug interacts with the enzyme and is converted by the *peroxidase* activity into a toxic derivative which acts at a second, as yet unknown, site (Zhang,Y., et al., 1992).

Clearly, further experiments are required to address this hypothesis. For a better understanding of the complex interrelations between *catalase* and *peroxidise,* the oxidation of phenols, in the present paper some studies on the kinetic characterization of this enzyme in bacterials has been carried out with the detection of KatG isoenzyme patterns. Whereas *C-ter* capping by positively charged M244 residues in *E-helix* is effective to neutralize the helix dipole which lead to destabilization of the helix through entropic effect. Since it may be less effective to neutralize the *B-helix* dipole, the electron can transfer to carbonyl oxygen atoms of *C-ter* capping with G99 and Y101 residues in *B-helix* from heme via the W95–Y218 covalent linkage adduct with M244 centred octahedral coordinate. Just as *O*DA analogous compounds are bound D125, E194 and E222 on mobile LL1 loop, as the destabilization or cleavage of the covalent adduct linkage between M244 and Y218 may be caused by the mobile Y218 which located on the upstream residues of LL1 loop in binding site. Since the competitive inhibition of *O*DA substrate may imply in binding SHA inhibitor (as hypothetical INH), the right candidates of the INH may be narrowed down and selected out in a great number of these lead compounds, if the higher binding and lower activation energies for ET during *peroxidase* reaction are predicted by further docking and additional *ab initio* MO calculations based on the structure of binding site of INH-sensitive (*peroxidase*) KatG variants. *Peroxidases* are highly polymorphic enzymes, and the functionality of each isoenzyme depends on its (acidic) nature and its persistent growth phase of the *Mt* clinical strains. In order to facilitate the *peroxidase* activity and to understand the metabolic functions that are needed for the persistence of *Mt,* compounds that could eradicate persisters effectively can be useful in the design of HIV/anti-tuberculosis drugs.

5. Acknowledgements

The author wish to thank Mr. W. Higuchi for excellent technical assistance, Dr. K. Yoshimatsu for preparing *Hm*KatG variants and Prof. T. Fujiwara for providing recombinant KatG from *Haroarcula marismortui* from the Department of Biological Sciences at Shizuoka University. I thank also Gorana Screrbe for her expert secretarial assistant and for helpful discussions during the preparation of this manuscript.

6. References

Albery, E.J., and Knowles, J.R. (1976) Evolution of enzyme function and the development of catalytic efficiency. *Biochemistry* Vol.15, pp. 5631-5640, ISSN 0006-2960

Bertrand, T., Eady, N.A.J., Jones, J.N., Jesmin, J.M., Nagy, J.M., Jamart-Gregoire, B., Raven, E.L., Brown,K.A. (2004) Crystal structure of Mycobacterium tuberculosis *catalase-peroxidase*. The *Journal of Biological Chemistry* Vol.279, pp.38991-38999, ISSN 0021-9258

Carpena, X., Wiseman, B., Deermagarn, T., Singh, R., Switala, J., Ivancich, A., Fita, I., and Loewen,P.C. (2005) A molecular switch and electronic circuit modulate *catalase* activity in *catalase-peroxidases EMBO Report* Vol.6, pp.1156-1162, ISSN 1469-221X

DeLano, W. L. (2002) The PyMOL Molecular Graphics System, DeLano Scientific, Palo Alto, CA

Dewar M.J.S., Zoebisch E.G., Healy E.F., and Stewart, J.J.P. (1985) Development and use of quantum mechanical molecular models. 76. AM1: a new general purpose quantum mechanical molecular model. *Journal of the American Chemical Society* Vol.107, pp.3902-3909, ISSN 0002-7863.

Dobner P., Rüsch-Gerdes S., Bretzel G., Feldmann K., Rifai M.,Löscher T., and Rinder H. (1997) Usefulness of *Mycobacterium tuberculosis* genomic mutations in the genes katG and inhA for the prediction of isoniazid resistance. Int J Tuberc Lung Dis Vol.1, pp365-369, ISSN 1027-3719

Fukui, K., Yonezawa, T., and Nagata, C. (1954) Theory of Substitution in Conjugated Molecules. *Bulletin of the Chemical Society of Japan*, Vol.27, pp.423-427, ISSN 0009-2673

Fukui, K., Yonezawa, T., and Nagata, C.(1957) MO-Theoretical Approach to the Mechanism of Charge Transfer in the Process of Aromatic Substitutions. *The Journal of chemical physics*, Vol.27, pp.1247-1259, ISSN 0021-9606

Fukuzawa, K., Kitaura, K., Nakata, K., Kaminuma, T., and Nakano, T. (2003) Fragment molecular orbital study of the binding energy of ligands to the estrogen receptor. *Pure and Applied chemistry*, Vol.75, pp.2405-2410, ISSN 1365-3075

Ghiladi, R.A., Cabelli, P.R., and Ortiz de Montellano,P.R. (2004) Superoxide reactivity of KatG: Insights into isoniazid resistance pathways in TB. *Journal of the American Chemical Society* Vol.126, pp.4772-4773, ISSN 0002-7863

Ghiladi, R.A., Medzihradszky,K.F., Rusnak,F.M., and Ortiz de Montellano,P.R. (2005a) Correlation between isoniazid resistance and superoxide reactivity in Mycobacterium tuberculosis KatG. *Journal of the American Chemical Society* Vol.127, pp.13428-13442, ISSN 0002-7863

Ghiladi, R. A., Knudsen, G. M., Medzihradszky, K.F., and Ortiz de Montellano, P. R. (2005b) The Met-Tyr-Trp Cross-link in *Mycobacterium tuberculosis Catalase-peroxidase* (KatG). *The Journal of Biological Chemistry* Vol. 280, pp. 22651–22663, ISSN 0021-9258

Ghiladi, R. A., Medzihradszky, K. F., and Ortiz de Montellano, P. R. (2005c) Role of the Met-Tyr-Trp Cross-Link in *Mycobacterium tuberculosis Catalase-Peroxidase* (KatG) As Revealed by KatG(M255I). *Biochemistry*, Vol.44, pp.15093-15105, ISSN 0006-2960

Heering, H. A., Indiani, C., Regelsberger, G., Jakopitsch, C., Obinger, C., and Smulevich, G. (2002) New Insights into the Heme Cavity Structure of *Catalase-Peroxidase*: A Spectroscopic Approach to the Recombinant *Synechocystis* Enzyme and Selected Distal Cavity Mutants., *Biochemistry* Vol.41, pp. 9237–9247, ISSN 0006-2960.

Haas W.H., Schilke K., Brand J., Amthor B., Weyer K., Fourie P.B., Bretzel G., Sticht- Groh V., and Bremer H.J. (1997) Molecular analysis of katG gene mutations in strains of *Mycobacterium tuberculosis* complex from Africa. *Antimicrob Agents Chemother* Vol.41, pp.1601-1603, ISSN 0066-4804

Heym, B., Zhang, Y., Poulet, S., Young, D., and Cole, S. (1993) Characterisation of the KatG gene encoding *catalase-peroxidase* required for the isoniazid susceptibility of Mycobacterium tuberculosis. *Journal of Bacteriology.*, Vol.175, pp.4255-4259, ISSN 0021-9193

Heym B., Alzari P.M., Honoré N., and Cole S.T. (1995) Missense mutations in the *catalase-peroxidase,* katG are associated with isoniazid resistance in *Mycobacterium tuberculosis. Molecular microbiology* Vol.15, pp.235-245, ISSN 0950-382X

Hillar, A., Peters, B., Pauls, R., Loboda, A., Zhang, H., Mauk, A.G., and Loewen, P.C. (2000) Modulation of the Activities of *Catalase–Peroxidase* HPI of *Escherichia coli* by Site-Directed Mutagenesis. *Biochemistry* Vol.39, pp.5868-5875, ISSN 0006-2960

Ikeda-Saito M., Shelly D.A., Lu L., Booth K.S., Caughey W.S. and Kimura S. (1991) Salicylhydroxamic acid inhibits myeloperoxidase. *J.Biol.Chem.* Vol.266, pp.3611-3616, ISSN 0021-9258

Itakura H., Oda Y., Fukuyama K. (1997) Binding mode of benzhydroxamic acid to arthromyces ramosus *peroxidase* shown by X-ray crystallographic analysis of the complex at 1.6 Å resolution. *FEBS Letters* Vol.412, pp.107-110, ISSN 0014-5793

Ivancich, A., Jakopitsch, C., Auer, M., Un, S., and Obinger, C. (2003) Protein-Based Radicals in the *Catalase-Peroxidase* of *Synechocystis* PCC6803: A Multifrequency EPR Investigation of Wild-Type and Variants on the Environment of the Heme Active Site. *Journal of the American Chemical Society* Vol. 125, pp. 14093–14102, ISSN 0002-7863

Jakopitsch, C., Auer, M., Regelsberger, G., Jantschko, W., Furtmüller, P.G., Ruker, F., and Obinger, J. (2003a) Distal site Aspartate is Essential in the *Catalase* Activity of *catalase-peroxidases*. *Biochemistry* Vol.42, pp.5292-5300, ISSN 0006-2960

Jakopitsch, C., Auer, M., Ivancich, A., Ruker, F., Furtmüller, P.G., and Obinger, J. (2003b) Total conversion of bifunctional *catalase-peroxidase* (KatG) to Monofunctional *peroxidase* by exchange of a conserved distal side tyrosine. *Journal of Biological Chemistry* Vol.278, pp.20185-20191, ISSN 0021-9258

Jakopitsch, C., Ivancich, A., Schmuckenschlager, A., Wanasinghe, A., Furtmüller, P.G., Ruker, F., and Obinger, C. (2004) Influence of the unusual covalent adduct on the kinetics and formation of radical intermediates in Synechocystis *catalase peroxidase*: A stopped-flow and EPR chracterization of the Met275, Tyr249, and Arg439 variants. *Journal of Biological Chemistry* Vol.279, pp.46082-46095, ISSN 0021-9258

Marttila H J., Soini H., Eerola E., Vyshnevskaya E., Vyshnevskiy B. I., Otten T. F., Vasilyef A. V., and Viljanen M. K. (1998) A Ser315Thr substitution in KatG is predominant in genetically heterogeneous multidrug-resistant *Mycobacterium tuberculosis* isolates originating from the St Petersburg area in Russia. Antimicrob Agents Chemother Vol.42, pp.2443-2445, ISSN 0066-4804

Modi S., Behere D.V., and Mitra S. (1989) Binding of aromatic donor molecules to lactoperoxidase: proton NMR and optical difference spectroscopic studies. *Biochim.Biophys.Acta* - Proteins and Proteomics Vol.996, pp.214-225, ISSN 1570-9639.

Musser J.M., Kapur V., Williams D. L., Kreiswirth B. N., van Soolingen D., van Embden J.D.A. (1996) Characterization of the *catalase-peroxidase* gene (katG) and inhA locus in isoniazid-resistant and susceptible strains of *Mycobacterium tuberculosis* by automated DNA sequencing: restricted array of mutations associated with drug resistance. *The Journal of Infectious Diseases* Vol.173, pp.196-202, ISSN 0022-1899.

Pearson R.G. (1986) Absolute electronegativity and hardness correlated with molecular orbital theory. *Proceedings of the National Academy of Sciences of the United States of America* Vol.83, pp.8440-8441, ISSN 0027-8424

Petersen M.R. and Csizmadia I.G. (1979) Determination and analysis of the formic acid conformation hypersurface. *J.Amer.Chem.Soc.* Vol.101, pp. 1076-1079, ISSN 0002-7863

Poulos, T.L., Edwards,S.L., Wariishi,H., and Gold,M.H. (1993) Crystallographic refinement of lignin *peroxidase* at 2Å. *Journal of Biological Chemistry*, Vol. 268, pp.4429-4440, ISSN 0021-9258

Regelsberger, G., Jakopitsch, C., Rüker, F., Krois, D., Peschek, G. A., and Obinger, C. (2000) Effect of Distal Cavity Mutations on the Formation of Compound I in Catalase-Peroxidases. *Journal of Biological Chemistry*, Vol. 275, pp.22854-22861, ISSN 0021-9258

Saint-joanis, B., Souchon, H., Wilming, M., Johnsson, K., Alzari, P.M., and Cole, S.T.(1999). Use of site-directed mutagenesis to probe the structure, function and isoniazid activation of the *catalase/peroxidase*, KatG, from *Mycobacterium tuberculosis. Biochemical Journal* Vol.338, pp.753-760, ISSN0264-6021.

Santoni, E., Jakopitsch, C., Obinger, C., and Smulevich, G. (2004) Manipulating the covalent link between distal side tryptophan, tyrosine, and methionine in *catalase-peroxidases*: An electronic absorption and resonance Raman study. *Biopolymers* Vol.74, pp. 46–50, ISSN 1097-0282.

Sato, T., Higuchi, W., Yoshimatsu K., and Fujiwara, T. *Catalase* Functional Implication Guided by Structure, Docking, Molecular Orbital and Kinetics Studies on Ser305Thr and Arg409Leu variants of *Hm*KatG (2011a; to be submitted)

Sato, T., Higuchi, W., Yoshimatsu K., and Fujiwara, T. Crystal Structures and *Peroxidatic* Function of Cyanide Arg409Leu (R409L) Variant and its Complexes with *o*-Dianisidine (*O*DA) in KatG from *Haloarcula Marismortui*. (2011b; to be submitted)

Sato, T., Higuchi, W., Yoshimatsu K., and Fujiwara, T. Crystal Structure and Kinetics Studies on Ser305Ala variant of KatG from *Haloarcula Marismortui* and its Complexes with Inhibitor SHA (2011c; to be submitted)

Sato, T., Ten-i, T., Higuchi, W., Yoshimatsu K., and Fujiwara, T. Crystal Structure and Kinetic Studies on Met244Ala [M244A] Variants of KatG from *Haloarcula Marismortui*. (2011d; to be submitted)

Singh, R., Wiseman,B., Deemagarn,T. Donald,L.J., Duckworth,H.W., Carpena,X., Fita,I., and Loewen,P.C. (2004) *Catalase-peroxidase* (KatG) exhibit NADH oxidase Activity. *Journal of Biological Chemistry* Vol.279, pp.43098-43106, ISSN 0021-9258

Slayden, R.A. and Barry III, C.E. (2000) The genetics and biochemistry of isoniazid resistance in *Mycobacterium tuberculosis. Microbes and Infection* Vol.2, pp.659-669, ISSN 1286-4579.

Stewart J.J.P., (1989a) Optimization of parameters for semiempirical methods. I. Method, *Journal Computational Chemistry*, Vol.10, pp209-220, ISSN 1096-987X.

Stewart J.J.P., (1989b) Optimization of parameters for semiempirical methods. II. Applications, *Journal Computational Chemistry*, Vol.10 , pp221-264, ISSN 1096-987X.

Stewart J.J.P., Mopac 2002 release 2.1 Fujitsu Limited,Tokyo, JAPAN(2002)

Stewart J.J.P., (2007) Optimization of parameters for semiempirical methods V: Modification of NDDO approximations and application to 70 elements. *Journal of Molecular Modeling* Vol.13, pp.1173-1213,ISSN 1610-2940)

Ten-i, T., Kumasaka, T., Higuchi, W., Tanaka, S., Yoshimatsu, K., Fujiwara, T. and Sato, T. (2007) Expression, purification, crystallization and preliminary X-ray analysis of Met244Ala variant of *catalase-peroxidase* (KatG) from the haloarchaeon Haloarcula marismortui. Acta Crystallographica Section F Vol. 63, pp.940-943, ISSN 1744-3091

Wei, C.J., Lei,B., Musser, J.M., and Tu, S.C. (2003) Isoniazid activation defects in recombinant Mycobactrium tuberculosis *Catalase-Peroxidase* (KatG) mutants evident in InhA inhibitor production. *Antimicrobial Agents and Chemotherapy*, Vol.47, pp.670-675, ISSN 0066-4804

Wengenack N. L., Uhl J. R., St. Amand A. L., Tomlinson A. J., Benson L. M., Naylor S., Kline B. C., Cockerill III F. R., and Rusnak F. (1997) Recombinant *Mycobacterium tuberculosis* KatG (S315T) is a competent *catalse-peroxidase* with reduced activity toward isoniazid. *J Infect Dis* Vol.176, pp.722-727, ISSN 0022-1899

Welinder, K.G. (1991) Bacterial *catalase-peroxidases* are gene duplicated members of the plant *peroxidase* superfamily. *Biochimica et Biophysica Acta - Proteins and Proteomics*Vol.1080, pp.215-220, ISSN 1570-9639.

Welinder, K.G. (1992) Superfamily of plant, fungal and bacterial *peroxidases*. *Current Opinion in Structural Biology* Vol.2, pp.388-393, ISSN 0959-440X

Yamade,Y., Fujiwara,T., Sato,T., Igarashi,N., and Tanaka,N. (2002) The 2.0Å crystal structure of *catalase-peroxidase* from *Haroarcula marismortui*. *Nature Structural Biology* Vol.9, pp691-695, ISSN10728368

Zhang, Y., Heym,B., Allen,B., Young,D., and Cole,S. (1992) The *catalase-peroxidase* gene and isoniazid resistance of Mycobacterium tuberculosis. *Nature* Vol.358, pp.591-593, ISSN 0028-0836

Part 2

Vitamins & Antioxidants

Effects of Dietary Fortification of Vitamin A and Folic Acid on the Composition of Chicken Egg

M.J. Ogundare and S.A. Bolu
Department of Animal Production, University of Ilorin, Ilorin Nigeria

1. Introduction

Micronutrient Malnutrition (MNM) is widespread in the industrialized nations, but more prominent in the developing regions of the world. It affects human of all age groups, but young children and women of reproductive age tend to be among the high-risk groups of micronutrient deficiencies. Micronutrient malnutrition has many adverse effects on human health, some of which, are clinically evident. At moderate levels of deficiency (which can be detected by biochemical or clinical measurements) MNM can have serious detrimental effects on human biological functions. Thus, in addition to the obvious and direct health effects, MNM has profound implications for economic development and productivity, particularly in terms of the potentially huge public health costs and the loss of human capital formation.

Fortification is the practice of deliberately increasing the content of an essential micronutrient, i.e. vitamins and minerals (including trace elements) in a food, so as to improve the nutritional quality of the food supply and provide a public health benefit with minimal risk to health (Lindsay et al., 2006). Fortification of food with micronutrients is a valid technology for reducing micronutrient malnutrition as part of a food-based approach. Food fortification is particularly imperative when and where existing food supplies and limited access fail to provide adequate levels of the respective nutrients in the diet. In such cases, food fortification reinforces and supports ongoing nutrition improvement programmes and should be regarded as part of a broader, integrated approach to prevent MNM, thereby complementing other approaches to improve micronutrient status. The public health significance of food fortification is a function of: helping the at-risk population to reach the biological efficacious threshold of micronutrient intake through the consumption of fortified food; and keeping the intake below levels that may cause adverse effects due to excesses (toxicity) on individuals who consume the fortified food in large amounts.

Food fortification is a deliberate addition of one or more micronutrients to particular foods, for the purpose of increasing the intake of these micronutrients and consequently, correct or prevent a demonstrated deficiency. It has the dual advantage of being able to deliver nutrients to large segments of the population without radical changes in food consumption patterns (Lofti, 1996). In developed countries, food fortification has proven an effective and low-cost way to increase the micronutrient supply and reduce the consequences of micronutrient deficiencies (Omar and Jose, 2002).

Food fortification, unlike other forms of MNM interventions does not necessitate a change in dietary patterns of the population, can deliver a significant proportion of the recommended dietary allowances for a number of micronutrients on a continuous basis, and does not demand for individual compliance. Nutrients to be fortified are often be incorporated into the existing food production and distribution system, and therefore, can be sustained over a long period of time. Food fortification is a more cost-effective and sustainable solution. It plays a major role in improving the diet and meeting the micronutrient needs of the population. It is also an integrated food-based strategy; other strategies include dietary diversification, homestead production, development of national food and improved food processing and storage.

The technical considerations in food fortification include selection of appropriate food vehicles that are consumed by a sizeable proportion of the population and lend themselves to centralized processing on an economical scale. The product should be one that is distributed through a wide network so that it reaches all parts of the country.

Food fortification has a long history of use in industrialized countries for the successful control of deficiencies of vitamin A and D, several B vitamins (thiamin, riboflavin and niacin), iodine and iron. Salt iodization was introduced in the early 1920s in both Switzerland (Burgi et al., 1990) and the United States of America (Marine and Kimball, 1920) and has since expanded progressively all over the world to the extent that iodized salt is now used in most countries.

From the early 1940s, the fortification of cereal products with thiamine, riboflavin and niacin became common practice. Margarine was fortified with vitamin A in Denmark and milk with vitamin D in the United States. Foods for young children were fortified with iron, a practice which has substantially reduced the risk of iron-deficiency anaemia in this age group.

In more recent years, folic acid fortification of wheat has become widespread in the Americas, a strategy adopted by Canada and the United States and about 20 Latin American countries (Lindsay et al., 2006). Being a food-based approach, food fortification offer a number of advantages over other interventions aimed at preventing and controlling micronutrient malnutrition (World Bank, 1994).

The burden of malnutrition in Africa and other developing economy is a major global concern. Malnutrition in its various manifestations has resulted loss of productivity. This is a direct is a direct consequence of the effects of lack of essential nutrients, vitamins and minerals in particular (WHO, 2002). Interest in micronutrient malnutrition has increased greatly over the last few years because it contributes substantially to the global burden of disease. More than 2 billion people in the world today suffer from micronutrient deficiencies caused largely by a dietary deficiency of vitamins.

In 2000, the World Health Report identified iron, vitamin A and zinc deficiencies as micronutrients that contributed to serious health risk factors (Lindsay et al., 2006).Vitamin A deficiency is a major global problem affecting populations in developing areas of more than seventy-five countries where clinical and subclinical conditions have been observed (Summer and Davidson, 2002; WHO,2003). It has also been reported that 41% of population under 5 years of age in developing countries suffers from inadequate vitamin A intake. This has resulted to about half a million of children going blind each year and thirteen and half million develop night-blindness yearly (WHO, 2000).

Populations in malaria-endemic areas are known to be at high risk of folate deficiency. In certain parts of Africa, malaria illness during pregnancy has been reported as the most common cause of megaloblastic erythropoiesis (Fleming, 1989). This is due to extensive haemolysis occasioned by malaria as well as, suppression of erythropoiesis. As a result of increased red cell turnover (i.e. erythroid hyperplasia), folate requirement is further increased, especially during pregnancy when folate requirements are already high for maternal and fetal tissue growth (O'Conner, 1994).

Supplements, usually in the form of pills, capsules or syrups had been used to provide relatively large doses of micronutrients. Nationwide supplementation programmes combined to supplement children once or twice a year with high-doses of Vitamin A are combined in at least 43 countries with National Immunization Days (Klemm et al., 1997).

Though supplementation is often the fastest way to control identified deficiencies, it however requires the procurement and purchase of micronutrients in relatively expensive pre-packaged forms, an effective distribution system and a high degree of consumer compliance (especially for water-soluble vitamins prescribed for use on a long term basis). Irregular supplies and poor compliance has been reported by many supplementation programme managers as being the major constraints to success (Lindsay et al., 2006; Maria and Bruno, 2002).

Among the different foods delivering essential nutrients to the human body, egg is important, because it supplies rich and balanced source of essential amino acids, fatty acids as well as some minerals and vitamins (Surai and Sparks, 2001). Research has shown that the levels of many nutrients found in hen's eggs can be altered by altering their level in the hen's diet (Gibson, 2000).

In this study, we manipulated vitamin A and folate fortification in laying hens diet to increase their eggs content. The success of this study may position the egg as an important source of these dietary vitamins and lead to an improvement in consumer acceptance of this commodity as a healthful product. This is of paramount relevance to the vulnerable human groups for these nutrients e.g. children and pregnant women.

2. Materials and methods

Two hundred (200) Black Harco pullets (23 weeks old) were fed a standard layers mash. The pullets were monitored for egg production and health status for 2 weeks. One hundred and forty (140) highest-producing and healthy laying hens were selected from this population for the study. The selected hens were divided into seven groups of five replicates per group. Each replicate containing 4 birds (20 birds per treatment) were randomly allocated to a two tier cage with cell dimensions of 38 cm x 40cm. Birds were given water and feed (Table 1) *ad libitum* for a study period of eight weeks. Vitamin A capsules containing water miscible retinyl palmitate beadlet called "25-CWS" (CWS-cold water soluble) was administered orally to individual birds with the aid of a sterilized syringe while the synthetic feed grade monoglutamate form of folate-folic acid (Dr Meyers) used was incorporated into the diet.

Feed intake and body weight gain were recorded weekly. Hen-day production, feed utilization efficiency were calculated. Weights of egg, albumen, yolk and shell samples were measured. Albumen height was measured with a spherometer, shell thickness with a micrometer screw gauge and egg length and breadth with a vernier caliper. Egg albumen quality (Haugh units) and yolk index were calculated (Bolu and Balogun, 1998)

Vitamin A analysis of four eggs per treatment was carried out using standard HPLC analytical procedures as outlined for vitamin A determination (Nielsen, 2002). The concentration of 5-methyltetrahydrofolate in egg yolk extracts was determined by reverse-phase HPLC with fluorescence detection, as described by Vahteristo et al., (1997).

At the end of the study, blood samples were taken from the wing vein of six (6) birds from each treatment into bijou bottles containing EDTA (anticoagulant). Packed cell volume (PCV), haemoglobin concentration (Hb), total RBC and total WBC were evaluated according to Dacie and Lewis (1977). Plasma obtained from centrifugation of was used to determine alkaline phosphatase (AP.EC.3.1.3.1) activity by the kinetic method of (Frajola et al., 1965). Proximate analysis of the basal diet was carried out according to the method described by (AOAC 1990). Response criteria were subjected to analysis of variance (ANOVA) using the SPSS statistical package (SPSS, Inc., 2000). Differences between treatment means were separated using Duncan Multiple Range Test (Duncan, 1995).

Ingredients	**Quantity (kg)**
Maize	45
Corn bran	10
Wheat offal	5.9
Groundnut cake	19
Brewers Dried Grain	5.0
Soybean Meal	2.0
Palm kernel cake	3.0
Fishmeal	1.0
Bone meal	0.27
Oyster Shell	8.0
Salt	0.3
Lysine	0.1
Methionine	0.15
Toximil	0.03
	100.00
Calculated nutrient composition	
Crude Protein (%)	17.94
Metabolizable energy (Kcal/kg)	2672.67

Table 1a. Composition of Basal Diet (/100kg feed)

Ingredients	Diets						
	1	2	3	4	5	6	7
Basal diet	+	+	+	+	+	+	+
3,750 i.u vit A/kg diet	+						
Commercial premix* (+ve control)			+				
0.5mg folate/kg diet			+				
1.0mg folate/kg diet				+			
7,500 i.u vit A/kg diet					+		
7,500 i.u.vit A+1.0mg folate/kg diet						+	
-ve control							+

*provided (per 2.5kg of diet), vitamin A (i.u) 10,000,000, vitamin D_3 (i.u) 2, 000, 000, vit. E (i.u) 12, 000, vit. K (mg) 2,000, Thiamin (mg) 1,500, Riboflavin (mg) 4,000, Pyridoxine (mg) 1,500, Niacin (mg)15,000, Cyanocobalamin (mcg) 10,Panthothenic acid (mg) 5,000, Folic acid(mg) 500, Biotin (mcg) 20,Choline chloride (mg) 100,000,Manganese (mg) 75,000, Zinc (mg) 50,000, Iron (mg) 20,000,Copper(mg) 5,000, Iodine (mg) 1,000,Selenium (mg) 200, Cobalt(mg) 500, Antioxidant (mg) 125,00

Table 1b. Composition of Experimental Diet

3. Results and discussion

The performance and egg quality parameters of laying pullets fed the different treatments did not vary (P>0.05) except for egg weight (Table 2). This observation corroborates the reports Hebert et al., (2004) and Lin et.al, 2002 that supplemental Folic acid and vitamin A respectively, did not affect body weight change of experimental hens. Folic acid levels have been reported not to affect feed consumption (Hebert et al. 2005). In the same vein, Lin et al., 2002 reported that neither the vitamin A supplemental levels nor the treatment regimens had a significant effect on feed intake or laying performance of experimental hens ($P > 0.05$).

Significant differences (P<0.05) observed in the values of egg weight was due to dietary treatments. Birds fed 100% dietary inclusion of folic acid had the highest egg weight (61.74g) while the lowest value of egg weight (56.83g) was recorded for birds fed 200% dietary inclusion of vitamin A. Herbert e al. (2005) observed in a study of diet × strain interaction that higher egg mass was a result of increased folic acid supplementation. March et al. (1972) reported a decrease in egg weight and a marked decline in the hen-day egg production at higher levels of vitamin A supplementation.

The results of the haematological indices and serum enzyme are shown on Table 3. Significant differences were observed among the dietary treatments, The values of the PCV of all the treatments were within the normal range (18.5 - 45.2%) reported by Mitruka and Rawnsely (1977) though treatments 4 and 6 were significantly lower than the others. The results obtained for haemoglobin concentration (Hb) followed the same pattern with that of PCV. The function of the blood in transporting hormones, metabolites, as thermo-regulators and general homeostasis has been clearly established (Duke, 1975). The haematological examination of the present study showed no adverse effect on the health status of the experimental birds as a result of the dietary fortification of vitamin A and Folic acid. Phosphatase enzymes are already detectable in the chick embryo as early as the first 5 days of incubation and it is an indirect assessment of bone mineralization (Needham, 1963). Values obtained from this study showed that dietary fortification of vitamins A and Folic acid does not impair calcium and phosphorus utilization for bone mineralization.

Egg laying index	Diets^{A}						
	1	2	3	4	5	6	7
Average initial weight (kg/bird)	1.46	1.51	1.55	1.43	4.50	1.50	1.55
Final body weight (kg/bird)	1.55	1.53	1.61	1.50	1.55	1.55	1.54
Weight change (g/bird/week)	1.63	1.35	1.13	1.28	1.17	0.96	0.64
Feed intake (g/d)	122.29	113.64	119.85	117.42	110.04	118.29	114.05
Feed efficiency (kg/dozen eggs)	2.37	2.43	2.50	2.40	2.36	2.57	2.51
Hen-day egg production (%)	14.48	14.92	15.32	14.72	14.54	15.77	15.41
Egg weight (g)	64.82	58.92	61.07	62.67	61.07	59.10	57.32
Albumen height (mm)	57.87^{ab}	57.54^{ab}	61.74^{c}	58.25^{ab}	56.83^{a}	58.24^{ab}	59.62^{bc}
Egg shell thickness (mm)	6.13	6.28	6.35	6.31	6.42	5.87	5.62
Yolk index^{B}	3.95	3.98	4.04	4.03	4.01	3.96	3.89
Haugh Unit** (%)	75.92	77.19	76.94	77.94	79.27	72.65	71.64
Liveability (%)	100.0	100.0	100.0	100.0	100.0	100.0	100.0

abc: Means in the same column with different superscripts are significantly different ($P < 0.05$).
^{A}1=basal diet+3,750 IU retinyl palmitate; 2: basal diet with premix alone (+ve control); 3: basal diet+ 0.5mg folic acid/kg; 4: basal diet+1.0mg folic acid; 5: basal diet+7,500 IU retinyl palmitate; 6: basal diet+ 7500 IU retinyl palmitate+ 1.0mg folic acid/kg; 7: basal diet without premix and fortification (- ve control).

Table 2. Effects of Dietary Fortification of Vitamin A and Folic Acid on Performance and Egg Quality Parameters of Laying Hens.

The results of egg yolk retinol and folate concentrations are shown on Tables 4 and 5 respectively. Regression analysis clearly showed a significant positive correlation between dietary retinyl palmitate and egg retinol (r = 0.813). The regression equation was $y = 0.1172x + 528.175$, where y is egg retinol (i.u/100g), and x is dietary retinyl palmitate (i.u/kg), indicating that egg retinol increased linearly as dietary vitamin A rose. These data are in agreement with previous reports indicating the influence of dietary vitamin A on its concentration in the egg yolk (Squires and Naber, 1993; Qui and Sim, 1998). The addition of increasing levels of dietary folic acid showed a trend to reduce the yolk retinol content, decreasing in 47.69, and 34.03%, respectively for 0.5, and 1.0 mg of supplemental folic acid/kg of diet, in comparison with the control group (Table 4). These results, however, proved to be not significant.

Haematological Indices	Diets/Treatments						
	1	2	3	4	5	6	7
Packed cell volume (%)	24[ab]	22[ab]	27[bc]	19[a]	22[ab]	19[a]	28[c]
Heamoglobin (g/l)	4.72[d]	3.79[b]	5.20[e]	4.00[c]	4.00[c]	3.5[a]	5.25[e]
Red Blood cell (10^{12}/l)	1.15[b]	1.16[b]	1.90[c]	1.00[a]	1.20[b]	1.13[ab]	1.93[c]
White Blood cell (10^{9}/l)	8.9[ab]	9.1[b]	10.3[d]	12.9[e]	8.7[a]	10.1[d]	9.7[c]
Neutrophils (%)	19[a]	26[b]	26[b]	32[d]	29[c]	27[bc]	20[a]
Lymphocytes (%)	79[d]	73[c]	74[c]	66[a]	70[b]	70[b]	80[d]
Monocytes (%)	2	0	0	1	1	3	0
Eosinophils (%)	0	1	0	1	0	0	0
SERUM ENZYME							
Plasma alkaline phosphatase	8[a]	16[c]	10[b]	20[d]	8[a]	22[e]	10[b]

abcd: Means in the same column with different superscripts are significantly different ($P < 0.05$).

Table 3. Effects of Dietary Fortification of Vitamin A and Folic Acid on the Haematology and Serum Enzyme of Laying Hens.

Treatments[1-7]				
Vitamin A supplementation (IU/kg of diet)	**Folic Acid supplementation (mg/kg of diet)**	**Yolk retino (IU/100g of yolk)**	**Yolk retinol (% change)**	**Correlation coefficient (r)**
1	3,750	0	604.02[a]	-8.2[B]
2	0	0	658.01[a]	------
3	0	0.5	592.98[a]	- 9.8[B]
4	0	1.0	915.97[b]	+39.2[B]
5	7,500	0	2597.9[c]	+294.8[B]
6	7,500	1.0	580 [a]	- 11.9[B]
7	0	0	675 [a]	+2.6[B]
Vitamin A supplementation (IU/kg of diet)				
0		710	------	
3.750		604.02	- 14.93[C]	0.813[E]
7.500		1588.99	+123.80[C]	
Folic Acid supplementation (mg/kg of diet)				
0		1133.75	------	
0.5		592.98	- 47.69[D]	------
1.0		747.99	- 34.03[D]	------

abc: Means in the same column with different superscripts are significantly different ($P < 0.05$).
[1-7]1=basal diet+3,750 IU retinyl palmitate; 2: basal diet with premix alone (+ve control); 3: basal diet+ 0.5mg folic acid/kg; 4: basal diet+1.0mg folic acid; 5: basal diet+7,500 IU retinyl palmitate; 6: basal diet+ 7500 IU retinyl palmitate+ 1.0mg folic acid/kg; 7: basal diet without premix and fortification (- ve control).
[B]Percentage change in comparison to the +ve control group.
[C]Percentage of yolk retinol in comparison with no Vitamin A supplementation.
[D] Percentage of yolk retinol in comparison with no Folic Acid supplementation.
[E]Correlation coefficient between yolk retinol and dietary Vitamin A supplementation, whole data, regardless of levels of supplemental Folic Acid.

Table 4. Egg yolk Retinol (iu/100g of yolk), Percentage of change, and Correlation Coefficients.

Significant and progressive incorporation of folate into the egg yolk was achieved by feeding laying hens a basal diet supplemented with increasing levels of folic acid, regardless of supplemental vitamin A. The regression analysis was significant (r = 0.971), and the regression equation was $y = 154x + 106.74$, where y is egg folate (μg/100g), and x is dietary Folic acid (mg/kg), indicating that egg folate increased linearly as dietary Folic acid increased. House et al., (2002) and the work of others (Sherwood et al., 1993), reported that egg folate concentrations responded to increasing levels of dietary folic acid supplementation of crystalline folic acid per kilogram of diet.

Treatments[1-7]					
	Vitamin A supplementation (IU/kg of diet)	**Folic Acid supplementation (mg/kg of diet)**	**Egg Folate (µ/100g of yolk)**	**Yolk Folate (% change)**	**Correlation coefficient (r)**
1	3,750	0	93.09[a]	-10.8[B]	
2	0	0	104.34[a]	------	
3	0	0.5	205.64[c]	+97.08[B]	
4	0	1.0	323.48[d]	+210.02[B]	
5	7,500	0	92.63[a]	-11.22[B]	
6	7,500	1.0	176.11[b]	+68.78[B]	
7	0	0	93.12[a]	-10.75[B]	
Vitamin A supplementation (IU/kg of diet)					
	0		181.65	------	
	3.750		93.09	-48.75[C]	------
	7.500		134.37	-26.02[C]	------
Folic Acid supplementation (mg/kg of diet)					
	0		95.79	------	
	0.5		205.64	+114.67[D]	------
	1.0		249.79	+160.61[D]	0.971 [E]

abcd: Means in the same column with different superscripts are significantly different ($P < 0.05$).
[1-7]1=basal diet+3,750 IU retinyl palmitate; 2: basal diet with premix alone (+ve control); 3: basal diet+ 0.5mg folic acid/kg; 4: basal diet+1.0mg folic acid; 5: basal diet+7,500 IU retinyl palmitate; 6: basal diet+ 7500 IU retinyl palmitate+ 1.0mg folic acid/kg; 7: basal diet without premix and fortification (- ve control).
[B]Percentage change in comparison to the +ve control group.
[C]Percentage of yolk folate in comparison with no Vitamin A supplementation.
[D] Percentage of yolk folate in comparison with no Folic Acid supplementation.
[E]Correlation coefficient between yolk folate and dietary Folic acid supplementation, whole data, regardless of levels of supplemental Vitamin A.

Table 5. Egg Yolk Folate (*m*g/100g Yolk), Percentage of Change, and Correlation Coefficients.

Supplementation with increasing amounts of dietary vitamin A produced significant reductions in average egg yolk folate concentrations of 48.75 and 26.02%, respectively, for eggs from hens fed 3,750 and 7,000 i.u of supplemental vitamin A/kg (Table 5).

Generally, the findings of this study further reinforce the need to profitably produce enriched eggs in a sustainable way through fortification. Individual fortifications of vitamin A and folic acid at 200% inclusion level of NRC recommendations which was significantly better in overall performance and egg yolk vitamin deposit would be recommended until further studies are carried out to elucidate the interactions between vitamin A and Folic acid.

4. References

A.O.A.C. (1990). Official Methods of Analysis. 15th ed. Association of Official Analytical Chemists. Arlington, VA.

Cavalli-Sforza, T. and Bosch, D. (2000). Food fortification as part of an integrated food and nutrition strategy. In: Manila Forum 2000: Strategies to fortify essential foods in Asia and the Pacific. Asian Development Bank, International Life Sciences Institute, Washington, DC, Micronutrient Initiative, Ottawa, Canada. pp. 351–52

Dacie, J.W. and Lewis, S.M. (1977). Practical haematology. 5th. Ed. Longman Group Ltd. Pg 21-68.

Duke H.H. (1975). Duke;s Physiology of Domestic Animal Textbook. 8th edition. Ithae and London constock publishing Associate.

Duncan, D.B. (1955). Multiple Range and F-Test Biometric: 11:1-42.

El-Husseiny, O.M., Soliman, A.Z., Abd-Elsamee, M.O. and Omara, I.I. (2005). Effect of dietary energy methionine, choline and folic acid levels on layers performance. Egy. Poult. Sci. J., 25: 931-956.

Fitzgerald, S. (1997). Fortification rapid assessment guidelines and tool (FRAT) Path Canada.

Fleming, A.F. (1989). The aetiology of pregnancy in tropical Africa. Trans R Soc Trop. Med Hyg 83: 441-448.

Frajola, W.J., William, R.D. and Austin, R.A. (1965). The kinetic spectrophotometric assay for serum Alkaline Phosphatase. Am. J. Clin. Path. 43:261-264.

Gibson, R. (2000). Literature and Patent Search to determine the feasibility of developing eggs with specific nutrients. RIRDC Completed Projects in 1999-2000.

Hebert, K., House, J.D. and Guenter, W. (2004). Efficiency of folate deposition in eggs through-out the production cycle of Hy-line W98 and W36 laying hens. Poult. Sci., 83: (suppl.1).

Herbert, K., House, J.D. and Guenter, W. (2005). Effect of dietary folic acid supplementation on egg folate content and the performance and folate status of two strains of laying hens. Poult.Sci. 84: 1533-1538.

House, J.D., Braun, K., Balance, D.M., O'Connor, C.P. and Guenter, W. (2002). The enrichment of eggs with folic acid through supplementation of the laying hen diet. Poult Sci.81: 1332-1337.

Klemm, R.D.W., Ungson, B.D., Lopez, C.T., Villate, E.E., Ramos, A.C., Serdooncillo, M., Paulino, L.S., Longaza, S.G., Guarin, R.M., Tirazona, A.M. and Borata, F.R. (1997). Coverage and factors associated with vitamin A, iron and iodine supplementation during the National Micronutrient Day 1996 in Samar and Albay. Mimeo. USAID - HKI, 1997.

Lindsay, A., Omar, D. and Richard, H. (2006). Guidelines on food fortification with micronutrients. WHO Library Catalouging- in- Publication.

Lin, H., Wang, L.F., Song, J.L., Xie, Y.M. and Yang, Q.Y. (2002). Effect of Dietary Supplemental Levels of Vitamin A on the Egg Production and Immune Responses of Heat-Stressed Laying Hens. Poultry Science 81:458–465

Lofti M. (1996). Micronutrient fortification of foods: current practices, research and opportunities. Ottawa, Micronutrient Initiative, International Agricultural Centre.

March, B. E., Coates, V and Goudie, C. (1972). Delayed hatching time of chicks from dams fed excess vitamin A and from eggs injected with vitamin A. Poult. Sci. 51:891–896.

Maria, A. and Bruno de Benoist. (2002). Vitamin A Deficiency Control. /UNICEF Strategy. Micronutrient and Trace Elements Unit Department of Nutrition for Health and Development World Health Organization Geneva

Micronutrient Initiative. (1997) Food fortification to end micronutrient malnutrition: Stateof-the-Art Symposium Report, 2 August 1997, Montreal, Canada. Ottawa, Micronutrient Initiative, International Agricultural Centre, 1998.

Miltruka, B.M and Rawnsley, H. (1977). In clinical, biochemical and heamatological values in normal experimental animals. 106-112. Masson Publishing. U.S.A Inc. new York.

Needham, J. (1963). Chemical Embryology. Hafner Publishing Comp, New York, London.

Neilsen, S.S. (2002). Introduction to the chemical analysis of food. Pg 253.CBS publishers. New Delhi.

O'Connor, D.L. (1994). Folate status during pregnancy and lactation. In: Nutrient regulation during pregnancy, lactation and infant growth. Allen L, King J, Lonnerdal B.(eds) Plenum Press, New York, pp. 157–172.

Omar, D. and Jose, O.M. (2002). Food Fortification to Reduce Vitamin A Deficiency: International Vitamin A Consultative Group Recommendations J. Nutr. 132:2927S-2933S, September 2002.

Qui, G. H and Sim, J. S. (1998). Natural tocopherol enrichment and its effect in n-3 fatty acid modified chicken eggs. J. Agric. Food Chem. 46:1920–1926

Sherwood, I.A., Alphin, R.L., Saylor, W.W. and White, H.B. (1993). Folate metabolism and deposition in eggs by laying hens.Archives Biochem. Biophys., 307; 66-72.

Sommer, A. and Davidson, F.R. (2002). Assessment and Control of Vitamin A Deficiency: The Annecy Accords. J.Nutr. 132(9): 2845S-2850.

SPSS, Inc. (2000). SigmaPlot 2000 Graphical Software Package for Windows Version 10.0. Statistics Canada, 2001. CANSIM II Database. http://www.statcan.ca/english/Estat/licence. htm.

Squires, M.W. and Naber, E.C. (1993). Vitamin profiles of eggs as indicators of nutritional status in the laying hen: Riboflavin study. Poultry Sci. 72:483-494.

Surai, P.F. and Sparks, H.C. (2001). Designer eggs: from improvement of egg consumption to functional food. Anim. Feed Sci. Tech.12:7-16.

Vahteristo, L.T., Ollilainen, V and Varo, P.J. (1997). Liquid chromatographic determination of folate monoglutamates in fish meat, egg, and dairy products consumed in Finland. J.AOAC Int.80:373–8.

WHO. (2000). Vitamin A deficiency. WHO website updated 17 Feb 2000 http://www.who.int/vaccinesdiseases/diseases/vitamin_a.htm]

World Health Report. (2002). Reducing risks, promoting healthy life: overview. Geneve. World Health Organization.2002(WHO/WHR/02.1).

WHO (2003). Combating vitamin A deficiency. WHO website, updated September 2003. World Health Organisation, Rome, Italy. [http://www.who.int/nut/vad.htm]

8

Mitochondria and Antioxidants: The Active Players in Islet Oxidative Stress

Tatyana V. Votyakova, Rita Bottino and Massimo Trucco
Children's Hospital of Pittsburgh of UPMC, University of Pittsburgh
USA

1. Introduction

Diabetes mellitus (DM) of type 1 (T1D) and type 2 (T2D) are characterized by persistently high glucose (HG) blood levels known as hyperglycemia. The preponderance of evidence points to a significant role of oxidative stress in the development of complications in patients with DM. In particular, the pathogenic increase in reactive oxygen (ROS) and nitrogen species (RNS) as well as accumulation of oxidation and nitration products has been well documented in cases of diabetes. ROS and RNS affect all types of biological molecules: they cause oxidation of membrane lipids, modification of protein amino groups as well as deoxynucleotites. Insulin producing beta-cells, which are part of pancreatic islets of Langerhans, perform the energetically demanding function of sensing blood glucose and releasing insulin to sustain metabolic homeostasis. They are highly specialized endocrinal cells with a complex system of signal transduction and insulin producing capacity. In pathology of diabetes beta-cells are the ones which are most susceptible to oxidative stress.

2. Sources of oxidative stress in islets

In islets, reactive species can originate from several sources. NAD(P)H-oxidase located in the plasma membrane produces molecule of O^{2-}. Several isoforms of this enzyme were found in islets (Uchizono et al. 2006; Newsholme et al. 2007; Newsholme et al. 2009) and they are considered as substantial producers of ROS. The enzyme can be activated by exposure to fatty acids and in normal conditions it is believed to participate in glucose-stimulated insulin secretion (GSIS) (Graciano et al. 2011; Santos et al. 2011). Islets also possess both types of nitric oxide synthases (NOS): constitutive cNOS (Nakada et al. 2003) and inducible (iNOS) (Darville and Eizirik 1998; Kutlu et al. 2003). Nitric oxide produced by cNOS is part of normal beta-cell physiology, while activation of iNOS is associated with beta-cell destruction, in particular via cytokines produced by immune cells.

There is plentiful evidence that hyperglycemic conditions cause rise in ROS and RNS in beta-cells reviewed in: (Newsholme et al. 2007; Acharya and Ghaskadbi 2010). It was shown that high glucose triggers generation of ROS in rodent and human islets as well as in insulinoma cell lines (Tanaka et al. 1999; Tanaka et al. 2002; Bindokas et al. 2003; Robertson et al. 2003). A definitive role for mitochondria in glucose-induced ROS signal was proposed in a number of papers (Maechler and Wollheim 2001; Brownlee 2003; Fridlyand and

Philipson 2004; Newsholme et al. 2007; Newsholme et al. 2009). Thus, mitochondria are considered as an important ROS generator in islets.

3. Islet mitochondria as a source of reactive oxygen species

The main role of mitochondria in a cell is the production of ATP molecules for cellular energetic needs using for that energy of metabolite oxidation. ATP production by mitochondria is central for glucose sensing and insulin release in beta-cells, the fact that directs attention to mitochondria in diabetes studies. In the process of metabolite oxidation, mitochondria are dealing with the transfer of electrons along respiratory Complexes I through IV. Eventually four electrons are combined with four protons, H^+, and one molecule of oxygen, producing two molecules of water; the reaction takes place within Complex IV (for reference see (Nicholls and Ferguson 2002). However, as a side reaction, a low portion of unpaired electrons leaks from the respiratory complexes and interacts with molecular oxygen producing a molecule of superoxide anion radical, a form of ROS. Currently, it is believed that the sites of ROS production in the mitochondrial respiratory chain are Complex I and Complex III (Turrens 2003; Rigoulet, Yoboue, and Devin 2011).

For proper mitochondrial functioning with low levels of ROS production, the availability of end-point electron acceptor, molecular oxygen, is absolutely necessary. Temporal hypoxia, which causes a halt in electron flow followed by reoxygenation and resuming of electron flow, results in an increase of mitochondrial ROS (Selivanov et al. 2009). This phenomenon is related to a widely known phenomenon of ischemia-reperfusion injury and is a critical factor in the process of islet isolation from donor pancreata for transplantation purposes, as donor tissues inevitably become hypoxic when blood circulation stops. Presence of antioxidants in preservation and isolation media detoxifies mitochondria-derived ROS and substantially improves viability of islets and their potency to normalize blood glucose in recipient diabetic animals (Bottino et al. 2002; Bottino et al. 2004; Sklavos et al. 2010).

Manipulation of electron flow in the respiratory chain by specific inhibitors of mitochondrial respiratory Complexes can either increase or decrease ROS generation. In normal conditions, with low levels of glucose and low base level of ROS, inhibitors of Complexes I and III, rotenone and antimycin, respectively, increase ROS production in rat islets (Armann et al. 2007; Leloup et al. 2009). This is consistent with data obtained on isolated mitochondria *in vitro* (Votyakova and Reynolds 2001; Starkov, Polster, and Fiskum 2002; Starkov and Fiskum 2003; Rigoulet, Yoboue, and Devin 2011).

An important question is whether mitochondrial inhibitors are capable of down regulating ROS signals originating from the respiratory chain under conditions of hyperglycemia and hyperlipidemia. The answer depends on the type of chemical agent and the locus it binds in the respiratory chain. Oxidation of succinate in Complex II results in a high ROS generation because this prosess initiates forward electron flow to Complex III as well as reverse electron flow upstream to Complex I (Selivanov et al. 2011). Thus, inhibition of succinate oxidation would lead to a decrease in overall ROS. This effect was observed in a work by Sakai and collaborators (Sakai et al. 2003) where an inhibitor of Complex II thenoyltrifluoroacetone (TTFA) decreased glucose-stimulated ROS in human islets and MIN-6 cells. The inhibitor of Complex III antimycin A blocks Q-cycle in a way that increases free radical forms of respiratory chain components (Votyakova and Reynolds 2001; Starkov,

Polster, and Fiskum 2002; Starkov and Fiskum 2003). Leloup and co-authors observed a similar effect of antimycin A on rat islets (Leloup et al. 2009).

Complex I, or NADH-oxidoreductase, is the largest and the most sophisticated segment in the respiratory chain; its architecture, functioning and mechanisms of ROS generation has been extensively studied (Magnitsky et al. 2002; Vinogradov 2008). In mammals it consists of 47 subunits with a number of redox centers to transfer electrons. Among them, there are flavine mononucleotide at the entrance of the Complex, which binds NADH, several sulfur-iron clusters and several Coenzyme Q binding sites in the middle of the Complex. According to the current consensus, all inhibitors, which bind to subunits located in the middle of Complex I and block electron flow within the complex, such as rotenone, piericidin and others, increase ROS (Grivennikova and Vinogradov 2006). The only known inhibitor that decreases ROS generation in Complex I is diphenylene iodonium, which binds at the very entrance at the flavin mononucleotide (FMN) site (Liu, Fiskum, and Schubert 2002). This compound, though, is unspecific and also inhibits NAH(P)H oxidase of the plasma membrane. In other words, if an inhibitor blocks the very entry of Complex I at the site of FMN and prevents the access of electrons into Complex I at all, it would prevent ROS generation. If an inhibitor blocks electron flow somewhere in the middle of Complex I and allows redox centers upstream of the block to be over-reduced, it would result in an increase of ROS (Genova et al. 2003). In accordance with this notion, inhibitory effects of diphenylene iodonium on ROS induced by hyperglycemia in MIN-6 cells were reported by (Tsubouchi et al. 2005), though the researchers attributed this fact solely to the plasma membrane NAD(P)H oxidase inhibition. Regarding the other Complex I inhibitors, there is some controversy in the literature. The same authors reported no effect of rotenone, the effect could be expected if ROS was already increased, while earlier work by Sakai and collaborators reported that rotenone twofold decreased hyperglycemia-induced ROS in human islets and MIN-6 cells (Sakai et al. 2003).

There is a way to modulate mitochondrial ROS generation without interference into the activity of respiratory Complexes. In the cell this function belongs to a special group of proteins located in the inner membrane of mitochondria, called uncoupling proteins (USPs), which in a highly controlled manner modulate membrane potential, basically working as proton conductors (Ricquier and Bouillaud 2000). Islets express an UCP2 isoform of uncoupling protein (Gimeno et al. 1997), and in a number of publications it was shown that overexpression of UCP2 downregulated the levels of ROS (Pi et al. 2009; Affourtit, Jastroch, and Brand 2011).

The same effect can be achieved by using chemical compounds called uncouplers, which, while different in structure, have two properties in common: they can penetrate into mitochondrial membrane and are capable to easily accept or dissociate H^+ ion. Uncouplers decrease membrane potential by transporting protons into mitochondrial matrix and, in principle, their function is similar to that of UCPs, though they are more powerful modulators (see (Nicholls and Ferguson 2002) for detailed mechanism). Uncouplers were shown to decrease ROS production in glucose-stimulated human and rat islets as well as in MIN-6 cell line (Sakai et al. 2003; Leloup et al. 2009). This is consistent with the data observed on isolated mitochondria *in vitro* (Votyakova and Reynolds 2001; Selivanov et al. 2008).

Overall, the amount of data on the mechanism of mitochondrial ROS generation in beta-cell is limited and this issue is underexplored, mainly because islets are precious and less available for research compared with other more abundant tissues, like liver, muscle, heart and even brain. Mostly intriguing is the fact that, despite extensive evidence of increased level of free radicals in conditions of high glucose or high lipids, there is still no satisfactory molecular mechanism explaining exactly how high levels of cellular energy metabolites cause beta-cell mitochondria to produce more free radicals.

4. Antioxidant capacity of insulin producing beta-cells

Antiradical defense systems in the cell consist of a gamut of small antioxidant molecules and enzymes capable of interaction with reactive oxygen and nitrogen species. Endogenous antioxidant molecules are vitamins A, C and E, sulfur-containing compounds like amino acid cysteine and tripeptide glutathione (GSH - reduced form, GSSG - oxidized form). Coenzyme Q may also act as an antioxidant in particular conditions. These compounds possess different efficiency in scavenging harmful oxidants, glutathione being the most potent. The main players in enzymatic antiradical protection are superoxide dismutases (SOD), catalase and glutathione peroxidases (GPx). Within the cell two isoforms of SOD are found: Zn,Cu-SOD (SOD1) which is located in cytoplasm and Mn-SOD (SOD2) located in mitochondrial matrix. They are the first line of defense as they convert superoxide anion radical into O_2 and H_2O_2 by dismutation. The hydrogen peroxide next can be converted into H_2O and O_2 by catalase or, to H_2O, by GPx; the latter will use reduced glutathione molecule as a substrate. Glutathione is a recyclable molecule: its oxidized form is reduced back by glutathione reductase. Mitochondria have both enzymes (about of 10% of total cellular activity) which, working in concert, effectively detoxify peroxides and recycle GSH (for reference see (Halliwell 2001)). The proteins UCPs, which downregulate ROS generation in mitochondria, can also be formally added to antioxidant proteins.

Glucose sensor function and production of insulin on demand are the two dominating functions in beta-cell physiology and, apparently, this comes at a cost of downregulating some other functions. High susceptibility of islets to oxidative insults was well established long ago, and this feature was utilized to specifically target these cells, thus, creating animal models of diabetes. Compared to cells from other tissues, islet cells have profoundly lower activity of enzymes involved into antiradical defense. As early as in 1979 Grankvist and co-authors shown that alloxan effectively destroys rat islet cells through ROS-mediated mechanisms (Grankvist et al. 1979). Two years later they reported that beta-cell super-sensitivity to oxidative agent is due to a deficiency of anti-radical defense capacity (Grankvist, Marklund, and Taljedal 1981). The activity of the main anti-oxidant enzymes was found to be about 30% for both types of SODs and only in single percentage range for catalase and glutathione peroxidase (1.2% and 1.8%, respectively), as compared to liver. This data was confirmed by estimating the levels of mRNA of the respective genes by Lenzen and co-workers (Lenzen, Drinkgern, and Tiedge 1996). It should be noted, though, that a more appropriate comparison of islet enzymes' activities would be with that of tissues with similar functions and/or intensity of metabolism. As Table 1 shows, when compared with hypophysis or brain tissues, islets' activities of anti-oxidant enzymes were only 50% or lower.

Gene Expression	Cu/Zn SOD	Mn SOD	Catalase	Glutathione Peroxidase
Pancreatic islets % of Liver*	38	30	n.d. **	15
Pancreatic islets % Pituitary gland	48	64	n.d.**	23
Pancreatic islets % of Brain	49	45	n.d.**	38

Table 1. The levels of anti-oxidant gene expression in islets compared to brain and pituitary gland. *Data adopted from Lenzen & al. (1996). **In (Tiedge et al. 1997) it was determined to be around 5%.

Expression of anti-oxidant enzymes in islets may vary within the same species reflecting genome variations. Zraika and co-authors have found that mRNA levels of Mn-SOD in islets of a diabetes-prone DBA/2 mice were twofold higher than in islets of C57BL/6 mice (Zraika et al. 2006), while the level of catalase was the same. This means that at the same level of superoxide anion radical beta-cells of DBA/2 mice would produce H_2O_2 twice as fast compared to that of C57BL/6 mice, and consequently, having the same level of catalase, DBA/2 islets would deal with higher levels of peroxide.

Among species the differences in activities of antioxidant enzymes can also vary substantially. It was shown that human islets have more active catalase and SOD than rodent ones and, consequently, are more resistant to oxidative stress (Welsh et al. 1995).

The expression of antiradical enzymes can be changed in diabetes conditions. In islets of Goto–Kakizaki/Paris rats, a model of T2D, expression of the whole spectrum of antiradical defense genes is increased along with an increased level of reduced glutathione compared with normal healthy animals (Lacraz et al. 2010). In patients with T2D expression of Cu,Zn-SOD was found reduced (Sakuraba et al. 2002). Diabetes is a complex and dynamic disease, in which epigenetic and environmental factors can differently affect the expression of antioxidant enzymes at particular time points of its development. Thus, it is often difficult to compare data on diabetes-related oxidative stress in short-lived rodents to long-lived humans.

It is important to note that beta-cells possess a substantial activity of enzymes dealing with superoxide anion radicals, while the activity of the enzymes decomposing H_2O_2 is very low, especially that of catalase. This feature is in line with increasing evidence that suggests a signaling role for H_2O_2 molecule in the process of insulin secretion (Pi et al. 2009; Affourtit, Jastroch, and Brand 2011).

5. Application of chemical compounds with antioxidative properties as a strategy to offset oxidative stress

Low level of antioxidant defense in beta-cells suggested that supplementation of antioxidants could be beneficial. A number of chemical compounds with antioxidant properties were employed to prevent or counteract oxidative stress in islets. The studies were conducted on diabetic animals and on isolated islets. Both preventive and curing actions of antioxidants were studied; the antioxidative agents were administered either prior to induction of diabetes of oxidation stress, or in the course of developing processes.

Historically, the first to study were vitamins A, C and E which are naturally present in the body and possess the ability to scavenge free radicals. Unfortunately they offered very limited or no protection (Kaneto et al. 2001). Hence, a broad spectrum of chemical compounds was tested in search for agents capable to counteract oxidative burden experiencing by islets. They differ in chemical nature and by the mechanism through which they protect from oxidative stress. The most important aspect in search of these compounds is to pay attention to their potential side effects. We will focus on the most prominent ones.

5.1 Antioxidant properties of N-acetyl-L-cysteine (NAC)

Antioxidant properties of N-acetyl-L-cysteine (NAC), an acetylated derivative of amino acid L-cysteine) were employed in a range of medical conditions such as neurodegeneration (Pocernich et al. 2011), cardiovascular diseases (De Rosa et al. 2010), gastroenterological diseases (Ramudo and Manso 2010; Jegatheeswaran and Siriwardena 2011), transplantation (Czubkowski, Socha, and Pawlowska 2011) and diabetes (Kaneto et al. 2001). There are several mechanisms by which NAC can modulate oxidative stress: (i) it is a precursor in the synthesis of glutathione, an important component of the cellular antiradical defense system in the cell; (ii) NAC, as a thiol-containing compound, is able to directly reduce free radicals as well as S-S bonds in proteins, thus modulating redox signaling (Parasassi et al. 2010)

It was reported that NAC is protective or partially protective in animal models of diabetes. It reduces levels of oxidative stress markers and preserves beta-cell mass and function in STZ-treated hamsters (Takatori et al. 2004). Intravenous co-infusion of NAC with high glucose into Wistar rats offered only partial protection, as it quenched ROS, but did not restore beta-cell functions (Tang et al. 2007). However, the same group reported that NAC, co-infused with free fatty acids, not only decreased beta-cells ROS caused by prolonged exposure to fatty acids, but preserved their insulin and C-peptide responses to hyperglycemic clamps (Oprescu et al. 2007). These two studies show that effectiveness of particular antioxidant molecules depends on the type of oxidative stimuli and the metabolic pathways that are intervened. A study by Kaneto and co-workers also demonstrated that NAC can be protective in T2D-like metabolic deregulations. Using the db/db diabetic mouse model, the researchers found that NAC improved glucose-stimulated insulin response in these animals and, on a molecular level, NAC increased expression of pancreatic and duodenal homeobox factor-1 (PDX-1) in islets, a beta-cell-specific transcription factor (Kaneto et al. 2001).

NAC protects against oxidative stress stimuli *in vitro*. LDL oxidation level, which is relevant to pathology of T2D, significantly changes the expression of genes involved in the production and secretion of insulin and in cell survival mechanisms. This effect was offset by NAC (Favre et al. 2011). NAC counteracts the damaging effect of human amylin (hA), a small fibrillogenic protein, which accumulates in beta-cells in most subjects with T2D (Konarkowska et al. 2005). It is believed that NAC offers protection as a reagent capable of reducing protein SH groups, rather than a general ROS scavenger.

NAC is also protective against oxidative damage caused by direct short exposure of islets to H_2O_2 and free fatty acids, but is inefficient or partially efficient against long exposure islets to high glucose, or cytokines (Khaldi et al. 2006; Oprescu et al. 2007; Michalska et al. 2010). In some cases NAC, while decreasing ROS, did not restore beta-cell functions completely (Wang et al. 2004; Tang et al. 2007).

5.2 (R)-alpha-Lipoic acid (ALA)

(R)-alpha-Lipoic acid (ALA), (3R)-1,2-dithiolane-3-pentanoic acid, is a cyclic disulfide, being an oxidized form of its dithiol congener, (6R)-6,8-dimercaptooctanoic acid, or (R)-dihydrolipoic acid. Two sulfur atoms in ALA, which are connected to each other by a disulfide bond, can undergo facile and highly reversible redox processes (Arner, Nordberg, and Holmgren 1996). Thus, this compound can feature either antioxidant or pro-oxidant properties depending on particular redox context (Haramaki et al. 1997).

ALA is an essential co-factor in several mitochondrial oxidative complexes; the most important of these are pyruvate dehydrogenase (PDH) complex, 2-oxoglutarate dehydrogenase (OGDH) complex, and the complex for oxidation of branched chain amino acids (BCDH) (Nelson 2005).

Administration of alpha-lipoic acid to non-obese diabetic (NOD) mice decreased incidence of diabetes induced by cyclophosphamide from 60% to 30%. It also reduced severe intraislet infiltration and increases the percentage of islets with mild per-insular and periductular infiltrates (from 8.4 to 29.6 and 25.9%, respectively, $P < 0.01$) (Faust et al. 1994). The authors concluded that the anti-inflammatory action of lipoic acid may be due to its ability to scavenge oxygen radicals and to suppress nitric oxide production.

In the alloxan-induced diabetic mouse ALA lowered blood glucose, increased insulin release and prevented loss of beta cells and their dysfunction (Zhang et al. 2009). ALA prevented development of diabetes mellitus in obese Otsuka Long-Evans Tokushima Fatty (OLETF) rats, a T2D animal model. It diminished glycosuria, reduced body weight and protected pancreatic beta-cells from destruction. ALA also reduced triglyceride accumulation in skeletal muscle and pancreatic islets (Song et al. 2005).

In vitro studies provided a deeper insight into the mechanism of ALA action on cellular and organellar levels. ALA counteracted oxidative stress induced by proinflamatory cytokines IL-1β, IL-6 and IFN-γ by preventing NF-κB activation (Zhang et al. 2009) and restoring insulin secretion (Schroeder et al. 2005). When a direct oxidative insult was applied in a form of H_2O_2 (Lee, Kwon, et al. 2009) or xantine-xantine oxidase (Burkart et al. 1993), pretreatment with ALA decreased cellular ROS and c-JNK activation, stabilized mitochondrial membrane potential and induced Akt phosphorylation, altogether offering protection to beta-cells from oxidative stress. In MIN6 cells and rat islets, this compound also offsets the deleterious actions of free fatty acids, which feature an *in vitro* model for conditions of T2D (Shen et al. 2008). In particular, it decreased levels of ROS, restored mitochondrial membrane potential, glucose-induced ATP and glucose stimulated insulin secretion.

It is worth noting that ALA, like many other redox active chemicals (Skulachev et al. 2009), has an optimum concentration for anti-radical actions, and, consequently, for protective activity. Optimum protective concentration for isolated mouse islets against oxidative stress induced by cytokine IL-6 was found to be 10^{-9} M; lower and higher concentrations were not effective (Schroeder et al. 2005). In the case of chemically-induced oxidative stress in INS-1 cells or isolated islets, ALA prevented apoptotic cell death in 150-300 μM concentrations, while higher concentrations caused apoptosis. Discrepancies in the effective concentration ranges reported in these studies may lie in different nature of oxidative stress stimuli and in different protective mechanisms by which ALA acted in each case. Concentration of 2 mM, which is not physiologically relevant, impaired functions of isolated rat islets and MIN6 cells (Targonsky et al. 2006).

The nature of protective effects exhibited by ALA and NAC suggests that these compounds can act by multiple mechanisms, working as direct scavengers of free radicals, or by controlling of the thiol-disulfide level of reduction in signaling protein molecules (Pietta 2000; Pandey and Rizvi 2009; Parasassi et al. 2010). The antioxidant properties of ALA can be boosted by mitochondrial thioreductase (Trx), which restores its reduced form (Packer, Witt, and Tritschler 1995).

5.3 Coenzyme Q (CoQ)

Coenzyme Q (CoQ), a quinone, is an essential component of mitochondrial respiratory complexes I and III. In general, quinones can undergo two reversible one-electron redox processes, converting them into semiquinones and hydroquinones. High stability of semiquinone radicals renders the redox reversibility and is the basis for antioxidant properties of the quinone-related systems; both quinones and hydroquinones can serve as radical-protecting agents.

In cells, exogenous CoQ can exhibit either antioxidant or pro-oxidant properties, depending on conditions. As it was shown (Schroeder et al. 2005), a very low concentration of 10^{-12} M CoQ10 restored insulin production by mouse islets which were impaired by exposure to cytokine IL-1β. However, higher concentrations showed no effect or were even harmful. Overdoses of CoQ (50-200μM), although able to stimulate insulin release, were toxic to human islets and INS-1 cells. At these concentrations, CoQ, being a strong electrophile, covalently binds to E2 components of pyruvate dehydrogenase and α-ketoglutarate dehydrogenase complexes in mitochondria causing a substantial inhibition of the complexes and eventually triggering cell apoptosis (MacDonald et al. 2004). This circumstance put into a question the possibility to utilize CoQ10 as antioxidant to counteract diabetes-related oxidative burden as it would be difficult to find the right dose, while overdose would greatly outweigh potential benefits.

Some synthetic and plant-derived compounds may possess profound antioxidant properties, actively reacting with free radicals and ROS. These compounds belong to different chemical types; the most known of them are stable nitroxyl free radicals, metalloporphyrins, and plant polyphenols.

Superoxide anion radical is the first product resulting from the passing of unpaired electrons to molecular oxygen in the chain of chemical reactions causing oxidative stress. Thus, its detoxification would seem strategically advantageous. This notion prompted the introduction of a number of compounds which belong to different classes, but can all act as superoxide scavengers as they are able to effectively disproportionate superoxide to dioxygen and hydrogen peroxide. These compounds have a common name of SOD-mimetics.

5.4 TEMPOL (4-hydroxy-2,2,6,6-tetramethylpiperidine-1-oxyl)

TEMPOL (4-hydroxy-2,2,6,6-tetramethylpiperidine-1-oxyl) is a nitroxyl stable radical. Being redox-active, TEMPOL facilitates the metabolism of many reactive oxygen and nitrogen species (see review by (Wilcox 2010), in particular, TEMPOL catalyzes $O_2^{\cdot-}$ disproportionation. Its activity is not limited to this reaction, though, as it also catalytically converts H_2O_2 into water and dioxygen in a catalase-like reaction and inhibits generation of $OH^{\cdot}$ from H_2O_2 in the presence of redox-active transition metals in the Fenton reaction (Soule et al. 2007; Wilcox and Pearlman 2008).

Application of TEMPOL in diabetic animal models and *in vitro* on isolated islets was found to be protective against a variety of oxidative stress stimuli. Intravenous co-infusion of TEMPOL with high glucose or with free fatty acids into rats prevented islet dysfunction caused by hyperglycemia (Tang et al. 2007) or by hyperlipidemia (Oprescu et al. 2007). In Zuker rats, an animal model for T2D that features hyperglycemia, hyperinsulinemia as well as renal oxidative stress and high blood pressure, TEMPOL administered in drinking water reduced blood glucose, insulin secretion, renal oxidative stress and blood pressure which implies normalization of islet function (Banday et al. 2005).

5.5 Metalloporphyrins

Metalloporphyrins represent another well-known class of organic redox-active compounds. Manganese metalloporphyrins feature high chemical stability in different oxidation states and coordination of substrates at the central metal atom (Patel and Day 1999). Manganese ion can coordinate additional water or H_2O_2 in its axial sites, thus, performing its catalytic action. The stability of the porphyrin ring enables a variety of reversible catalytic metal-centered redox processes. Depending on substitutents, Mn-porphyrin compounds vary in their overall electric charge, redox potential and lipophic-hydrophylic properties. Eventually these physico-chemical properties translate into different catalytic activities and different abilities to penetrate into cells and cellular compartments. These compounds are capable of catalyzing not only the reaction of $O_2^{\cdot-}$ dismutation, but also the reduction or disproportionation of other reactive oxygen and nitrogen species, like H_2O_2, $HO^{\cdot}$, NO and $ONOO^-$ (Patel and Day 1999).

Manganese complexes of porphyrins were employed in studies of a number of biological models of oxidative stress (Batinic-Haberle et al. 2011). In an animal model of STZ-diabetic rats, Mn-porphyrin MnTM-2-PyP^{5+}, administered after the dose of STZ, counteracted the oxidative stress as judged by decreased levels of lipid peroxidation in blood plasma and erythrocytes (MDA products). Though it did not normalize blood glucose, it still decreased mortality of STZ-treated animals and increased their life span (Benov and Batinic-Haberle 2005). The data suggested that application of this antioxidant after a major STZ-induced oxidative stress did not restore islet function, but rather ameliorated following hyperglycemia-related complications. In the other experimental settings, the animals were preconditioned with Mn-porphyrin (MnTE-2PyP^{4+}, FBC-007), possessing a higher catalytic activity and a better ability to penetrate into cells. The animals were also given regular injections after STZ administration. Such treatment prevented development of diabetes in mice as monitored up to 120 days (Sklavos et al. 2010). It appears, therefore, that a preventive measure may play an important role in successful diabetes treatment.

In vitro, Mn-porphyrins showed a protective action also on the models of cultured isolated human and rodent islets as well as on insulin producing cell lines (INS-1 cells) subjected to oxidative insult. Bottino and co-workers demonstrated that the presence of MnTE-2PyP^{5+}

(AEOL10113) and MnTDE-2-ImP^{5+} (AEOL10150) as supplement to media during islet isolation resulted in up to three-fold increase of the viable mass of human islets, the fact of vital importance for transplantation medicine (Bottino et al. 2002). In a consequent publication a cascade of stressful events triggered by the procedure of islet isolation from the whole pancreas was studied in detail (Bottino et al. 2004). The islet isolation is a lengthy procedure that causes, apart of hypoxia-reoxygenation, mechanical and chemical stress to the cells. This results in activation of stress-related signals NF-κB and poly(ADP-ribose) polymerase (PARP) as well as increased levels of proinflamatory cytokines (Bottino et al. 2004). Mn-porphyrin effectively decreased NF-κB binding to DNA, PARP activation and release of cytokines and chemokines in islet cells, eventually resulting in higher survival and better insulin release(Bottino et al. 2004). Compound MnTE-2PyP^{5+} protected human islets from STZ-induced cell death and ensured better islet function after transplantation into immunodeficient diabetic mice. This holds true for transplantation of islets from the same mice strain (syngeneic), different strain (allogeneic) or in case of transplantation of human islets into mice (xenogeneic) (Sklavos et al. 2010).

Due to the complexity of metabolic and signal pathways, which vary in different cell types and physiological conditions, a possible protective effect of a Mn-porphyrin may depend on the cell type and the type of oxidative insult. Thus, it was reported that MnTMPyP preserved INS-1 cell viability and insulin secretion upon exposure to both NO and O_2^-, while human islets were protected by this compound only from NO, but not from superoxide anion radical (Moriscot et al. 2007).

5.6 Polyphenolic compounds

Polyphenolic compounds, naturally occurring in plants, were extensively studied in recent years as potential remedies against many diseases like cancer, cardio-vascular disorders and diabetes. They can be found in numerous dietary and medicinal plants and comprise an important part of the human diet, though they are generally viewed as nonnutrients. Particularly rich in polyphenols are red grapes, berries, tea leaves and some spices. These polyphenols comprise several types of compounds, i.e. phenolic acids, stilbenes, lignans and flavonoids, the latter are oxygen heterocycles (chromenes, for structures and more detailed classification see review by (Pandey and Rizvi 2009)). Flavonoids are formed in plants from aromatic amino acids phenylalanine and tyrosine, and malonate (Pietta 2000).

Polyphenols act as mild reductants in alkali and neutral pH by reducing common inorganic and organic oxidants and react with radicals both in reversible and irreversible ways, depending on their particular structure, thus, exerting direct antioxidant capacity (Pietta 2000; Pandey and Rizvi 2009). There are also other ways how polyphenols can chemically intervene into oxidative processes as shown in experiments *in vitro*: polyphenols can chelate transition metals like copper and iron, which catalyze propagation of radical chain reactions (Afanas'ev et al. 1995; Korkina and Afanas'ev 1997; Brown et al. 1998) and inhibit the enzymes responsible for superoxide anion production, such as xanthine oxidase (Arimboor et al. 2011) and protein kinase C (Ursini et al. 1994). However, interaction of flavonoids with isolated beef heart mitochondria in vitro caused an additional production of ROS due to inhibition of respiratory complex I (Hodnick et al. 1986; Hodnick et al. 1988; Hodnick, Duval, and Pardini 1994).

Several papers reported the protective effects of polyphenols *in vitro* on islets or cultured insulin-producing cells against oxidative challengers. It was also shown that polyphenols from olive leaves protected INS-1 cells from H_2O_2 toxicity (Cumaoglu et al. 2011). Cells pre-incubated with whole leaf extract or individual polyphenol compound oleuropein, followed by peroxide treatment, showed a lower percentage of necrotic and apoptotic death compared to untreated controls. Polyphenols stimulated activity of catalase, which resulted in a lower level of cellular ROS, along with improvement in insulin production. In this study it was suggested that polyphenols act through a redox-modulating mechanism rather than through direct free radical scavenging. Supplementation of cultured media with polyphenols from green tea increased recovery rates of isolated human and nonhuman primate islets. Polyphenols from tea extracts preserved islets by increasing the level of anti-apoptotic Bcl-2 and decreasing level of pro-apoptotic BAX (Zhang et al. 2004).

In animal models of both T1D and T2D, polyphenols of different origin were reported to lower blood glucose levels (Al-Awwadi et al. 2004; Su, Hung, and Chen 2006; Ciocoiu et al. 2009; Dixit and Kar 2010; Ong et al. 2011). The hypoglycemic effect can be caused by multiple mechanisms. Polyphenols can decrease glucose absorption from intestine as they inhibit amylase, a polysaccharide-hydrolyzing enzyme (Ong et al. 2011). On the other hand, some studies show that polyphenols enhance glucose uptake by muscle cells via increasing expression of glucose transporter Glut4 (Cao et al. 2007; Ong et al. 2011). The third possible mechanism of lowering blood glucose by these compounds could be an inhibition of gluconeogenesis in liver by inhibiting glucose-6-phosphatase, a key enzyme in this process (Ong et al. 2011). Lowering blood glucose itself alleviates excessive metabolic burden on islets and undoubtedly plays a positive role in preserving their mass and function, a notion which was confirmed by (Coskun et al. 2005; Hahm, Park, and Son 2011).

Administration of polyphenols to diabetic animals resulted in a lower level of markers of oxidative stress as judged by lower levels of products of lipid peroxidation in pancreatic homogenates (Coskun et al. 2005), kidneys (Lee, Wang, et al. 2009) and blood plasma (Ciocoiu et al. 2009; Hininger-Favier et al. 2009) of STZ-rats. This may be a result of direct antioxidant activity of polyphenols, as well as a result of their ability to upregulate expression of antioxidant-defense enzymes: glutathione peroxidase (GSHPx) superoxide dismutase and catalase activities (Ciocoiu et al. 2009; Lee, Wang, et al. 2009; Dixit and Kar 2010). It is well known that oxidative stress plays a substantial role in the destruction of beta-cells by infiltrated self macrophages and lymphocytes. The fact that administration of polyphenols to NOD mice, a model for autoimmune diabetes, significantly decreased incidence of the disease is evidence of a direct redox modulation by polyphenols (Zunino, Storms, and Stephensen 2007).

Numerous publications were devoted to study effects of polyphenols on a number of health-related functions. However, it should be noted that the effective therapeutic dose of natural polyphenols is rather high varying from 15 to 500mg/kg of animal body weight. This translates into 1 to 20g of polyphenols a day for a human patient of 65 kg of weight; the amount seems to be unfeasible to get as a part of regular diet or acute therapeutic treatment. However, studies on the effect of polyphenol-rich diets with much lower doses on human patients mostly of T2D report improvement in several metabolic responses, like blood glucose and lipid content, insulin sensitivity and markers of oxidative stress (Banini et al. 2006; Dembinska-Kiec et al. 2008; Stote and Baer 2008; Zunino 2009; Fenercioglu et al. 2010).

6. Concluding remarks

According to the current hypothesis, the lack of anti-ROS defense capacity in pancreatic beta-cells is related to the signaling role of oxidants in glucose sensing and insulin release. In normal physiological conditions the levels of oxidants are low and mitochondria play an essential role in the chain of signaling events by releasing ROS in a controlled manner. Redox reactions in beta-cells are in fine balance, but this balance can be destroyed by persistent hyperglycemia and hyperlipidemia. In conditions of diabetes beta-cells are imposed to a burden of free radicals not considered in evolution and, therefore, antioxidants administered in a proper way can alleviate the oxidative burden and offset the destruction of beta-cells.

7. References

Acharya, J. D., and S. S. Ghaskadbi. 2010. Islets and their antioxidant defense. *Islets* 2 (4):225-35.

Afanas'ev, I. B., E. A. Ostrachovitch, N. E. Abramova, and L. G. Korkina. 1995. Different antioxidant activities of bioflavonoid rutin in normal and iron-overloading rats. *Biochem Pharmacol* 50 (5):627-35.

Affourtit, C., M. Jastroch, and M. D. Brand. 2011. Uncoupling protein-2 attenuates glucose-stimulated insulin secretion in INS-1E insulinoma cells by lowering mitochondrial reactive oxygen species. *Free Radic Biol Med* 50 (5):609-16.

Al-Awwadi, N., J. Azay, P. Poucheret, G. Cassanas, M. Krosniak, C. Auger, F. Gasc, J. M. Rouanet, G. Cros, and P. L. Teissedre. 2004. Antidiabetic activity of red wine polyphenolic extract, ethanol, or both in streptozotocin-treated rats. *J Agric Food Chem* 52 (4):1008-16.

Arimboor, R., M. Rangan, S. G. Aravind, and C. Arumughan. 2011. Tetrahydroamentoflavone (THA) from Semecarpus anacardium as a potent inhibitor of xanthine oxidase. *J Ethnopharmacol* 133 (3):1117-20.

Armann, B., M. S. Hanson, E. Hatch, A. Steffen, and L. A. Fernandez. 2007. Quantification of basal and stimulated ROS levels as predictors of islet potency and function. *Am J Transplant* 7 (1):38-47.

Arner, E. S., J. Nordberg, and A. Holmgren. 1996. Efficient reduction of lipoamide and lipoic acid by mammalian thioredoxin reductase. *Biochem Biophys Res Commun* 225 (1):268-74.

Banday, A. A., A. Marwaha, L. S. Tallam, and M. F. Lokhandwala. 2005. Tempol reduces oxidative stress, improves insulin sensitivity, decreases renal dopamine D1 receptor hyperphosphorylation, and restores D1 receptor-G-protein coupling and function in obese Zucker rats. *Diabetes* 54 (7):2219-26.

Banini, A. E., L. C. Boyd, J. C. Allen, H. G. Allen, and D. L. Sauls. 2006. Muscadine grape products intake, diet and blood constituents of non-diabetic and type 2 diabetic subjects. *Nutrition* 22 (11-12):1137-45.

Batinic-Haberle, I., Z. Rajic, A. Tovmasyan, J. S. Reboucas, X. Ye, K. W. Leong, M. W. Dewhirst, Z. Vujaskovic, L. Benov, and I. Spasojevic. 2011. Diverse functions of cationic Mn(III) N-substituted pyridylporphyrins, recognized as SOD mimics. *Free Radic Biol Med* 51 (5):1035-53.

Benov, L., and I. Batinic-Haberle. 2005. A manganese porphyrin suppresses oxidative stress and extends the life span of streptozotocin-diabetic rats. *Free Radic Res* 39 (1):81-8.

Bindokas, V. P., A. Kuznetsov, S. Sreenan, K. S. Polonsky, M. W. Roe, and L. H. Philipson. 2003. Visualizing superoxide production in normal and diabetic rat islets of Langerhans. *J Biol Chem* 278 (11):9796-801.

Bottino, R., A. N. Balamurugan, S. Bertera, M. Pietropaolo, M. Trucco, and J. D. Piganelli. 2002. Preservation of human islet cell functional mass by anti-oxidative action of a novel SOD mimic compound. *Diabetes* 51 (8):2561-7.

Bottino, R., A. N. Balamurugan, H. Tse, C. Thirunavukkarasu, X. Ge, J. Profozich, M. Milton, A. Ziegenfuss, M. Trucco, and J. D. Piganelli. 2004. Response of human islets to isolation stress and the effect of antioxidant treatment. *Diabetes* 53 (10):2559-68.

Brown, J. E., H. Khodr, R. C. Hider, and C. A. Rice-Evans. 1998. Structural dependence of flavonoid interactions with Cu2+ ions: implications for their antioxidant properties. *Biochem J* 330 (Pt 3):1173-8.

Brownlee, M. 2003. A radical explanation for glucose-induced beta cell dysfunction. *J Clin Invest* 112 (12):1788-90.

Burkart, V., T. Koike, H. H. Brenner, Y. Imai, and H. Kolb. 1993. Dihydrolipoic acid protects pancreatic islet cells from inflammatory attack. *Agents Actions* 38 (1-2):60-5.

Cao, H., I. Hininger-Favier, M. A. Kelly, R. Benaraba, H. D. Dawson, S. Coves, A. M. Roussel, and R. A. Anderson. 2007. Green tea polyphenol extract regulates the expression of genes involved in glucose uptake and insulin signaling in rats fed a high fructose diet. *J Agric Food Chem* 55 (15):6372-8.

Ciocoiu, M., A. Miron, L. Mares, D. Tutunaru, C. Pohaci, M. Groza, and M. Badescu. 2009. The effects of Sambucus nigra polyphenols on oxidative stress and metabolic disorders in experimental diabetes mellitus. *J Physiol Biochem* 65 (3):297-304.

Coskun, O., M. Kanter, A. Korkmaz, and S. Oter. 2005. Quercetin, a flavonoid antioxidant, prevents and protects streptozotocin-induced oxidative stress and beta-cell damage in rat pancreas. *Pharmacol Res* 51 (2):117-23.

Cumaoglu, A., L. Rackova, M. Stefek, M. Kartal, P. Maechler, and C. Karasu. 2011. Effects of olive leaf polyphenols against HO toxicity in insulin secreting beta-cells. *Acta Biochim Pol* 58 (1):45-50.

Czubkowski, P., P. Socha, and J. Pawlowska. 2011. Oxidative stress in liver transplant recipients. *Ann Transplant* 16 (1):99-108.

Darville, M. I., and D. L. Eizirik. 1998. Regulation by cytokines of the inducible nitric oxide synthase promoter in insulin-producing cells. *Diabetologia* 41 (9):1101-8.

De Rosa, S., P. Cirillo, A. Paglia, L. Sasso, V. Di Palma, and M. Chiariello. 2010. Reactive oxygen species and antioxidants in the pathophysiology of cardiovascular disease: does the actual knowledge justify a clinical approach? *Curr Vasc Pharmacol* 8 (2):259-75.

Dembinska-Kiec, A., O. Mykkanen, B. Kiec-Wilk, and H. Mykkanen. 2008. Antioxidant phytochemicals against type 2 diabetes. *Br J Nutr* 99 E Suppl 1:ES109-17.

Dixit, Y., and A. Kar. 2010. Protective role of three vegetable peels in alloxan induced diabetes mellitus in male mice. *Plant Foods Hum Nutr* 65 (3):284-9.

Faust, A., V. Burkart, H. Ulrich, C. H. Weischer, and H. Kolb. 1994. Effect of lipoic acid on cyclophosphamide-induced diabetes and insulitis in non-obese diabetic mice. *Int J Immunopharmacol* 16 (1):61-6.

Favre, D., G. Niederhauser, D. Fahmi, V. Plaisance, S. Brajkovic, N. Beeler, F. Allagnat, J. A. Haefliger, R. Regazzi, G. Waeber, and A. Abderrahmani. 2011. Role for inducible cAMP early repressor in promoting pancreatic beta cell dysfunction evoked by oxidative stress in human and rat islets. *Diabetologia* 54 (9):2337-46.

Fenercioglu, A. K., T. Saler, E. Genc, H. Sabuncu, and Y. Altuntas. 2010. The effects of polyphenol-containing antioxidants on oxidative stress and lipid peroxidation in Type 2 diabetes mellitus without complications. *J Endocrinol Invest* 33 (2):118-24.

Fridlyand, L. E., and L. H. Philipson. 2004. Does the glucose-dependent insulin secretion mechanism itself cause oxidative stress in pancreatic beta-cells? *Diabetes* 53 (8):1942-8.

Genova, M. L., M. M. Pich, A. Biondi, A. Bernacchia, A. Falasca, C. Bovina, G. Formiggini, G. Parenti Castelli, and G. Lenaz. 2003. Mitochondrial production of oxygen radical species and the role of Coenzyme Q as an antioxidant. *Exp Biol Med (Maywood)* 228 (5):506-13.

Gimeno, R. E., M. Dembski, X. Weng, N. Deng, A. W. Shyjan, C. J. Gimeno, F. Iris, S. J. Ellis, E. A. Woolf, and L. A. Tartaglia. 1997. Cloning and characterization of an uncoupling protein homolog: a potential molecular mediator of human thermogenesis. *Diabetes* 46 (5):900-6.

Graciano, M. F., L. R. Santos, R. Curi, and A. R. Carpinelli. 2011. NAD(P)H oxidase participates in the palmitate-induced superoxide production and insulin secretion by rat pancreatic islets. *J Cell Physiol* 226 (4):1110-7.

Grankvist, K., S. L. Marklund, and I. B. Taljedal. 1981. CuZn-superoxide dismutase, Mn-superoxide dismutase, catalase and glutathione peroxidase in pancreatic islets and other tissues in the mouse. *Biochem J* 199 (2):393-8.

Grankvist, K., S. Marklund, J. Sehlin, and I. B. Taljedal. 1979. Superoxide dismutase, catalase and scavengers of hydroxyl radical protect against the toxic action of alloxan on pancreatic islet cells in vitro. *Biochem J* 182 (1):17-25.

Grivennikova, V. G., and A. D. Vinogradov. 2006. Generation of superoxide by the mitochondrial Complex I. *Biochim Biophys Acta* 1757 (5-6):553-61.

Hahm, S. W., J. Park, and Y. S. Son. 2011. Opuntia humifusa stems lower blood glucose and cholesterol levels in streptozotocin-induced diabetic rats. *Nutr Res* 31 (6):479-87.

Halliwell, B and Gutteridge, J.M.C. 2001. *Free Radicals in Biology and Medicine*. Oxford: University Press.

Haramaki, N., D. Han, G. J. Handelman, H. J. Tritschler, and L. Packer. 1997. Cytosolic and mitochondrial systems for NADH- and NADPH-dependent reduction of alpha-lipoic acid. *Free Radic Biol Med* 22 (3):535-42.

Hininger-Favier, I., R. Benaraba, S. Coves, R. A. Anderson, and A. M. Roussel. 2009. Green tea extract decreases oxidative stress and improves insulin sensitivity in an animal model of insulin resistance, the fructose-fed rat. *J Am Coll Nutr* 28 (4):355-61.

Hodnick, W. F., D. L. Duval, and R. S. Pardini. 1994. Inhibition of mitochondrial respiration and cyanide-stimulated generation of reactive oxygen species by selected flavonoids. *Biochem Pharmacol* 47 (3):573-80.

Hodnick, W. F., F. S. Kung, W. J. Roettger, C. W. Bohmont, and R. S. Pardini. 1986. Inhibition of mitochondrial respiration and production of toxic oxygen radicals by flavonoids. A structure-activity study. *Biochem Pharmacol* 35 (14):2345-57.

Hodnick, W. F., E. B. Milosavljevic, J. H. Nelson, and R. S. Pardini. 1988. Electrochemistry of flavonoids. Relationships between redox potentials, inhibition of mitochondrial respiration, and production of oxygen radicals by flavonoids. *Biochem Pharmacol* 37 (13):2607-11.

Jegatheeswaran, S., and A. K. Siriwardena. 2011. Experimental and clinical evidence for modification of hepatic ischaemia-reperfusion injury by N-acetylcysteine during major liver surgery. *HPB (Oxford)* 13 (2):71-8.

Kaneto, H., G. Xu, K. H. Song, K. Suzuma, S. Bonner-Weir, A. Sharma, and G. C. Weir. 2001. Activation of the hexosamine pathway leads to deterioration of pancreatic beta-cell function through the induction of oxidative stress. *J Biol Chem* 276 (33):31099-104.

Khaldi, M. Z., H. Elouil, Y. Guiot, J. C. Henquin, and J. C. Jonas. 2006. Antioxidants N-acetyl-L-cysteine and manganese(III)tetrakis (4-benzoic acid)porphyrin do not prevent beta-cell dysfunction in rat islets cultured in high glucose for 1 wk. *Am J Physiol Endocrinol Metab* 291 (1):E137-46.

Konarkowska, B., J. F. Aitken, J. Kistler, S. Zhang, and G. J. Cooper. 2005. Thiol reducing compounds prevent human amylin-evoked cytotoxicity. *FEBS J* 272 (19):4949-59.

Korkina, L. G., and I. B. Afanas'ev. 1997. Antioxidant and chelating properties of flavonoids. *Adv Pharmacol* 38:151-63.

Kutlu, B., A. K. Cardozo, M. I. Darville, M. Kruhoffer, N. Magnusson, T. Orntoft, and D. L. Eizirik. 2003. Discovery of gene networks regulating cytokine-induced dysfunction and apoptosis in insulin-producing INS-1 cells. *Diabetes* 52 (11):2701-19.

Lacraz, G., F. Figeac, J. Movassat, N. Kassis, and B. Portha. 2010. Diabetic GK/Par rat beta-cells are spontaneously protected against H2O2-triggered apoptosis. A cAMP-dependent adaptive response. *Am J Physiol Endocrinol Metab* 298 (1):E17-27.

Lee, B. W., S. J. Kwon, H. Y. Chae, J. G. Kang, C. S. Kim, S. J. Lee, H. J. Yoo, J. H. Kim, K. S. Park, and S. H. Ihm. 2009. Dose-related cytoprotective effect of alpha-lipoic acid on hydrogen peroxide-induced oxidative stress to pancreatic beta cells. *Free Radic Res* 43 (1):68-77.

Lee, W. C., C. J. Wang, Y. H. Chen, J. D. Hsu, S. Y. Cheng, H. C. Chen, and H. J. Lee. 2009. Polyphenol extracts from Hibiscus sabdariffa Linnaeus attenuate nephropathy in experimental type 1 diabetes. *J Agric Food Chem* 57 (6):2206-10.

Leloup, C., C. Tourrel-Cuzin, C. Magnan, M. Karaca, J. Castel, L. Carneiro, A. L. Colombani, A. Ktorza, L. Casteilla, and L. Penicaud. 2009. Mitochondrial reactive oxygen species are obligatory signals for glucose-induced insulin secretion. *Diabetes* 58 (3):673-81.

Lenzen, S., J. Drinkgern, and M. Tiedge. 1996. Low antioxidant enzyme gene expression in pancreatic islets compared with various other mouse tissues. *Free Radic Biol Med* 20 (3):463-6.

Liu, Y., G. Fiskum, and D. Schubert. 2002. Generation of reactive oxygen species by the mitochondrial electron transport chain. *J Neurochem* 80 (5):780-7.

MacDonald, M. J., R. D. Husain, S. Hoffmann-Benning, and T. R. Baker. 2004. Immunochemical identification of coenzyme Q0-dihydrolipoamide adducts in the E2 components of the alpha-ketoglutarate and pyruvate dehydrogenase complexes partially explains the cellular toxicity of coenzyme Q0. *J Biol Chem* 279 (26):27278-85.

Maechler, P., and C. B. Wollheim. 2001. Mitochondrial function in normal and diabetic beta-cells. *Nature* 414 (6865):807-12.

Magnitsky, S., L. Toulokhonova, T. Yano, V. D. Sled, C. Hagerhall, V. G. Grivennikova, D. S. Burbaev, A. D. Vinogradov, and T. Ohnishi. 2002. EPR characterization of ubisemiquinones and iron-sulfur cluster N2, central components of the energy coupling in the NADH-ubiquinone oxidoreductase (complex I) in situ. *J Bioenerg Biomembr* 34 (3):193-208.

Michalska, M., G. Wolf, R. Walther, and P. Newsholme. 2010. Effects of pharmacological inhibition of NADPH oxidase or iNOS on pro-inflammatory cytokine, palmitic acid or H2O2-induced mouse islet or clonal pancreatic beta-cell dysfunction. *Biosci Rep* 30 (6):445-53.

Moriscot, C., S. Candel, V. Sauret, J. Kerr-Conte, M. J. Richard, M. C. Favrot, and P. Y. Benhamou. 2007. MnTMPyP, a metalloporphyrin-based superoxide dismutase/catalase mimetic, protects INS-1 cells and human pancreatic islets from an in vitro oxidative challenge. *Diabetes Metab* 33 (1):44-53.

Nakada, S., T. Ishikawa, Y. Yamamoto, Y. Kaneko, and K. Nakayama. 2003. Constitutive nitric oxide synthases in rat pancreatic islets: direct imaging of glucose-induced nitric oxide production in beta-cells. *Pflugers Arch* 447 (3):305-11.

Nelson, D.L. and Cox, M.M 2005. *Lehninger Principles of Biochemistry*. Forth Edition ed. New York, NY: W.H. Freeman and Co.

Newsholme, P., E. P. Haber, S. M. Hirabara, E. L. Rebelato, J. Procopio, D. Morgan, H. C. Oliveira-Emilio, A. R. Carpinelli, and R. Curi. 2007. Diabetes associated cell stress and dysfunction: role of mitochondrial and non-mitochondrial ROS production and activity. *J Physiol* 583 (Pt 1):9-24.

Newsholme, P., D. Morgan, E. Rebelato, H. C. Oliveira-Emilio, J. Procopio, R. Curi, and A. Carpinelli. 2009. Insights into the critical role of NADPH oxidase(s) in the normal and dysregulated pancreatic beta cell. *Diabetologia* 52 (12):2489-98.

Nicholls, D.G. , and F.J Ferguson. 2002. Bioenergetics 3. Academic Press.

Ong, K. W., A. Hsu, L. Song, D. Huang, and B. K. Tan. 2011. Polyphenols-rich Vernonia amygdalina shows anti-diabetic effects in streptozotocin-induced diabetic rats. *J Ethnopharmacol* 133 (2):598-607.

Oprescu, A. I., G. Bikopoulos, A. Naassan, E. M. Allister, C. Tang, E. Park, H. Uchino, G. F. Lewis, I. G. Fantus, M. Rozakis-Adcock, M. B. Wheeler, and A. Giacca. 2007. Free fatty acid-induced reduction in glucose-stimulated insulin secretion: evidence for a role of oxidative stress in vitro and in vivo. *Diabetes* 56 (12):2927-37.

Packer, L., E. H. Witt, and H. J. Tritschler. 1995. alpha-Lipoic acid as a biological antioxidant. *Free Radic Biol Med* 19 (2):227-50.

Pandey, K. B., and S. I. Rizvi. 2009. Plant polyphenols as dietary antioxidants in human health and disease. *Oxid Med Cell Longev* 2 (5):270-8.

Parasassi, T., R. Brunelli, G. Costa, M. De Spirito, E. Krasnowska, T. Lundeberg, E. Pittaluga, and F. Ursini. 2010. Thiol redox transitions in cell signaling: a lesson from N-acetylcysteine. *ScientificWorldJournal* 10:1192-202.

Patel, M., and B. J. Day. 1999. Metalloporphyrin class of therapeutic catalytic antioxidants. *Trends Pharmacol Sci* 20 (9):359-64.

Pi, J., Y. Bai, K. W. Daniel, D. Liu, O. Lyght, D. Edelstein, M. Brownlee, B. E. Corkey, and S. Collins. 2009. Persistent oxidative stress due to absence of uncoupling protein 2 associated with impaired pancreatic beta-cell function. *Endocrinology* 150 (7):3040-8.

Pietta, P. G. 2000. Flavonoids as antioxidants. *J Nat Prod* 63 (7):1035-42.

Pocernich, C. B., M. L. Bader Lange, R. Sultana, and D. A. Butterfield. 2011. Nutritional approaches to modulate oxidative stress in Alzheimer's disease. *Curr Alzheimer Res* 8 (5):452-69.

Ramudo, L., and M. A. Manso. 2010. N-acetylcysteine in acute pancreatitis. *World J Gastrointest Pharmacol Ther* 1 (1):21-6.

Ricquier, D., and F. Bouillaud. 2000. Mitochondrial uncoupling proteins: from mitochondria to the regulation of energy balance. *J Physiol* 529 Pt 1:3-10.

Rigoulet, M., E. D. Yoboue, and A. Devin. 2011. Mitochondrial ROS generation and its regulation: mechanisms involved in H(2)O(2) signaling. *Antioxid Redox Signal* 14 (3):459-68.

Robertson, R. P., J. Harmon, P. O. Tran, Y. Tanaka, and H. Takahashi. 2003. Glucose toxicity in beta-cells: type 2 diabetes, good radicals gone bad, and the glutathione connection. *Diabetes* 52 (3):581-7.

Sakai, K., K. Matsumoto, T. Nishikawa, M. Suefuji, K. Nakamaru, Y. Hirashima, J. Kawashima, T. Shirotani, K. Ichinose, M. Brownlee, and E. Araki. 2003. Mitochondrial reactive oxygen species reduce insulin secretion by pancreatic beta-cells. *Biochem Biophys Res Commun* 300 (1):216-22.

Sakuraba, H., H. Mizukami, N. Yagihashi, R. Wada, C. Hanyu, and S. Yagihashi. 2002. Reduced beta-cell mass and expression of oxidative stress-related DNA damage in the islet of Japanese Type II diabetic patients. *Diabetologia* 45 (1):85-96.

Santos, L. R., E. Rebelato, M. F. Graciano, F. Abdulkader, R. Curi, and A. R. Carpinelli. 2011. Oleic Acid Modulates Metabolic Substrate Channeling during Glucose-Stimulated Insulin Secretion via NAD(P)H Oxidase. *Endocrinology* 152 (10):3614-21.

Schroeder, M. M., R. J. Belloto, Jr., R. A. Hudson, and M. F. McInerney. 2005. Effects of antioxidants coenzyme Q10 and lipoic acid on interleukin-1 beta-mediated inhibition of glucose-stimulated insulin release from cultured mouse pancreatic islets. *Immunopharmacol Immunotoxicol* 27 (1):109-22.

Selivanov, V. A., T. V. Votyakova, V. N. Pivtoraiko, J. Zeak, T. Sukhomlin, M. Trucco, J. Roca, and M. Cascante. 2011. Reactive oxygen species production by forward and reverse electron fluxes in the mitochondrial respiratory chain. *PLoS Comput Biol* 7 (3):e1001115.

Selivanov, V. A., T. V. Votyakova, J. A. Zeak, M. Trucco, J. Roca, and M. Cascante. 2009. Bistability of mitochondrial respiration underlies paradoxical reactive oxygen species generation induced by anoxia. *PLoS Comput Biol* 5 (12):e1000619.

Selivanov, V. A., J. A. Zeak, J. Roca, M. Cascante, M. Trucco, and T. V. Votyakova. 2008. The role of external and matrix pH in mitochondrial reactive oxygen species generation. *J Biol Chem* 283 (43):29292-300.

Shen, W., K. Liu, C. Tian, L. Yang, X. Li, J. Ren, L. Packer, E. Head, E. Sharman, and J. Liu. 2008. Protective effects of R-alpha-lipoic acid and acetyl-L-carnitine in MIN6 and isolated rat islet cells chronically exposed to oleic acid. *J Cell Biochem* 104 (4):1232-43.

Sklavos, M. M., S. Bertera, H. M. Tse, R. Bottino, J. He, J. N. Beilke, M. G. Coulombe, R. G. Gill, J. D. Crapo, M. Trucco, and J. D. Piganelli. 2010. Redox modulation protects islets from transplant-related injury. *Diabetes* 59 (7):1731-8.

Skulachev, V. P., V. N. Anisimov, Y. N. Antonenko, L. E. Bakeeva, B. V. Chernyak, V. P. Erichev, O. F. Filenko, N. I. Kalinina, V. I. Kapelko, N. G. Kolosova, B. P. Kopnin, G. A. Korshunova, M. R. Lichinitser, L. A. Obukhova, E. G. Pasyukova, O. I. Pisarenko, V. A. Roginsky, E. K. Ruuge, Senin, II, Severina, II, M. V. Skulachev, I. M. Spivak, V. N. Tashlitsky, V. A. Tkachuk, M. Y. Vyssokikh, L. S. Yaguzhinsky, and D. B. Zorov. 2009. An attempt to prevent senescence: a mitochondrial approach. *Biochim Biophys Acta* 1787 (5):437-61.

Song, K. H., W. J. Lee, J. M. Koh, H. S. Kim, J. Y. Youn, H. S. Park, E. H. Koh, M. S. Kim, J. H. Youn, K. U. Lee, and J. Y. Park. 2005. alpha-Lipoic acid prevents diabetes mellitus in diabetes-prone obese rats. *Biochem Biophys Res Commun* 326 (1):197-202.

Soule, B. P., F. Hyodo, K. Matsumoto, N. L. Simone, J. A. Cook, M. C. Krishna, and J. B. Mitchell. 2007. The chemistry and biology of nitroxide compounds. *Free Radic Biol Med* 42 (11):1632-50.

Starkov, A. A., and G. Fiskum. 2003. Regulation of brain mitochondrial H2O2 production by membrane potential and NAD(P)H redox state. *J Neurochem* 86 (5):1101-7.

Starkov, A. A., B. M. Polster, and G. Fiskum. 2002. Regulation of hydrogen peroxide production by brain mitochondria by calcium and Bax. *J Neurochem* 83 (1):220-8.

Stote, K. S., and D. J. Baer. 2008. Tea consumption may improve biomarkers of insulin sensitivity and risk factors for diabetes. *J Nutr* 138 (8):1584S-1588S.

Su, H. C., L. M. Hung, and J. K. Chen. 2006. Resveratrol, a red wine antioxidant, possesses an insulin-like effect in streptozotocin-induced diabetic rats. *Am J Physiol Endocrinol Metab* 290 (6):E1339-46.

Takatori, A., Y. Ishii, S. Itagaki, S. Kyuwa, and Y. Yoshikawa. 2004. Amelioration of the beta-cell dysfunction in diabetic APA hamsters by antioxidants and AGE inhibitor treatments. *Diabetes Metab Res Rev* 20 (3):211-8.

Tanaka, Y., C. E. Gleason, P. O. Tran, J. S. Harmon, and R. P. Robertson. 1999. Prevention of glucose toxicity in HIT-T15 cells and Zucker diabetic fatty rats by antioxidants. *Proc Natl Acad Sci U S A* 96 (19):10857-62.

Tanaka, Y., P. O. Tran, J. Harmon, and R. P. Robertson. 2002. A role for glutathione peroxidase in protecting pancreatic beta cells against oxidative stress in a model of glucose toxicity. *Proc Natl Acad Sci U S A* 99 (19):12363-8.

Tang, C., P. Han, A. I. Oprescu, S. C. Lee, A. V. Gyulkhandanyan, G. N. Chan, M. B. Wheeler, and A. Giacca. 2007. Evidence for a role of superoxide generation in glucose-induced beta-cell dysfunction in vivo. *Diabetes* 56 (11):2722-31.

Targonsky, E. D., F. Dai, V. Koshkin, G. T. Karaman, A. V. Gyulkhandanyan, Y. Zhang, C. B. Chan, and M. B. Wheeler. 2006. alpha-lipoic acid regulates AMP-activated protein kinase and inhibits insulin secretion from beta cells. *Diabetologia* 49 (7):1587-98.

Tiedge, M., S. Lortz, J. Drinkgern, and S. Lenzen. 1997. Relation between antioxidant enzyme gene expression and antioxidative defense status of insulin-producing cells. *Diabetes* 46 (11):1733-42.

Tsubouchi, H., T. Inoguchi, M. Inuo, M. Kakimoto, T. Sonta, N. Sonoda, S. Sasaki, K. Kobayashi, H. Sumimoto, and H. Nawata. 2005. Sulfonylurea as well as elevated glucose levels stimulate reactive oxygen species production in the pancreatic beta-cell line, MIN6-a role of NAD(P)H oxidase in beta-cells. *Biochem Biophys Res Commun* 326 (1):60-5.

Turrens, J. F. 2003. Mitochondrial formation of reactive oxygen species. *J Physiol* 552 (Pt 2):335-44.

Uchizono, Y., R. Takeya, M. Iwase, N. Sasaki, M. Oku, H. Imoto, M. Iida, and H. Sumimoto. 2006. Expression of isoforms of NADPH oxidase components in rat pancreatic islets. *Life Sci* 80 (2):133-9.

Ursini, F., M. Maiorino, P. Morazzoni, A. Roveri, and G. Pifferi. 1994. A novel antioxidant flavonoid (IdB 1031) affecting molecular mechanisms of cellular activation. *Free Radic Biol Med* 16 (5):547-53.

Vinogradov, A. D. 2008. NADH/NAD+ interaction with NADH: ubiquinone oxidoreductase (complex I). *Biochim Biophys Acta* 1777 (7-8):729-34.

Votyakova, T. V., and I. J. Reynolds. 2001. DeltaPsi(m)-Dependent and -independent production of reactive oxygen species by rat brain mitochondria. *J Neurochem* 79 (2):266-77.

Wang, X., H. Li, D. De Leo, W. Guo, V. Koshkin, I. G. Fantus, A. Giacca, C. B. Chan, S. Der, and M. B. Wheeler. 2004. Gene and protein kinase expression profiling of reactive oxygen species-associated lipotoxicity in the pancreatic beta-cell line MIN6. *Diabetes* 53 (1):129-40.

Welsh, N., B. Margulis, L. A. Borg, H. J. Wiklund, J. Saldeen, M. Flodstrom, M. A. Mello, A. Andersson, D. G. Pipeleers, C. Hellerstrom, and et al. 1995. Differences in the expression of heat-shock proteins and antioxidant enzymes between human and rodent pancreatic islets: implications for the pathogenesis of insulin-dependent diabetes mellitus. *Mol Med* 1 (7):806-20.

Wilcox, C. S. 2010. Effects of tempol and redox-cycling nitroxides in models of oxidative stress. *Pharmacol Ther* 126 (2):119-45.

Wilcox, C. S., and A. Pearlman. 2008. Chemistry and antihypertensive effects of tempol and other nitroxides. *Pharmacol Rev* 60 (4):418-69.

Zhang, G., S. Matsumoto, S. H. Hyon, S. A. Qualley, L. Upshaw, D. M. Strong, and J. A. Reems. 2004. Polyphenol, an extract of green tea, increases culture recovery rates of isolated islets from nonhuman primate pancreata and marginal grade human pancreata. *Cell Transplant* 13 (2):145-52.

Zhang, Z., J. Jiang, P. Yu, X. Zeng, J. W. Larrick, and Y. Wang. 2009. Hypoglycemic and beta cell protective effects of andrographolide analogue for diabetes treatment. *J Transl Med* 7:62.

Zraika, S., K. Aston-Mourney, D. R. Laybutt, M. Kebede, M. E. Dunlop, J. Proietto, and S. Andrikopoulos. 2006. The influence of genetic background on the induction of oxidative stress and impaired insulin secretion in mouse islets. *Diabetologia* 49 (6):1254-63.
Zunino, S. 2009. Type 2 diabetes and glycemic response to grapes or grape products. *J Nutr* 139 (9):1794S-800S.
Zunino, S. J., D. H. Storms, and C. B. Stephensen. 2007. Diets rich in polyphenols and vitamin A inhibit the development of type I autoimmune diabetes in nonobese diabetic mice. *J Nutr* 137 (5):1216-21.

9

Nitric Oxide Signaling During Senescence and Programmed Cell Death in Leaves

Meena Misra[1,2,*], Amarendra Narayan Misra[1,2] and Ranjeet Singh[1,2]
[1]*Post-Graduate Department of Biosciences & Biotechnology*
Fakir Mohan University, Balasore
[2]*Centre for Life Sciences, School of Natural Sciences, Central University of Jharkhand*
Ratu-Lohardaga Road, Brambe, Ranchi
India

1. Introduction

Senescence is a ubiquitous developmental process that leads to the death of a cell, an organ, or an organism and occurs at the final stage of their development. There is a striking divergence and convergence between plants and animals regarding senescence regulation (Kenyon, 2001). The mechanisms of regulation of ageing in animals including p53, telomerase and telomere dynamics, DNA damage sensing and repair, and transcriptional activation and inactivation by histone acetylation/deacetylation are either not present or do not appear to play an important role in plant ageing. On the contrary, plants have evolved their own unique senescence-regulating mechanisms. These include the modulation of senescence by phytohormones, photosynthetic machinery, and protein degradation. In plants, the chloroplast is reported to be the first origin and target for initiating senescence (Misra and Biswal, 1980, 1981; 1982a, b, c, Biswal et al., 2001; Dilnawaz et al. 2001; Misra et al. 2011a), whereas in animals the mitochondrion serves as the initiator (Thomas, 2002). Thus, we may infer that plants and animals have evolved conserved strategies for the regulation of senescence, while employing diverse molecular mechanisms that have been shaped during the long history of evolution. Senescence is a genetically-controlled developmental programme, but it has no adaptive advantages in animals, except that in plants leaf senescence is a recruited nutrient recycle programme and hence is considered to have a strong adaptive advantage (Bleecker, 1998). This is the final developmental phase of a leaf which starts with nutrient salvage and ends with cell death. However, until late in senescence the process requires cell viability and is often reversible (Thomas *et al.*, 2003). There has been some debate about the degree of overlap of senescence and PCD (Thomas *et al.*, 2003; van Doorn & Woltering, 2004, 2008). van Doorn & Woltering (2004, 2008) identified three positions regarding the overlap in senescence and PCD. Some authors assumed total overlap (Noodén, 2004), but van Doorn & Woltering (2004, 2008) postulated that overlap is complete and senescence and PCD are synchronous. However a minority, argued that there was no overlap (Delorme *et al.*, 2000; Thomas *et al.*, 2003). van Doorn & Woltering (2008)

* Corresponding Author

defined PCD as 'the process that leads to the moment of death and the degradation that goes on after this moment'. Senescent cells are being actively recycled, which means that the cells are subjected to nucleases, proteases and photosynthetic breakdown along with various other senescent-related stresses. Normally one might expect that a cell which is subjected to this type of stress would activate PCD.

2. Symptoms associated with leaf senescence

Chloroplasts are the first organelles in leaves to show the symptoms of leaf senescence (Misra and Biswal, 1980, Dilnawaz et al., 2001; Misra et al., 2002; Misra et al. 2011a). Chlorophyll (Chl) is the key photosynthetic pigment required for the absorption of sunlight and photo-energy transduction by plants for primary productivity on earth. Absorbed energy can also be transferred from Chl to oxygen, resulting in the production of reactive oxygen species (ROS) (Apel and Hirt,2004). Likewise, inhibition of Chl biosynthesis or degradation can lead to the accumulation of phototoxic intermediates and ROS production (Apel and Hirt,2004; Pruzinska et al.2007).

There is a stepwise degradation of Chl (Hortensteiner 2006) during leaf senescence or accelerated cell death caused by various biotic or abiotic stresses (Matile et al., 1999). Hence, leaf yellowing is regarded as an external symptom of programmed cell death processes in senescing leaf cells (Nooden et al., 1997; Hortensteiner and Matile, 2004). The first visible event during senescence is leaf yellowing (Misra and Biswal, 1980; Quirino et al., 2000). Photosynthesis gradually slows down (Dilnawaz et al., 2001; Misra et al., 2002; 2010; 2011a; Misra and Misra, 1987; 2010; Misra and Biswal, 1982) and finally stops, but mitochondria and nucleus remain functional and respiration continues until the end of the senescence (Srivastava, 2002). Besides leaf age, darkness is more commonly considered to be an inducer of leaf senescence and induces loss of chlorophyll (Misra and Biswal, 1980, 1981), nucleic acid and protein (Misra and Biswal, 1982b), increase in the photosynthetic excitation pressure (Misra et al., 2011a) and alterations in the thylakoid membrane function such as photosystem I (PS I) and PS II activities (Biswal et al., 2001; Misra and Biswal, 1982a; Misra, 1993a). The expression of various photosynthesis associated genes was reduced (Misra, 1993b; Kleber-Janke and Krupinska, 1997; Weaver et al., 1998).

A decrease in nucleic acids occurs during leaf senescence (Misra and Biswal, 1982b). Total RNA levels are rapidly reduced along with the progression of senescence (Misra, 1993b; Taylor et al., 1993). Their initial decline is apparent in the chloroplast rRNAs and cytoplasmic rRNAs. This decrease in rRNAs contents is followed by that of the cytoplasmic mRNA and tRNA, and is accompanied by enhanced activity of several RNases. The earliest structural evidence for senescence appears in the chloroplast, manifested as changes in the grana and the formation of lipid droplets. Polysomes and ribosomes generally decrease fairly early, reflecting diminished protein synthesis. In comparison, the mitochondria and nucleus, both essential to energy production and gene expression, remain intact until the last stages (Quirino et al., 2000). This is because the senescing cells must be functional for progression of senescence until a late stage of senescence, possibly for the efficient re-utilization of cellular materials. In the final stage of senescence, when the leaves have turned almost completely yellow, typical symptoms of PCD, such as controlled vacuolar collapse, chromatin condensation, and DNA laddering have been reliably detected in naturally senescing leaves from a variety of plants including rice, tobacco, and five trees (Buchanan-

Wollaston et al. 2003). Eventually, disintegration of the plasma and vacuolar membranes become apparent, and the loss of integrity in the plasma membrane then leads to a disruption in cellular homeostasis, ending the life of a cell (Buchanan-Wollaston et al. 2003). Lipid-degrading enzymes, such as phospholipase D, phosphatidic acid phosphatase, lytic acyl hydrolase, and lipooxygenase, appear to be involved in the hydrolysis and metabolism of membrane lipids (Thompson et al., 2000). The majority of fatty acids is either oxidized to provide energy for the senescence process or processed to α-ketoglutarate via the glyoxylate cycle. This α-ketoglutarate can be converted into phloem-mobile sugars through gluconeogenesis or else used to mobilize the amino acids released during leaf protein degradation (Buchanan-Wollaston, 1997).

3. Factors regulating senescence in leaves

Leaf senescence is controlled by various internal and external factors including leaf age, light conditions, nutrient supply and environmental stress (Lim and Nam 2005). Plants integrate these factors through endogenous signaling molecules and coordinate the senescence process. Senescence is under genetic control and requires differential expression of specific genes. Expression of photosynthesis-associated genes (*PAGs*) is downregulated, while many other genes, designated as senescence-associated genes (*SAGs*), are upregulated during senescence (Lim and Nam, 2005; Lim et al., 2003).

During plant growth light dosage has an effect on ageing; high light intensity results in premature senescence when compared with growth under standard light intensities, while low light intensities delay the senescence process (Misra 1995; Misra and Misra, 1987; 1989a, b; 1995; Nooden et al., 1996).

Many other stress-inducing conditions such as drought, darkness, high temperature, high light, ozone, and pathogen attack can hasten leaf senescence as well (Biswal et al., 2001; Misra et al., 2001a, b; Misra and Misra 1986a, b; 2002; Misra and Biswal 1980, 1981, 1982a, b, c; 1987; 2000; Lim and Nam, 2005).

3.1 Hormonal regulation of leaf senescence

Plant hormones play an important roles in the regulation of the onset of senescence (Misra and Biswal 1980; Misra and Misra 1989a; 1991). All the identified phytohormones are reported to be involved in leaf senescence. Among the five classic hormones, the roles of ethylene and cytokinin in leaf senescence have long been established. Besides, jasmonic acid, salicylic acid, nitric oxide, and brassinosteroid are also implicated in regulating leaf senescence.

Cytokinins play a master regulatory role in leaf senescence (Misra and Biswal, 1980). While increasing cytokinin production could delay leaf senescence (Gan and Amasino, 1995), reducing endogenous cytokinin levels resulted in accelerated senescence (Masferrer et al., 2002). The components involved in cytokinin signalling is reported by Kieber and Skallar (2010) and Hwang et al. (2002). Among the genes characterised, only the receptor CKI1 and the response regulator ARR2 appear to be involved in regulating leaf senescence (Hwang et al., 2002).

Auxin, which regulates many aspects of plant growth and development, also influence leaf senescence, and is reported to delay leaf senescence (Misra and Biswal, 1980; Ellis et al., 2005).

The role of ethylene in senescence has been demonstrated by several studies. Both ethylene-insensitive mutants *etr1-1* and *ein2* show increased leaf longevities (Grbic and Bleecker, 1995; Johnson and Ecker, 1998) and antisense suppression of the tomato ACC oxidase resulted in delayed leaf senescence (John et al., 1995). Exogenously applied ethylene induces premature leaf senescence in *Arabidopsis* (Weaver et al., 1998). However, constitutive application of ethylene does not change the longevity of the leaves. Both *ctr1* mutants and *Arabidopsis* plants grown in the continuous presence of exogenous ethylene did not show premature senescence (Grbic and Bleecker, 1995). These studies suggest that ethylene does not directly regulate the onset of leaf senescence, but it acts to modulate the timing of leaf senescence (Grbic and Bleecker, 1995; Jing et al., 2002). The combined physiological and genetic studies by Buchanan-Wollaston et al. (2003) using *Arabidopsis* mutants exhibited a dual function of ethylene, both as an inducer and a repressor, in the induction of leaf senescence and that such a role of ethylene was differentially modulated by multiple genetic loci.

Salicylic acid (SA) is also involved in the regulation of this senescence. Its levels increase in senescing leaves, possibly accounting for the enhanced expression of *SAG* genes (Morris et al., 2000). Mutant analysis in *Arabidopsis* plants showed that the SA pathway has a very specific role in natural senescence, possibly in the final death phase or PCD (Morris et al., 2000; Buchanan-Wollaston et al., 2005).

Jasmonates (JAs) were proposed to play a regulatory role in leaf senescence. Early experiments, involving treating leaves or cell cultures with jasmonates, showed that a loss of chlorophyll was induced and the expression of photosynthetic-associated genes was suppressed (Creelman and Mullet, 1997). Jasmonates could rapidly induce the expression of chlorophyllase (Tsuchiya et al., 1999) and several *SAG*s (He et al., 2002). However, mutants that are impaired in JA biosynthesis and signalling (Berger, 2001), were not aberrant in phenotypic expression for leaf senescence, suggesting that jasmonates are not essential for leaf senescence. In addition, transgenic plants that either overexpress allene oxide synthase, jasmonic acid carboxyl methyltransferase, or underexpress lipoxygenase, did not show abnormal leaf senescence. Thus, molecular genetic analysis of jasmonate-related mutants did not generate any crucial link between jasmonate action and leaf senescence, and the role of jasmonates in leaf senescence is still a question of debate. It is possible that other factors might induce leaf senescence in the absence of JA.

Brassinosteroids could promote leaf senescence and mutants deficient in brassinosteroids showed altered leaf senescence indicating their involvement in leaf senescence (Clouse and Sasse, 1998; Yin et al., 2002). Nevertheless, a systematic study is needed to dissect the regulatory functions of these hormones in leaf senescence.

Abscisic acid (ABA) is a key hormone that mediates plant responses to environmental stresses. ABA levels rise in senescing leaves, and exogenously applied ABA induces the expression of several SAGs (Weaver et al., 1998), consistent with an effect on leaf senescence. Environmental stresses, including drought, high salinity, and low temperatures, have positive influences on leaf senescence and, under those conditions, leaf ABA contents rise. Concurrent with this increase (Gepstein and Thimann, 1980), genes encoding the key enzyme in ABA biosynthesis show greater expression (van der Graaff et al., 2006). Physiological analysis has shown that ABA (abscisic acid) could promote leaf senescence, but to date molecular genetic analysis has not generated a crucial link between ABA and leaf senescence (Fedoroff, 2002). But there is an interaction and crosstalk between plant hormones, regulating the signaling network, which in turn regulates almost all processes in plants. This is likely the case in regulating leaf senescence by plant hormones.

4. Degradation of chloroplast protein during senescence

Protein synthesis and protein degradation are equally important for changes in the protein pattern and are of fundamental importance for the normal development, homeostasis and final death of a plant cell (Vierstra, 1996). Proteolysis in plants is a complex process involving many enzymes and multifarious proteolytic pathways in various cellular compartments (Sakamoto, 2006; Misra et al., 1991). ATP-independent and ATP-dependent proteolytic pathways are involved in plant proteolysis (Callis, 1995). It has been proposed that chloroplast proteins may be degraded by vacuolar proteases, via the ubiquitin pathway in the cytosol and also by the plastidial Clp system (Vierstra, 1996). The protein and RNA degradation parallels a loss in photosynthetic activity (Misra and Biswal, 1981a; 1982; Buchanan-Wollaston et al., 2003). The degradation products are transported out of the leaves to other parts of the plant, and so the senescing leaf continues to function as a source of nutrients to the whole plant. This nutrient salvage leads to the hydrolysis of macromolecules and subsequent remobilization requires a complex array of metabolic pathways. The initial stage of senescence symptoms is a breakdown in membrane structure within the chloroplasts, where more than 50% of the leaf protein and more than 70% of its lipids are present (Misra and Biswal, 1982; Misra and Misra, 1987; Hortensteiner and Feller, 2002). Chloroplasts are a major site of protein degradation during leaf senescence (Misra et al. 2002; Feller et al., 2008). Chloroplast degeneration is accompanied by chlorophyll degradation and a progressive loss of chloroplast proteins, e.g., ribulose bisphosphate carboxylase (Rubisco) and chlorophyll *a/b* binding protein (CAB) (Misra 1993a, b; Misra and Biswal, 1982b; Dilnawaz et al. 2001). The complete hydrolysis of proteins to free amino acids depends on the actions of several endo- and exopeptidases (Misra 1993a, b). Rubisco is the most abundant protein on earth and contributes a high percentage to the total leaf nitrogen remobilized to the developing parts of the plant by leaf senescence (Feller et al., 2008). Roberts et al. (2003) reported a serine protease activity in senescing wheat leaves for which Rubisco was a target protein. LSU (Large Subunit Rubisco) proteolysis is most likely catalyzed enzymatically by a metalloendopeptidase or a cysteine endopeptidase inside the chloroplasts (Thoenen et al., 2007) and/or non-enzymatically cleaved by reactive oxygen species (Dilnawaz et al., 2001; Nakano et al., 2006). Reactive oxygen species may directly cleave the plastid proteins or modify it in a manner making it more susceptible to proteolytic cleavage (Misra 1993a). Increased levels of reactive oxygen species during leaf senescence is reported (Tunc-Ozdemir et al., 2009; Misra et al., 2011a).

4.1 Regulation of chlorophyll degradation

Chlorophyll (chl), the most abundant pigment on earth, is a key component of photosynthesis required for the absorption of sunlight. A common and stepwise Chl degradation pathway in higher plants is elucidated by Hortensteiner (2006). Plant leaves generally change in color from green to yellow or red as a result of the breakdown of the green pigment chlorophyll (Chl) combined with carotenoid retention or anthocyanin accumulation, as a result of leaf senescence or accelerated cell death caused by various biotic or abiotic stresses (Matile et al.,

1999, Nooden et al., 1997). Chl catabolism is a multistep pathway. Chls in thylakoid membranes are degraded to nonfluorescent Chl catabolites and are accumulated in the vacuoles of senescing cells (Matile et al., 1988; Hortensteiner, 2006). There are three steps in Chl catabolism before the porphyrin ring is cleaved:

Step I - chlorophyllase converts Chl a into chlorophyllide a (Chlide a),

Step II - Mg-chelating substance converts Chlide a into pheophorbide a (Pheide a),

Step III - Pheide a oxygenase (PaO) converts Pheide a into red Chl catabolite.

Subcellular fractionation experiments show that chlorophyllase activity is present in the inner envelope membrane of chloroplasts (Matile et al., 1997). However, the substrate Chl, is tightly bound to the light-harvesting chlorophyll binding protein I (LHCPI) and II complexes in the thylakoid membranes. The spatial separation between Chlase and Chl has led to the hypothesis for a possible Chl carrier in the chloroplast stroma that shuttles between thylakoid and inner envelope membranes for Chl transport (Matile et al., 1997, 1999; Hortensteiner and Matile, 2004). Satoh et al. (1998) proposed the water-soluble chlorophyll protein (WSCP) as a feasible candidate for a Chl carrier. However, a recent report indicated that WSCP might act as a Chlide transporter during Chl synthesis in developing leaves rather than during Chl degradation in senescing leaves (Reinbothe et al., 2004). In higher plants, Chlorophyllase (CLH) genes encode soluble proteins that are predicted to localize in the cytoplasm, vacuole, or chloroplast stroma (Tsuchiya et al., 1999; Takamiya et al., 2000; Okazawa et al., 2006). In Arabidopsis thaliana, At CLH1 (At1g19670) encodes a putative cytosolic chlorophyllase and is upregulatedin response to stress and/or senescence-related hormones such as wounding, methyl jasmonate, and coronatine (Benedetti et al., 1998; Tsuchiya et al., 1999; Benedetti and Arruda, 2002). On the other hand, At CLH2 (At5g43860), which encodes a putative chloroplast chlorophyllase, is constitutively expressed at a low level throughout leaf development, and this expression is unaffected by either by stress or by senescence (Tsuchiya et al., 1999; Benedetti and Arruda, 2002). However, the gene(s) encoding the inner envelope membrane-bound chlorophyllase has not yet been identified, and it is still unknown which chlorophyllases are involved in the first step of Chl catabolism during leaf senescence. In this respect, the stay-green (also called nonyellowing) mutants isolated from several plants have been of great interest in elucidating the genetic and biochemical mechanisms of Chl breakdown during leaf senescence (Thomas and Smart, 1993; Thomas and Howarth, 2000).

The nonyellowing *sid* mutant of *F. pratensis* accumulates significant amounts of Chlide a and Pheide a in the senescing leaves and has no PaO activity (Roca et al., 2004), suggesting that Chl dephytylation by Chlases is suppressed by Pheide a accumulation in senescing leaves. In Arabidopsis, the PaO-impaired mutants, pao1 and AsACD1, which were induced by T-DNA insertion and antisense silencing of ACCELERATED CELL DEATH1 (ACD1) (Greenberg and Ausubel, 1993), respectively, maintained leaf greenness only during dark-induced senescence (Tanaka et al., 2003; Pruzinska et al., 2005). However, the decreased PaO activity in the mutants resulted in age- and light-dependent cell death in mature leaves, possibly due to the accumulation of the photodynamic Chl catabolite Pheide a (Pruzinska et al.,2005; Tanaka et al., 2003). Thus, high levels of Pheide a in the AsACD1 leaves did not exhibit persistent greenness during natural senescence (Tanaka et al., 2003). Another PaO-impaired mutant in maize (*Zea mays*), lethal leaf spot-1 (lls1), forms several necrotic spots that spread continuously until all of the mature leaves are wilted and bleached (Gray et al., 2002). The red Chl catabolite reductase-impaired acd2 mutants in Arabidopsis also exhibit spontaneous cell death lesions in mature leaves (Mach et al., 2001). These defective phenotypes demonstrate that genetic lesions associated with the Chl catabolic pathway will ultimately result in cell death in green mature leaves, thus demonstrating that Chl catabolism is tightly regulated throughout plant development. Since the staygreen(Sgr) mutants do not show any age- and/or light-dependent cell death syndrome under natural growth conditions, it has been proposed that, during leaf senescence, the stay-green genes may encode the regulatory proteins triggering Chl catabolism rather than one of the Chl catabolic enzymes (Pruznska et al., 2005; Tanaka et al., 2003; Hortensteiner, 2006). It was recently reported that the stay-green y mutant in Festuca/Lolium forage plants also resulted from a frameshift mutation of a Sgr homolog (Armstead et al., 2006). Sgr is highly senescence-inducible and encodes a previously uncharacterized chloroplast protein whose amino acid sequence is extremely conserved in higher plants. The overexpression of Sgr in transgenic rice and the reduced expression of Sgr homologs in the Arabidopsis pao1 and acd1-20 mutants demonstrate that Chl degradation is regulated by Sgr at the transcriptional level and that Sgr transcription is repressed by either an increase of Pheide a or a lack of PaO activity in the senescing leaves.

5. Dark induced senescence in leaves

Leaf senescence is an active process regulated by exogenous and endogenous factors especially light quality and quantity (Nooden et al., 1996). Darkness is more commonly considered to be an inducer of senescence. Besides leaf age, darkness is more commonly considered to be an inducer of leaf senescence and induces loss of chlorophyll (Misra and Biswal, 1980, 1981), nucleic acid and protein (Misra and Biswal, 1982b), and alterations in the thylakoid membrane function such as photosystem I (PS I) and PS II activities (Biswal et al., 2001; Misra and Biswal, 1982a; Misra, 1993a). The expression of various photosynthesis associated genes was reduced (Misra, 1993b; Kleber-Janke and Krupinska, 1997; Weaver et al., 1998). The dark induced senescence is supposed to be be reversed by returning plants to the light, at least within a thresh hold dark treatment (Kleber-Janke and Krupinska, 1997). The *SAG* gene expression in response to both whole-plant darkness and natural senescence, showed varied responses, with some genes responding similarly to both treatments and others responding very differently (Kleber-Janke and Krupinska, 1997; Weaver et al., 1998). When whole plants were darkened, in contrast, by all criteria except loss of total protein and chlorophyll (which still occurred less quickly and less strongly than with individual leaf darkness), senescence was not induced (Weaver and Amasino,2001).

6. Regulation of senescence by cytokinins, sugers and light

During the process of leaf senescence, chlorophyll and photosynthetic proteins are degraded (Humbeck et al., 1996). There are several factors that can accelerate or delay this breakdown of the photosynthetic apparatus. Whereas the plant growth regulators ABA and ethylene accelerate the symptoms of senescence (Smart, 1994), exogenous application of cytokinins inhibits the degradation of chlorophyll and photosynthetic proteins (Richmond and Lang, 1957; Badenoch-Jones et al., 1996). Senescence is also delayed in transgenic plants producing cytokinin by expression of a bacterial gene encoding IPT, the enzyme catalyzing the first step of cytokinin synthesis (Smart et al., 1991; Gan and Amasino, 1995).

Sugar signaling has emerged as an important regulator of leaf senescence (Rolland et al., 2002). Several lines of evidence suggest that a high concentration of sugars lowers photosynthetic activity and induces leaf senescence (Quirino et al., 2000; Moore et al., 2003). Senescence is then triggered when those concentrations go above an acceptable level. Hexokinases are involved in sugar sensing in higher plants. Studies using their over expressors have demonstrated that increased hexokinase levels stimulate a rise in sugar content that is associated with reduced photosynthetic activity (Jang et al., 1997; Dai et al., 1999). One notable phenotype found in transgenic plants is accelerated leaf senescence, supporting the idea that lower photosynthetic activity may be related to premature leaf senescence via hexokinase. Moreover, a glucose-insensitive *Arabidopsis* mutant (gin2), with a lesion in one of the hexokinases shows delayed senescence (Moore et al., 2003). However, The *hysl (hypersenescencel)* mutant has increased sensitivity to exogenously applied sugars as well as an accelerated leaf senescence phenotype (Yoshida et al., 2002b). Therefore, one might suggest that an enhanced sugar signal in that mutant causes diminished photosynthesis and induces premature senescence, likely via hexokinase. Sugar-signaling pathways interact intimately with those signaling pathways regulated by hormones, e.g., auxin, cytokinin, or abscisic acid, during plant development. This is likely the case in the regulation of *Arabidopsis* leaf senescence. Such control through sugar signaling probably is also affected by other factors, such as nitrogen status and developmental stage. Integration of these factors into a senescence program might be important to the proper regulation of timing for its onset and progression.

The photoautotrophic nature of plants makes them fundamentally different from animals. Their energy input depends on the available photosynthetic activity, light and CO_2 and altering the available sources could substantially change the process of leaf senescence. Miller et al. (1997) found that elevated CO_2 could accelerate the shift of leaf development from the photosynthetic activity increase phase to the decrease phase. Ludewig and Sonnewald (2000) subsequently showed that this was caused by the earlier onset of leaf senescence. Leaf senescence was also examined in plants with reduced available sources. In *Rubisco* antisense tobacco plants, less dry weight and chlorophyll content was achieved than the wild type at maturity, while the leaf ontogeny was not altered (Miller et al., 2000). The most striking feature of the *Rubisco* antisense plants is that the senescence was markedly prolonged resulting in extended longevity. This pattern is similar to one of the stay-green mutants described in pea (Thomas and Howarth, 2002). More recently, the *Arabidopsis* delayed leaf senescence mutant *ore4-1*, was shown to contain a T-DNA insertion in the plastid ribosomal small subunit protein 17 (*PRPS17)* gene (Woo et al., 2002). The *ore4-1* mutants achieved less dry weight and contained less chlorophyll contents as in the *Rubisco* antisense plants, and more importantly, the photosynthetic system I activity of the *ore4-1* mutants was impaired. These results suggest that disruption of *PRPS17* resulted in reduced chloroplast function and energy input, perhaps mimicking the effect of calorie restriction in animals. Thus, increased energy input (mimicking over feeding in animals?) could accelerate leaf senescence, whereas reduced energy input had an opposite effect. It has been proposed that leaf senescence is initiated when photosynthetic activity drops below a certain threshold level (Hensel et al., 1993). This threshold could be related to leaf sugar levels. Indeed, leaf soluble sugar content increases with leaf age, and growth on media supplemented with sugars could repress photosynthesis associated gene (*PAG)* transcription and translation (Dijkwel et al., 1997; Jang et al., 1997; Wingler et al., 1998).

Sugars could specifically inhibit the expression of several *SAG*s associated with dark induction (Fujiki et al., 2001). However, in *SAG12-ipt* (isopentenyl transferase) transgenic tobacco the sugar levels were not different from *SAG12-GUS* plants, although senescence in the former one was substantially delayed (Ludewig and Sonnewald, 2000). In the senescent leaves the soluble sugars showed higher levels than in the non-senescent leaves presumably due to the breakdown of chloroplast and cell wall compounds (Quirino et al., 2000). This suggests that increased sugar levels are a consequence rather than a signal to initiate senescence. Exogenous sugars also had different effects on the expression profile of *SAG*s. While enhancing the expression of *SAG21* and *SAG13*, sugars inhibited the expression of *SAG12* (Noh and Amasino, 1999; Xiao et al., 2000). Taken together, the absolute level of sugars appears not to be directly involved in the regulation of leaf senescence. On the other hand, compelling evidence shows that sugar sensing and signalling can influence senescence. In *Arabidopsis* plants overexpressing sense and antisense hexokinase genes (*AtHXK1* and *AtHXK2*), the greening process and the expression profile of *PAG*s and *SAG21* were directly correlated with *AtHXK* expression levels (Jang et al., 1997; Xiao et al., 2000). Similar results were observed in transgenic tomato plants overexpressing *Arabidopsis AtHXK1* (Dai et al., 1999). The *gin2* mutant that has a lesion in the *AtHXK1* gene shows delayed leaf senescence as well as reduced glucose sensitivity (Quirino et al., 2000; Rolland et al., 2002). The *cpr5* mutant that was originally isolated based on the altered pathogen resistance was shown to have sugar hypersensitivity and early leaf senescence (Bowling et al., 1997; Yoshida et al., 2002a). Thus, altered energy intake or sensing can substantially influence senescence. However, more studies are needed to elucidate the precise molecular mechanisms. It is known that sugars can interact with several distinct signalling pathways such as ABA, ethylene, light and cytokinins, all of which are implicated in regulation of leaf senescence (Smeekens, 2000; Rolland et al., 2002). The effect of sugars on leaf senescence may depend on these interactions.

Leaf senescence is influenced by environmental conditions, including light, temperature, ion, salts, nutrient or water stress, or oxidative stresses induced by ozone or UV-B and biotic stresses like pathogen infection. Low light intensities or darkness results in the reduced expression of light-dependent genes and the disappearance of photosynthetic proteins and chlorophyll (Thomas, 1978). Exposure to extremely high or low temperatures, pathogen attack, or water/nutrient deficiency can also trigger leaf yellowing. Thus, parts of the signaling pathways that are associated with environmental stresses would be predicted to regulate leaf senescence. Expression profiles of 402 potential stress-related genes that encode known or putative transcription factors from *Arabidopsis* have been monitored in various organs, at different developmental stages, and under several biotic and abiotic stresses (Chen et al., 2002). Among the 43 transcription factor genes that are reportedly induced during senescence, 28 are also induced by stress treatment, suggesting extensive overlapping responses. Downstream genes for senescence-enhanced transcription factors might play a role either in executing leaf senescence or in protecting the cellular functions required for proper progression or completion of that senescence.

Light, perceived by a variety of photoreceptors, affects developmental processes over the entire life span, and may also play a role in leaf senescence (Cherry et al., 1991; Thiele et al., 1999). For example, transgenic plants over expressing phytochrome A *(PhyA)* or phytochrome B *(PhyB)* exhibit greater longevity. Although a mechanism for delayed senescence has not yet been proposed, one cause might be a slower chlorophyll degradation and leaf yellowing in these mutants. Since phytochrome acts as the light receptor for the expression of many photosynthetic genes, a lower red/far-red ratio reaching the lower leaves of a plant can also accelerate the senescence of these leaves (Rousscaux et al., 1996). In nonsenescent leaves sugar accumulation can lead to a decline in chlorophyll and photosynthetic proteins (Krapp and Stitt, 1994). Glc and Suc repress the transcription of photosynthetic genes, probably acting via hexokinase as a sugar sensor (Jang et al., 1997). The involvement of sugar-mediated repression of genes in the regulation of natural senescence is less clear (Feller and Fischer, 1994). The concentration of leaf sugars can increase during leaf senescence and accumulation of sugars, induced by removal of sinks or phloem interruption, can both accelerate and delay senescence (Frohlich and Feller, 1991). The response of leaves to the accumulation of sugars must therefore also depend on other factors, such as the C:N status of the leaf (Paul and Driscoll, 1997), light (Dijkwel et al., 1997) and plant growth regulators (Koch, 1996). For example, it has been suggested that cytokinin, in addition to delaying senescence, could block some of the responses to sugars (Jang et al., 1997). The interactions of cytokinins, light, and sugars during senescence in transgenic tobacco (*Nicotiana tabacum* L.) plants with autoregulated synthesis of cytokinin (Gan and Amasino, 1995). The transgenic tobacco plants express the gene for IPT under control of the senescence-specific SAG 12 promoter (Lohman et al., 1994). This promoter is activated at the onset of senescence, leading to the synthesis of cytokinin. Because of the inhibition of senescence by cytokinin, the promoter is actively attenuated. This results in an autoregulatory loop, preventing the overproduction of cytokinin and confining expression solely to those tissues that have initiated senescence. Apart from a delay in senescence, these plants therefore develop normally (Gan and Amasino, 1995). In cotyledons of curcurbits the synthesis of HPR is induced by cytokinin (Chen and Leisner, 1985; Andersen et al., 1996) and light (Bertoni and Becker, 1993), and the activity of HPR decreases during senescence (De Bellis and Nishimura, 1991).

It was shown that leaves of soybean (Guiamet et al., 1989) and sunflower (Rousseaux et al., 1996) senesced more quickly when the red:far-red ratio of the light they received was decreased, and that far-red light induces chlorophyll loss in tobacco leaves (Rousseaux et al., 1997). It also has been shown that tobacco and overexpressing oat phytochrome A display both delayed leaf senescence (Cherry et al., 1991) and an inhibited response to the senescence promoting effects of far-red light (Rousseaux et al., 1997). Weaver and Amasino (2001) examined a *hy2/hy3* phytochrome double mutant line, in which there is no phytochrome B and low levels of all other phytochromes (Parks and Quail, 1991; Somers et al., 1991), and observed it to behave much like wild-type controls, after both whole-plant and individual leaf dark treatments. One should note, however, that if phytochrome is involved in inhibiting senescence it might not be surprising that covered leaf senescence would continue to occur in phytochrome mutants, and in some instances might occur more strongly.

7. Senescence-associated PCD

Senescence is a complex, highly ordered process, during which plant organs undergo a series of biochemical and physiological changes ultimately resulting in the death of the organ (Smart, 1994; Buchanan-Wollaston, 1997). However, until late in senescence the process requires cell viability and is often reversible (Thomas *et al.*, 2003). There has been some debate about the degree of overlap of senescence and PCD (Thomas *et al.*, 2003; van Doorn & Woltering, 2004). van Doorn & Woltering (2004) identified various types of overlap in senescence and PCD.

i. Total overlap: Senescence and PCD overlap is complete and are synchronous (Noodén, 2004; van Doorn & Woltering (2004, 2008).
ii. No overlap: PCD either operated as the final instalment of senescence, or equally plausible, operated following senescence (Delorme *et al.*, 2000; Thomas *et al.*, 2003). van Doorn & Woltering (2008) defined PCD as 'the process that leads to the moment of death and the degradation that goes on after this moment'. Senescent cells are being actively recycled, which means that the cells are subjected to nucleases, proteases and photosynthetic breakdown along with various other senescent-related stresses. Normally one might expect that a cell which is subjected to this type of stress would activate PCD. However, in order to complete senescent recycling, PCD may have to be actively suppressed during the senescence process and only activated when recycling has been completed. For example, Bax inhibitor-1 (BI-1) has been shown to suppress PCD in both plant and animal cells (Watanabe & Lam, 2006). Bax inhibitor-1 is upregulated during harvest-induced senescence in broccoli (Coupe *et al.*, 2004). Another PCD suppressor, defender against apoptotic death (DAD1), is upregulated during leaf senescence (Dong *et al.*,1998), whereas it is down regulated before the onset of apoptotic like cell death in shorter-lived petals of pea (Orzaez & Granell,1997), *Alstroemeria* (Wagstaff *et al.*, 2003) and gladiolus(Yamada *et al.*, 2004). It is hypothysed that senescence may indeed be a rich source of PCD genes, but genes involved in suppression, rather than activation, of the process, and indeed activation of PCD, may be a result of senescence having terminated and consequently a cessation of transcription of PCD-suppressing genes and gene products.

8. Altered membrane lipase expression delays leaf senescence

Membrane deterioration leading to leakiness and loss of selective permeability is an early and ubiquitous feature of senescence (Thompson et al, 1998). It affects both the plasma membrane and intracellular membranes, and results in loss of ionic and metabolite gradients that are essential for normal cell function (Paliyath and Thompson, 1990) initiating programmed cell death. As the study of membrane lipids in green leaves is relatively cumbersome, the informations regarding senescence induced changes in membrane lipids are deried from floral senescence and PCD. Lipases are involved in leakiness of cellular membranes in senescing plant tissues (Borochov et al, 1982), resulting in an increase in the nonesterified/esterified fatty acid ratio (Thompson et al,1998). Moreover, the onset of membrane leakiness appears to be attributable, at least in part, to lateral phase separations of non-esterified fatty acids within the plane of the membrane bilayer. The resulting mixture of lipid phases causes the membranes to become leaky because of packing imperfections at the phase boundaries (Thompson et al, 1998). In addition, de-esterified polyunsaturated fatty acids in senescing membranes serve as substrates for lipoxygenase and ensuing lipid peroxidation, and the formation of peroxidized lipids in membranes also contributes to the onset of leakiness (Thompson et al, 1998). The abundance of the lipase mRNA increases just as carnation flowers begin to senesce, and expression of the gene is also induced by treatment with ethylene (Hong, 2000).

During leaf senescence, reactive oxygen species (ROS) and oxidative damage increases, whereas the levels of antioxidant enzymes such as SOD, catalase, and ascorbate peroxidase decreases (Orendi et al.,, 2001; Munne-Bosch and Alegre, 2002). Leaf senescence and the expression of various *SAGs* were promoted in old leaves upon exposure to UV-B or ozone, which are known oxidative damage inducing treatments (Miller et al., 1999; John et al., 2001). Mutant analysis and studies on transgenic plants provided a more straightforward support for the role of ROS in senescence (Orvar and Ellis, 1997; Willekens et al., 1997). Thus the molecular analysis substantiates the direct involvement of ROS in leaf senescence. ROS have a tight relationship with membrane and lipid dynamics since the membrane associated NAD(P)H oxidases can sense both endogenous and exogenous stresses and are one of the major generators of ROS (Mittler, 2002). The involvement of lipid metabolism in leaf senescence was demonstrated (He and Gan, 2002). Lipids are produced by fatty acid biosynthesis pathways, hence mutations in this pathway were also shown to change leaf senescence (Mou et al., 2000; Wellesen et al., 2001). Thus, ROS-induced membrane shuffling and lipid metabolism is not a passive wear and tear process but actively involved in leaf senescence. There is an intrinsic link between oxidative damage and leaf senescence, and the free radical theory of ageing seems to apply to plant senescence. Senescent tissues are stressed and subjected to gradually increasing oxidative damage. But, leaf cells continue to functionally operate transcriptional and translational activities along the progression of leaf senescence. So, plants have inherent mechanisms to safe guard the genome stability until the last stage of leaf senescence i.e. through PCD. (Liu et al. 2007).

9. Senescence in plants: Conserved strategies and novel pathways

Senescence is a universal phenomenon in living organisms and in higher plants it is manifested by the senescence of leaves. Constituting the last part of leaf development, leaf senescence has evolved as an indispensable process to maximise the reutilization of nutrients accumulated in the senescing leaves (Leopold, 1961; Bleecker, 1998). Molecular dissection of senescence process will provide information about the regulation of developmental cell death in plants or leaves, which can be utilized in crops for agricultural productivity. The expression of *SAGs* are up-regulated during senescence. To date over 100 *SAGs* have been identified in diverse plant species, and the list of *SAGs* is still increasing. Their expression profiles have been examined during development and under various induction conditions (Smart, 1994; Buchannan-Wollaston, 1997; Gan and Amasino, 1997; Nam, 1997). Many clones of *SAGs* showing up-regulation of their expression during senescence are also been reported to be overexpressed with abitic and biotic stress. Plant leaf senescence is modulated at a large array of genetic loci (Bleecker and Patterson, 1997; Buchanan-Wollaston, 1997; Quirino et al., 2000). However, a single gene mutation can substantially alter the lifespan in yeast, worms, fruit flies and some mammals (Kirkwood and Austad, 2000). Emerging evidences allowed the analysis of pathways involved and to compare the molecular strategies and the conserved nature of the senescence processes in plants and animals.

10. Nitric oxide evolution during leaf senescence

More recently, NO, a mediator of various plant developmental and (patho) physiological processes (Neill et al.2003; Crawford and Guo 2005; Mur et al. 2006), has been implicated in plant senescence and maturation. For instance, the temporal progress of fruit maturation and floral senescence is associated with a significant decrease in NO emission, and application of NO donating compounds retards flower senescence and extends the post-harvest life of fruits and vegetables (Leshem et al. 1998). Similarly, NO emission from *Arabidopsis* plants decreases significantly when plants mature and leaves start to senesce (Magalhaes et al. 2000). In addition, exogenous NO counteracts the promotion of leaf senescence caused by ABA and methyl jasmonate in rice (Hung and Kao, 2004).

To date, two major biochemical means of NO production have been identified in plants. On one hand, NO can be produced through enzymatic or non-enzymatic reduction from nitrite (Stöhr *et al.* 2001; Rockel *et al.* 2002; Bethke *et al.*, 2004). In fact,a major source of NO in plants originates from nitrite mediated by the action of nitrate reductase (Yamasaki and Sakihama 2000; Kaiser *et al.* 2002).

Interestingly, several factors, among them cytokinin, light and nitrate treatment, do simultaneously stimulate the expression or activity of NR (Crawford 1995; Yu, Sukumaran and Márton 1998), enhance the *in planta* production of NO (Magalhaes *et al.* 2000; Tun et al. 2001; Planchet *et al.* 2006) and retard the progress of plant senescence (Smart 1994). On the other hand, NO can be generated by NOS from L-arginine, and a corresponding plant NOS gene (*AtNOS1*) has been cloned and characterized in *Arabidopsis* (Guo *et al.*, 2003).The AtNOS1 protein is localized to mitochondria, and is involved in ABA-induced stomatal closure, the control of flowering and defense responses towards bacterial lipopolysaccharide elicitors (Guo *et al.* 2003; He *et al.* 2004; Zeidler *et al.* 2004; Guo and Crawford 2005). Moreover, NOS activity appears to represent one enzymatic means to influence plant senescence, as dark-induced leaf senescence occurs more rapidly in *Atnos1* knockout mutants compared with WT plants (Guo and Crawford 2005). In addition, a NOS-like activity of pea peroxisomes is down-regulated during the senescence process in pea leaves (Corpas *et al.* 2004).

In an attempt to study the (patho)physiological effects of reduced endogenous NO levels in plants, Mishina et al.,(2007) expressed a bacterial NOD under the control of an inducible promoter in *Arabidopsis* and observed that the resulting NO-deficient plants undergo a senescence-like process several days after activation of the NOD. They showed that this NOD-induced senescence process shares many similarities with natural senescence at the molecular level, and report effects of exogenous NO treatment on the progression of NOD- and dark-induced senescences andsupport the hypothesis that NO acts as a negative regulator of leaf senescence.

Arabidopsis Col-0 plants expressing the bacterial flavohemoglobin Hmp (Zeier *et al.* 2004) that functions as a NOD in *Escherichia coli*, converting NO to nitrate by using NAD(P)H and O2 (Poole and Hughes 2000). Leaf yellowing as a consequence of chlorophyll degradation represents the first visible symptom of senescence (Quirino *et al.* 2000). During natural senescence and senescence induced by artificial treatments, like shading of detached leaves, many photosynthetic genes are actively down-regulated (Lohman *et al.* 1994). When following the expression patterns of the typical photosynthesis-related genes *CAB* and *RBCS* [small subunit of ribulose 1·5-bisphosphate carboxylase/ oxygenase (Rubisco)] were down-regulated after NOD induction and this decline in expression of photosynthetic genes just occurred before visible signs of leaf yellowing (Mishina et al., 2007).

In addition to the down-regulation of photosynthesis, leaf senescence is characterized by an increase in expression of a multitude of genes that are often referred to as ***SAGs*** (Buchanan-

Wollaston 1997; Gepstein *et al.* 2003). Recently, Vladkova et al. (2011) reported that NO directly affects photosynthesis by enhancing it at low concentrations and is inhibitory at higher concentrations. This concentration dependent regulation of photosynthesis could inturn regulate leaf senescence (Misra et al. 2010a, b, 2011b)

Visible leaf yellowing in response to NOD expression occurred faster and was more pronounced in old than in young Hmp leaves. The findings that the expression of a NO degrading enzyme in *Arabidopsis* and the concomitant decrease in plant NO levels (Zeier *et al.* 2004) lead to a yellowing phenotype and changes in gene expression similar to senescence. These findings suggest that the senescence-like effect in Hmp plants was a result of NO deficiency that could be compensated by external NO. Induced expression of the *E. coli* flavohemoglobin Hmp in *Arabidopsis*, an enzyme functioning as a NOD (Poole and Hughes 2000), reduced both the *in planta* detection of NO as well as the emission of NO from plants, and furthermore increased the NO degrading capacity of leaf extracts (Zeier *et al.* 2004). The cysteine protease gene *SAG12* is known to be expressed exclusively during senescence, and its expression is therefore used as a senescence-specific marker (Zimmermann et al., 2006). The up-regulation of *SAG12* during the NOD-induced senescence phenotype indicates a high similarity of the process to natural senescence.

Reduction of nitrite by NR represents a major enzymatic plant NO source (Yamasaki and Sakihama 2000; Kaiser *et al.* 2002). Consequently, factors like nitrate feeding, high light treatment or cytokinin application, all of which stimulate NR activity (Crawford 1995; Yu *et al.* 1998), were able to enhance NO formation in plants (Magalhaes *et al.* 2000; Tun *et al.* 2001; Planchet *et al.* 2006). The nitrate remobilization out of the senescing leaves might cause a perturbation in the NO synthesis and regulate the process of senescence and associated PCD.

11. Regulation of leaf senescence by nitric oxide

A negative regulatory role of NO in plant senescence and maturation is studied extensively (Leshem *et al.* 1998; Magalhaes *et al.* 2000; Hung and Kao 2004; Guo and Crawford 2005). The findings that leaf yellowing, the down-regulation of *CAB* and SAG expression during NOD-induced senescence are attenuated by exogenous application of NO. During natural ageing of *Arabidopsis*, plant NO emission decreases continuously and reaches a minimum value during senescence (Magalhaes *et al.* 2000). Considering the processes observed in NO-deficient Hmp and *Atnos1* plants, we speculate that during leaf development, NO levels might fall below a certain threshold that in turn contributes to the induction of natural senescence.

Thus, application of NO releasing chemicals or NO gas extended the post-harvest life of fruits and vegetables and retarded the senescence of flowers (Leshem *et al.* 1998). Further, NO donors counteracted methyl jasmonate and ABA promoted senescence of rice leaves (Hung and Kao 2003, 2004), and attenuated dark-induced leaf senescence in *Arabidopsis Atnos1* mutants (Guo and Crawford 2005). The dose-dependent action of NO as a double-edged sword (Colasanti and Suzuki 2000), providing anti-senescent properties at lower and damaging effects at higher concentrations.

The integration and balance of internal and external factors is thought to be important in controlling the induction of leaf senescence (Yoshida 2003). Leaf age represents a significant internal variable influencing senescence. For instance, Weaver and Amasino (2001) have shown that dark induced senescence in individual leaves occurs more rapidly and strongly in older leaves than in younger ones. The older plants exhibit lower NO emission than young ones (Magalhaes *et al.* 2000).

NO emission is generally much more pronounced in light-situated plants as compared with darkened plants, and higher light intensities give rise to stronger NO emission signals than lower light levels (Magalhaes *et al.* 2000; Planchet *et al.* 2006). Similarly, nitrate-grown plants emit considerably higher amounts of NO than plant grown on ammonium (Planchet *et al.* 2006), and cytokinins give rise to increased NO releases in plant cell cultures (Tun *et al.* 2001). Light, nitrate and cytokinin are thus capable to stimulate endogenous NO production, and an internal NO generation caused by the previous treatments may have consequently counter balanced the NO diminishing action of NOD in Hmp plants. Considering the postulated antisenescent properties of NO, this in turn would explain the observed attenuation of NOD-induced senescence. NR may play a key role in mediating these effects, as NR expression and activity are known to be positively regulated by the three treatments applied (Crawford 1995; Yu *et al.* 1998), and at the same time, NR represents a major NO source in plants (Yamasaki and Sakihama 2000; Kaiser *et al.* 2002).A similar scenario would be feasible in plant development, during which light conditions, nitrogen nutrition and other environmental factors could be integrated through cytokinin action, NR activity and NO levels to influence the regulation of natural senescence.

NO has been recognized to possess both pro- and antioxidant effects in plants (Misra et al. 2010a, b, 2011), and this antagonism may be based on the relative ratios of ROS and NO levels in different physiological situations. By inhibition of antioxidant enzymes like catalase and ascorbate peroxidase, NO may contribute to elevated ROS levels and oxidative stress under certain circumstances (Clark *et al.* 2000). During the oxidative burst, NO ensures prolonged H2O2 levels at the site of pathogen challenge. Contrastingly, NO has been shown to act as an antioxidant in other situations. NO donors protect from oxidative damage caused by methylviologen herbicides, and counteract ROS-mediated programmed cell death in barley aleurone layers (Beligni and Lamattina 1999; Beligni *et al.* 2002).

During ABA- and jasmonate-induced senescence in rice, NO-releasing substances prevent an increase in H2O2 levels and lipid peroxidation (Hung and Kao, 2004). Accelerated dark-induced senescence in *Atnos1* mutants is accompanied with increased ROS levels and protein oxidation (Guo and Crawford 2005). As a free radical, NO reacts with superoxides in a diffusion limited reaction to form peroxynitrite (Huie and Padmaja 1993), and a subsequent fast isomerization of this toxic compound to a harmless end product like nitrate represents a possible mechanism to reduce ROS levels and cell damage through oxidative stress.

Although oxidative stress and ozone application induce various SAGs, these treatments do not initiate expression of the specific senescence marker *SAG12* (Miller *et al.* 1999; Navabpour *et al.* 2003).Thus, ROS elevation is not sufficient to trigger a full and coordinated execution of the natural senescence programme, and attenuation of *SAG12* expression by NO must therefore occur by ROS independent mechanisms. The plant hormone ethylene is capable of promoting senescence in plants (Smart 1994; Grbic and Bleecker 1995), and ethylene levels rise when plants start to senesce (Aharoni *et al.* 1979; Magalhaes *et al.* 2000). Considering the negative correlation of ethylene and NO emission during plant ageing (Magalhaes *et al.* 2000) and the up-regulation of ACC synthase during NOD induced senescence, it is plaussible that falling NO levels also contribute to senescence regulation by initiating ethylene biosynthesis.

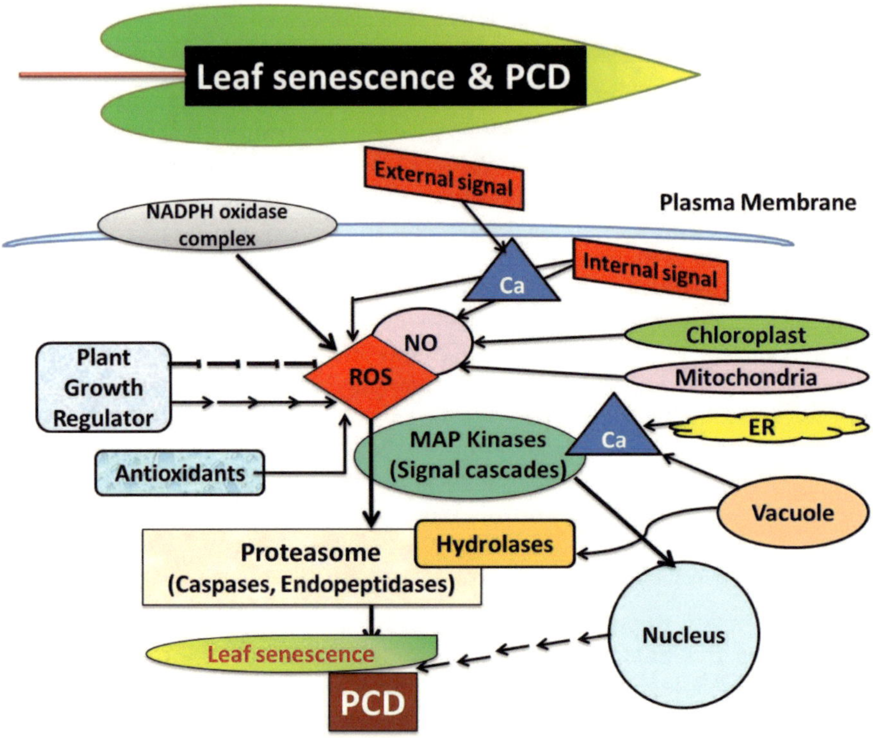

Fig. 1. A schematic diagram of the process and control of leaf senescence and the associated programmed cell death (PCD). The external and internal signals regulating leaf senescence and PCD is shown. The compartmentation of and the origin of different signals or signaling molecules are shown in the right panel. The central panel indicates the regulatory pathway. The leaft panel shows the physiological and biochemical regulatory mechanism for the control of various signaling and oxidative network pathways during leaf senescence or PCD in plants.

On one hand, NO can be produced through enzymatic or non-enzymatic reduction from nitrite (Misra et al. 2010a, b, 2011b and references there in). In fact, a major source of NO in plants originates from nitrite mediated by the action of nitrate reductase. Interestingly, several factors, among them cytokinin, light and nitrate treatment, do simultaneously stimulate the expression or activity of NR (Misra et al. 2010a, b, 2011b), enhance the *in planta* production of NO and retard the progress of plant senescence. On the other hand, NO can be generated by NOS from L-arginine, and a corresponding plant NOS gene (*AtNOS1*) has been cloned and characterized in *Arabidopsis* (Guo et al., 2003). The AtNOS1 protein is localized to mitochondria, and is involved in ABA-induced stomatal closure, the control of flowering and defense responses towards bacterial lipopolysaccharide elicitors (Zeidler *et al.* 2004; Guo and Crawford 2005). Moreover, NOS activity appears to represent one enzymatic means to influence plant senescence, as dark-induced leaf senescence occurs more rapidly in *Atnos1*knockout mutants compared with WT plants (Guo and Crawford 2005). In addition, a NOS-like activity of pea peroxisomes is down-regulated during the senescence process in pea leaves (Corpas *et al.* 2004).

12. Conclusion and future perspectives

This review focuses on the physiological, biochemical and molecular aspects of leaf senescence. Studies on these aspects are extensive and there is an wealth of knowledge on the senescence processes in plants. Although senescence and programmed cell death in animal tissue is well established, however the temporal and spatial correlation between these two events in plants and leaves are not well defined. This review tried to make a synthesis of the upto date information regarding these two aspects and bring about a common mechanism for both processes being regulated through a network of inter-related regulatory network such as reactive oxygen species and kinase caskades (Fig. 1). Also, the possible pathways of synthesis, cellular compartmentation and regulation of NO during leaf senescence and/or PCD is illucidated. However, this is the beginning of a new era of NO signaling in leaf senescence and PCD. The journey starts now for the elucidate of the mechanism of NO signaling in leaf senescence and PCD in plants.

13. Acknowledgements

This work was supported by funds from UGC-MRP No.36-302/2008(SR) and DST- INT/ BULGARIA/ B70/06 to ANM. MM acknowledges the award of DST-WoS and UGC PDF for Women.

14. References

Aharoni N, Anderson JD, Lieberman M (1979). Production and action of ethylene in senescing leaf discs. Effects of indoleacetic acid, kinetin, silver ion, and carbon dioxide. *Plant Physiol* 64: 805–809.

Andersen BR, Jin G, Chen R, Ertl JR, Chen CM (1996). Transcriptional regulation of hydroxypyruvate reductase gene expression by cytokinin in etiolated pumpkin cotyledons. *Planta* 198: 1–5.

Apel K and Hirt H (2004). Reactive oxygen species: metabolism, oxidative stress, and signal transduction. *Annu. Rev. Plant Biol.* 55, 373–399.

Armstead I, Donnison I, Aubry S, Harper J, Hörtensteiner S, James C, Mani J, Matt Moffet, Ougham H, Roberts L, Thomas A, Weeden N, Thomas H, King I (2006). From crop to model to crop: Identifying the genetic basis of the staygreen mutation in the Lolium/Festuca forage and amenity grasses. *New Phytol.* 172: 592–597.

Badenoch-Jones J, Parker CW, Letham DS, Singh S (1996). Effect of cytokinins supplied via the xylem at multiples of endogenous concentrations on transpiration and senescence in derooted seedlings of oat and wheat. *Plant Cell Environ* 19: 504–516.

Beligni MV, Fath A, Bethake PC, Lamattina L, Jones RL (2002). Nitric oxide acts as an antioxidant and delays programmed cell death in barley aleurone layers. *Plant Physiol* 129: 1462–1650.

Beligni MV and Lamattina L (1999 a). Nitric oxide counteracts cytotoxic processes mediated by reactive oxygen species in plant tissues. *Planta* 208: 337–344.

Beligni MV and Lamattina L (1999 b). Nitric oxide protects against cellular damage produced by methyl violgen herbicides in potato plants. *Nitric Oxide Biol Chem* 3: 199–208.

Benedetti CE, and Arruda P (2002). Altering the expression of the chlorophyllase gene ATHCOR1 in transgenic *Arabidopsis* caused changes in the chlorophyll-to-chlorophyllide ratio. *Plant Physiol* 128: 1255–1263.

Benedetti CE, Costa CL, Turcinelli SR, Arruda P (1998). Differential expression of a novel gene in response to coronatine, methyl jasmonate, and wounding in the Coi1 mutant of *Arabidopsis*. *Plant Physiol* 116: 1037–1042.

Berger S (2001). Jasmonate-related mutants of *Arabidopsis* as tools for studying stress signals. *Planta* 214, 497-504.

Bertoni GP and Becker WM (1993). Effects of light fluence and wavelength on expression of the gene encoding cucumber hydroxypyruvate reductase. *Plant Physiol* 103: 933–941

Bethke PC, Badger MR, Jones RL (2004). Apoplastic synthesis of nitric oxide by plant tissues. *Plant Cell* 16: 332-341.

Biswal AK, Ramaswamy NK, Mathur M, Misra AN (2001a). Light regulated protein kinase activity in thylakoid membranes of NaCl salt stressed seedlings. Photosynthesis. PS2001. CSIRO Publ., Melbourne, Australia.

Biswal AK, Dilnawaz F, David KAV, Ramaswamy NK, Misra AN (2001b). Increase in the intensity of thermoluminescence Q-band during leaf ageing is due to a block in the electron transfer from Q_A to Q_B. *Luminescence* 16: 309-313.

Bleecker AB (1998). The evolutionary basis of leaf senescence: method to the madness? *Curr Opin Plant Biol* 1: 73-78.

Bleecker AB and Patterson SE (1997). Last exit: senescence, abscission, and meristem arrest in *Arabidopsis*. *Plant Cell* 9: 1169-1179.

Borochov A, Halevy AH, Shinitzky M (1982). Senescence and the fluidity of rose petal membranes. Relationship to phospholipid metabolism. *Plant Physiol* 69: 296-299.

Bowling SA, Clarke JD, Liu Y, Klessig DF, Dong X (1997). The cpr5 mutant of *Arabidopsis* expresses both NPR1-dependent and NPR1-independent resistance. *Plant Cell* 9: 1573-1584.

Breeze E, Harrison E, Mc Hattie S, Hughes L, Hickman RDG, Hill C, Kiddle SJ, Kim YS, Penfold CA, Jenkins DJ, Zhang C, Morris K, Jenner CE, Jackson SD, Thomas B, Tabrett A, Legaie R, Moore JD, Wild DL, Ott S, Rand DA, Beynon J, Denby KJ, Mead A, Buchanan-Wollaston V (2011). 'High-resolution temporal profiling of transcripts during *Arabidopsis* leaf senescence reveals a distinct chronology of processes and regulation'. *Plant Cell* 23(3): 873 - 894 (1040-4651)

Buchanan-Wollaston V, Page T, Harrison E, Breeze E, Lim P O, Nam H G, Lin J F, Wu S H, Swidzinski J, Ishizaki K, Leaver C (2005). Comparative transcriptome analysis reveals significant differences in gene expression and signalling pathways between developmental and dark/starvation induced senescence in *Arabidopsis*. *Plant J* 42: 567 - 585.

Buchanan-Wollaston V (1997). The molecular biology of leaf senescence. *J Exp Bot* 48: 181-199.

Buchanan-Wollaston V, Earl S, Harrison E, Mathas E, Navabpour S, Page T, Pink D (2003). The molecular analysis of leaf senescence – a genomics approach. *Plant Biotech J* 1: 3-22.

Callis J (1995). Regulation of protein degradation. *Plant Cell* 7: 845-857.

Chen CM and Leisner SM (1985). Cytokinin-modulated gene expression in excised pumpkin cotyledons. *Plant Physiol* 77: 99–103.

Chen W, Powart NJ, Glazebroo JH, Katagiri F, Chang HS, Eulgem T, Mauch F, Luan T, Zou G, Whitham S A, Budworth PR, Tao Y, Xie Z, Chen X, Lam S, Kreps JA, Harper JF,

Si-Ammour A, Mauch-Mani B, Heinlein M, Kobayashi K, Hohn T, Dangl JL, Wang X, Zhu T (2002). Expression profile matrix of *Arabidopsis* transcription factor genes suggest their putative functions in response to environmental stress. *Plant Cell* 14: 559-574.

Cherry JR, Hershey HP, Vierstra RD (1991). Characterization of tobacco expressing functional oat phytochrome: Domains Responsible for the Rapid Degradation of Pfr Are Conserved between Monocots and Dicots. *Plant Physiol* 96(3):775–785.

Clark D, Dunar J, Navarre DA, Klessig DF (2000). Nitric oxide inhibition of tobacco catalase and ascorbate peroxidase. *Mol Plant-Microbe Interact* 13: 1380–1384.

Clouse SD and Sasse JM (1998). Brassinosteroids: essential regulators of plant growth and development. *Annu Rev Plant Physiol Plant Mol Biol* 49: 327-451.

Corpas FJ, Barroso JB, del Rıo LA (2004). Enzymatic sources of nitric oxide in plant cells: beyond one protein-one function. *New Phytol* 162: 246–248.

Coupe SA, Watson LM, Ryan DJ, Pinkney TT, and Eason JR. 2004. Molecular analysis of programmed cell death during senescence in *Arabidopsis thaliana* and *Brassica oleracea*: cloning broccoli LSD1, Bax inhibitor and serine palmitoyltransferase homologues. *J Exp Bot* 55: 59–68.

Crawford NM and Guo FQ (2005). New insights into nitric oxide metabolism and functions. *Trends Plant Sci* 259: 1360-85.

Crawford NM (1995). Nitrate: nutrient and signal for plant growth. *Plant Cell* 7: 859–868.

Creelman RA and Mullet JE (1997). Biosynthesis and action of jasmonates in plants. *Annu Rev Plant Physiol Plant Mol Biol* 48: 355-381.

Dai P, Akimaru H, Tanaka Y, Maekawa T, Nakafuku M, andIshii S (1999). Sonic hedgehog-induced activation of the Gli1 promoter is mediated by GLI3. *J Biol Chem* 274: 8143–8152.

De Bellis L and Nishimura M (1991). Development of enzymes of the glyoxylate cycle during senescence of pumpkin cotyledons. *Plant Cell Physiol* 32: 555–561.

Delorme V, McCabe PF, Kim DJ, Leaver CJ (2000). A matrix metalloproteinase gene is expressed at the boundary of senescence and programmed cell death in cucumber. *Plant Physiol* 123: 917–927.

Dijkwel PP, Huijser C, Weisbeck PJ, Chua NH, Smeekens SCM (1997). Sucrose control of phytochrome A signaling in *Arabidopsis*. *Plant Cell* 9: 583–595.

Dilnawaz F, Mohapatra P, Misra M, Ramaswamy NK, Misra AN (2001). The distinctive pattern of photosystem 2 activity, photosynthetic pigment accumulation, and ribulose-1, 5-bisphosphate carboxylase/oxygenase content of chloroplasts along the axis of primary wheat leaf lamina. *Photosynthetica* 39: 557-563.

Dong YH, Zhan XC, Kvarnheden A, Atkinson RG, Morris BA, Gardner RC (1998). Expression of a cDNA from apple encoding a homologue of DAD1, an inhibitor of programmed cell death. *Plant Sci* 139: 165–174.

Ellis CM, Nagpal P, Young JC, Hagen G, Guilfoyle TJ, Reed JW (2005). AUXIN RESPONSE FACTOR1 and AUXIN RESPONSE FACTOR2 regulate senescence and floral organ abscission in *Arabidopsis thaliana*. *Development* 132: 4563–4574.

Fedoroff NV (2002). Cross-talk in abscisic acid signaling. *Sci STKE* 140.

Feller U, Anders I, Demirevska K (2008). Degradation of Rubisco and other chloroplast protein under abiotic stress. *Gen Appl P Physiol* 34 (1-2): 5-18.

Feller U and Fischer A (1994). Nitrogen metabolism in senescing leaves. *Crit Rev Plant Sci* 13: 241–273.

Frohlich V and Feller U (1991). Effect of phloem interruption on senescence and protein remobilization in the flag leaf of field grown wheat. *Biochem Physiol Pflanzen* 187: 139–147.

Fujiki Y, Yoshikawa Y, Sato T, Inada N, Ito M, Nishida I, Watanabe A (2001). Dark inducible genes from *Arabidopsis thaliana* are associated with leaf senescence and repressed by sugars. *Physiol Plant* 111: 345-352.

Gan S and Amasino RM (1995). Inhibition of leaf senescence by autoregulated production of cytokinin. *Sci* 270: 1986–1988.

Gan S and Amasino RM (1997). Making sense of senescence (molecular genetic regulation and manipulation of leaf senescence). *Plant Physiol* 113: 313-319.

Gepstein S and Thimann KV (1980). Changes in the abscisic acid content of oat leaves during senescence. *Proc Natl Acad Sci USA* 77: 2050-2053.

Gepstein S, Sabehi G, Carp MJ, Hajouj T, Nesher MFO, Dor IYC, Bassani M (2003). Large-scale identification of leaf senescence-associated genes. *Plant J* 36:629–642.

Gray J, Janick-Bruckner D, Bruckner B, Close PS, Johal GS (2002). Light-dependent death of maize lls1 cells is mediated by mature chloroplasts. *Plant Physiol* 130: 1894–1907.

Grbic V and Bleecker AB (1995). Ethylene regulates the timing of leaf senescence in *Arabidopsis*. *Plant J* 8: 595-602.

Greenberg JT and Ausubel FM (1993). *Arabidopsis* mutants compromised for the control of cellular damage during pathogenesis and aging. *Plant J* 4: 327–341.

Guiamet JJ, Willemoes JG, Montaldi ER (1989). Modulation of progressive leaf senescence by the red far-red ratio of incident light. *Bot Gaz* 150: 148–151.

Guo F, Okamoto M, Crawford NM (2003). Identification of a plant nitric oxide synthase gene involved in hormonal signaling. *Sci* 302: 100–103.

Guo FQ and Crawford NM (2005). *Arabidopsis* nitric oxide synthase1 is targeted to mitochondria and protects against oxidative damage and dark-induced senescence. *Plant Cell* 17: 3436–3450.

He Y, Fukushige H, Hildebrand DF, Gan S (2002). Evidence supporting a role of jasmonic acid in *Arabidopsis* leaf senescence. *Plant Physiol* 128: 876–884.

He Y, Tang RH, Hao Y, Stevens RD, Cook CW, Ahn SM, Jing L, Yang Z, Chen L, Guo F, Fiorani F, Jackson RB, Crawford NM, Pei ZM (2004). Nitric oxide represses the Arabidopsis floral transition. *Sci* 305: 1968-1971.

He Y and Gan S (2002). A gene encoding an acyl hydrolase is involved in leaf senescence in *Arabidopsis*. *Plant Cell* 14: 805-815.

Hensel LL, Grbic V, Baumgarten DA, Bleecker AB (1993). Developmental and agerelated processes that influence the longevity and senescence of photosynthetic tissues in *Arabidopsis*. *Plant Cell* 5: 553-564.

Hong Y, Wang TW, Hudak KA, Schade F, Froese CD, Thompson JE (2000). An ethylene-induced cDNA encoding a lipase expressed at the onset of senescence. *Proc Natl Acad Sci USA* 97: 8717–8722.

Hortensteiner S (2006). Chlorophyll degradation during senescence. *Annu Rev Plant Biol* 57: 55–77.

Hortensteiner S and Feller U (2002). Nitrogen metabolism and remobilization during senescence. *J Exp Bot* 53: 927–937.

Hortensteiner S and Matile P (2004). How leaves turn yellow: Catabolism of chlorophyll. In Plant Cell Death Processes, *L.D. Nooden, ed (San Diego, CA: Academic Press)*, pp. 189-202.

Huie RE and Padmaja S (1993). The reaction of NO with O_2^-. *Free Radic Res Commun* 235: 264-267.

Humbeck K, Quast S, Krupinska, K (1996). Functional and molecular changes in the photosynthetic apparatus during senescence of flag leaves from field-grown barley plants. *Plant Cell Environ* 19: 337-344.

Hung KT and Kao CH (2004). Nitric oxide acts as an antioxidant and delays methyl jasmonate-induced senescence of rice leaves. *J Plant Physiol* 161: 43-52.

Hung KT and Kao CH (2003). Nitric oxide counteracts the senescence of rice leaves induced by abscisic acid. *J Plant Physiol* 160: 871-879.

Kieber JJ and Schaller GE (2010). The perception of cytokinin: a story 50 years in the making. *Plant Physiol* 154: 487-492.

Hwang I, Chen HC, Sheen J (2002). Two-component signal transduction pathways in *Arabidopsis*. *Plant Physiol* 129: 500-515.

Jang JC, Leon P, Zhou L, Sheen J (1997). Hexokinase as a sugar sensor in higher plants. *Plant Cell* 9: 5-19.

Jing HC, Sturre MJ, Hille J, Dijkwel PP (2002). *Arabidopsis* onset of leaf death mutants identify a regulatory pathway controlling leaf senescence. *Plant J* 32: 51-63.

John CF, Morris K, Jordan BR, Thomas B, A-H-Mackerness S (2001). Ultraviolet-B exposure leads to upregulation of senescence- associated genes in *Arabidopsis thaliana*. *J Exp Bot* 52: 1367-1373.

Johnson R and Ecker R (1998). The ethylene gas signal transduction pathway: A molecular perspective. *Annu Rev Genet* 32: 227-254.

Kaiser A, Sell S, Hehl R (2002). Heterologous expression of a bacterial homospermidine synthase gene in transgenic tobacco: effects on the polyamine pathway. *Arch Pharm (Weinheim)* 335: 143-151.

Kenyon C (2001). A conserved regulatory system for aging. *Cell* 105 (2): 165-168.

Kirkwood TBL and Austad SN (2000). Why do we age? *Nature* 408: 233-238.

Kleber-Janke T and Krupinska K (1997). Isolation of cDNA clones for genes showing enhanced expression in barley leaves during dark-induced senescence as well as during senescence under field conditions. *Planta* 203: 332-340.

Koch KE (1996). Carbohydrate-modulated gene expression in plants. *Annu Rev Plant Physiol Plant Mol Biol* 47: 509-540.

Krapp A and Stitt M (1994). Influence of high carbohydrate content on the activity of plastidic and cytosolic isoenzyme pairs in photosynthetic tissues. *Plant Cell Environ* 17: 861-866.

Leopold AC (1961). Senescence in plant development. *Sci* 134: 1727-1734.

Leshem Y, Wills RBH, Ku VVV (1998). Evidence for the function of the free radical gas-nitric oxide (NO) as an endogenous maturation and senescence regulating factor in higher plant. *Plant Physiol Biochem* 36: 825-835.

Lim PO and Nam HG (2005). The molecular and genetic control of leaf senescence and longevity in *Arabidopsis*. *Curr Top Dev Biol* 67: 49-83.

Lim PO, Woo HR, Nam HG (2003). Molecular genetics of leaf senescence in *Arabidopsis*. *Trends Plant Sci* 8: 272-278.

Liu YD, Ren DT, Pike S, Pallardy S, Gassmann W, Zhang S Q (2007). Chloroplast-generated reactive oxygen species are involved in hypersensitive response-like cell death mediated by a mitogen-activated protein kinase cascade. *Plant J* 51: 941–954.

Lohman KN, Gan S, John MC, Amasino RM (1994). Molecular analysis of natural leaf senescence in *Arabidopsis thaliana*. *Physiol Plant* 92: 322–328.

Ludewig F and Sonnewald U (2000). High CO_2-mediated downregulation of photosynthetic gene transcripts is caused by accelerated leaf senescence rather than sugar accumulation. *FEBS Lett* 479: 19–24.

Mach JM, Castillo AR, Hoogstraten R, Greenberg JT (2001). The *Arabidopsis*-accelerated cell death gene ACD2 encodes red chlorophyll catabolite reductase and suppresses the spread of disease symptoms. *Proc Natl Acad Sci USA* 98: 771–776.

Magalhães JR, Monte DC, Durzan D (2000). Nitric oxide and ethylene emission in *Arabidopsis thaliana*. *Physiol Mol Biol Plants* 6: 117-127.

Masferrer A, Arro M, Manzano D, Schaller H, Fernandez-Busquets X, Moncalean P, Fernandez B, Cunillera N, Boronat A, Ferrer A (2002). Over expression of *Arabidopsis thaliana* farnesyl diphosphate synthase (FPS1S) in transgenic *Arabidopsis* induces a cell death/senescence like response and reduced cytokinin levels. *Plant J* 30: 123-132.

Matile P, Ginsburg S, Schellenberg M, Thomas H (1988). Catabolites of chlorophyll in senescing barley gerontoplasts. *Plant Physiol Biochem* 34: 55–59.

Matile P, Hortensteiner S, Thomas H (1999). Chlorophyll degradation. *Annu Rev Plant Physiol Plant Mol Biol* 50: 67–95.

Matile P, Schellenberg M, Vicentini F (1997). Localization of chlorophyllase in the chloroplast envelope. *Planta* 201: 96–99.

Miller A, Schlagnhaufer C, Spalding M , Rodermel S (2000). Carbohydrate regulation of leaf development: Prolongation of leaf of leaf senescence in Rubisco antisense mutants of tobacco. *Photosyn Res* 63: 1–8.

Miller A, Tsai CH, Hemphill D, Endres M, Rodermel S and Spalding M (1997). Elevated CO_2 effects during leaf ontogeny (A new Perspective on Acclimation). *Plant Physiol* 115(3): 1195–1200.

Miller JD, Arteca RN, Pell EJ (1999). Senescence-associated gene expression during ozone-induced leaf senescence in *Arabidopsis*. *Plant Physiol* 120: 1015–1023.

Mishina TE, Lamb C, Zeier J (2007). Expression of a NO degrading enzyme induces a senescence programme in *Arabidopsis*. *Plant Cell Environ* 30: 39–52.

Misra AN (1993a). CAB gene expression during chloroplast senescence. *In* DAE Symposium on Photosynthesis and Plant Molecular Biology, Department of Atomic Energy, Govt. of India, Bombay pp. 246-151.

Misra AN (1993b). Molecular mechanism of turn-over of 32 kDa-herbicide binding protein of photosyntem II reaction center. *In* Advances in Plant Biotechnology & Biochemistry (Lodha M L, Mehta S L, Ramagopal S and Srivastava G P, eds.), IARI, New Delhi pp. 73-78.

Misra AN (1995). Assimilate partitioning in pearl millet (*Pennisetum glaucum* L.R.Br.). *Acta Physiol Plant* 17: 41-46.

Misra AN and Biswal AK (2000). Thylakoid membrane protein kinase activity as a signal transduction pathway in chloroplasts. *Photosynthetica* 38: 323-332.

Misra AN, Biswal AK, Misra M (2002). Physiological, biochemical and molecular aspects of water stress responses in plants, and the biotechnological applications. *Proc Nat Acad Sci (India)* 72.B (2): 115-134.

Misra AN and Biswal UC (1980). Effect of phytohormones on the chlorophyll degradation during aging of chloroplasts *in vivo* and *in vitro*. *Protoplasma* 105: 1-8.

Misra AN and Biswal UC (1981). Changes in photosynthetic pigments during aging of attached and detached leaves, and of isolated chloroplasts. *Photosynthetica* 15: 75-79.

Misra AN and Biswal UC (1982a). Differential changes in electron transport properties of chlorplasts during aging of attached and detached leaves and of isolated chloroplasts. *Plant Cell Env* 5: 27-30.

Misra AN and Biswal UC (1982b). Changes in plastid macromolecules during aging of attached and detached leaves, and of isolated chloroplasts. *Photosynthetica* 16: 22-26.

Misra AN and Biswal UC (1987). Senescenece induced changes in chlorplast absorption spectra of attached and detached leaves, and of isolated chlorplasts in wheat (*Triticum aestivum* L.). *Plant Physiol Biochem* 14: 159-164.

Misra A N, Hall S, Barber J (1991). The isolated DI/D2/Cyt b559 complex of photosystem two reaction centre possesses a serine type endopeptidase activity. *Biochim Biophys Acta* 1059: 239-242.

Misra AN, D Latowski and K Strzalka (2011a). Violaxanthin de-epoxidation in aging cabbage (*Brassica oleracea* L.) leaves play as a sensor for photosynthetic excitation pressure. *J Life Sci* 5: 182-191.

Misra AN and Misra M (1986a). Effect of temperature on senescing rice leaves. I. Photoelectron transport activity of chloroplasts. *Plant Sci* 46: 1-4.

Misra AN and Misra M (1987). Effect of age and rehydration on greening of wheat leaves. *Plant Cell Physiol* 28: 47-51.

Misra AN, Misra M, Singh R (2010a). Nitric oxide biochemistry, mode of action and signaling in plants. *J Med Plant Res* 4(25): 2729-2739.

Misra AN, Misra M, Singh R (2010b). Nitric oxide: A ubiquitous signaling molecule with diverse role in plants. *African J Plant Sci* 5: 57-74.

Misra AN, Misra M, Singh R (2011b) Nitric oxide ameliorates stress responses in plants. *Plant Soil Env* 57 (3): 95-100.

Misra M and Misra AN (1986b). Catalase activity and pigment content in senescing rice leaves. *Agric Biol Res* 3: 107-113.

Misra M and Misra AN (1989a). Role of axis on the interaction of light and kinetin on retardation of wheat leaf senescence. *Environ Ecol* 7: 906-910.

Misra M and Misra AN (1989b). Effect of tempeature on chloroplat absorption spectra during senescence of detatched rice leaves. *Adv Biosci* 7: 906-910.

Misra M and Misra AN (1995). Photosynthetic pigment and protein content of pearl millet seedling grown at high and low photon flux densities. *Indian J Plant Physiol* 2: 148-150.

Misra M and Misra AN (1991). Hormonal regulation of detatched rice leaf senescene. *In* Environmental Contamination and Hygien (Praksh R and Ali A, eds.), Jagmandir Books, New Delhi, India pp. 47-55.

Misra AN, Dilnawaz F, Misra M, Biswal AK (2001a). Thermoluminescence in chloroplasts as an indicator of alterations in photosystem II reaction center by biotic and abiotic stress. *Photosynthetica* 39: 1-9.

Misra AN, Srivastava A, Strasser RJ (2001b). Utilisation of fast Chlorophyll *a* fluorescence technique in assessing the salt/ion sensitivity of mung bean and brassica seedlings. *J Plant Physiol* 158: 1173-1181.

Mittler R (2002). Oxidative stress, antioxidants and stress tolerance. *Trends in Plant Sci* 7: 405–410.

Moore B, Zhou L, Rolland F, Hall Q, Cheng WH, Liu YX, Hwang I, Jones T, Sheen J (2003). Role of the *Arabidopsis* glucose sensor HXK1 in nutrient, light, and hormonal signaling. *Sci* 300: 332-336.

Morris K, MacKerness SA, Page T, John CF, Murphy AM, Carr JP, Buchanan-Wollaston V (2000). Salicylic acid has a role in regulating gene expression during leaf senescence. *Plant J* 23: 677–685

Mou Z, He Y, Dai Y, Liu X, Li J (2000). Deficiency in fatty acid synthase leads to premature cell death and dramatic alterations in plant morphology. *Plant Cell* 12: 405–417.

Munne-Bosch S and Alegre L (2002). The function of tocopherols and tocotrienols in plants. *Crit Rev Plant Sci* 21: 31–57

Mur LAJ, Craver TLW, Prats E (2006). NO way to live; the various roles of nitric oxide in plant–pathogen interactions.*J Exp Botany* 57: 489–505.

Nakano RH, Ishida A, Makino T, Mae T (2006). *In vivo* fragmentation of the large subunit of ribulose-1,5-bisphosphate carboxylase by reactive oxygen species in an intact leaf of cucumber under chilling-light conditions. *Plant Cell Physiol* 47: 270-276.

Nam HG (1997). The molecular genetic analysis of leaf senescence. *Curr Opin Biotech* 8: 200-207.

Navabpour S, Morris K, Allen R, Harrison E, A-H-Mackerness S, Buchanan-Wollaston V (2003). Expression of senescence enhanced genes in response to oxidative stress. *J Exp Bot* 54: 2285–2292.

Neill S, Barros R, Bright J, Desikan R, Hancock J, Harrison J, Morris P, Ribeiro D, Wilson I (2008). Nitric oxide, stomatal closure, and abiotic stress. *J Exp Bot* 59: 165–176.

Neill SJ, Desikan R, Hancock JT (2003). Nitric oxide signaling in plants. *New Phytol* 159: 11-35.

Noh YS and Amasino RM (1999). Identification of a promoter region responsible for the senescence-specific expression of SAG12. *Plant Mol Biol* 41: 181-194.

Noodén LD (2004). Introduction. In: Noodén LD. *Plant cell death processes.* Amsterdam, the Netherlands: Elsevier, 1–18.

Nooden LD, Guiamet JJ, John I (1997). Senescence mechanisms. *Physiol Plant* 101: 746–753.

Nooden LD, Hillsberg JW, Schneider MJ (1996). Induction of leaf senescence in *Arabidopsis thaliana* by long days through a light-dosage effect. *Physiol Plant* 96: 491-495.

Okazawa A, Tang L, Itoh Y, Fukusaki E, Kobayashi A (2006). Characterization and subcellular localization of chlorophyllase from *Ginkgo biloba. Z Naturforsch* 61c: 111–117.

Orendi G, Zimmermann P, Baar C, Zentgraf U (2001). Loss of stressinduced expression of catalase3 during leaf senescence in *Arabidopsis thaliana* is restricted to oxidative stress. *Plant Sci* 161: 301–314.

Orvar BL, and Ellis BE (1997). Transgenic tobacco plants expressing antisense RNA for cytosolic ascorbate peroxidase show increased susceptibility to ozone injury. *Plant J* 11(6): 1297-1305.

Orzaez D and Granell A (1997). DNA fragmentation is regulated by ethylene during carpel senescence in *Pisum sativum. Plant J* 11: 137–144.

Paliyath G and Thompson JE (1990). Evidence for early changes in membrane structure during post-harvest development of cut carnations. *New Phytol* 144: 555-562.

Parks BM and Quail PH (1991). Phytochrome-deficient hy1 and hy2 long hypocotyl mutants of *Arabidopsis* are defective in phytochrome chromophore biosynthesis. *Plant Cell* 3: 1177–1186.

Paul MJ and Driscoll SP (1997). Sugar repression of photosynthesis: the role of carbohydrates in signalling nitrogen deficiency through source:sink imbalance. *Plant Cell Environ* 20: 110–116.

Planchet E, Sonoda M, Zeier J, Kaiser WM (2006). Nitric oxide (NO) as an intermediate in the cryptogein-induced hypersensitive response-a critical re-evaluation. *Plant Cell Environ* 29: 59–69.

Poole RK and Hughes MN (2000). New functions for the ancient globin family: bacterial responses to nitric oxide and nitrosative stress. *Mol Microbiol* 36: 775-783.

Pruzinska A, Anders E, Aubry S, Schenk N, Tapernoux-Lüthi E, Müller, T, Kräutler B, Hörtensteiner S (2007). In vivo participation of red chlorophyll catabolite reductase in chlorophyll breakdown. *Plant Cell* 19: 369–387.

Pruzinska A, Tanner G, Aubry S, Anders I, Moser S, Muller T, Ongania KH, Krautler B, Youn JY, Liljegren SJ, Hortensteiner S (2005). Chlorophyll breakdown in senescent Arabidopsis leaves. Characterization of chlorophyll catabolites and chlorophyll catabolic enzymes involved in the degreening reaction. *Plant Physiol* 139: 52–63.

Quirino BF, Noh YS, Himelblau E, Amasino RM (2000). Molecular aspects of leaf senescence. *Trends Plant Sci* 5: 278-282.

Reinbothe C, Satoh H, Alcaraz JP, Reinbothe S (2004). A novel role of water-soluble chlorophyll proteins in the transitory storage of chlorophyllide. *Plant Physiol* 134: 1355–1365.

Richmond AE and Lang A (1957). Effect of kinetin on protein content and survival of detached *Xanthium* leaves. *Sci* 125: 650–651.

Roberts IN, Murray PF, Caputo CP, Passeron S, Barneix AJ (2003). Purification and characterization of a subtilisin-like serine protease induced during the senescence of wheat leaves. *Physiol Plant* 118: 483-490.

Roca M, James C, Pruzinska A, Hortensteiner S, Thomas H, Ougham H (2004). Analysis of the chlorophyll catabolism pathway in leaves of an introgression senescence mutant of *Lolium temulentum*. *Phytochem* 65: 1231–1238.

Rockel P, Strube F, Rockel A, Wildt J, Kaiser WM (2002). Regulation of nitric oxide (NO) production by plant nitrate reductase in vivo and in vitro. *J Exp Bot* 53:103-110.

Rolland F, Moore B, Sheen J (2001). Sugar Sensing and Signaling in Plants. *Plant Cell* S185-S205.

Rousseaux MC, Ballare CL, Jordan ET, Vierstra RD (1997). Directed overexpression of PHYA locally suppresses stem elongation and leaf senescence responses to far-red radiation. *Plant Cell Environ* 20: 1551–1558.

Rousseaux MC, Hall AJ, Sanchez RA (1996). Far-red enrichment and photosynthetically active radiation level influence leaf senescence in field-grown sunflower. *Physiol Plant* 96: 217–224.

Sakamoto W (2006). Protein degradation machineries in plastids. *Annu Rev Plant Biol* 57: 599-621.

Satoh H, Nakayama K, Okada M (1998). Molecular cloning and functional expression of a water-soluble chlorophyll protein, aputative carrier of chlorophyll molecules in cauliflower. *J Biol Chem* 273: 30568–30575.

Smart CM (1994). Gene expression during leaf senescence. *New Phytol* 126: 419–448.

Smart CM, Scofield SR, Bevan MW, Dyer TA (1991). Delayed leaf senescence in tobacco plants transformed with *tmr*, a gene for cytokinin production in *Agrobacterium*. *Plant Cell* 3: 647–656.

Smeekens S (2000). Sugar-induced signal transduction in plants. *Annu Rev Plant Physiol Plant Mol Biol* 51: 49–81.

Somers DE, Sharrock RA, Tepperman JM, Quail PH (1991). The hy3 long hypocotyl mutant of *Arabidopsis* is deficient in phytochrome B. *Plant Cell* 3: 1263–1274.

Srivastava LM (2002). Plant growth and development hormones and environment. *Academic Press California* 0-12-660570-X.

Stöhr C and Ullrich WR (2002). Generation and possible roles of NO in plant roots and their apoplastic space. *J Exp Bot* 53: 2293-2303.

Takamiya K, Tsuchiya T, Ohta H (2000). Degradation pathway(s) of chlorophyll: What has gene cloning revealed? *Trends Plant Sci* 5: 426–431.

Tanaka R, Hirashima M, Satoh S, Tanaka A (2003). The *Arabidopsis*-accelerated cell death gene ACD1 is involved in oxygenation of pheophorbide a: Inhibition of the pheophorbide a oxygenase activity does not lead to the "stay-green" phenotype in *Arabidopsis*. *Plant Cell Physiol* 44: 1266–1274.

Taylor KC, Hammer CU, Alley RB, Clausen HB, Dahl-Jensen D,Gow AJ, Gundestrup NS, Kipfstrep J, Moore JC, Waddington ED (1993). Electrical conductivity measurements from the GISP2 and GRIP Greenland ice cores. *Nature* 366: 549 - 552.

Thiele A, Herold M, Lenk I, Quail PH, Gatz C (1999). Heterologous expression of *Arabidopsis* phytochrome B in transgenic potato influences photosynthetic performance and tuber development. *Plant Physiol* 120: 73–82

Thoenen M, Herrmann B, Feller U (2007). Senescence in wheat leaves: is a cysteine endopeptidase involved in the degradation of the large subunit of Rubisco? *Acta Physiol Plant* 29: 339-350.

Thomas H (1978). Enzymes of nitrogen mobilization in detached leaves of *Lolium temulentum* during senescence. *Planta* 142: 161–169.

Thomas H, Ougham HJ, Wagstaff C, Stead AD (2003). Defining senescence and death. *J Exp Bot* 54: 1127–1132.

Thomas H (2002). Ageing in plants. *Mechanisms Ageing Develop* 123: 747–53.

Thomas H and Howarth CJ (2000). Five ways to stay green. *J Exp Bot* 51: 329–337.

Thomas H and Smart CM (1993). Crops that stay green. *Ann Appl Biol* 123: 193–219.

Thompson J, Taylor C, Wang TW (2000). Altered membrane lipase expression delays leaf senescence. *Biochem Soc Trans* 28(6): 775-777.

Thompson JE, Legge RL, Barber RF (1987). The role of free radical in senescence and wounding. *New Phytol* 105: 317–344.

Thompson JE, Froese CD, Madey E, Smith MD, Hong Y (1998). *Prog Lipid Res* 37: 119-141.

Tsuchiya T, Ohta H, Okawa K, Iwamatsu A, Shimada H, Masuda T, Takamiya K (1999). Cloning of chlorophyllase, the key enzyme in chlorophyll degradation: Finding of a lipase motif and the induction by methyl jasmonate. *Proc Natl Acad Sci USA* 96: 15362–15367.

Tun NN, Holk A, Scherer FE (2001).Rapid increase of NO release in plant cell cultures induced by cytokinin. *FEBS Lett.* 509:174–176.

Tunc-Ozdemir M, Miller G, Song L, Kim J, Sodek A, Koussevitzky S, Misra AN, Mittler R, Shintani D (2009). Thiamin confers enhanced tolerance to oxidative stress in *Arabidopsis*. *Plant Physiol 151*: 421–432.

van der Graaff E, Schwacke R, Schneider A, Desimone M, Flugge UI, Kunze R (2006). Transcription analysis of Arabidopsis membrane transporters and hormone pathways during developmental and induced leaf senescence. *Plant Physiol* 141: 776-792

van Doorn WG and Woltering EJ (2004). Senescence and programmed cell death: substance or semantics? *J Exp Bot* 55: 2147–2153.

van Doorn WG and Woltering EJ (2008). Physiology and molecular biology of petal senescence. *J Exp Bot* 59: 453–480.

Vierstra RD (1996). Proteolysis in plants: mechanisms and functions. *Plant Mol Biol* 32: 275-302.

Wagstaff C, Malcolm P, Rafiq A, Leverentz M, Griffiths G, Thomas B, Stead A, Rogers H (2003). Programmed cell death (PCD) processes begin extremely early in *Alstroemeria* petal senescence. *New Phytol* 160: 49–59.

Vladkova R, Dobrikova AG, Singh R, Misra AN, Apostolova E (2011). Photoelectron transport ability of chloroplast thylakoid membranes treated with NO donor SNP: Changes in flash oxygen evolution and chlorophyll fluorescence. *Nitric Oxide Biol Chem* 24: 84-90.

Watanabe N and Lam E (2006). Arabidopsis Bax inhibitor-1 functions as an attenuator of biotic and abiotic types of cell death. *Plant J* 45: 884–894.

Weaver LM and Amasino RM (2001). Senescence Is Induced in Individually Darkened *Arabidopsis* Leaves, but Inhibited in Whole Darkened Plants. *Plant Physiol* 127: 876–886.

Weaver LM, Gan S, Quirino B, Amasino RM (1998). A comparison of the expression patterns of several senescence-associated genes in response to stress and hormone treatments. *Plant Mol Biol* 37: 455–469.

Wellesen K, Durst F, Pinot F, Benveniste I, Nettesheim K, Wisman E, Steiner-Lange S, Saedler H, Yephremov A (2001). Functional analysis of the LACERATA gene of Arabidopsis provides evidence for different roles of fatty acid ω-hydroxylation in development. *Proc Natl Acad Sci USA* 98: 9694–9699.

Willekens H, Chamnongpol S, Davey M, Schraudner M, Langebartels C, Van Montagu M, Inze D, Van CampW (1997). Catalase is a sink for H_2O_2 and is indispensable for stress defence in C3 plants. *EMBO J* 16: 4806–4816

Wingler A, von Schaewen A, Leegood RC, Lea PJ, Quick WP (1998). Regulation of leaf senescence by cytokinin, sugars, and light. Effects on NADH-dependent hydroxypyruvate reductase. *Planta* 116: 329 –335.

Woo HR, Goh CH, Park JH, Teyssendier de la Serve B, Kim JH, Park YI, Nam HG (2002). Extended leaf longevity in the *ore4-1* mutant of *Arabidopsis* with a reduced expression of a plastid ribosomal protein gene. *Plant J* 31: 331-340.

Xiao W, Sheen J, Jang JC (2000). The role of hexokinase in plant sugar signal transduction and growth and development. *Plant Mol Biol* 44: 451–461.

Yamada T, Takatsu Y, Kasumi M, Marubashi W, Ichimura K (2004). A homolog of the defender against apoptotic death gene (DAD1) in senescing gladiolus petals is down-regulated prior to the onset of programmed cell death. *J Plant Physiol* 161: 1281–1283.

Yamasaki H and Sakihama Y (2000). Simultaneous production of nitric oxide and peroxynitrite by plant nitrate reductase: In vitro evidence for the NR-dependent formation of active nitrogen species. *FEBS Lett* 468: 89-92.

Yin Y, Wang ZY, Garcia SM, Li J, Yoshida S, Asami T, Chory J (2002). BES1 Accumulates in the Nucleus in Response to Brassinosteroids to Regulate Gene Expression and Promote Stem Elongation. *Cell* 109(2): 181-191.

Yoshida S, Ito M, Nishida I, Watanabe A (2002b). Identification of a novel gene HYS1/CPR5 that has a repressive role in the induction of leaf senescence and pathogen-defence responses in *Arabidopsis thaliana. Plant J* 29: 427 - 437

Yoshida S, Ito M, Callis J, Nishida I, Watanabe A (2002a). A delayed leaf senescence mutant is defective in arginyl-tRNA:protein arginyltransferase, a component of the N-end rule pathway in *Arabidopsis. Plant J* 32:129-137

Yoshida S (2003). Molecular regulation of leaf senescence. *Curr Opin Plant Biol* 6: 1–6.

Yu IC, Parker J, Bent AF (1998). Gene-for-gene disease resistance without the hypersensitive response in *Arabidopsis* dnd1 mutant. *Proc Natl Acad Sci USA* 95: 7819–7824.

Zeidler D, Zahringer U, Gerber I, Dubery I, Hartung T, BorsW, Hutzler P, Durner J (2004). Innate immunity in *Arabidopsis thaliana*: lipopolysaccharides activate nitric oxide synthase (NOS) and induce defense genes. *Proc Natl Acad Sci USA* 101: 15811–15816.

Zeier J, Delledonne M, Mishina T, Severi E, Sonoda M, Lamb C (2004). Genetic elucidation of nitric oxide signaling in incompatible plant–pathogen interactions. *Plant Physiol* 136: 2875–2886.

Zimmermann P, Heinlein C, Orendi G, Zentgraf U (2006). Senescence-specific regulation of catalases in *Arabidopsis thaliana* (L.) Heynh. *Plant Cell Environ* 29: 1049-1060.

10

Menaquinone as Well as Ubiquinone as a Crucial Component in the *Escherichia coli* Respiratory Chain

Naoko Fujimoto[1], Tomoyuki Kosaka[2] and Mamoru Yamada[1,2]
[1]*Graduate School of Medicine, Yamaguchi University, Ube*
[2]*Faculty of Agriculture, Yamaguchi University, Yamaguchi*
Japan

1. Introduction

Isoprenoid quinones, which are found as membrane-bound compounds in almost all living organisms, generally have functions as electron carriers or antioxidants. Of the various isoprenoid quinones, ubiquinones (UQs) and menaquinones (MKs) have been extensively studied to reveal not only their physiological functions but also their biosynthesis at gene level.

In molecular structure, both UQs and MKs consist of a polar head group and a hydrophobic side chain. The latter part provides the molecules with a lipid-soluble character to allow them to perform vital functions in membrane lipid bilayers, whereas the former group enables interaction with membrane proteins. The quinone ring of the head group has a crucial activity by a two-step reversible reduction reaction to form a quinol structure (Fig. 1). There are three

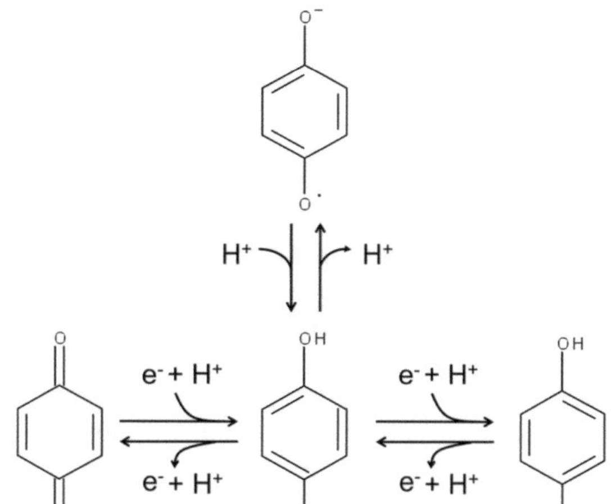

Left, oxidized form; middle, semiquinone intermediate; right, reduced form; upper, semiquinone radical ion.

Fig. 1. Oxidation states of quinone ring.

oxidized states of the quinone ring. The addition of one electron and one proton to the fully oxidized form results in the semiquinone form. The addition of a second electron and proton to generates the fully reduced form. When the semiquinone loses a proton, it becomes a semiquinone radical anion. The reduced form of isoprenoid quinones becomes more polar, and thus the quinol portion may be preferentially located in the boundary region between hydrophobic and hydrophilic micro-environments (Fiorini et al., 2008; Lenaz et al., 2007).

Such chemical and functional features of quinones make them suitable mediators of electron transfer between different protein complexes in biological membranes. Most of UQ or MK molecules function as mobile carriers of electrons in membranes, but some of them are tightly associated with protein molecules to function in intramolecular electron transfer (Goodwin, 1977; Lichtenthaler, 1977). In addition, UQs physiologically possess an antioxidant capacity against lipid peroxidation (Landi et al., 1984). However, little attention has been given to MKs as natural antioxidants. Phylloquinone (PhQ) and MK have been shown to prevent lipid peroxidation as dose UQ, but menadione has no such inhibitory effect (Vervoort et al., 1997).

1.1 Occurrence and structure of menaquinones

MKs, which have a low redox potential compared to that of UQs, function as the most widespread respiratory quinones. They are assumed to be evolutionarily the most ancient type of isoprenoid quinones and are found in many microorganisms such as archaea and bacteria (Goodwin, 1977; Lubben, 1995; Nitschke et al., 1995; Schoepp-Cothenet et al., 2009; Soballe & Poole, 1999).

The degree of saturated or hydrogenated polyprenyl side-chains of MKs is sometimes different depending on these microorganisms. *Archaea* contains mainly MKs, which are often dehydrogenated. The major MK in *Mycobacterium phlei* is MK-9(H_2) (Gale et al., 1963), whereas the major MK in *Corynebacterium diphtheriae* is MK-8(H_2) (Scholes & King, 1965). Such dihydromenaquinones are widespread in corynebacteria and mycobacteria, whereas even more highly saturated MKs occur in certain actinomycetes (Collins et al., 1977; Lancaster, 2003).

MKs mostly possess methylnaphthoquinone as a head group, but different modifications of the head group are present in some prokaryotes, which provide them with different redox potentials (Fig. 2). Demethylmenaquinones (DMKs) in some bacteria (Baum & Dolin, 1965; Collins & Jones, 1979; Lester et al., 1964) lack the ring methyl substitutant (C-2). DMKs with different sizes of polyprenyl side chains from one to nine isoprene units have been reported (Hammond & White, 1969). Methylmenaquinone (MMK) with an additional methylation has a redox potential similar to that of MK. Thermoplasmaquinone (TPQ) is one kind of MMK. Methionaquinone (MTK) contains a methylthio group instead of a methyl group in the naphtoquinone (Hiraishi et al., 1999; Ishii et al., 1987). Both MTK and TPQ occur in Thermoplasma (Lubben, 1995; Shimada et al., 2001). DMKs and MMK are often found in proteobacteria and Gram-positive bacteria (Biel et al., 2002; Collins & Jones, 1981). Chlorobiumquinone (ChQ), which is found in photosynthetic green sulfur bacteria, is an isoprenoid naphthoquinone containing a carbonyl group in its side chain (Collins & Jones, 1981).

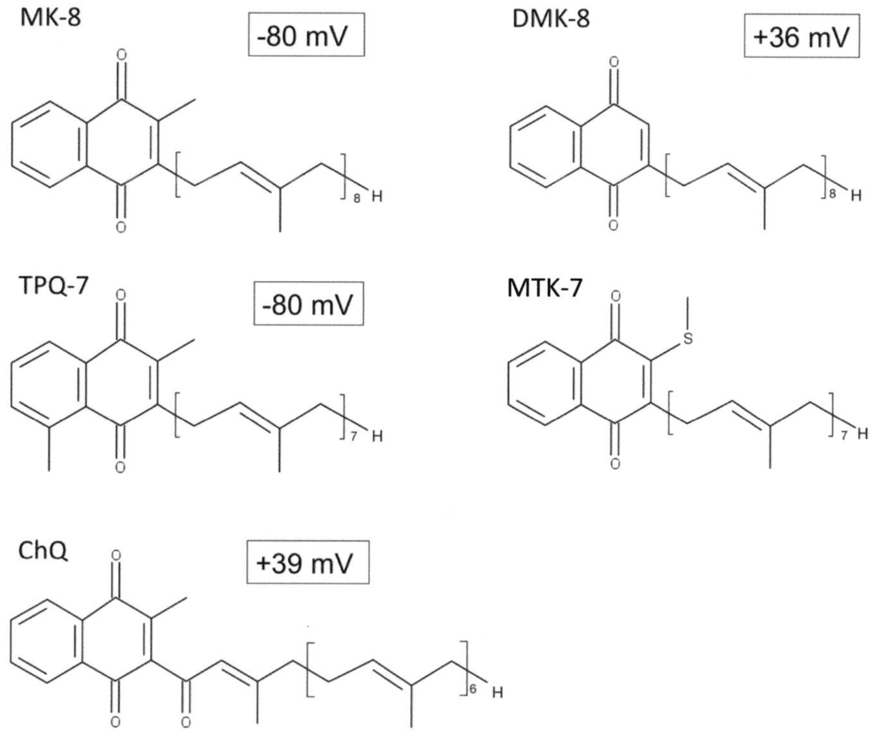

Menaquinone-8 (MK-8) and demethylmenaquinone-8 (DMK-8) occur in *E. coli*. Thermoplasmaquinone-7 (TPQ-7) and methionaquinone-7 (MTK-7) occur in *Thermoplasma acidphilum*. ChQ, chlorobiumquinone (1′-oxo-menaquinone-7) is present in *Chlorobium tepidum*. Adapted from work of Nowicka and Kruk (Nowicka & Kruk, 2010). The redox potential of MTK-7 is not available.

Fig. 2. Structures and redox potentials of isoprenoid naphthoquinones.

Most Gram-positive bacteria and anaerobic Gram-negative bacteria contain MK as a sole isoprenoid quinone (Collins et al., 1981). Most *Bacteroides* and *Bacillus spp.* produce MK as a major isoprenoid quinone. *Thermus thermophilus, Actinobacillus actinoides, Thermoplasma acidphilum, Lactobacillus mali, Lactobacillus yamanashiensis, Streptococcus cremoris, Planococcus, Staphylococcus, Corynebacterium, Brevibacterium, Arthrobacter, Chlorobium thiosulfatophium* and *Chloropseudomonas ethylicum* produce MKs as their sole isoprenoid quinones. *Halobacterium* and *Halococcus* have an unsaturated MK (MK-8) and a dihydrogenerated MK (MK-8(H_2)) (Kushwaha et al., 1974). Typical strains of species within the genera *Actinobacillus* and *Pasteurella* produce DMKs as major isoprenoids. *Actinobacillus actinomycetemcomitans, Actinobacillus suis, Pasteurella bettii, Pasteurella pneumotropica, Streptococcus faecalis, Haemophilus influenza* and *Haemophilus aergyptilus* produce only DMKs. Gram-negative facultatively anaerobic rods and several genera within the family *Enterobacteriaceae* such as *Escherichia, Klebsiella* and *Promonas* contain mixtures of MK and DMK. Certain species of the genera *Aeromonas* and *Erwinia* also contain a mixture of MKs.

The appearance of MKs on Earth is assumed to be associated with the reducing character of the atmosphere before the occurrence of oxygenic photosynthesis and dramatic increase in environmental oxygen concentration (Schoepp-Cothenet et al., 2009). It is thought that various quinones with higher redox potentials had diverged independently from naphthoquinones in a few groups of prokaryota along with acquisition of aerobic metabolism. The reduced form of MKs is highly reactive with molecular oxygen and is subjected to non-catalytic oxidation, and it is therefore thought to function inefficiently in an oxygen-containing atmosphere (Schoepp-Cothenet et al., 2009). The isoprenoid side chain of most MKs consists of 6-10 prenyl units, but MKs bearing 1-5 or 11-14 prenyl units are found in some species (Collins & Jones, 1981). The side chain is generally unsaturated, but it is also partially or fully saturated in some organisms (Collins & Jones, 1981). The degree of saturation of the side chain appears to be dependent on growth temperature.

1.2 Biosynthesis of menaquinones

1.2.1 Biosynthesis of the isoprenoid side chain of menaquinones

The head group and the isoprenoid side chain of MK are separately synthesized. Both parts are then combined together by an enzyme in the prenyltransferase family and then subjected to further modifications. The isoprenoid side chain is synthesized from dimethylallyl diphosphate (DMAPP) and isopentenyl diphosphate (IPP). There are two distinct pathways to build IPP precursors: the mevalonate (MVA) pathway and the deoxyxylulose 5-phosphate (DXP) pathway (Lange et al., 2000). *Archaea* are thought to have acquired the MVA pathway, based on evidence that species in this group possess genes homologous to those encoding enzymes in this pathway but not in the DXP pathway. On the other hand, all bacteria except for *Myxococcus fulvus, Chloroflexus aurantiacus, Streptomycetes, Rickettsia prowazekii* and *Mycoplasma genitalium* have acquired the DXP pathway (Lange et al., 2000). All animals and fungi utilize the MVA pathway. Higher plants utilize both pathways in different organelles, the DXP pathway for plastids and the MVA pathway for the cytosolic compartment. In higher plants, the biosynthesis of prenyl chains of phylloquinone, plastoquinone, tocopherolquinone, chlorophyll and tocopherol thus proceed by the DXP pathway, whereas UQ synthesis is carried out by the MVA pathway (Lichtenthaler, 1999). Algae and *Protista* also show distinctive utilization of both pathways. *Euglena gracilis* utilizes only the MVA pathway, *Chlamydomonas reinhardtii* utilizes only the DXP pathway, and the rhodophyte *Cyanidium caldarum* uses both pathways (Lange et al., 2000). This complication in utilization of the pathways might be due to horizontal gene transfer and/or endosymbiotic origin of organelles. The isoprenoid side chain is synthesized by a specific prenyl diphosphate synthase through a series of condensation reactions of precursors.

1.2.2 Biosynthesis of the head group of menaquinones

Two pathways are known to be involved in biosynthesis of the head group of MK (Fig. 3a and b). In both pathways, the precursor of the head group is 1,4-dihydroxy-2-naphthoate (DHNA), which is derived from chorismate in the shikimate pathway. In one of the two pathways, there are 7 enzymes required for formation of a quinone portion, DHNA, in which 6 *men* genes are involved (Widhalm et al., 2009) (Table 1). The quinone portion synthesized by the successive reactions is combined with the isoprenoid side chain by the condensation reaction of DHNA prenyltransferase. The number of prenyl units composing the isoprenoid side chain, which differs from species to species, may be determined by the length in the tertiary structure of isoprenoid diphosphatease. For example, *E. coli* and

Geobacter metallireducens produce MK-8, whereas, *Bacillus firmus* produces MK-7 (Hedrick et al., 2009; Soballe & Poole, 1999). Following the condensation reaction, the naphthoate group is methylated by C-metyltransferase (Lee et al., 1997).

In the other pathway called the futalosine pathway, which has recently been discovered in some microorganisms, orthologues of certain *men* genes are absent (Hiratsuka et al., 2008) (Fig. 3B). Four enzymes specific for this pathway are involved at the beginning of DHNA biosynthesis followed by the same reactions as those in the former pathway (Hiratsuka et al., 2009) (Table 2).

Gene	Product
menF	Isochorismate synthase
menD	2-succinyl-6-hydroxy-2,4-cyclohexadiene-1-carboxylate synthase
menH	(1R, 6R)-2-succinyl-6-hydroxy-2,4-cyclohexadiene-1-carboxylic acid synthase
menC	*o*-succinyl benzoate synthase
menE	*o*-succinyl benzoate-CoA synthase
menB	1,4-dihydroxy-2-naphthoyl-CoA synthase
ybgC	1,4-dihydroxy-2-naphthoate thioesterase
menA	1,4-dihydroxy-2-naphthoate octaprenyltransferase
ubiE	2-DMK methyltransferase

Table 1. Genes and their products related to the biosynthetic pathway of menaquinone-8 in *E. coli.*

Gene	Function
mqnA	Condensation of chrismate, inosine and C2 unit to form futalosine
mqnB	Futalosin hydrolase
mqnC	Cyclization of dehypoxanthinylfutalosine to form cyclic dehypoxanthinylfutalosine
mqnD	Cleavage of cyclic dehypoxanthinylfutalosine to form 1,4-dihydroxy-2-naphthoate

Table 2. Genes and their functions related to the futalosine pathway.

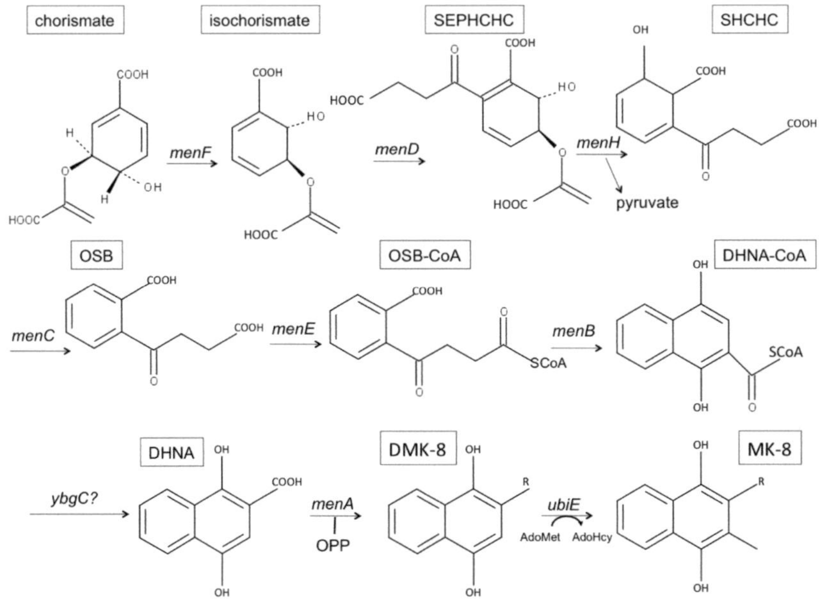

(a)

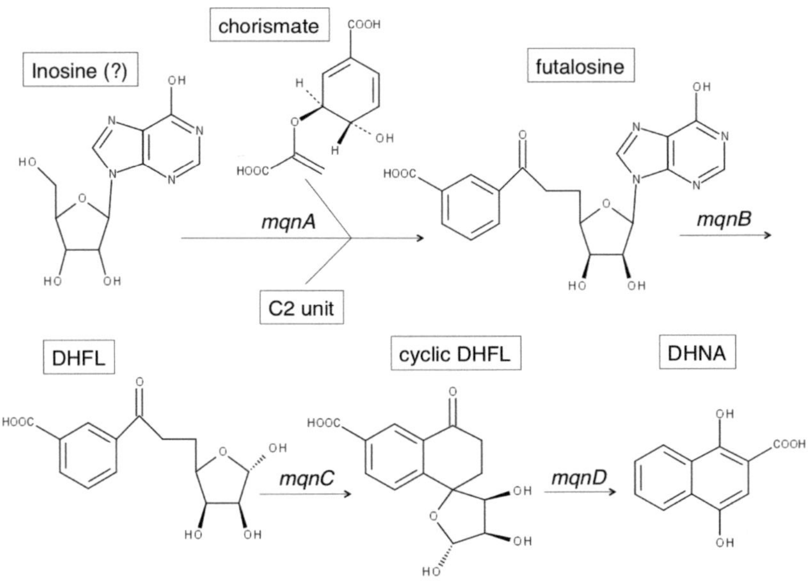

(b)

(a) The pathway to form menaquinone-8 in *E. coli*. DHNA, 1,4-dihydroxy-2-naphthoate; R, octaprenyl side chain; OSB, *o*-succinylbenzoate; SEPHCHC, 2-succinyl-5-enolpyruvyl-6-hydroxy-3-cyclohexene-1-carboxylic acid; SHCHC, (1R, 6R)-2-succinyl-6-hydroxy-2,4-cyclohexadiene-1-carboxylic acid; OPP, octaprenyl diphosphate; AdoMet, S-adenosylmethionine; AdoHcy, S-adenosylhomocysteine.
(b) The alternative pathway to form DHNA. DHFL, dehypoxanthinylfutalosine. Adapted from works of Hiratsuka et al. (Hiratsuka et al., 2008) and Nowicka and Kruk (Nowicka & Kruk, 2010).

Fig. 3. Biosynthetic pathways of menaquinone.

1.3 Functions of menaquinones

1.3.1 Menaquinones in the bacterial photosynthetic electron transport chain

MKs or UQs are known to engage in photosynthetic electron transport of several photosynthetic microorganisms (Nowicka & Kruk, 2010). MKs are found in photosystem II (PSII)-type reaction centers (RCs) of purple bacteria and green filamentous bacteria and in photosystem I (PSI)-type RCs of green sulfur bacteria and heliobacteria (Ke, 2001). On the other hand, UQs are involved in photosynthesis of many purple bacterial species. *Halorhodospira halophila* seems to distinctively utilize the two quinones, MK-8 for the photosynthetic electron transfer reaction and UQ-8 for the respiratory reaction (Schoepp-Cothenet et al., 2009). Green filamentous bacteria dominantly utilize MK-10 both for photosynthetic and respiratory electron transport reactions (Hale et al., 1983). MK-4 is present in the PSI of the cyanobacterium *Gloeobacter violaceus* or *Synechococcus* PCC 7002 diatoms and primitive red alga *C. caldarum* (Ikeda et al., 2008; Mimuro et al., 2005; Sakuragi et al., 2005). Green sulfur bacteria contain MK-7 in their PSI-type RCs (Hauska et al., 2001), similar to that of PSI of green filamentous bacteria. The involvement of MK in the reaction of their RCs is not obvious because the electron transfer activity in their RCs was not hampered without MK. MK-9 is found at the A1 site of the RC of *Helicobacterium chlorum*, but its physiological function remains to be clarified (Neerken & Amesz, 2001; Oh-oka, 2007).

MK occurs in chlorosomes of green photosynthetic bacteria (Frigaard et al., 1997; Frigaard et al., 1998; Kim et al., 2007). A small amount of MK-7 with a large amount of chlorobiumquinone is found in chlorosomes of the green sulfur bacterium *Chlorobium tepidum* (Frigaard et al., 1997). On the other hand, MK is present as a major quinone in chlorosomes of the thermophilic green bacterium *Chloroflexus auranticus*, which lacks chlorobiumquinones. It has been reported that chlorosomes of *C. tepidum* exhibited high fluorescence, which rapidly decreased under aerobic conditions (Frigaard et al., 1997). The authors speculated that chlorobiumquinone senses a redox state and inhibits electron transfer to the RC under aerobic conditions to avoid possible oxidative stress (Frigaard et al., 1997).

1.3.2 Menaquinone in the prokaryotic respiratory chain

MK mediates electron transport reactions as does UQ in respiratory chains of prokaryotes. Both quinones contribute to the formation of transmembrane potential via electron-accepting or -donating reactions with other respiratory components. The transmembrane potential that consists of a proton gradient and an electron gradient across membranes is used for ATP synthesis and for transport of many materials between the inside and outside of cells. There are two mechanisms for the formation of the proton gradient: proton pump mechanism and redox loop mechanism. In the latter mechanism, during electron transfer reactions, oxidation of one substrate and proton(s) release simultaneously occur on the inside of membranes and then reduction of the other substrate and proton(s) binding simultaneously occur on the outside of the membrane (Richardson & Sawers, 2002). MK as well as UQ is thus crucial components in the respiratory chain of prokaryotic cells for energy production and cell maintenance.

Reduction of MK is coupled with oxidation of reductants, such as NADH, succinate, sulfide, thiocyanate, ammonium, formate and hydrogen, and oxidation of reduced MK (MKH_2) is coupled with reduction of oxidants such as oxygen, nitrogen dioxide, nitrite, sulfate, sulfide, thiosulfate, polysulfide, elemental sulfur and fumarate. These reduction/oxidation reactions are carried out by specific enzymes, reductases and oxidases that participate in the reaction of MK and oxidation of MKH_2, respectively. A number of MK reductases called as oxidoreductases or dehydrogenases for NADH, succinate, formate, hydrogen, malate, pyruvate and glycerol-3-phosphate have been studied (Lancaster & Simon, 2002; Richardson, 2000).

NDH-1-type dehydrogenases (NDH-1) functioning as an NADH:MK oxidoreductase in many microorganisms are able to pump protons across the membrane (Yagi et al., 1998). They are complexes consisting of 13–14 subunits and analogous to mitochondrial complex I. Their detailed structure and the relationship between their structure and function, however, are still poorly understood. NDH-2-type dehydrogenases also acting as an NADH:MK oxidoreductase are also ubiquitous in bacteria but are simpler and smaller in structure than NDH-1. The NDH-2-type dehydrogenases lack proton pumping activity (Nantapong et al., 2005). This type of dehydrogenase might be crucial for substrate oxidation under relatively transmembrane potential-rich conditions.

Succinate dehydrogenase as a succinate:MK oxidoreductase, which is also widely spread in microorganisms, is analogous to mitochondrial complex II but has no proton-pumping activity (Azarkina & Konstantinov, 2010; Fernandes et al., 2005; Kurokawa & Sakamoto, 2005; Madej et al., 2006; Xin et al., 2009). The tertiary structure of succinate dehydrogenase in *E. coli*, which consists of 4 subunits, has been resolved (Iverson et al., 1999). However, there are controversial opinions regarding the site for quinone reduction in the dehydrogenase molecule (Fernandes et al., 2005). The dehydrogenase utilizes both quinones to perform coupled reaction of oxidation of succinate to fumarate and reduction of quinone and also catalyzes the coupled reaction of reduction of fumarate to succinate and oxidation of the reduced form of quinone. Since MK has a redox potential lower that that of UQ, the reduction of MK is endergonic and that of UQ is exergonic under standard conditions (Madej et al., 2006). Therefore, no succinate oxidation coupled with MK reduction occurs without energy supply from outside. In Gram-positive aerobic bacteria having only MK, such as *Bacillus subtilis* and *Bacillus licheniformis*, energization of the cellular membrane is necessary for succinate:MK oxidoreductase activity (Azarkina & Konstantinov, 2010; Madej et al., 2006). Therefore, MK reduction reaction of the enzyme is proposed to be proton-driven.

Formate dehydrogenase, which oxidizes formate to carbon dioxide, can generate a proton gradient by a redox loop mechanism. As described above for the redox loop mechanism, the formate oxidation site of the enzyme is orientated towards the periplasm, whereas the MK reduction site is located close to the cytoplasmic side of the membrane (Jormakka et al., 2002). Similarly, hydrogenase that catalyzes hydrogen oxidation coupled with MK reduction is assumed to form a proton gradient by the redox loop mechanism (Kröer et al., 2002).

$F_{420}H_2$:quinone oxidoreductase as an MK reductase in the strictly anaerobic, sulfate-reducing archeon *Archaeoglobus fulgidus* is thought to be a functional equivalent of NADH:quinone oxidoreductase (Bruggemann et al., 2000). This enzyme is a multi-subunit complex similar in structure to bacterial NDH-1 and mitochondrial complex I and may be composed of three subcomplexes. The significant homology of archaebacterial enzyme subunits to those of bacterial NDH-1 and complex I (Bruggemann et al., 2000) allows us to consider the evolutionary relationship as a biochemical unity.

On the other hand, MKH_2 oxidation reactions are known to be performed by several MKH_2:oxidant oxidoreductases. Coupling oxidants for the reactions are nitrate (nitrite), sulfate, fumarate, and oxygen as an terminal electron acceptor. Membrane-bound nitrate reductase (Nir) in *E. coli,* which is expressed under anaerobic or microaerobic conditions, catalyzes the reduction reaction of nitrate to nitrite coupled with oxidation of MKH_2 and concomitantly generates a proton gradient (Pinho et al., 2005). *E. coli* Nir, which is expressed at a high concentration of nitrate, can utilize both MKH_2 and UQH_2 as electron donors (Brondijk et al., 2002; Giordani & Buc, 2004). Enzymes similar in structure to nitrate reductase have been postulated to participate in sulfate and selenate reduction in some bacteria (Ma et al., 2009; Pires et al., 2003). Fumarate respiration in anaerobic bacteria is achieved by using fumarate as a terminal electron acceptor and formate and hydrogen as electron donors into the chain (Lancaster, 2003). Since the reaction of MKH_2:fumarate oxidoredactase is exergonic, the enzyme is assumed to be able to generate a proton gradient by a redox loop mechanism (Madej et al., 2006).

Other MK/MKH_2-involved reactions are found in cytochrome bc_1 complexes (Schutz et al., 2000; Trumpower, 1990) and quinol oxidases (Ingledew & Poole, 1984) in respiratory chains. Quinone like UQ may form a Q cycle in the cytochrome bc_1 complex, which has been proposed to form a proton gradient by a mechanism similar to a redox loop mechanism. A model of the Q cycle allows us to speculate that there are at least two quinone-binding sites in the complex. The tertiary structure of the ubiquinol oxidase from *E. coli* has been resolved and the binding sites of quinone have been predicted (Abramson et al., 2000).

One of quinol-oxidizing enzymes is ubiquinol oxidase as a terminal oxidase in bacteria. Terminal oxidases utilizing oxygen as an electron acceptor in bacterial respiratory chains can be classified into two families: heme:copper oxidases and cytochrome *bd* oxidases. The former family catalyzes quinol oxidation reaction coupled with the reduction of oxygen as a terminal electron acceptor and generates a proton gradient partially by a proton pump mechanism (Kusumoto et al., 2000; Unden & Bongaerts, 1997), whereas the latter family transfers electrons from quinol to oxygen and generates a proton gradient by a redox loop mechanism (Junemann, 1997).

Another family of MKH_2-oxidizing enzymes is the NapC/NirT family. Each enzyme in this family occurs as a membrane-bound complex of tetraheme or pentaheme *c*-type cytochromes and catalyzes quinol oxidation reaction coupled with reduction of periplasmic proteins in cytoplasmic membranes of Gram-negative bacteria. The membrane-bound NrfH in *Wolinella succinogenes* reduces the periplasmic complex of nitrite reductase NrfA (Simon et al., 2001). NapC in *E. coli* may participate in electron transfer to the periplasmic complex of nitrate reductase NapA-NapB (Brondijk et al., 2002). CymA in *Shewanella oneidensis* is assumed to transfer electrons to a wide range of reductases (Schwalb et al., 2003).

Facultative anaerobic bacteria are known to be able to utilize both MK and UQ in respiratory chains. *E. coli* possesses two distinct but structurally similar enzymes, succinate dehydrogenase and fumarate reductase. The former is involved in oxidation reaction of succinate coupled with the reduction of UQ under aerobic conditions, and the latter functions in reducing fumarate and oxidizing reduced MK under anaerobic conditions. The cells might perform an efficient reaction by a combination of these enzymes and types of isoprenoid quinones, which are differently synthesized or produced under the two different conditions (Maklashina et al., 2006).

Strictly aerobic bacteria and strictly anaerobic bacteria possess only UQ and MK, respectively (Soballe & Poole, 1999). Facultative anaerobic proteobacteria including *E. coli* have both UQ and MK (Soballe & Poole, 1999; Wallace & Young, 1977). Such bacteria that are able to synthesize both quinones seem to control the relative production ratio of the two quinones depending on growth conditions. The oxygen level appears to affect relative amounts of these two kinds of quinones. When *E. coli* grows under aerobic conditions, UQ level is 4–5-times higher than the sum of MK and DMK levels, whereas under anaerobic conditions, the amount of UQ is three-times smaller than the sum of the amounts of both MKs (Wallace & Young, 1977; Wissenbach et al., 1990). Since the redox potential of MKs is lower than that of UQs, the former isoprenoid quinones are more suitable for respiratory chains with lower-potential electron acceptors (Nitschke et al., 1995; Soballe & Poole, 1999). When *E. coli* lacks one of the two quinones, the remaining one can functionally replace the missing one, but this is not the case for all respiratory pathways (Wallace & Young, 1977). Dependence of the quinone chosen at different oxygen concentrations has been also reported in Archaea. The facultative anaerobic, thermophilic archaeon *Thermoplasma acidophilum* has three isoprenoid naphthoquinones of MK-7, TPQ-7 and MTK-7 with different redox potentials. Under anaerobic conditions, TPQ-7 constitutes 97% of the total quinone pool, whereas under aerobic conditions, all the three quinones exist in nearly equal amounts Shimada et al., 2001).

In addition to the function in electron transfer chains, MK may have a crucial function in gene or cellular regulation of some microbes. MK has been shown to participate in the regulation of nitrogen fixation in *Klebsiella pneumoniae*. NifL, which acts as a co-repressor in the regulation of expression of nitrogen fixation genes, is reduced by the reduced form of MK under anaerobic conditions and becomes incompetent by its sequestration on the membrane (Thummer et al., 2007). A derivative of MK-9(H_2) with modification of a sulfate group at its isoprenoid chain has recently been identified in *Mycobacterium tuberculosis* and has been shown to act as a negative regulator of virulence in mice, suggesting its involvement in the regulation of host-pathogen interactions (Holsclaw et al., 2008).

2. Role of menaquinone as an electron acceptor and as a prosthetic group for dehydrogenases in *E. coli*

2.1 Menaquinone has a crucial role as an electron acceptor like ubiquinone in *E. coli*

2.1.1 Menaquinone is involved in activation of ArcB in a two-component system

Two-component systems that are widely spread in prokaryotes and plants are crucial for adaptation to changes in environmental and intracellular conditions. The two-component system of ArcB and ArcA as a transmembrane sensor kinase and a cognate response regulator, respectively, allows *E. coli* to sense aerobiosis to control over 30 operons (Fig. 4) (Iuchi & Lin, 1988; Iuchi et al., 1989; Iuchi et al., 1990). ArcBA is involved in sensing oxygen availability (Rolfe et al., 2011) or the redox state of the quinone pool (Bekker et al., 2010) and the concomitant transcriptional regulation of oxidative and fermentative catabolism. ArcB possesses an elaborate cytosolic structure that comprises three catalytic domains, each of which has a specific function of a primary transmitter with a conserved His292, a receiver with a conserved Asp576, or a secondary transmitter with a conserved His717. ArcA consists of an N-terminal receiver domain with a conserved Asp54 and a C-terminal helix-turn-helix DNA-binding domain (Kwon et al., 2000).

Two ArcB molecules form two reversible disulfide bridges at Cys180 and Cys241, which are located in the so-called PAS domain of the protein at the cytoplasmic side of *E. coli*. The PAS domain is the site to interact with UQ-UQH_2 couple or MK-MKH_2 couple (Malpica et al., 2004). The kinase activity of ArcB is highly dependent on the covalent linkage. Under signal perception, ArcB undergoes autophosphorylation at His292, which is enhanced by several fermentation metabolites such as D-lactate, pyruvate and acetate. The phosphoryl group is then sequentially transferred to ArcA via a His292 to Asp576 to His717 to Asp54 phosphorely (Kwon et al., 2000).

Bekker et al. (2010) have demonstrated that the ArcBA two-component system is regulated by the redox state of both the UQ and the MK pool in membranes (Fig. 4). The ArcBA

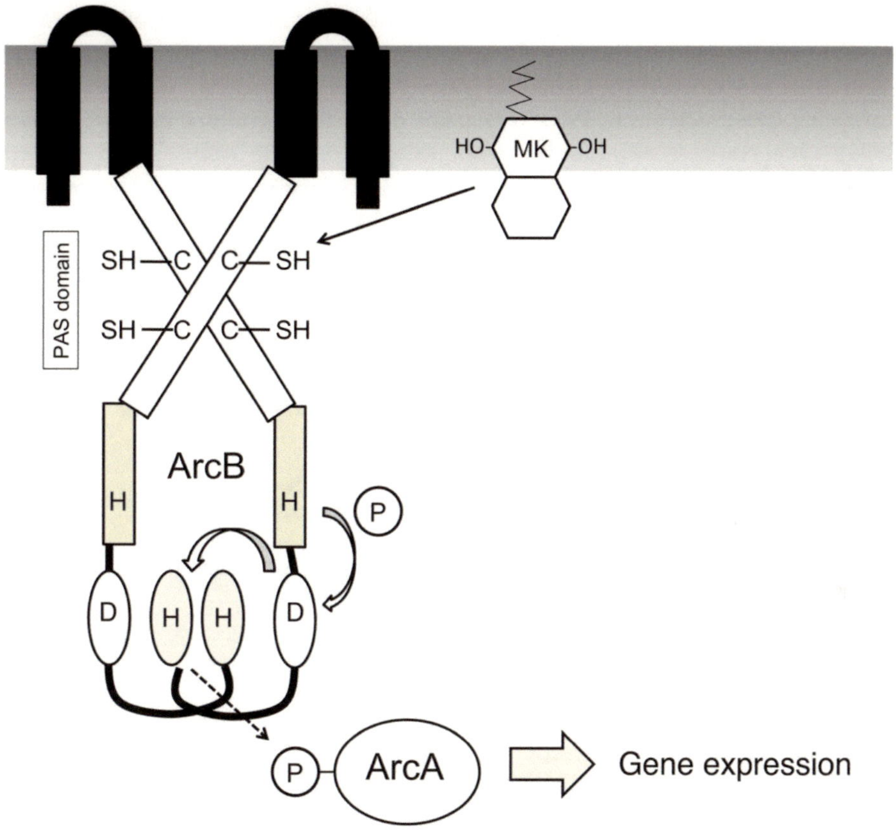

Two ArcB molecules form reversible disulfide bridges via two cysteine residues at the PAS domain of the protein. ArcB is activated by phosphorylation, depending on the reduction level of the disulfide bridges, which causes an intramolecular relay of the phosphoryl group and finally the phosphoryl group is transferred to ArcA, which in turn expresses its regulon. Recently, Bekker et al. (Bekker et al., 2010) indicate that the MK pool plays an important role in ArcB activation. Adapted from works of Bekker et al. (Bekker et al., 2010) and Malpica et al. (Malpica et al., 2004). H, histidine kinase domain; D, receiver domain.

Fig. 4. Involvement of MK in the signal transduction pathway via ArcBA two-component system.

system exhibits high and low activities under anaerobic and aerobic conditions, respectively. *In vitro* experiments have revealed that the residues in disulfide bonds of the ArcB complex can be oxidized by UQ (Malpica et al., 2004). Bekker et al. (2010) has shown that the deletion of *ubiC,* encoding an enzyme for the UQ biosynthesis, causes no effect on regulation of ArcB in the anaerobic-aerobic transition, but the deletion of *menB*, encoding an enzyme for the MK biosynthesis, leads to inactivation of ArcB. Therefore, MKs seems to play a major role in ArcB activation.

2.1.2 Menaquinone works as an electron acceptor in the DsbA-DsbB system

Formation of a disulfide bridges is crucial for the maturation process for envelope and secretary proteins, which intrinsically depends on distinct functions of several Dsb proteins (Collet & Bardwell, 2002; Kadokura et al., 2003). DsbA is involved in the oxidative folding of proteins newly synthesized in the periplasm of *E. coli* using its own disulfide bridge (Cys30-Cys33), and then re-oxidized by DsbB, an inner membrane protein to recycle the catalytic activity of DsbA (Bardwell et al., 1993; Missiakas et al., 1993; Guilhot et al., 1995; Kishigami et al., 1995). Oxygen is a ultimate electron acceptor for the DsbA/DsbB system under aerobic conditions (Bader et al., 1999; Kobayashi & Ito, 1999), where UQ receives electrons from DsbB (Bader et al., 1999; Bader et al., 2000) and the further electron transfer process to oxygen is mediated by the respiratory chain (Fig. 5) (Kobayashi et al., 1997).

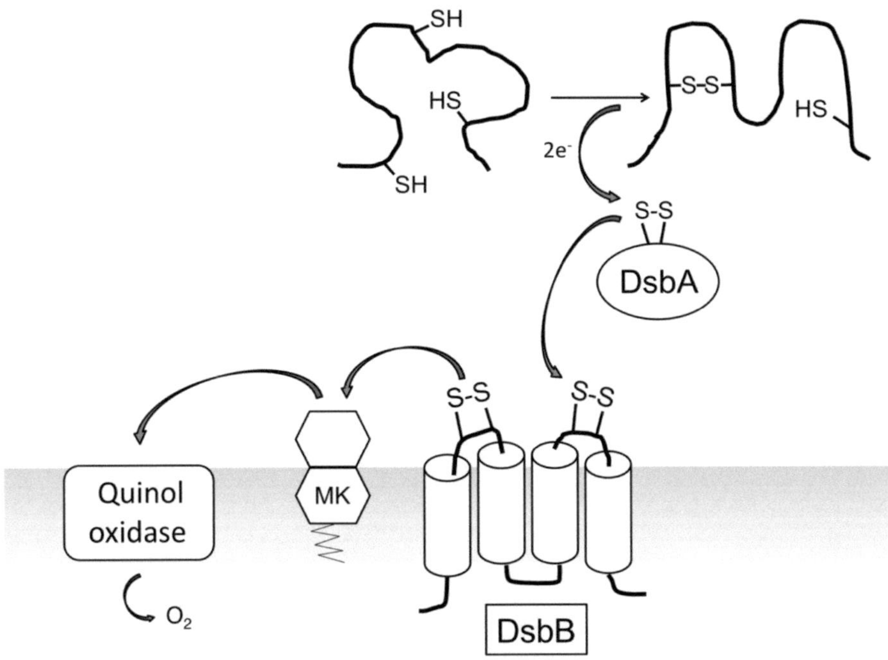

A disulfide bridge is introduced by DsbA, and the extracted electrons are transferred from DsbA to DsbB. The electrons are further transferred to oxygen via the respiratory chain, in which MK or UQ are involved. The arrows indicate the flow of electrons in the DsbB-DsbA oxidative pathway. Adapted from work of Inaba and Ito (Inaba & Ito, 2008).

Fig. 5. Involvement of MK in the DsbB-DsbA oxidative pathway.

In the inner membrane, UQ receives electrons from DsbB to restore the DsbA-oxidizing activity of DsbB (Bader et al., 2000). DsbB possesses two pairs of essential cysteines in its periplasmic domains as shown in Fig. 5, which are Cys41 and Cys44 in the N-terminal loop and Cys104 and Cys130 in the C-terminal loop (Jander et al., 1994). The former pair is oxidized by respiratory components (Kobayashi & Ito, 1999). The interaction of DsbB at the region around its Cys44 with UQ has been suggested, though the UQ-binding site on DsbB has not been directly determined (Xie et al., 2002). Mutations of residues 42 and 43 affect the redox potential and reactivity of DsbB, and those of the short segment between Cys44 and the second transmembrane region impair UQ-8-dependent DsbB oxidation *in vivo* (Inaba & Ito, 2002; Kadokura et al., 2000; Kobayashi et al., 2001). The DsbA/DsbB system is also functioning in disulfide bridge formation under anaerobic conditions in *E. coli* (Bader et al., 1999; Bader et al., 2000; Kadokura et al., 2000), where MK occurs the major quinone species (Wallace & Young, 1977). According to the assumption that MK plays an electron acceptor like UQ in the DsbA/DsbB system (Kadokura et al., 2000; Kishigami et al., 1995), mutational analysis was performed, which revealed that impairment of both the UQ and MK biosynthetic pathways results in dysfunction of the DsbA/DsbB system (Kobayashi et al., 1997). Arg48 substitutions in DsbB result in a low-activity enzyme that can no longer utilize the MK analog menadione as an *in vitro* electron-accepting substrate (Kadokura et al., 2000). MK-8 has been shown to associate with DsbB, similar to that of UQ, by spectroscopic analysis (Takahashi et al., 2004). The *in vitro* reaction of DsbA oxidation with MK-8 has been shown to be slower than the UQ-dependent reaction.

2.2 Menaquinone as well as UQ as a bound quinone is crucial for catalytic activity and intramolecular electron transfer in *E. coli* membrane-bound glucose dehydrogenase

Membrane-bound glucose dehydrogenase (mGDH) is a good model for primary dehydrogenases in terms of its occurrence as a single protein and as an apo-protein (Yamada et al., 1993a; Matsushita et al., 1997; Elias & Yamada, 2003), which allows study with both forms of apo- and holo-enzymes (Ameyama et al., 1985). It has been demonstrated that mGDH has two quinone (Q)-binding sites, one (Q_I) for bound Q and the other (Q_{II}) for bulk Q (Elias et al., 2004), which is near the membrane surface rather than in the hydrophobic interior (Miyoshi et al., 1999), and that intramolecular electron transfer following the catalytic reaction occurs from $PQQH_2$ directly to Q in the Q_{II} site or via bound Q (Fig. 6). Pulse radiolysis analysis revealed that the two redox centers are closely located at a distance of 11-13 Å (Kobayashi et al., 2005) and that Asp466 and Lys493 are involved in proton donation to the semiquinone anion radical of bound Q and in electron transfer from bound UQ to PQQ, respectively (Elias et al., 2000; Mustafa et al., 2008a). Recent mGDH analysis provided the first evidence that the primary dehydrogenase in respiratory chains utilizes both MK and UQ as a bound Q and suggest that bound MK occurs in a fashion similar to that of bound UQ in the mGDH molecule and functions as an electron acceptor from PQQ (Mustafa et al., 2008b). We also presented the data for the first time that suggest the requirement of bound Q for catalytic reaction in quinoprotein dehydrogenases (Mustafa et al., 2008b).

mGDH is expressed not only under aerobic conditions but also under anaerobic conditions, although the expression level is relatively low (Yamada et al., 1993b). mGDH expressed in the *ubiA* mutant defective in UQ biosynthesis was found to contain MK-8, and its MK content in purified mGDH protein was estimated to be 0.9 ± 0.03 mol/mol of mGDH (Mustafa et al., 2008b). Functional activities of purified bound MK-containing mGDH (MK-mGDH) and non-bound quinone-containing mGDH (Q-free mGDH) purified from *ubiA menA* cells were compared with those of bound UQ-containing mGDH (UQ-mGDH) from *menA* mutant cells. Both PMS reductase activity as glucose dehydrogenase activity and UQ-2 reductase activity of MK-mGDH were found to be equivalent to those of UQ-mGDH. The latter activity reflects the total ability of catalytic reaction and the successive intramolecular electron transfer from PQQ to UQ-2 at the Q_{II} site. The Q-free enzyme, however, exhibited only 18% of the dehydrogenase activity and 6% of the UQ-2 reductase activity of UQ-bearing mGDHs (Mustafa et al., 2008b).

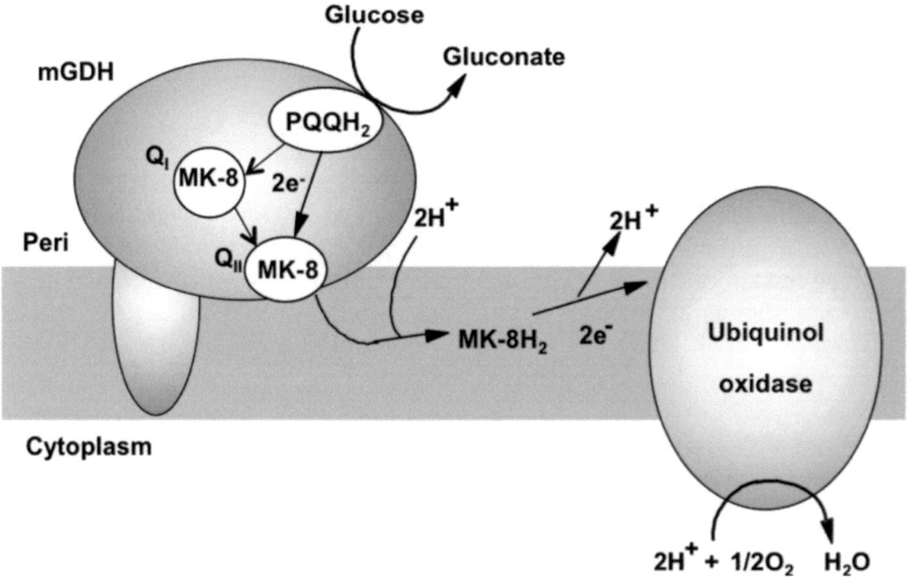

mGDH contains bound Q at the Q_I site and interacts with bulk Q at the Q_{II} site. Electrons from reduced PQQ following D-glucose oxidation are transferred to bound Q and then bulk Q or directly to bulk Q. Both MK and UQ as a Q can be incorporated into the Q_I site of mGDH molecule and interact at the Q_{II} site.

Fig. 6. Involvement of MK-8 in the intramolecular electron transfer in mGDH of *E. coli*.

It has been reported that UQ-1 incorporated into Q-free DsbB functions in a manner similar to that of bound Q (Inaba et al., 2005). External addition of UQ-1 showed UQ-1 dose-dependent increase in PMS reductase activity in Q-free mGDH. A similar level of increase in UQ-2 reductase activity was observed in the presence of UQ-1 in Q-free mGDH (Mustafa et al., 2008b). A radiolytically generated hydrated electron reacted with the bound MK to form a semiquinone anion radical with an absorption maximum at 400 nm. Subsequently, decay of the absorbance at 400 nm was accompanied by an increase in the absorbance at 380 nm with a first order rate constant of 5.7×10^3 s^{-1}. This indicates an intramolecular electron transfer from the bound MK to the PQQ. EPR analysis revealed that characteristics of the semiquinone radical of bound MK are similar to those of the semiquinone radical of bound UQ and indicated an electron flow from PQQ to MK as in the case of UQ. Taken together, the results suggest that MK is incorporated into the same pocket as that for UQ to perform a function almost equivalent to that of UQ and that bound quinone is involved at least partially in the catalytic reaction and primarily in the intramolecular electron transfer of mGDH (Mustafa et al., 2008b).

2.3 Possible interaction of menaquinone with ubiquinone-binding protein in *E. coli*

It is generally accepted that almost 90% of UQ is distributed freely in membranes as a UQ pool (Lenaz, 2001). Recently, Barros et al. (2005) speculated that Coq10 as a mitochondrial UQ-binding protein, which is not associated with succinate- and NADH-UQ reductase or the bc_1 complex in *Saccharomyces cerevisiae*, is involved in transport of UQ from its synthetic site to its functional site. Cui and Kawamukai (2009) reported that Coq10 binds to CoQ10 and is required for 11 proper respiration activity in Schizosaccharomyces pombe. These findings not only expand our knowledge of the regulation of UQ but also allow us to challenge the longstanding notion that UQ molecules are freely moving in the hydrophobic environment of membranes. Coq10 possesses 13 highly conserved hydrophobic amino acid residues presumably involved in the binding to CoQ_{10}. The presumption was partially proven by alanine substitution of these amino acid residues, in which L63A and W104A caused defective respiration and growth retardation on minimal medium. A human Coq10 ortholog has been shown to functionally compensate for a *coq10* null mutant. Therefore, Coq10 seems to be crucial for respiratory activity in various organisms.

Coq10 orthologs are also present in various microorganisms including *E. coli,* though their physiological functions remain to be established. Our preliminary experiments showed that an *E. coli* YfjG plays a function similar to that of Coq10 and allowed us to speculate that it interacts with UQ or MK to enhance the activity of electron transfer between dehydrogenase and ubiquinol oxidase. The alignment of YfjG orthologs in Gram-negative bacteria exhibited 11 highly conserved amino acid residues, which are partially overlapping with the conserved residues in Coq10 orthologs, some of which are involved in UQ binding (Cui & Kawamukai, 2009). A *yfjG*-disrupted mutant exhibited different properties in respiration activity of *E. coli*. Dehydrogenase activities of mGDH and NADH dehydrogenase and oxidase activity of ubiqionol oxidase in the *yfjG*-disrupted mutant were nearly the same as those in the parental strain. However, glucose and NADH oxidase activities were significantly decreased. Notably, the catalytic sites of mGDH and NADH dehydrogenase face the periplasm and cytoplasm, respectively. The mutation thus seems to affect the electron transfer ability between dehydrogenases and ubiquinol oxidase. Therefore, it is speculated that YfjG binds to UQ or MK to stimulate electron transfer in the respiratory chain.

3. Conclusion

In this chapter, first, we have shown the fundamental information on MKs including their structures, biosynthesis and physiological functions. We have also introduced several proteins intrinsically interacting not only with UQ-8 but also with MK-8 in *E. coli*. These proteins can associate with both quinones that have different ring structures and different redox potentials. In some cases, both quinones function as a mediator of electrons via the respiratory chain or as a sensor of the redox state of cells. In other cases, both quinones are incorporated into protein molecules and indispensable for enzyme activity. Such microbes that are capable of synthesizing both types of quinones might have evolved to adapt different growth conditions only by changing quinone, but not by changing the cognate protein(s). We thus notice one of ingenious strategies of microbes in utilization of the quinones.

4. Acknowledgements

This work was partially supported by a Grant-in-Aid for Basic Research from the Ministry of Education, Science and Culture of Japan, the Program for Promotion of Basic Research Activities for Innovative Biosciences and the Special Coordination Funds for Promoting Science and Technology, Ministry of Education, Culture, Sports, Science and Technology.

5. References

Abramson, J., Riistama, S., Larsson, G., Jasaitis, A., Svensson-Ek, M., Laakkonen, L., Puustinen, A., Iwata, S. & Wikstrom, M. (2000). The structure of the ubiquinol oxidase from *Escherichia coli* and its ubiquinone binding site, *Nat Struct Biol* 7(10): 910-917.

Ameyama, M., Nonobe, M., Hayashi, M., Shinagawa, E., Adachi, O. & Matsushita, K. (1985). Mode of binding of pyrroloquinoline quinone to apo-glucose dehydrogenase, *Agric Biol Chem* 49: 1227-1231.

Azarkina, N. & Konstantinov, A.A. (2010). Energization of *Bacillus subtilis* membrane vesicles increases catalytic activity of succinate: Menaquinone oxidoreductase, *Biochemistry (Moscow)* 75(1): 50-62.

Bader, M., Muse, W., Ballou, D., Gassner, C. & Bardwell, J. (1999). Oxidative protein folding is driven by the electron transport system, *Cell* 98(2): 217-227.

Bader, M., Xie, T., Yu, C.A. & Bardwell, J. (2000). Disulfide bonds are generated by quinone reduction, *J Biol Chem* 275(34): 26082-26088.

Bardwell, J., Lee, J.O., Jander, G., Martin, N.L., Belin, D. & Beckwith, J. (1993). A pathway for disulfide bond formation *in vivo*, *Proc Natl Acad Sci U S A* 90(3): 1038-1042.

Baum, R. & Dolin, M.I. (1965). Isolation of 2-Solanesyl-1,4-Naphthoquinone from *Streptococcus faecalis*, *J Biol Chem* 240(8): 3425-3433.

Bekker, M., Alexeeva, S., Laan, W., Sawers, G., Teixeira de Mattos, J. & Hellingwerf, K. (2010). The ArcBA two-component system of *Escherichia coli* is regulated by the redox state of both the ubiquinone and the menaquinone pool, *J Bacteriol* 192(3): 746-754.

Biel, S., Simon, J., Gross, R., Ruiz, T., Ruitenberg, M. & Kroger, A. (2002). Reconstitution of coupled fumarate respiration in liposomes by incorporating the electron transport enzymes isolated from *Wolinella succinogenes*, *Eur J Biochem* 269(7): 1974-1983.

Brondijk, T., Fiegen, D., Richardson, D.J. & Cole, J. (2002). Roles of NapF, NapG and NapH, subunits of the *Escherichia coli* periplasmic nitrate reductase, in ubiquinol oxidation, *Mol Microbiol* 44(1): 245-255.

Bruggemann, H., Falinski, F. & Deppenmeier, U. (2000). Structure of the $F_{420}H_2$:quinone oxidoreductase of *Archaeoglobus fulgidus* identification and overproduction of the $F_{420}H_2$-oxidizing subunit, *Eur J Biochem* 267(18): 5810-5814.

Collet, J. & Bardwell, J. (2002). Oxidative protein folding in bacteria, *Mol Microbiol* 44(1): 1-8.

Collins, M.D., Pirouz, T., Goodfellow, M. & Minnikin, D.E. (1977). Distribution of menaquinones in actinomycetes and corynebacteria, *J Gen Microbiol* 100(2): 221-230.

Collins, M.D. & Jones, D. (1979). The distribution of isoprenoid quinones in streptococci of serological groups D and N, *J Gen Microbiol* 114(1): 27-33.

Collins, M.D. & Jones, D. (1981). Distribution of isoprenoid quinone structural types in bacteria and their taxonomic implication, *Microbiol Rev* 45(2): 316-354.

Collins, M.D., Ross, H.N.M., Tindall, B.J. & Grant, W.D. (1981). Distribution of isoprenoid quinones in halophilic bacteria, *J Appl Microbiol* 50(3): 559-565.

Cui, T.Z. & Kawamukai, M. (2009). Coq10, a mitochondrial coenzyme Q binding protein, is required for proper respiration in *Schizosaccharomyces pombe, FEBS J* 276(3): 748-759.

Elias, M.D., Tanaka, M., Izu, H., Matsushita, K., Adachi, O. & Yamada M. (2000) Functions of amino acid residues in the active site of *Escherichia coli* PQQ-containing quinoprotein glucose dehydrogenase. *J Biol Chem* 275(10): 7321-7326

Elias, M.D., Nakamura, S., Migita, C.T., Miyoshi, H., Toyama, H., Matsushita, K., Adachi, O. & Yamada, M. (2004). Occurrence of a bound ubiquinone and its function in *Escherichia coli* membrane-bound quinoprotein glucose dehydrogenase, *J Biol Chem* 279(4): 3078-3083.

Fernandes, A.S., Konstantinov, A.A., Teixeira, M. & Pereira, M.M. (2005). Quinone reduction by *Rhodothermus marinus* succinate: menaquinone oxidoreductase is not stimulated by the membrane potential, *Biochem Biophys Res Commun* 330(2): 565-570.

Fiorini, R., Ragni, L., Ambrosi, S., Littarru, G.P., Gratton, E. & Hazlett, T. (2008). Fluorescence studies of the interactions of ubiquinol-10 with liposomes, *Photochem Photobiol* 84(1): 209-214.

Frigaard, N.U., Takaichi, S., Hirota, M., Shimada, K. & Matsuura, K. (1997). Quinones in chlorosomes of green sulfur bacteria and their role in the redox-dependent fluorescence studied in chlorosome-like bacteriochlorophyll c aggregates, *Arch Microbiol* 167(6): 343-349.

Frigaard, N.U., Matsuura, K., Hirota, M., Miller, M. & Cox, R.P. (1998). Studies of the location and function of isoprenoid quinones in chlorosomes from green sulfur bacteria, *Photosynth Res* 58(1): 81-90.

Gale, P.H., Arison, B., Trenner, N.R., Page, A.C.J. & Folkers, K. (1963). Characterization of vitamin K_9(H) from *Mycobacterium phlei, Biochemistry* 2(1): 200-203.

Giordani, R. & Buc, J. (2004). Evidence for two different electron transfer pathways in the same enzyme, nitrate reductase A from *Escherichia coli, Eur J Biochem* 271(12): 2400-2407.

Goodwin, T.W. (1977) The prenyllipids of the membranes of higher plants. In: *The prenyllipids of the membranes of higher plants.* Tevini, M. & Lichtenthaler, H.K., pp. (29-7), Springer.

Guilhot, C., Jander, G., Martin, N.L. & Beckwith, J. (1995). Evidence that the pathway of disulfide bond formation in *Escherichia coli* involves interactions between the cysteines of DsbB and DsbA, *Proc Natl Acad Sci U S A* 92(21): 9895-9899.

Hale, M.B., Blankenship, R. & Fuller, R.C. (1983). Menaquinone is the sole quinone in the facultatively aerobic green photosynthetic bacterium *Chloroflexus aurantiacus, Biochim Biophys Acta* 723(3): 376-382.

Hammond, R.K. & White, D.C. (1969). Separation of vitamin K2 isoprenologues by reversed-phase thin-layer chromatography, *J Chromatogr* 45(3): 446-452.

Hauska, G., Schoedl, T., Remigy, H. & Tsiotis, G. (2001). The reaction center of green sulfur bacteria(1), *Biochim Biophys Acta* 1507(1-3): 260-277.

Hedrick, D.B., Peacock, A.D., Lovley, D.R., Woodard, T.L., Nevin, K.P., Long, P.E. & White, D.C. (2009). Polar lipid fatty acids, LPS-hydroxy fatty acids, and respiratory quinones of three *Geobacter* strains, and variation with electron acceptor, *J Ind Microbiol Biot* 36(2): 205-209.

Hiraishi, A., Yamamoto, H., Kato, K. & Maki, Y. (1999). A new structural type of methionaquinones isolated from hot spring sulfur-turf bacterial mats, *J Gen Appl Microbiol* 45(1): 39-41.

Hiratsuka, T., Furihata, K., Ishikawa, J., Yamashita, H., Itoh, N., Seto, H. & Dairi, T. (2008). An alternative menaquinone biosynthetic pathway operating in microorganisms, *Science* 321(5896): 1670-1673.

Hiratsuka, T., Itoh, N., Seto, H. & Dairi, T. (2009). Enzymatic properties of futalosine hydrolase, an enzyme essential to a newly identified menaquinone biosynthetic pathway, *Biosci Biotechnol Biochem* 73(5): 1137-1141.

Holsclaw, C.M., Sogi, K.M., Gilmore, S.A., Schelle, M.W., Leavell, M.D., Bertozzi, C. & Leary, J.A. (2008). Structural characterization of a novel sulfated menaquinone produced by *stf3* from *Mycobacterium tuberculosis, ACS chemical biology* 3(10): 619-624.

Ikeda, Y., Komura, M., Watanabe, M., Minami, C., Koike, H., Itoh, S., Kashino, Y. & Satoh, K. (2008). Photosystem I complexes associated with fucoxanthin-chlorophyll-binding proteins from a marine centric diatom, *Chaetoceros gracilis, Biochim Biophys Acta* 1777(4): 351-361.

Inaba, K. & Ito, K. (2002). Paradoxical redox properties of DsbB and DsbA in the protein disulfide-introducing reaction cascade, *EMBO J* 21(11): 2646-2654.

Inaba, K., Takahashi, Y.H. & Ito, K. (2005). Reactivities of quinone-free DsbB from *Escherichia coli, J Biol Chem* 280(38): 33035-33044.

Inaba, K. & Ito, K. (2008). Structure and mechanisms of the DsbB-DsbA disulfide bond generation machine, *Biochim Biophys Acta* 1783(4): 520-529.

Ingledew, W.J. & Poole, R.K. (1984). The respiratory chains of *Escherichia coli, Microbiol Mol Biol R* 48(3): 222-271.

Ishii, M., Kawasumi, T., Igarashi, Y., Kodama, T. & Minoda, Y. (1987). 2-Methylthio-1, 4-naphthoquinone, a unique sulfur-containing quinone from a thermophilic hydrogen-oxidizing bacterium, *Hydrogenobacter thermophilus, J Bacteriol* 169(6): 2380–384.

Iuchi, S. & Lin, E.C. (1988) *arcA* (*dye*), a global regulatory gene in *Escherichia coli* mediating repression of enzymes in aerobic pathways. Proc Natl Acad Sci USA, 85(6): 1888–1892.

Iuchi, S., Cameron, D. & Lin, E.C. (1989). A second global regulator gene (*arcB*) mediating repression of enzymes in aerobic pathways of *Escherichia coli, J Bacteriol* 171(2): 868-873.

Iuchi, S., Matsuda, Z., Fujiwara, T. & Lin, E.C. (1990). The *arcB* gene of *Escherichia coli* encodes a sensor-regulator protein for anaerobic repression of the arc modulon, *Mol Microbiol* 4(5): 715-727.

Iverson, T.M., Luna-Chavez, C., Cecchini, G. & Rees, D.C. (1999). Structure of the *Escherichia coli* fumarate reductase respiratory complex, *Science* 284(5422): 1961-1966.

Jander, G., Martin, N.L. & Beckwith, J. (1994). Two cysteines in each periplasmic domain of the membrane protein DsbB are required for its function in protein disulfide bond formation, *EMBO J* 13(21): 5121-5127.

Jormakka, M., Tönroth, S., Byrne, B. & Iwata, S. (2002). Molecular basis of proton motive force generation: structure of formate dehydrogenase-N, *Science* 295(5561): 1863-1868.

Junemann, S. (1997). Cytochrome *bd* terminal oxidase, *Biochim Biophys Acta* 1321(2): 107-127.

Kadokura, H., Bader, M., Tian, H., Bardwell, J. & Beckwith, J. (2000). Roles of a conserved arginine residue of DsbB in linking protein disulfide-bond-formation pathway to the respiratory chain of *Escherichia coli, Proc Natl Acad Sci U S A* 97(20): 10884-10889.

Kadokura, H., Katzen, F. & Beckwith, J. (2003). Protein disulfide bond formation in prokaryotes, *Annu Rev Biochem* 72: 111-135.

Ke, B. (2001). (Ed.). Photosynthesis: Photobiochemistry and Photobiophysics, Advances of Photosynthesis, Vol.10, Kluwer Academic Publishers, Dordecht.

Kim, H., Li, H., Maresca, J.A., Bryant, D. & Savikhin, S. (2007). Triplet exciton formation as a novel photoprotection mechanism in chlorosomes of *Chlorobium tepidum, Biophys J* 93(1): 192-201.

Kishigami, S., Akiyama, Y. & Ito, K. (1995). Redox states of DsbA in the periplasm of *Escherichia coli, FEBS Lett* 364(1): 55-58.

Kobayashi, T., Kishigami, S., Sone, M., Inokuchi, H., Mogi, T. & Ito, K. (1997). Respiratory chain is required to maintain oxidized states of the DsbA-DsbB disulfide bond formation system in aerobically growing *Escherichia coli* cells, *Proc Natl Acad Sci U S A* 94(22): 11857-11862.

Kobayashi, T. & Ito, K. (1999). Respiratory chain strongly oxidizes the CXXC motif of DsbB in the *Escherichia coli* disulfide bond formation pathway, *EMBO J* 18(5): 1192-1198.

Kobayashi, T., Takahashi, Y.H. & Ito, K. (2001). Identification of a segment of DsbB essential for its respiration-coupled oxidation, *Mol Microbiol* 39(1): 158-165.

Kobayashi, K., Mustafa, G., Tagawa, S. & Yamada, M. (2005). Transient formation of a neutral ubisemiquinone radical and subsequent intramolecular electron transfer to pyrroloquinoline quinone in the *Escherichia coli* membrane-integrated glucose dehydrogenase, *Biochemistry* 44(41): 13567-13572.

Kröer, A., Biel, S., Simon, J., Gross, R., Unden, G. & Lancaster, C.R. (2002). Fumarate respiration of *Wolinella succinogenes*: enzymology, energetics and coupling mechanism, *Biochim Biophys Acta* 1553(1-2): 23-38.

Kurokawa, T. & Sakamoto, J. (2005). Purification and characterization of succinate: menaquinone oxidoreductase from *Corynebacterium glutamicum*, *Arch Microbiol* 183(5): 317-324.

Kushwaha, S.C., Gochnauer, M.B., Kushner, D.J. & Kates, M. (1974). Pigments and isoprenoid compounds in extremely and moderately halophilic bacteria, *Can J Microbiol* 20(2): 241-245.

Kusumoto, K., Sakiyama, M., Sakamoto, J., Noguchi, S. & Sone, N. (2000). Menaquinol oxidase activity and primary structure of cytochrome *bd* from the amino-acid fermenting bacterium *Corynebacterium glutamicum*, *Arch Microbiol* 173(5-6): 390-397.

Kwon, O., Georgellis, D. & Lin, E.C. (2000) Phosphorelay as the sole physiological route of signal transmission by the arc two-component system of *Escherichia coli*. *J Bacteriol* 182(13): 3858-3862.

Lancaster, C.R. & Simon, J. (2002). Succinate:quinone oxidoreductases from epsilon-proteobacteria, *Biochim Biophys Acta* 1553(1-2): 84-101.

Lancaster, C.R. (2003). *Wolinella succinogenes* quinol: fumarate reductase and its comparison to *E. coli* succinate: quinone reductase, *FEBS Lett* 555(1): 21-28.

Landi, L., Cabrini, L., Sechi, A.M. & Pasquali, P. (1984). Antioxidative effect of ubiquinones on mitochondrial membranes, *Biochem J* 222(2): 463-466.

Lange, B.M., Rujan, T., Martin, W. & Croteau, R. (2000). Isoprenoid biosynthesis: the evolution of two ancient and distinct pathways across genomes, *Proc Natl Acad Sci U S A* 97(24): 13172-13177.

Lee, P.T., Hsu, A.Y., Ha, H.T. & Clarke, C. (1997). A C-methyltransferase involved in both ubiquinone and menaquinone biosynthesis: isolation and identification of the *Escherichia coli ubiE* gene, *J Bacteriol* 179(5): 1748-1754.

Lenaz, G. (2001). A critical appraisal of the mitochondrial coenzyme Q pool, *FEBS Lett* 509(2): 151-155.

Lenaz, G., Fato, R., Formiggini, G. & Genova, M.L. (2007). The role of Coenzyme Q in mitochondrial electron transport, *Agric Biol Chem* 7: 8-33.

Lester, R.L., Whilte, D.C. & Smith, S.L. (1964). The 2-desmethyl vitamin K2's. a new group of naphthoquinones isolated from *Hemophilus parainfluenzae*, *Biochemistry* 3(7): 949-954.

Lichtenthaler, H.K. (1977) Regulation of prenylquinone synthesis in higher plants. In: *Regulation of prenylquinone synthesis in higher plants*. Tevini, M. & Lichtenthaler, H.K., pp. (231–58), Springer.

Lichtenthaler, H.K. (1999). The 1-deoxy-D-xylulose-5-phosphate pathway of isoprenoid biosynthesis in plants, *Annu Revi Plant Biol* 50(1): 47-65.

Lubben, M. (1995). Cytochromes of archaeal electron transfer chains, *Biochim Biophys Acta* 1229(1): 1-22.

Ma, J., Kobayashi, D.Y. & Yee, N. (2009). Role of menaquinone biosynthesis genes in selenate reduction by *Enterobacter cloacae* SLD1a-1 and *Escherichia coli* K12, *Environ Microbiol* 11(1): 149-158.

Madej, M.G., Nasiri, H.R., Hilgendorff, N.S., Schwalbe, H., Unden, G. & Lancaster, C.R. (2006). Experimental evidence for proton motive force-dependent catalysis by the diheme-containing succinate: menaquinone oxidoreductase from the Gram-positive bacterium *Bacillus licheniformis*, *Biochemistry* 45(50): 15049-15055.

Maklashina, E., Hellwig, P., Rothery, R.A., Kotlyar, V., Sher, Y., Weiner, J.H. & Cecchini, G. (2006). Differences in protonation of ubiquinone and menaquinone in fumarate reductase from *Escherichia coli*, *J Biol Chem* 281(36): 26655-26664.

Malpica, R., Franco, B., Rodriguez, C., Kwon, O. & Georgellis, D. (2004). Identification of a quinone-sensitive redox switch in the ArcB sensor kinase, *Proc Natl Acad Sci U S A* 101(36): 13318-13323.

Matsushita, K., Arents, J., Bader, R., Yamada, M., Adachi, O. & Postma, P.W. (1997). *Escherichia coli* is unable to produce pyrroloquinoline quinone (PQQ), *Microbiology* 143(Pt 10): 3149-3156.

Mimuro, M., Tsuchiya, T., Inoue, H., Sakuragi, Y., Itoh, Y., Gotoh, T., Miyashita, H., Bryant, D. & Kobayashi, M. (2005). The secondary electron acceptor of photosystem I in *Gloeobacter violaceus* PCC 7421 is menaquinone-4 that is synthesized by a unique but unknown pathway, *FEBS Lett* 579(17): 3493-3496.

Missiakas, D., Georgopoulos, C. & Raina, S. (1993). Identification and characterization of the *Escherichia coli* gene *dsbB*, whose product is involved in the formation of disulfide bonds *in vivo*, *Proc Natl Acad Sci U S A* 90(15): 7084-7088.

Miyoshi, H., Niitome, Y., Matsushita, K., Yamada, M. & Iwamura, H. (1999). Topographical characterization of the ubiquinone reduction site of glucose dehydrogenase in *Escherichia coli* using depth-dependent fluorescent inhibitors, *Biochim Biophys Acta* 1412(1): 29-36.

Mustafa, G., Ishikawa, Y., Kobayashi, K., Migita, C.T., Elias, M.D., Nakamura, S., Tagawa, S. & Yamada, M. (2008a) Amino acid residues interacting with both the bound quinone and coenzyme, pyrroloquinoline quinone, in *Escherichia coli* membrane-bound glucose dehydrogenase, *J Biol Chem* 283(32): 22215-22221

Mustafa, G., Migita, C.T., Ishikawa, Y., Kobayashi, K., Tagawa, S. & Yamada, M. (2008b) Menaquinone as well as ubiquinone as a bound quinone crucial for catalytic activity and intramolecular electron transfer in *Escherichia coli* membrane-bound glucose dehydrogenase, *J Biol Chem* 283(42): 28169-28175

Mustafa, G., Ishikawa, Y., Kobayashi, K., Migita, C.T., Tagawa, S. & Yamada, M. (2008c). Function of a bound ubiquinone in *Escherichia coli* quinoprotein glucose dehydrogenase, *Biofactors* 32(1-4): 23-29.

Nantapong, N., Otofuji, A., Migita, C.T., Adachi, O., Toyama, H. & Matsushita, K. (2005). Electron transfer ability from NADH to menaquinone and from NADPH to oxygen of type II NADH dehydrogenase of *Corynebacterium glutamicum*, *Biosci Biotechnol Biochem* 69(1): 149-159.

Neerken, S. & Amesz, J. (2001). The antenna reaction center complex of heliobacteria: composition, energy conversion and electron transfer, *Biochim Biophys Acta* 1507(1-3): 278-290.

Nitschke, W., Kramer, D.M., Riedel, A. & Liebl, U. (1995) From naphtho- to benzoquinones—(r)evolutionary reorganisations of electron transfer chain. In: *From naphtho- to benzoquinones—(r)evolutionary reorganisations of electron transfer chain.* Mathis, P., pp. (945–50), Kluwer Academic Publishers, Dordrecht.

Nowicka, B. & Kruk, J. (2010). Occurrence, biosynthesis and function of isoprenoid quinones, *Biochim Biophys Acta* 1797(9): 1587-1605.

Oh-oka, H. (2007). Type 1 reaction center of photosynthetic heliobacteria, *Photochem Photobiol* 83(1): 177-186.

Pinho, D., Besson, S., Silva, P.J., de Castro, B. & Moura, I. (2005). Isolation and spectroscopic characterization of the membrane-bound nitrate reductase from *Pseudomonas chlororaphis* DSM 50135, *Biochim Biophys Acta* 1723(1-3): 151-162.

Pires, R.H., Lourenç, A.I., Morais, F., Teixeira, M., Xavier, A.V., Saraiva, L.M. & Pereira, I.A.C. (2003). A novel membrane-bound respiratory complex from *Desulfovibrio desulfuricans* ATCC 27774, *Biochim Biophys Acta* 1605(1-3): 67-82.

Richardson, D.J. (2000). Bacterial respiration: a flexible process for a changing environment, *Microbiology* 146(Pt 3): 551-571.

Richardson, D.J. & Sawers, G. (2002). Structural biology. PMF through the redox loop, *Science* 295(5561): 1842-1843.

Sakuragi, Y., Zybailov, B., Shen, G., Bryant, D., Golbeck, J.H., Diner, B.A., Karygina, I., Pushkar, Y. & Stehlik, D. (2005). Recruitment of a foreign quinone into the A1 site of photosystem I. Characterization of a *menB rubA* double deletion mutant in *Synechococcus sp.* PCC 7002 devoid of FX, FA, and FB and containing plastoquinone or exchanged 9,10-anthraquinone, *J Biol Chem* 280(13): 12371-12381.

Schoepp-Cothenet, B., Lieutaud, C., Baymann, F., Vermélio, A., Friedrich, T., Kramer, D.M. & Nitschke, W. (2009). Menaquinone as pool quinone in a purple bacterium, *Proc Natl Acad Sci U S A* 106(21): 8549-8554.

Scholes, P.B. & King, H.K. (1965). Isolation of a naphthaquinone with partly hydrogenated side chain from *Corynebacterium diphtheriae, Biochem J* 97(3): 766-768.

Schutz, M., Brugna, M., Lebrun, E., Baymann, F., Huber, R., Stetter, K.O., Hauska, G., Toci, R., Lemesle-Meunier, D., Tron, P., Schmidt, C. & Nitschke, W. (2000). Early evolution of cytochrome *bc* complexes, *J Mol Biol* 300(4): 663-675.

Schwalb, C., Chapman, S. & Reid, G.A. (2003). The tetraheme cytochrome CymA is required for anaerobic respiration with dimethyl sulfoxide and nitrite in *Shewanella oneidensis, Biochemistry* 42(31): 9491-9497.

Shimada, H., Shida, Y., Nemoto, N., Oshima, T. & Yamagishi, A. (2001). Quinone profiles of *Thermoplasma acidophilum* HO-62, *J Bacteriol* 183(4): 1462-1465.

Simon, J., Pisa, R., Stein, T., Eichler, R., Klimmek, O. & Gross, R. (2001). The tetraheme cytochrome c NrfH is required to anchor the cytochrome *c* nitrite reductase (NrfA) in the membrane of *Wolinella succinogenes, Eur J Biochem* 268(22): 5776-5782.

Soballe, B. & Poole, R.K. (1999). Microbial ubiquinones: multiple roles in respiration, gene regulation and oxidative stress management, *Microbiology* 145(Pt 8): 1817-1830.

Takahashi, Y.H., Inaba, K. & Ito, K. (2004). Characterization of the menaquinone-dependent disulfide bond formation pathway of *Escherichia coli, J Biol Chem* 279(45): 47057-47065.

Thummer, R., Klimmek, O. & Schmitz, R.A. (2007). Biochemical studies of *Klebsiella pneumoniae* NifL reduction using reconstituted partial anaerobic respiratory chains of *Wolinella succinogenes*, *J Biol Chem* 282(17): 12517–2526.

Trumpower, B.L. (1990). Cytochrome bc_1 complexes of microorganisms, *Microbiol Rev* 54(2): 101-129.

Unden, G. & Bongaerts, J. (1997). Alternative respiratory pathways of *Escherichia coli*: energetics and transcriptional regulation in response to electron acceptors, *Biochim Biophys Acta* 1320(3): 217-234.

Vervoort, L.M., Ronden, J.E. & Thijssen, H.H. (1997). The potent antioxidant activity of the vitamin K cycle in microsomal lipid peroxidation, *Biochem Pharmacol* 54(8): 871-876.

Wallace, B.J. & Young, I.G. (1977). Role of quinones in electron transport to oxygen and nitrate in *Escherichia coli*. Studies with a *ubiA-menA*-double quinone mutant, *Biochim Biophys Acta* 461(1): 84-100.

Widhalm, J.R., van Oostende, C., Furt, F. & Basset, G. (2009). A dedicated thioesterase of the Hotdog-fold family is required for the biosynthesis of the naphthoquinone ring of vitamin K1, *Proc Natl Acad Sci U S A* 106(14): 5599-5603.

Wissenbach, U., Kröer, A. & Unden, G. (1990). The specific functions of menaquinone and demethylmenaquinone in anaerobic respiration with fumarate, dimethylsulfoxide, trimethylamine N-oxide and nitrate by *Escherichia coli*, *Arch Microbiol* 154(1): 60-66.

Xie, T., Yu, L., Bader, M., Bardwell, J. & Yu, C.A. (2002). Identification of the ubiquinone-binding domain in the disulfide catalyst disulfide bond protein B, *J Biol Chem* 277(3): 1649-1652.

Xin, Y., Lu, Y.K., Fromme, R., Fromme, P. & Blankenship, R. (2009). Purification, characterization and crystallization of menaquinol: fumarate oxidoreductase from the green filamentous photosynthetic bacterium *Chloroflexus aurantiacus*, *Biochim Biophys Acta* 1787(2): 86-96.

Yagi, T., Yano, T., Di Bernardo, S. & Matsuno-Yagi, A. (1998). Procaryotic complex I (NDH-1), an overview, *Biochim Biophys Acta* 1364(2): 125-133.

Yamada, M., Sumi, K., Matsushita, K., Adachi O., & Yamada, Y. (1993a) Topological analysis of quinoprotein glucose dehydrogenase in *Escherichia coli* and its ubiquinone-binding site, *J Biol Chem* 268(17):12812-12817

Yamada, M., Asaoka, S., Saier, M.H.J. & Yamada, Y. (1993b) Characterization of the *gcd* gene from *Escherichia coli* K-12 W3110 and regulation of its expression, *J Bacteriol* 175(2): 568-571.

Part 3

Biotransformation & Gene Delivery

11

Genetically Modified Baker's Yeast *Saccharomyces cerevisiae* in Chemical Synthesis and Biotransformations

Ewa Białecka-Florjańczyk and Agata Urszula Kapturowska
Warsaw University of Life Sciences, Faculty of Food Sciences, Department of Chemistry
Poland

1. Introduction

Yeast *Saccharomyces cerevisiae* has been associated with human beings for more than 6000 years due to its use in food production, baking, wine and beer making. Potable and industrial ethanol production constitutes the majority of use of *S. cerevisiae* in biotechnological applications. However, baker's yeast also plays an important role as a model organism in the field of biochemistry, genetics and molecular biology. *S. cerevisiae* was the first eukaryotic organism to be sequenced in 1996 [Goffeau et al., 1996], and is clearly the most ideal eukaryotic microorganism for biological studies. Furthermore, the ease of genetic manipulation of yeast allows it to be used for analyzing and functionally dissecting gene products from other eukaryotes. The field of metabolic engineering, which utilizes genetic tools to manipulate microbial metabolism to enhance the production of compounds of interest has a particularly strong impact by providing new platforms for chemical production and facilitating the expansion of industrial (white) biotechnology [Nevoigt, 2008]. Baker's yeast can also be used as host organism for novel production of some industrially relevant chemicals. Our review focuses on the progress that has been achieved in the production of fine chemicals, bulk chemicals and fuels by genetic manipulation of the enzymatic activity of yeast, the combining of enzyme pathways from different microorganisms into *S. cerevisiae* and expressing genes from *S. cerevisiae* in other hosts. Attention to biotransformations catalyzed by genetically modified yeast will also be considered.

2. Bioethanol production

Bioethanol is usually obtained from the conversion of carbon based renewable feedstock and can be used as a fuel for vehicles in its pure form or as a gasoline additive to increase octane rating and improve vehicle emissions. Bioethanol is primarily produced by fermentation of sugar or the sugar components of starch. However, there has been constant research on its production from fibrous substances such as cellulose and hemicelluloses, which make up the bulk of most plant matter. Two chemical reactions take place during biomass conversion to ethanol: the hydrolysis of complex polysaccharides in the raw feedstock to simple sugars followed by their subsequent fermentation to ethanol. The second step of bioethanol production is caused by yeast or bacteria which feed upon the sugars. Therefore high ethanol yields from lignocellulosic biomass from agricultural and agro-industrial residues are dependent upon efficient hydrolysis of sugar polymers and utilization of all the available sugars including D-glucose, D-xylose, L-arabinose and other fermentable compounds.

S. cerevisiae, which plays a traditional and major role in industrial bioethanol production, has several advantages due to its high ethanol productivity as well as its high ethanol tolerance. However, baker's yeast cannot hydrolyze cellulose and is not able to use pentoses, which constitute up to 20% of lignocellulosic biomass. Many studies regarding the use of *S. cerevisiae* in metabolic engineering for xylose utilization have been reported and several reviews have been published [Chu & Lee, 2007; Hahn-Hägerdal et al., 2007; Matsushika et al., 2009]. The first step of D-xylose metabolism is its direct isomerization to D-xylulose catalyzed by bacterial xylose isomerase XR (EC 5.3.1.5) or stepwise transformation in yeast cells, firstly to xylitol (xylose reductase XR EC 1.1.1.21) and then to D-xylulose (xylitol dehydrogenase XDH EC 1.1.1.19). After phosphorylation of D-xylulose to D-xylulose-5-phosphate (xylulokinase XK EC 2.7.1.17) further metabolizm proceeds via a pentose phosphate pathway. Different strategies have been applied to engineering yeast including the introduction of initial xylose metabolism and xylose transport. However, these change the intracellular redox balance and result in over-expression of xylulokinase and further metabolism via a pentose phosphate pathway. Nevertheless they are insufficient for industrial bioprocesses mainly due to a low rate of reaction as compared with glucose fermentation [Kondo et al. 2010; Young et al., 2010].

3. Yeast cells as microbial chemical factories

Many plant secondary metabolites (i.e. small molecules, with complicated structures which are not involved in basic metabolic pathways and are not directly essential for photosynthetic or respiratory processes) have been identified as having beneficial effects on human health or nutrition, but their chemical synthesis is generally laborious and their isolation from natural sources is a difficult process with low yields. Furthermore, there is a wide range of chemicals essential in many industries as substrates in production processes or compounds necessary for appropriate process progress, that are produced by many microorganisms with low efficiency or during chemical synthesis, which is nowadays considered as non-ecological.

Baker's yeast can produce a diverse array of secondary metabolites [Pscheidt & Glieder, 2008], therefore the metabolic engineering of microorganisms is a new area of effective biosynthesis of these compounds and yeast is an important and attractive host for the heterologous and function expression of foreign genes encoding many important secondary metabolites of plants. Moreover baker's yeast possesses GRAS status, which is an advantage in the production of compounds that are intended for human consumption (e.g. the formation of antioxidants and aroma compounds during wine fermentation). A few examples in which modified *S. cerevisiae* were used as a whole-cell factories in production of chosen chemical compounds are presented in this chapter.

3.1 Alcohols (other than ethanol)

3.1.1 Butan-1-ol

Butan-1-ol can be regarded as superior biofuel to ethanol because of its greater hydrophobicity, higher energy density and the possibility of mixing with gasoline and transporting through existing pipeline infrastructure. Microbial butanol production is performed by members of the genus *Clostridium* [Inui et al. 2008]. The butanol pathway is expressed in *S. cerevisiae* from a range of organisms (*Escherichia coli, Clostridium beijerinckii* and *Ralstonia eutropha*) and owing to the tolerance of baker's yeast to alcohols, the production of butanol is raised ten-fold to 2.5 mg dm^{-3}. The most productive strains harbour the *C. beijerinckii* 3-hydroxybutyryl-CoA dehydrogenase, which uses NADH as a co-factor, and the acetoacetyl-CoA transferase from *S. cerevisiae* or *E. coli* [Steen et al., 2008].

3.1.2 Glycerol

Glycerol is considered to have a positive effect on the sensory properties of wine. Its concentration in wine varies between 1 - 15 g/dm^3. Many growth and environmental factors have been reported to influence the amount of glycerol produced by yeast in wine. In *S. cerevisiae* metabolism glycerol is a byproduct of the fermentation of sugar to ethanol. It is synthesized in the cytosol from dihydroxyacetone phosphate in two steps that are catalyzed by glycerol-3-phosphate dehydrogenase (GPDH) and glycerol-3-phosphatase (GPP) respectively. The former is the key enzyme in glycerol production [Wang et al., 2001]. More recently, genetic engineering approaches have been successful in redirecting the carbon flux towards glycerol. GPDH, a limiting enzyme for glycerol formation, is encoded by *GPD1* and *GPD2* genes. Overexpression of *GPD1* in a laboratory strain of baker's yeast and in a haploid strain V5 resulted in a marked increase in glycerol production at the expense of ethanol. Up to 28 g of glycerol per liter was formed by an engineered *S. cerevisiae* strain under conditions simulating wine fermentation [Remize et al., 1999].

Nowadays, the demand for glycerol is restricted due to the large quantity of glycerol generated during biodiesel production, so this direction of study is impractical.

3.1.3 Propane-1,3-diol

Propane-1,3-diol (1,3-PD) has numerous applications in production of polymers for cosmetics, foods, lubricants and medicines. Recently, there has been a strong industrial interests in a new kind of polyester, poly(trimethylene terephtalate), with 1,3-PD as a monomer. Nevertheless, its availability is restricted owing to its expensive chemical synthesis.

Non-modified *S. cerevisiae* can produce glycerol from D-glucose but cannot synthesize 1,3-propanediol. By taking advantage of genetically engineered *S. cerevisiae* 1,3-PD production is possible in two ways [Celińska, 2010]. One is to clone the yqhD gene from *E. coli* and dhaB gene from *K. pneumonia* required for the production of propane-1,3-diol from glycerol, and to integrate them into the chromosome W303-1A of *S. cerevisiae* by the *Agrobacterium tumefaciens* genetic transfer system. Both the yqhD and dhaB genes function in the engineered *S. cerevisiae* and lead to the production of 1,3-propanediol from D-glucose, but the amount of 1,3-PD is relatively small, due to low availability of glycerol in the reaction medium [Rao et al., 2008]. An alternative method of microbial 1,3-PD synthesis is described by Mendes et. al. using two recombinant microorganisms. The first step of the process is the conversion of sugar into glycerol by the metabolic engineered *Saccharomyces cerevisiae* strain HC42 adapted to high glucose concentrations (> 200 g dm^{-3}). The second step, carried out in the same bioreactor, is performed by the engineered strain *Clostridium acetobutylicum* DG1 (pSPD5) which converts glycerol to propane-1,3-diol [Mendes et al., 2011].

3.2 Organic acids – Pyruvic, lactic and malic

Saccharomyces cerevisiae does not naturally produce organic acids in large amounts, but its robustness and pH tolerance make it an excellent microorganism for researches in this field [Abbott et al., 2009]. The first attempts involved blocking the ethanol formation via deletion of four structural genes for alcohol dehydrogenase. Better results were obtained by eliminating the pyruvate decarboxylase (PDC) activity. Deletion of all three PDC genes completely eliminated alcoholic fermentation. Aerobic fermentation at pH 5 yielded pyruvic acid (fig. 1) at a concentration of aproximately 135 g dm^{-3} [Ischida et al., 2006]. The strategy for lactic acid (fig. 1) production consisted of two steps - deletion of one or more of three functional genes encoding pyruvate decarboxylase and the introduction of the bovine lactate dehydrogenase (L-LDH) in the genome under the control of PDC1 promoter.

Two enzymes, pyruvate carboxylase and malate dehydrogenase, produced by *S. cerevisiae* are involved in L-malic acid accumulation (Fig. 1). Overexpression of cytosolic malate dehydrogenase (MDH2) caused up to a 3.7 fold increase in L-malic acid production and an elevated accumulation of fumaric and citric acids [Pines et al., 1997]. Alternative way of L-malic production, including biotransformations, will be described in the chapter 4.4.

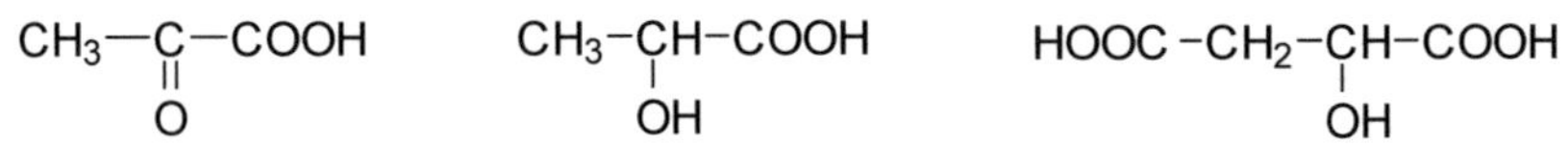

pyruvic acid lactic acid malic acid

Fig. 1. Organic acids produced by genetically modified *S. cerevisiae*

3.2.1 Sugar alcohols

Sugar alcohols (polyols, Fig. 2) can be synthesized from carbohydrates as a result of carbonyl group reduction to a hydroxyl one and can be used as a sweetners. Xylitol is the most popular one, due to its lower energy value than sucrose and the comparable sweetening power.

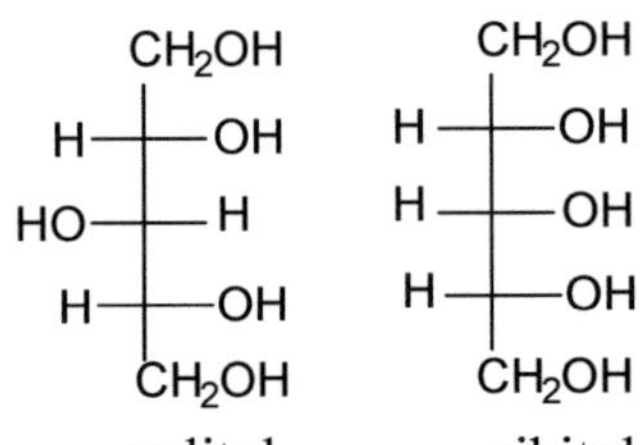

xylitol ribitol

Fig. 2. Sugar alcohols

Toivari et al described recombinant *Saccharomyces cerevisiae* strains that produce xylitol and ribitol from D-glucose in a single fermentation step. 8.5-fold enhancement of the total amount of the excreted sugar alcohols was achieved via expression of the xylitol dehydrogenase-encoding gene XYL2 of *Pichia stipitis* in the transketolase-deficient strain of *S. cerevisiae*. The additional introduction of the 2-deoxy-glucose 6-phosphate phosphatase-encoding gene DOG1 into the transketolase-deficient strain expressing the XYL2 gene induced further 1.6-fold increase in ribitol production [Toivari et al., 2007]

3.3 Lipid compounds

3.3.1 Fatty acids

Fatty acids are of considerable interest due to their pharmaceutical and nutritional values. These compounds are also necessary for cellular functions such as regulation of membrane fluidity. *Saccharomyces cerevisiae* are able to synthesize *de novo* only some saturated and monounsaturated fatty acids, mainly the C-16 and C-18 acids [Daum et al., 1998]. To reconstitute other long-chain and polyunsaturated fatty acids it is necessary to introduce genes of suitable enzymes such as desaturases and elongases eg. *A. thaliana* oleate desaturase gene (*FAD2*) or fatty acid desaturase from the fungus *M. alpina*. The simultaneous expression of Δ^{12}-desaturase and Δ^{6}-desaturase from *M. alpina* resulted in an increase in the content of γ-linolenic acid (18:3) to 8% of total fatty acids in yeast cells [Veen & Lang, 2004].

Due to the important roles of polyunsaturated fatty acids (PUFAs) in human health and nutrition the effect of overexpression of cytochrome b5 genes on fatty acid desaturation has been explored. The modification does not affect the fatty acid synthesis very much, but significantly enhances the synthesis of PUFA at 30 °C [Yazawa et al., 2010]. A number of desaturases from different sources (e.g. *Mucor rouxii, M. alpine*) have been functionally expressed in *S. cerevisiae* with a view to attain PUFAs formation [Chemler et al., 2006].

3.3.2 Isoprenoids

Isoprenoids (terpenoids) are ubiquitous in nature. They are a structurally diverse group of compounds and range from essential cell components to unique secondary metabolites. Terpenoids are based on combinations of isoprene units (C_5H_8), and their carbon skeleton is multiples of five carbon atoms e.g. monoterpenoids (10-carbon), sesquiterpenoids (15-carbon), diterpenoids (20-carbon) and carotenoids (40-carbon). Biosynthesis of these compounds proceeds via formation of isopentenyl pyrophosphate (IPP), dimethylallyl pyrophosphate (DMAPP), geranyl pyrophosphate (GPP), farnesyl pyrophosphate (FPP), and geranylgeranyl pyrophosphate (GGPP) (Fig. 3).

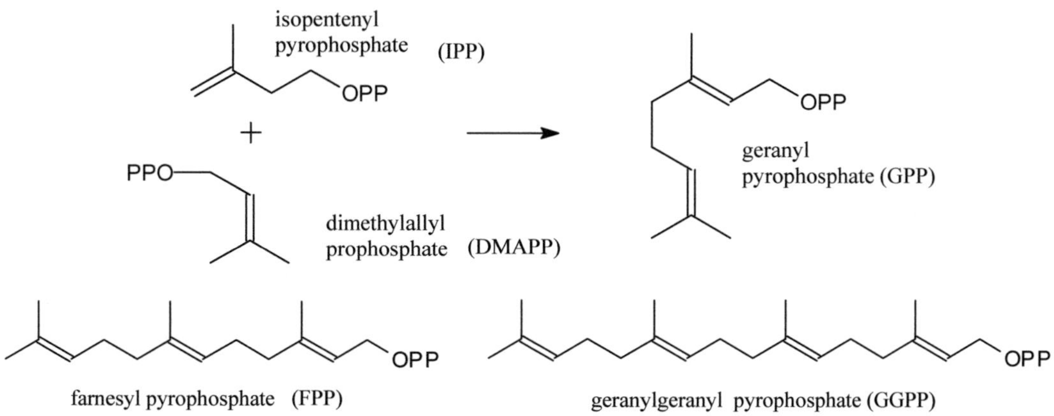

Fig. 3. Some terpene alcohols pyrophosphates - precursors of other terpenoids (OPP-pyrophosphate group)

Isoprenoids exist in low concentrations in their host organisms. They are complex molecules with a specified stereochemical structure, which makes them difficult to synthesize by chemical methods. Therefore, these compounds are of particular interest for microbial production. *E. coli, A. thaliana* and *S. cerevisiae* are the organisms with the most advanced genetic tools and are applied as heterologous hosts for terpenoids production, but they usually have very limited isoprenoid secondary metabolism minimizing competing metabolic fluxes.

Phosphorylated derivatives of geraniol (monoterpene) and geranylgeraniol (diterpene) are important molecules in the synthesis of various isoprenoids. (*E,E,E*)-geranylgeranyl pyrophosphate (GGPP) is an intermediate in carotenoids and ubiquinone biosynthesis [Tokuhiro et al., 2009] and was synthesized in engineered *S. cerevisiae* as a result of enhancement of the mevalonate pathway and redirection of carbon flux through overexpression of several key enzymes. The production achieved was 70.9 mg g^{-1} of dry cell weight (3.31 g dm^{-3}), which is very high in comparison to the production of other isoprenoids reported in the literature (e.g. 5.9 mg g^{-1} for carotene) [Verwaal et al., 2007].

Commercially important terpenoids are artemisinin and paclitaxel, so much work has focused on the introduction of exogenic metabolism in yeast to synthesize the precursors of these compounds [Chang & Keasling, 2006; Huang et al., 2008].

Artemisinin, an endoperoxide sesquiterpene lactone (Fig 4), was originally isolated from the chinese plant *Artemisia annua*. Its derivatives are now a group of drugs used worldwide for the treatment of malaria by means of artemisinin-combination therapies (ACTs). Engineered yeast systems are able to produce artemisinin or its precursor (artemisinic acid), which can be converted to artemisinin via chemical reactions (Fig. 4) [Shiba et al., 2007; Arsenault et al., 2008].

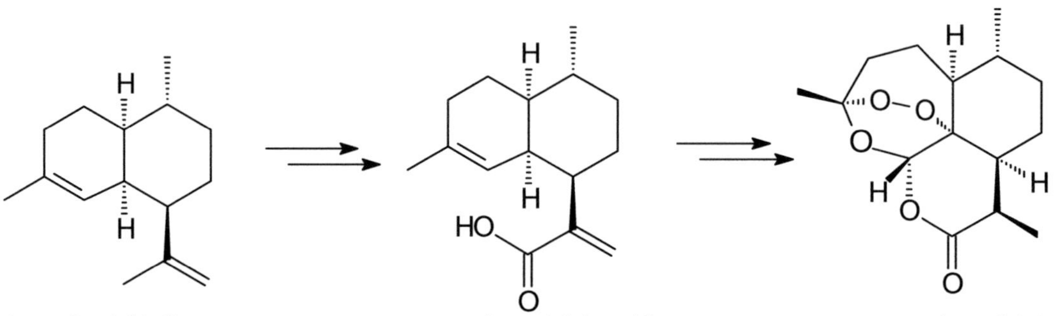

Amorpha-4,11-diene Artemisinic acid Artemisinin

Fig. 4. Synthesis of artemisinin [Ro et al., 2006]

Ro *et al.* cloned a gene from *A. annua,* which codes a cytochrome P 450, enabling the transformation of amorpha-4,11-diene to artemisinic acid, which in turn was converted to artemisinin by chemical synthesis [Ro et al., 2006]. In order to achieve a higher yield of amorpha-4,11-diene (Fig. 4), which is synthesized by cyclization of farnesyl pyrophosphate Shiba *et al.* engineered the pyruvate dehydrogenase bypass in *S. cerevisiae* [Shiba et al., 2007].

A partial biosynthesis of paclitaxel (trade name taxol, Fig. 5) – terpenoid, which is widely used in cancer therapy and produced as a secondary metabolite of yew trees (*Taxus* sp.), was constructed in *S. cerevisiae*. Dejong reported the functional expression of eight taxoid biosynthetic genes from *Taxus brevifolia* in yeast, elevating the accumulation of intermediate taxadiene [Dejong et al., 2006].

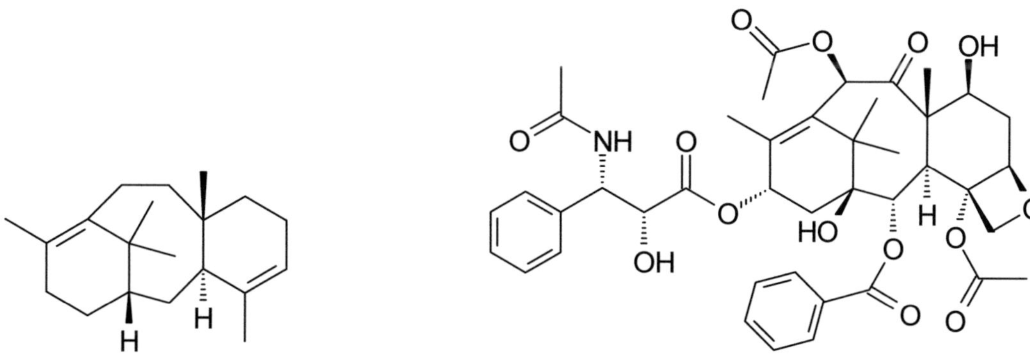

Taxadiene Paclitaxel

Fig. 5. Taxadiene and paclitaxel [Dejong et al., 2006]

Another way of synthesis of taxol building blocks is the biotransformation catalyzed by modified whole cell *S. cerevisiae* catalysts. This will be described in chapter 4.1.

3.3.3 Steroids

Steroids are the other group of compounds originating from the terpenoids precursors dimethylallyl pyrophosphate and isoprenylpyrophosphate (Fig. 3). One of them, hydrocortisone (Fig. 6), is an important starting material for synthesis of drugs with anti-inflammatory effect.

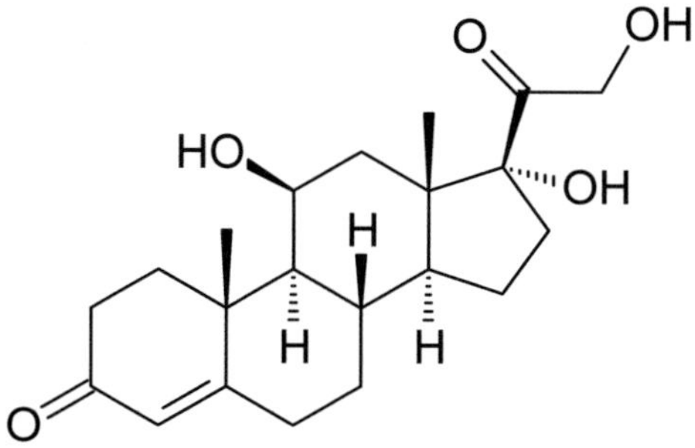

Fig. 6. Hydrocortisone

Hydrocortisone can be synthesized in *Saccharomyces cerevisiae* cells from simple carbon sources [Szczebara et al., 2003]. To optimize its yield the natural yeast pathway for sterols synthesis was rerouted after expressing genes from one plant enzyme and eight mammalian proteins including four of the P450 superfamily oxidases, three electron carriers and 3β-hydroxysterol oxidase/isomerase (3β-HSD). Under optimum conditions, a 70% yield of hydrocortisone was produced from the total steroid formation from glucose [Dumas et al., 2006].

3.4 Flavonoids

Flavonoids are a diverse class of plant secondary metabolites derived from the phenylpropanoid pathway. There has been increasing interest in flavonoids because many of them exhibit antioxidative activity due to free-radical scavenging. This is an attractive feature for drugs in coronary heart disease prevention and anticancer activity.

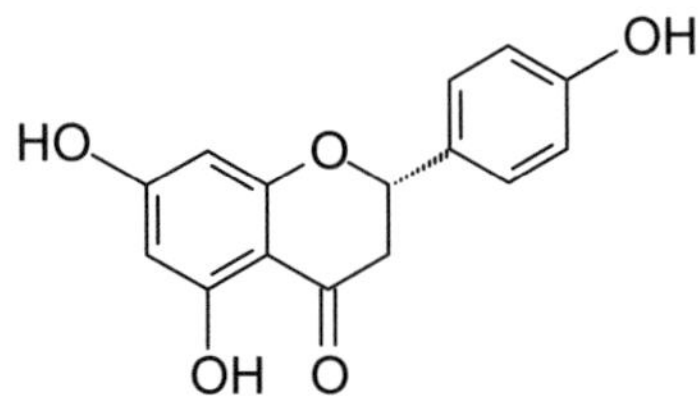

Fig. 7. Naringenin - flavanone

The biosynthesis of flavanone naringenin (Fig. 7), the central precursor of many flavonoids, is accomplished by introducing the phenylpropanoid pathway into *Saccharomyces cerevisiae* strains. It is achieved with the genes for phenylalanine ammonia lyase (PAL) from *Rhodosporidium toruloides*, 4-coumarate:coenzyme A (CoA) ligase (4CL) from *Arabidopsis thaliana* and chalcone synthase (CHS) from *Hypericum androsaemum* [Jiang et al., 2005] or with plant genes from heterologous origin such as 4-coumarate:coenzyme A ligase 4CL from *Petroselium crispum*, chalcone synthetases CHS from *Medicago sativa* and *Petunia x hybrida* and chalcone reductase CHR and chalcone isomerase from *M. sativa* [Yan et al., 2005].

Another method to produce flavonoid or isoflavonoid compounds in engineered yeast is to clone five soybean (*Glycine max*) chalcone isomerases (CHIs), key enzymes in the phenylpropanoid pathway [Ralston et al., 2005].

4. Biotransformations catalyzed by yeast cells

Baker's yeast may be considered as a producer of wide range of chemical compounds, a kind of microbial cell factory. However, *Saccharomyces cerevisiae* is a commonly applied to whole-cell biocatalysis in biotransformation, reactions based on enzymatic transformations of chemical compounds. Biotransformations, known as a branch of "white" biotechnology, provide efficient procedures in organic synthesis owing to the high chemo- and stereoselectivity of enzymes and offer a viable alternatives to chemical methods. Today, biotransfomation is a commonly accepted method for generating optically pure substances and for developing efficient synthesis of target compounds. In general, biotransformations are performed by the hydrolases or oxidoreductases. The remaining classes (transferases, isomerases, liases and ligases) are of lower, but increasing utility [Faber 2004].

Biotransformations may be carried out using isolated enzymes or microorganisms cells producing enzymes (whole-cell biocatalysis). The application of isolated and purified enzymes is profitable since the formation of undesired byproducts is avoided, whereas in cellular biotransformation systems undesirable product formation is possible due to the presence of other enzymes or simultaneous catalysis of several reactions. However the whole-cell biocatalysis has two important advantages: it is particularly beneficial when the regeneration of the cofactor is necessary (e.g. in redox reactions) and is favorable due to the cost effectiveness.

Baker's yeast has a great potential as a catalysts in organic chemistry owing to ease of handling, broad substrate acceptability and production of enzymes belonging to different classes. *S. cerevisiae* may be used in dry and pressed form, as raw yeast or lyophilized biomass and is capable of catalyzing many reactions in water or in organic media. These features are very important for chemists, because chemical laboratories are not usually equipped with the microbiological apparatus required for yeast cultivation [Csuk & Glanzer 1991; Servi, 1990; Białecka-Florjańczyk & Majewska, 2006].

The full sequencing of the *S. cerevisiae* genome, accompanied with the achievements in genetic engineering, have allowed new strains of yeast to be designed with specific high conversion yields and reaction selectivity. There is a growing interest in application of modified yeast in biotransformation reactions. Modern directions to improve catalytic abilities of baker's yeast include the use of surface display technology for enzymes and optimization or increase in availability of cofactor (required for bio-reduction reactions) or gene knock-out, to eliminate the activity of enzymes with conflicting and unwanted stereoselectivities. As commonly used technique is overexpression of the desired protein or expression of heterologous enzymes in yeast cells.

4.1 Reduction of carbonyl compounds

Baker's yeast is considered to produce over 20 different reductases. At present most of the characterized carbonyl (and dicarbonyl)-reducing enzymes of *S. cerevisiae* are grouped into two distinct protein categories - the aldo-keto reductase (AKR) and the short-chain dehydrogenase/reductase (SDR) superfamilies. They have been shown to reduce a broad array of ketones and ketoesters with different enantioselectivity and to have overlapping activities [Katz et al., 2003]. Reductions catalyzed with baker's yeast require the presence of nicotinoamide cofactors (mostly NADPH) [Johanson et al., 2005], which can be regenerated during the growth of microorganisms.

Enantioselectivity (or diastereoselectivity) of carbonyl compounds reduction is connected with the formation of an asymmetric center in the product (alcohol) molecule (Fig. 8). According to Csuk *et al.* and Servi, reduction of aliphatic methyl ketones using baker's yeast leads to synthesis of secondary alcohols with the *(S)* configuration [Csuk, Glanzer 1991; Servi, 1990].

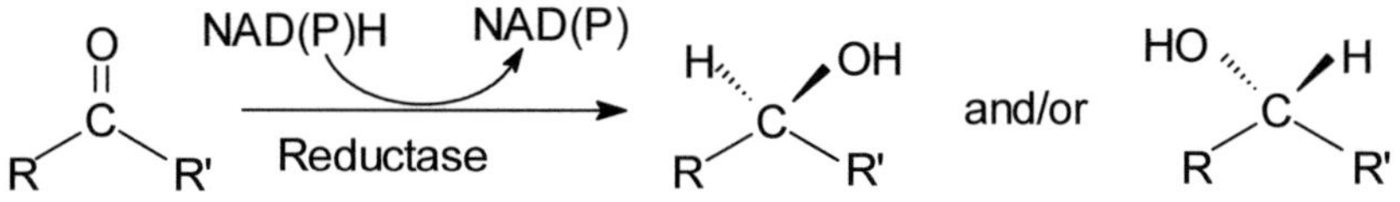

Fig. 8. Two enantiomeric alcohols formed during the reduction of carbonyl compound

Practical difficulties associated with the use of yeast as a chiral reducing agent arise from the presence of multiple enzymes overlapping substrate characteristics, but with differing enantio- and diastereoselectivities. Nakamura *et al.* isolated and characterized four oxidoreductases of raw baker's yeast in the reduction of 4-chloro-3-oxobutanoate [Nakamura et al., 1999]. The enantioselectivity of all of the enzymes examined was >99%, but two of them produced *(R)* and the others *(S)* isomers. Thus the different specificity of enzymes produced by *S. cerevisiae* is the reason for the low specificity of unmodified cells in this biotransformation. A recombinant *Saccharomyces cerevisiae* strain over-expressing the fatty acid synthase of *S. cerevisiae* (FAS) and the glucose dehydrogenase of *Bacillus subtilis* was applied by Engelking *et al.* for enantioselective reduction of ethyl 4-chloro-3-oxobutanoate to ethyl (*S*)-4-chloro-3-hydroxybutanoate (Fig. 9) as well as the reduction of ethyl benzoylacetate to ethyl (*S*)-3-hydroxy-3-phenylpropanoate. The enantiomeric excess (ee) was 90% in the case of the former and >97% in the case of the latter product [Engelking et al., 2006].

Fig. 9. Bioreduction of ethyl 4-chloroacetylacetate with *S. cerevisiae* yeast [Engelking et al., 2006]

A method performed by Rodriguez *et al.* involved modified baker's yeast for β-ketoesters reduction by using recombinant DNA techniques such as overexpression of carbonyl reductase with desirable stereoselectivities. Manipulating the levels of three proteins, fatty acid synthase, (Fasp); aldo-ketoreductase, (Ypr1p); α-acetoxy ketone reductase, (Gre2p) the stereoselectivity of β-ketoester reduction was improved and reached > 85 % in all cases [Rodriguez et al., 2001]. Conversely, the strain lacking aldo-ketoreductase produced hydroxyester with reversed stereoselectivity from unmodified baker yeast [Rodriguez et al., 1999].

One example of efficient genetic modification concerns the asymmetric reduction of the bicyclic diketone (Fig. 10) bicyclo[2.2.2]octane-2,6-dione (1) to (1*R*,4*S*,6*S*)-6-hydroxy-bicyclo[2.2.2]octane-2-one ((-)-2), which is used as an intermediate in the synthesis of transition metal based chiral chemical catalysts [Sarvary et al., 2001; Katz et al., 2002]. When carrying out the experiment with baker's yeast, ketoalcohol (-)-2 was obtained in 92–97% enantiomeric excess with 92% yield [Almqvist et al., 1993], but the diastereoselectivity of reduction of bicyclo[2.2.2]octane-2,6-dione catalyzed by whole cells of a modified strain of *S. cerevisiae* TMB4110 overexpressing the reductase gene YDR368w, was significantly improved. The product was achieved with 97% diastereomeric excess (de) and >99 enantiomeric excess (ee) [Johanson et al., 2008; Parachin et al., 2009].

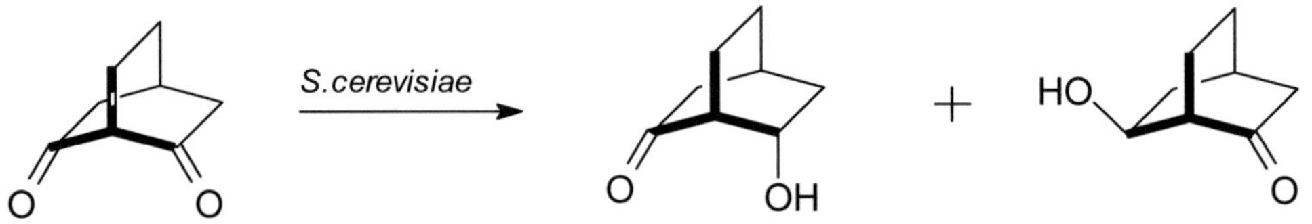

1 (-)-2 3

Fig. 10. Reduction of bicyclo[2.2.2]octane-2,6-dione (1) [Katz et al., 2003]

Another practical solution of this problem was uncovered by expression of appropriate endogenous enzymes in host cells, which do not produce these proteins. For example, expression of GCY1 and GRE3 genes from *S. cerevisiae* in *E. coli* cells achieved chiral βhydroxyesters with high optical purities with over 98% enantiomeric excess [Rodriguez et al., 2000].

Genetically modified cells of *S. cerevisiae* can catalyze biotransformation reaction, producing chiral building blocks important in the pharmaceutical industry (especially those obtained by reduction of α- and β -oxoesters or amides). Optically pure alcohols are important chiral blocks in the synthesis of valuable pharmaceuticals such as carnitine (required for the transport of fatty acids from the cytosol into the mitochondria during the breakdown of lipids) or paclitaxel [Stewart, 2000]. Using the structural analogy between substrate and molecules of β-oxothioesters participating in biosynthesis of fatty acids, the *S. cerevisiae* strain with punctual mutation in the FAS-2 gene (fatty acid synthase) catalyzed the stereoselective reduction of the ketone group in the β-lactam molecule (Fig. 11), eliminating *trans*-isomer [Kayser et al., 1999]. This reduction was applied in the synthesis of the paclitaxel (after coupling with readily available 10-deacetylbaccatin III) with a C_{13} side chain - (2*R*,3*S*)-*N*-benzoyl-3-phenylisoserine.

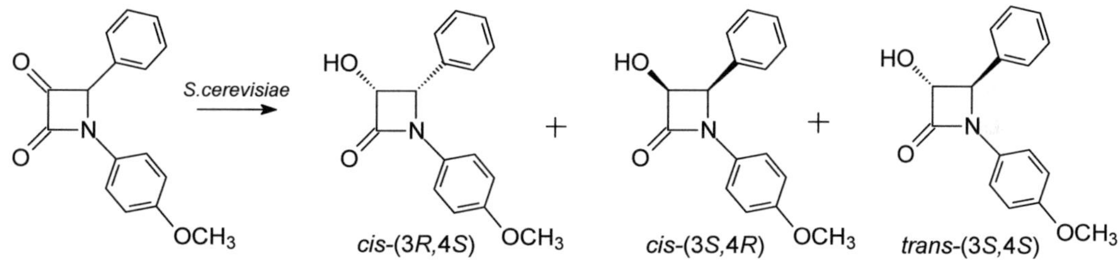

Fig. 11. Baker's yeast reduction of β-keto-α-lactam [Kayser et al., 1999]

Application of pure enzymatic preparations in organic chemistry generates a demand for coenzymes, which have to be added to the reaction medium. The use of baker's yeast in oxidoreductions is interesting, due to the ability of *S. cerevisiae* cells to synthesize not only the molecules of the specific enzymes, but also an essential coenzymes [Faber, 2004].

Although improvement of enantiomeric excess is critical for the product yield, regeneration of the co-factor is also important. This can be improved by genetic engineering. *S. cerevisiae* dicarbonyl reductases are NADPH-dependent and the reduced co-factor NADPH requires regeneration through the assimilation of a co-product, usually glucose, sucrose or ethanol. Glucose can be used as a co-substrate under both aerobic and anaerobic conditions. Unfortunately yeast generates large amounts of by-products from glucose, which may cause problems during downstream processing, with large amounts of glucose required to reduce small amounts of substrate. Ethanol is an interesting alternative to glucose as a co-substrate since it has a much lower utilisation rate and generate less CO_2. However, ethanol is toxic to the cells at high concentrations and NADPH cannot be regenerated from ethanol under anaerobic conditions.

Strain engineering for the design of efficient biocatalysts using glucose as a co-substrate has two potential objectives, to redirect the carbon flow towards NADPH-regenerating pathways and to slow down the rate of co-substrate utilisation in order to balance the reduction rate. In the *S. cerevisiae* strain the activity of phosphoglucose isomerase (PGI) was decreased and the baker's yeast strain alcohol dehydrogenase activity was deleted. In both cases the glucose consumption was limited without lost of reductase activity [Johanson et al., 2005]. The yeast phosphoglucose isomerase activity was decreased, and the short-chain dehydrogenase/reductase encoded by YMR226c was overexpressed in the genetically engineered *Saccharomyces cerevisiae* strain TMB4100 (1% PGI, YMR226c).

The whole cell biocatalyst was used for the kinetic resolution of racemic bicyclo[3.3.1]nonane-2,6-dione (rac-1, Fig. 12). This framework is a commonly occurring motif amongst natural products, displaying a wide scope of biological activities. Genetic modification reduced the demand for the glucose to regenerate NADPH, resulted in a greater reaction rate and produced higher selectivity towards the (+)-1 stereoisomer, reaching an enantiomeric excess of 100% after 75% conversion of the isomer (Fig.12) [Carlquist et al., 2008].

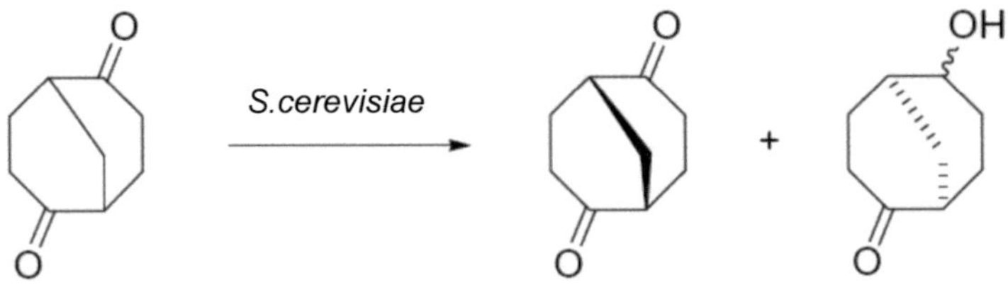

rac-**1** (+)-**1** **2**

Fig. 12. Kinetic resolution of bicyclo[3.3.1]nonane-2,6-dione by genetically modified baker's yeast [Carlquist et al., 2008]

In addition (+)-5,6-epoxybicyclo[2.2.1]heptane-2-one, ((+)-**1**, Fig. 12), and *endo*-(−)-5,6-epoxybicyclo [2.2.1]heptane-2-ol, *endo*-(−)-**2**, were obtained by asymmetric bioreduction catalyzed by the same *S. cerevisiae* yeast strain. R*ac*-**1** was kinetically resolved to give (+)-**1** with 95% enantiomeric excess and *endo*-(−)-**2** with 74% enantiomeric excess (Fig. 13) [Carlquist et al., 2009].

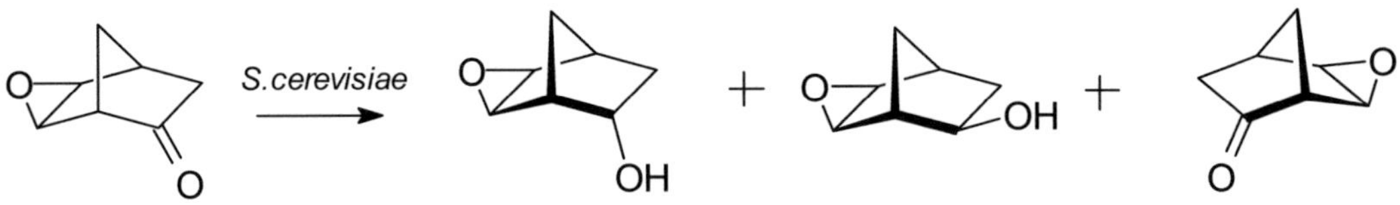

rac-1 *endo*-(-)-2 *exo*-2 (+)-1

Fig. 13. Kinetic resolution of 5,6-epoxybicyclo[2.2.1]heptan-2-one [Carlquist et al., 2009]

4.2 Baeyer–Villiger oxidation

"Designer yeast" is a new catalyst for Baeyer-Villiger (BV) oxidation, an organic reaction in which a ketone is oxidized to an ester by treatment with peroxy acids (Fig. 14).

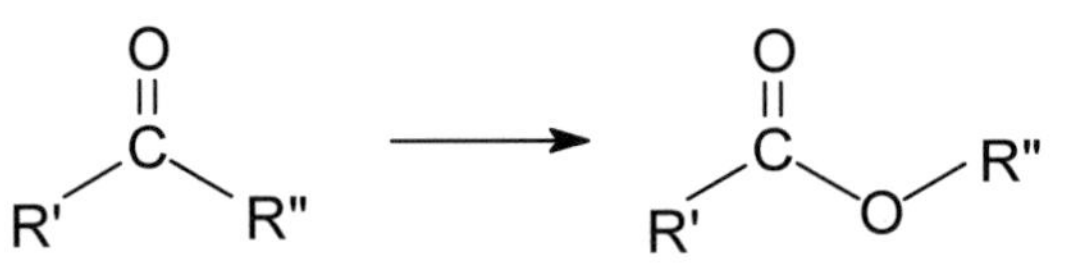

Fig. 14. Baeyer-Villiger reaction

The enzymatic Baeyer–Villiger oxidation represents an efficient approach to the asymmetric synthesis of chiral lactones and can be performed by biocatalysis with flavoenzymes called Baeyer-Villiger monooxygenases (BVMOs). BVMOs catalyze nucleophilic oxidation of ketones as well as electrophilic oxidation of heteroatoms such as boron, sulfur, selenium, nitrogen or phosphorus in organic compounds. Depending on the type of cofactor necessary for running the reaction BVMOs are distinguished between FAD- and NADPH-dependent enzymes and FMN- and NADH-dependent enzymes. Baeyer-Villiger monooxygenases are produced by the bacteria *Acinetobacter* sp. and *Pseudomonas* sp. and fungi such as *Aspergillus* sp. [Alphand et al., 2003]. Although a number of Baeyer–Villiger monooxygenases (EC 1.14.13.22) have been isolated from a variety of organisms, the NADPH-dependent cyclohexanone monooxygenase from the bacterium *Acinetobacter* sp. NCIB 9871 has been studied in the most detail. This enzyme catalyzes the second step in a catabolic pathway that allows the cells to utilize cyclohexanol as their sole source of carbon and energy. The ability of the purified enzyme or whole *Acinetobacter* cells to oxidize a diverse array of cyclic ketones to the corresponding lactones with high stereoselectivity is well established [Stewart et al., 1998].

NADPH-dependent cyclohexanone monooxygenase from *Acinetobacter* sp. has been expressed in baker's yeast. These genetically modified yeast cells can be used as biocatalysts in 4-alkilcyclohexanone oxidation to produce caprolactones (Fig. 15) with high enantiomeric excess [Stewart et al., 1998].

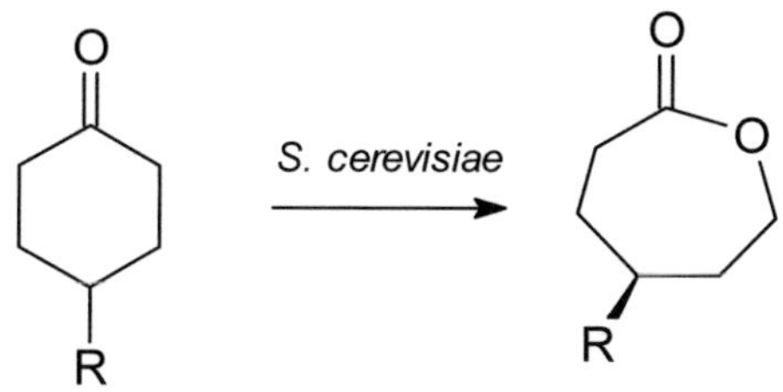

Fig. 15. 4-alkilcyclohexanone oxidation with baker's yeast [Stewart et al., 1998]

The same recombinant yeast can completely convert 2- and 3-substituted cyclopentanones and 2 and 3- substitued cyclohexanones. These reactions are found to be highly enantioselective, allowing the possibility to synthesize different stereoisomers depending upon the size of the ring and the sort of alkyl group. As a result of these modifications, reduction of the carbonyl group is minimized [Kayser et al., 1999; Stewart et al., 1998]. Chiral lactones are useful building blocks in the pharmaceutical industry since bicyclic and polycyclic lactones have received considerable attention as antitumor compounds, cardiac sarcoplasmic reticulum Ca^{2+}-pumping ATPase activators and as useful intermediates in the synthesis of potent drugs for the treatment of glaucoma and hypertension [Alphand et al., 2003].

4.3 Hydrolysis and esterification

About two thirds of reported practical biotransformations may be categorized as hydrolytic reactions involving ester and amide bonds using proteases, esterases or lipases [Loughlin, 2000]. Lipases, in addition to their biological significance, hold tremendous potential for exploitation in biotechnology. They display exquisite chemo-, regio-, and stereoselectivity, do not usually require cofactors and are readily available from microbial organisms [Jaeger &

Eggert, 2002]. The crystal structure of many lipases has been solved, facilitating considerably the design of rational engineering strategy. For these reasons lipases are commonly used in ester synthesis, acylglycerols modifications and in biodiesel production. Consequently, lipase genes are often integrated with host cell genomes of other microorganisms.

"True" lipases are defined as carboxylesterases, catalyzing the bioconversion of long-chain acylglycerols. The essential difference between lipases and esterases is the former act at the water-lipid interface, requiring a micelle formation by a water-insoluble substrate [Gill & Parish, 1997]. Although *S. cerevisiae* produces several hydrolytic enzymes such as esterases and lipases [Białecka-Florjańczyk et al., 2010] (some of which have been isolated), it is generally used as a host organism to express other microbial lipases of special catalytic activity.

4.3.1 Cell surface engineered baker's yeast in esterification reactions

The whole-cell catalysed reactions are often limited by barrier functions of the cell wall or membrane and thus the localization of enzyme, (extracellular, intracellular and membrane bound) plays an important role in lipase activity [Deive et al., 2009]. By the means of molecular engineering, the cell surface properties can be designed by displaying various functional proteins, especially enzymes. This technique provides to avoid mass transport problems of substrate and/or product across the cell membrane as the enzyme, necessary for catalysis, is displayed on the cells surface. Besides the anchoring to the cell surface usually results in increased biocatalyst stability. A cell surface engineering system of yeast *Saccharomyces cerevisiae* has been established and novel yeasts displaying lipases in their active form on the cell surface were constructed [Ueda & Tanaka, 2000]. Lipase-displaying whole-cell yeast biocatalysts have recently attracted attention for their use in biodiesel synthesis.

Biodiesel, a fuel for diesel engines, represents an alternative environmentally-friendly source of energy obtained from renewable materials. It is produced via tranesterification of vegetable oils with alcohols (methanol or ethanol) and comprises fatty acid methyl (FAME, Fig. 16) or ethyl esters (FAEE). For ecological reasons, the enzymatic transesterification is becoming of increasing interest, yet the high cost of enzymes (lipases) obstructs its full industrial application [Akoh et al., 2007, Ribeiro et al., 2011]. Much research has focused on methods which allow increased whole-cell biocatalytic activity and stability through changes in microorganism culture conditions, their immobilization and application of genetic engineering techniques [Kucharski et al., 2009].

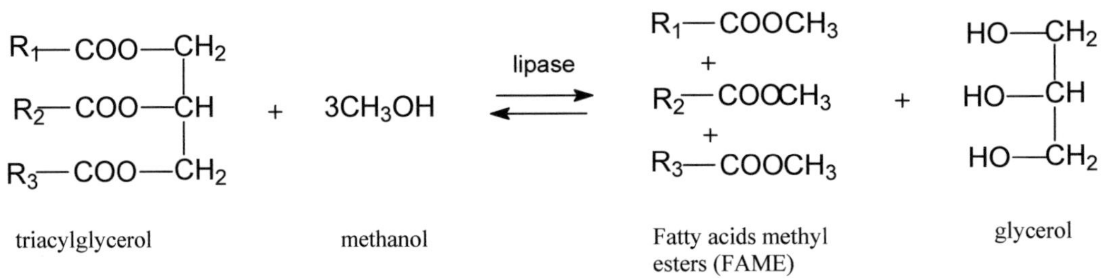

Fig. 16. Biodiesel production scheme

It has been shown that the application of recombinant baker's yeast reduces the number of operations required in biodiesel production and simplifies the removal of glycerol [Fukuda et al., 2008; Takahashi et al., 1998, 1999]. The possibilities of heterologous expression of nonspecific lipases such as *Candida rugosa, Pseudomonas cepacia* and *Pseudomonas fluorescens* has been reported, all of which exhibit relatively high conversion rates and methanol tolerance [Fukuda et al., 2008]. Whole-cell yeast biocatalysts, which intracellularly overproduce a recombinant lipase with a pro-sequence from *Rhizopus oryzae* (rProROL) were constructed and the content of active lipase in *S. cerevisiae* cells was maximized by optimizing the cultivation procedure. Lipase from *Rhizopus oryzae* (ROL) was chosen because its secretory production has been accomplished in *S. cerevisiae.* Overproducing ROL lipase baker's yeast strain MT8–1 was permeabilized and used for the synthesis of FAME with 71 % efficiency [Takahashi et al., 1999].

The cell surface engineering of yeasts by which functional proteins are displayed on the cell surface has enabled the development of a novel strategy for whole-cell biocatalysts. To enhance lipase activity and preserving the conformation of the active site near the C-terminal portion a linker peptide (spacer) consisting of the Gly/Ser repeated sequence was inserted at the C-terminal position. The display of active lipase from *Rhizopus oryzae* on the *S. cerevisiae* cell surface resulted in production of 2,3-dimercaptopropan-1-ol tributyl ester and insoluble triolein [Washida et al., 2001]. This eliminates the cell disintegration step, allowing the production of intracellular enzyme and subsequent ease of separation of the products from the catalysts. Both these factors influences the costs of the process, making it more cost effective [Takahashi et al., 1998]. There are other examples of enzymes displayed on baker's yeast surface such as the lipase CaLIP4 from *Candida albicans* [Breinig et al., 2002], lipase A from *Bacillus subtilis* [Mormeneo et al., 2008; Roustan et al., 2005] and lipase L1 from *Bacillus stearothermophilus* [Breinig et al., 2002].

The other tendency in *S. cerevisiae* modifications reflects the broadening of kinds of carbon sources that can be utilized by yeast cells for biodiesel production purposes

Though biodiesel synthesis is the main area of cell wall engineered yeast application nevertheless it was employed in some other esterifications. To improve cost-efficiency a whole-cell biocatalysts CALB-displaying was constructed using the Flo1p short (FS) anchor system. Lyophilized yeast cells were applied to an ester synthesis reaction at 60°C using adipic acid and butanol as substrates [Tanino et al., 2007]. Similar whole cell catalyst have been applied in the esterification of hexanoic acid with ethanol, yielding 98.2% of the ester under optimum conditions [Han et al., 2009].

Recombinant *R. oryzae* lipase (ROL) displayed on the *S. cerevisiae* cell surface was used in the resolution of enantiomers of (*R,S*)-1-phenylethanol, which serves as one of the important chiral building blocks. During enantioselective transesterification of this alcohol with vinyl acetate (Fig. 17) the yield of (*R*)-1-phenylethyl acetate reached 97 % with 93 % enantiomeric excess [Matsumoto et al., 2004].

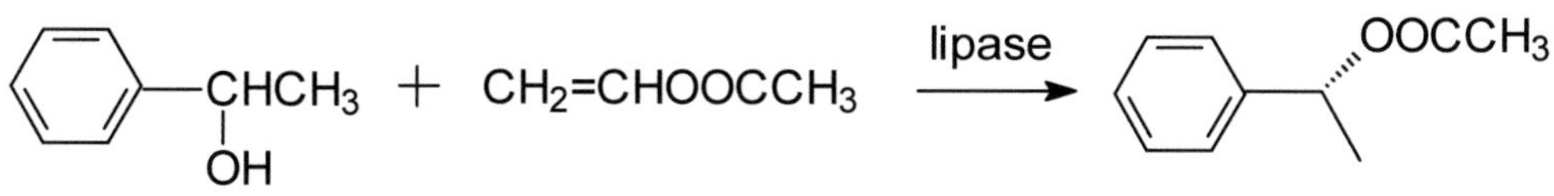

(*RS*)-1-phenylethanol vinyl acetate (*R*)-1-phenylethylacetate

Fig. 17. Synthesis of (*R*)-1-phenylethylacetate

The ROL-displaying yeast whole-cell biocatalyst catalyzed the stereospecific hydrolysis of the pharmaceutical precursor (*R,S*)-1-benzyloxy-3-chloro-2-propyl monosuccinate. The isomer (*R*) was isolated with the ee up to 95,5% (Fig. 18). *Rhizopus oryzae* lipase was displayed on the cell surface of *Saccharomyces cerevisiae* via the Flo1 N-terminal region (1100 amino acids), which corresponds to a flocculation functional domain [Nakamura Y., 2006].

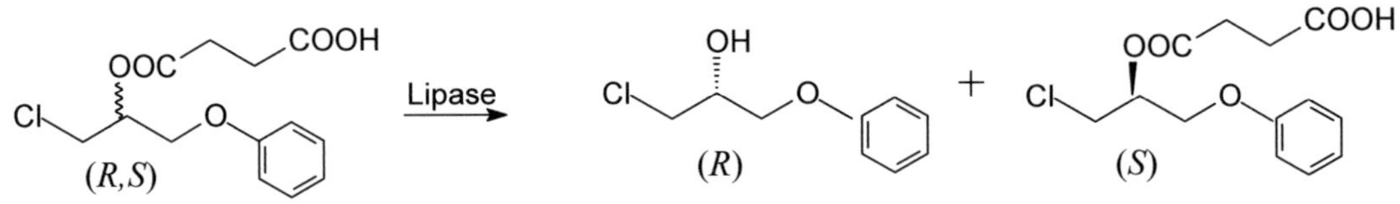

Fig. 18. Stereospecific hydrolysis of of (*R,S*)-1-benzyloxy-3-chloro-2-propyl monosuccinate

4.4 L-malic acid production

Biotransformation, using whole cells of modified *S. cervisiae* can be applied in the synthesis of L-malic acid (Fig. 19). Besides the food industry (acidulant E 296), malic acid has extensive application in the pharmaceutical and cosmetic industries. L-malic acid is enzymatically produced via the hydration of maleic or fumaric acid (FA) catalysed by fumarase (EC 4.2.1.2) from bacteria such as *Brevibacterium ammoniagenses* and *B. flavum* [Yazawa et al., 2010].

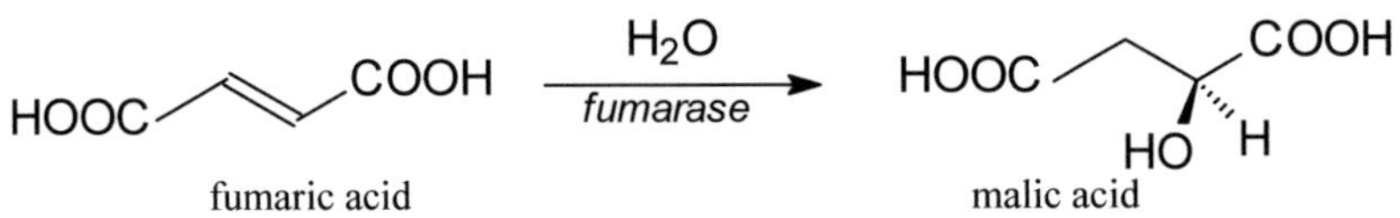

Fig. 19. Biotransformation of fumaric acid to L-malic acid

In the 1990s, production of L-malic acid from fumaric acid using *S. cerevisiae* cells was studied extensively with the amount of by-products minimized using modified *S. cerevisiae* cells with many copies of the fumarase gene as a biocatalysts [Presečki et al., 2007]. Conversion of fumaric acid to L-malic acid was carried out also in a bioreactor divided into three compartments by two supported liquid membranes. The yeast was immobilized in small glasslike beads of an alginate-silicate solgel matrix, allowing the reaction to reach almost 100% the conversion with only small amount of succinic acid produced. This was much higher than the 70% conversion rate found in industrial processes [Peleg et al., 1990].

5. Conclusions

S. cerevisiae has an enormous potential for the production of low and high molecular weight compounds by acting as an heterologous host and expressing biosynthetic enzymes or pathways. By coupling multiple enzymes, metabolic pathways in a single cell are created. This eliminates the need for purification of the chemical intermediates and the desired products can be prepared from simple, inexpensive and renewable materials. In the future the metabolic engineering of yeast may lead to an alternative production systems, helping overcome the limited availability of biologically active, commercially valuable and nutritionally important plant secondary metabolites compounds. It should be possible to significantly reduce development and costs of these microbial cell factories.

S. cerevisiae may be used as a whole-cell biocatalysts in the biotransformations, mostly in reduction or hydrolysis reactions. Modern research has lead to improvements in the catalytic ability of baker's yeast, including the surface display of enzymes, optimization or increase of the cofactor availability in bioreduction reactions and gene knock-out, which eliminates the activity of enzymes with conflicting, unwanted stereoselectivity.

It was shown that genetically engineered baker's yeast has a great potential as a biocatalysts in many branches of chemistry. The fact is that it is impossible to enumerate all of the published application of modified *S. cerevisiae* cells as to improve valuable chemicals production. It is worth mentioning that there are still problems not solved. Many studies regarding *S. cerevisiae* metabolic engineering for lignocellulosic biomass utilization have been provided, but this process involving baker's yeast is still insufficient for industrial bioprocesses mainly due to low rate of reaction. Nevertheless it can be predicted that engineered baker's yeast would be an efficient tool in chemical processes as to improve people lives.

6. Acknowledgements

This work was supported by Grant No N N209 107 639 from the State Committee for Scientific Research, Ministry of Scientific Research and Information Technology, Poland.

7. References

Abbott, D.A.; ZellC, R.M.; Pronk, J.T. & Maris A.J. (2009), Metabolic engineering of *Saccharomyces cerevisiae* for production of carboxylic acids: current status and challenges, *FEMS Yeast Research*, Vol.9, No.8, (December 2009), pp. 1123-1136, ISSN 1567-1356

Almqvist, F.; Eklund, L. & Frejd, T. (1993), An improved procedure for the synthesis of bicyclo[2.2.2]octane-2,6-dione, *Synthetic Communications*, Vol.6, (1993), pp. 957–960, ISSN 0039-7911

Alphand, V.; Carrea, G.; Wohlgemuth, R.; Furstoss, R. & Woodley, J.M. (2003), Towards large-scale synthetic applications of Baeyer-Villiger monooxygenases, *Trends in Biotechnology*, Vol. 21, No. 7, (July 2003), pp. 318-323, ISSN 0167-7799

Akoh, C.C.; Chang, S.W.; Lee, G.C. & Shaw J.F. (2007). Enzymatic approach to biodiesel production, *J Agric Food Chem.* Vol. 55, No.22, pp 8995-9005, ISSN 1520-5118

Arsenault, P.R.; Wobbe, K.K. & Weathers P.J. (2008), Recent advances in artemisinin production through heterologous expression, *Current Medicinal Chemistry*, Vol.15, No.27, (February 2008), pp. 2886-2896, ISSN 0929-8673

Białecka-Florjańczyk, E. & Majewska, E. (2006), Biotransformacje z udziałem drożdży *Saccharomyces cerevisiae*, *Biotechnologia*, Vol.3, (2006), pp. 113-133, ISSN 0860-7796

Białecka-Florjańczyk, E.; Krzyczkowska, J. & Stolarzewicz I. (2010), Catalytic activity of baker's yeast in ester hydrolysis, *Biocatalysis and Biotransformation*, Vol.28, No.4, (July 2010), pp. 288-291, ISSN 1024-2422

Breinig, F. & Schmitt, M.J. (2002), Spacer-elongated cell wall fusion proteins improve cell surface expression in the yeast *Saccharomyces cerevisiae, Applied Microbiology and Biotechnology,* Vol.58, No.5, (April 2002), pp. 637-640, ISSN 0175-7598

Carlquist, M.; Wallentin, C.; Warnmark, K. & Gorwa-Grauslund, M.F. (2008), Genetically engineered *Saccharomyces cerevisiae* for kinetic resolution of racemic bicyclo[3.3.1]nonane-2,6-dione, *Tetrahedron: Asymmetry,* Vol.19, No.19, (October 2008), pp. 2293-2295, ISSN 0957-4166

Carlquist, M.; Olsson, C.; Bergdahl, B.; Niel, E.W.J.; Gorwa-Grauslund, M.F. & Frejd, T. (2009), Kinetic resolution of racemic 5,6-epoxy-bicyclo[2.2.1]heptane-2-one using genetically engineered *Saccharomyces cerevisiae, Journal of Molecular Catalysis B,* Vol.58, No.1-4, (June 2009), pp. 98-102, ISSN 1381-1177

Celińska E. (2010), Debottlenecking the 1,3-propanediol pathway by metabolic engineering, *Biotechnology Advances,* Vol.28, No.4, (July-August 2010), pp. 519-530, ISSN 0734-9750

Chang, M.C. & Keasling, J.D. (2006), Production of isoprenoid pharmaceuticals by engineered microbes. *Nature Chemical Biology,* Vol.2, No.12, (December 2006), pp. 674-681, ISSN 1552-4450

Chemler, J.A.; Yan, Y. & Koffas, M.A.G. (2006), Biosynthesis of isoprenoids, polyunsaturated fatty acids and flavonoids in *Saccharomyces cerevisiae, In: Microbial Cell Factories,* Vol.5, 23.05.2006, Available from:
http://www.ncbi. nlm. nih.gov/ pmc/articles/PMC1533850/

Chu, B.C. & Lee, H. (2007), Genetic improvement of *Saccharomyces cerevisiae* for xylose fermentation, *Biotechnology Advances,* Vol. 25, No. 5, (September-October 2007), pp. 425-441, ISSN 0734-9750

Csuk, R. & Glanzer, B. (1991), Baker`s yeast mediated transformations in organic chemistry, *Chemical Reviews,* Vol. 91, No.1, (January 1991) pp. 49-97, ISSN 0009-2665

Daum, G.; Lees, N.D., Bard, M. & Dickson R. (1998), Biochemistry, cell biology and molecular biology of lipids of *Saccharomyces cerevisiae, Yeast,* Vol.14, No.16, (December 1998), pp. 1471–1510, ISSN 0749-503X

Deive, F.J.; Carvalho, E.; Pastrana, L.; Rua, M.L.; Longo, M.A. & Sanroman, M.A. (2009), Strategies for improving extracellular lipolytic enzyme production by *Thermus thermophilus* HB27, *Bioresource Technology,* Vol.100, No.14, (July 2009), pp. 3630-3637, ISSN 0960-8524

Dejong, J.M.; Liu, Y.; Bollon, A.P.; Long, R.M.; Jennewein, S.; Williams, D. & Croteau, R.B. (2006), Genetic engineering of taxol biosynthetic genes in *Saccharomyces cerevisiae,* Biotechnology and Bioeng*ineering,* Vol.93, No.2, (February 2006), pp. 212-224, ISSN 0006-3592

Dumas, B.; Brocard-Masson, C.; Assemat-Lebrun, K. & Achstetter, T. (2006), Hydrocortisone made in yeast: Metabolic engineering turns a unicellular microorganism into a drug-synthesizing factory, *Biotechnology Journal,* Vol.1, No.3, (March 2006), pp. 299–307, ISSN 1860-6768

Engelking, H.; Pfaller, R.; Wich, G. & Weuster-Botz, D. (2006), Reaction engineering studies on β-ketoester reductions with whole cell recombinants *Saccharomyces cerevisiae,*

Enzyme and Microbial Technology, Vol.38, No.3-4, (February 2006), pp.536-544, ISSN 0141-0229

Faber, K. (2004), *Biotransformations in organic chemistry,* Springer-Verlag, New York, ISBN 3-540-6334-7

Fukuda H., Hama S., Tamalampudi S. & Noda H. (2008). Whole-cell biocatalysts for biodiesel fuel production, *Trends in Biotechnology,* Vol.26, No.12, (December 2008), pp. 668–672, ISSN 0167-7799

Gill, J. & Parish, J. (1997), Lipases - Enzymes at an interface, *Biochemical Education,* Vol.25, No.1, (January 1997), pp. 2-5, ISSN 0307-4412

Goffeau, A.; Barrell, B.G.; Bussey, H.; Davis, R.W.; Dujon, B.; Feldmann, H.; Galibert, F.; Hoheisel, J.D.; Jacq, C.; Johnston, M.; Louis, E.J.; Mewes, H.W.; Murakami, Y.; Philippsen, P.; Tettelin, H. & Oliver, S.G.,(1996) Life with 6000 genes, *Science,* Vol. 74, No.546, (October 1996), pp. 546, 563-567, ISSN 0036-8075

Hahn-Hägerdal, B.; Karhumaa, K.; Jeppsson, M. & Gorwa-Grauslund, M.F. (2007), Metabolic engineering for pentose utilization in *Saccharomyces cerevisiae, Advances in Biochemical Engineering/ Biotechnology,* Vol.108, pp. 147-177, ISSN 0724-6145

Han, S.; Pan, Z.; Huang D.; Ueda, M.; Wang, X & Lin, Y. (2009), Highly efficient synthesis of ethyl hexanoate catalyzed by CALB-displaying *Saccharomyces cerevisiae* whole-cells in non-aqueous phase, *Journal of Molecular Catalysis: B,* Vol.59, No.1-3, (July 2009), pp. 168-172, ISSN 1381-1177

Huang, B.; Guo, J.; Yi, B.; Yu, X.; Sun L. & Chen, W. (2008), Heterologous production of secondary metabolites as pharmaceuticals in *Saccharomyces cerevisiae, Biotechnology Letters,* Vol.30, No.7, (July 2008), pp.1121-1137, ISSN 0141-5492

Inui, M.; Suda, M.; Kimura, S.; Yasuda, K.; Suzuki, H.; Toda, H.; Yamamoto, S.; Okino, S.; Suzuki, N. & Yukawa, H. (2008), Expression of *Clostridium acetobutylicum* butanol synthetic genes in *Escherichia coli, Applied Microbiology and Biotechnology,* Vol.77, No. 6 (January 2008), pp. 1305-1316, ISSN 0175-7598

Ishida, N.; Saitoh, S.; Ohnishi, T.; Tokuhiro, K.; Nagamori, E.; Kitamoto, K. & Takahashi, H. (2006), Metabolic engineering of *Saccharomyces cerevisiae* for efficient production of pure L-(+)-lactic acid, *Applied Biochemistry and Biotechnology,* Vol.131, No.1-3, (March 2006), pp. 795-807, ISSN 0273-2289

Jaeger, K.E. & Eggert, T. (2002), Lipases for biotechnology, *Current opinions in biotechnology,* Vol.13, No. 4, (August 2002) ,pp. 390-397, ISSN 0958-1669

Jiang, H.X.; Wood, K.V. & Morgan, J.A. (2005), Metabolic engineering of the phenylpropanoid pathway in *Saccharomyces cerevisiae, Applied and Environmental Microbiology,* Vol.71, No.6, (June 2005), pp. 2962-2969, ISSN 0099-2240

Johanson, T.; Katz, M. & Gorwa-Grauslund, M.F. (2005), Strain engineering for stereoselective bioreduction of dicarbonyl compounds by yeast reductases, *FEMS Yeast Research,* Vol.5, No.6-7, (January 2005), pp. 513-525, ISSN 1567-1356

Johanson, Y.; Carlquist, M.; Olsson, C.; Rudolf, A.; Frejd, T. & Gorwa-Grauslund, M.F. (2008), Reaction and strain engineering for improved stereoselective whole-cell reduction of a bicyclic diketone, *Applied Microbiology and Biotechnology,* Vol.77, No.5, (January 2008), pp. 1111-1118, ISSN 0175-7598

Katz, M.; Sarvary, I.; Frejd, T.; Hahn-Hagerdal, B. & Gorwa-Grauslund, B. (2002), An improved stereoselective reduction of a bicyclic diketone by *Saccharomyces cerevisiae* combining process optimization and strain engineering, *Applied Microbiology and Biotechnology*, Vol.59, No.6, (September 2002), pp. 641-648, ISSN 0175-7598

Katz M., Hahn-Hagerdal B. & Gorwa-Grauslund M. F., (2003), Screening of two complementary collections of *Saccharomyces cerevisiae* to identify enzymes involved in stereo-selective reductions of specific carbonyl compounds: an alternative to protein purification, *Enzyme and Microbial Technology*, Vol.33, No.3 (August 2003), pp. 163-172, ISSN 0141-0229

Kayser, M. M.; Chen, G. & Stewart, J. D. (1999), "Designer Yeast": an Enantioselective Oxidizing Reagent for Organic Synthesis *Synlett.*, No. 1, pp. 153-158. ISSN: 0936-5214

Kayser, M.M.; Mihovilovich, M.D.; Kearns, J.; Feicht, A. & Stewart, J.D. (1999), Baker's Yeast-Mediated Reductions of α-Keto Esters and an α-Keto-β-Lactam. Two Routes to the Paclitaxel Side Chain, *Journal of Organic Chemistry*, Vol.64, No.18, (3 September 1999), pp. 6603-6608, ISSN 0022-3263

Kondo, A.; Tanaka, T.; Hasunuma, T. & Ogino, C. (2010), Applications of yeast cell-surface display in bio-refinery, *Recent Patents on Biotechnology,* Vol.4, No.3, (November 2010), pp. 226-234, ISSN 1872-2083

Kucharski D.; Białecka-Florjańczyk E. & Stolarzewicz I. (2009) Mikrobiologiczne metody otrzymywania biodiesla, *Biotechnologia*, Vol.4 No.87, pp. 74-87, ISSN 0860-7796

Loughlin, W.A. (2000), Biotransformations in organic synthesis, *Biosource Technology,* Vol.74, No.1, (August 2000), pp. 49-62, ISSN 0960-8524

Matsumoto, T.; Ito, M.; Fukuda, H. & Kondo, A. (2004), Enantioselective transesterification using lipase-displaying yeast whole-cell biocatalyst, *Applied Microbiology and Biotechnology*, Vol.64, No.4, (May 2004), pp. 481-485, ISSN 0175-7598

Matsushika, A.; Inoue, H.; Kodaki, T. & Sawayama, S.; Ethanol production from xylose in engineered *Saccharomyces cerevisiae* strains: current state and perspectives. *Applied Microbiology and Biotechnology,* Vol.84, No.1, (August 2009), pp. 37-53, ISSN 0175-7598

Mendes, F.S.; González-Pajuelo, M.; Cordier, H.; François, J.M. & Vasconcelos I. (June 2011), 1,3-Propanediol production in a two-step process fermentation from renewable feedstock, In: *Applied Microbiology and Biotechnology,* 9.06.2011, Available from: http://www.springerlink.com/content/27r256n7r1638r5l

Mormeneo, M.; Andres, I.; Bofill, C.; Diaz, P. & Zuezo, J. (2008), Efficient secretion of *Bacillus subtilis* lipase A in *Saccharomyces cerevisiae* by translational fusion to the Pir4 cell wall protein, *Applied Microbiology and Biotechnology*, Vol. 80, No.3,(September 2008), pp. 437-445, ISSN 0175-7598

Nakamura, K.; Inoue, Y.; Shibahara, J.; Oka, S. & Ohno, A. (1988), Asymmetric reduction of β- and γ-nitro ketones by bakers' yeast. *Tetrahedron Letters*, Vol.29, No.7, (1988), pp. 4769-4770, ISSN 0040-4039

Nakamura,Y.; Matsumoto, T.; Nomoto, F.; Ueda, M.; Fukuda, H.& Kondo, A. (2006) Enhancement of activity of lipase-displaying yeast cells and their application to optical resolution of (*R*,*S*)-1-benzyloxy-3-chloro-2-propyl monosuccinate, *Biotechnology Progress*, 22, 998-1002. ISSN 1520-6033

Nevoigt, E. (2008). Progress in metabolic engineering of *Saccharomyces cerevisiae*, *Microbiology and Molecular Biology Reviews*, Vol. 72, No. 3, (September 2008), pp. 379-412, ISSN 1092-2172

Parachin, N.S.; Carlquist, M. & Gorwa-Grauslund, M. F. (2009), Comparison of engineered *Saccharomyces cerevisiae* and engineered *Escherichia coli* for the production of an optically pure ketoalkohol, *Applied Microbiology and Biotechnology*, Vol.84, No.3, (September 2009), pp. 487-497, ISSN 0175-7598

Peleg, Y.; Rokem, J.S.; Goldberg, I. & Pines, O. (1990), Inducible overexpression of the FUM1 gene in *Saccharomyces cerevisiae*: localization of fumarase and efficient fumaric acid bioconversion to L-malic acid., *Applied and Environmental Microbiology*, Vol.56, No.9, (September 1990), pp. 2777-2783, ISSN 0099-2240

Pines, O.; Shemesh, S.; Battat, E. & Goldberg, I. (1997), Overexpression of cytosolic malate dehydrogenase (MDH2) causes overproduction of specific organic acids in *Saccharomyces cerevisiae*, *Applied Microbiology and Biotechnology*, Vol.48, No.2, (August 1997), pp. 248-255, ISSN 0175-7598

Presečki, A.V.; Zelić, B. & Vasić-Rački, D. (2007), Comparison of the l-malic acid production by isolated fumarase and fumarase in permeabilized baker's yeast cells, *Enzyme and Microbial Technology*, Vol.41, No.5, (October 2007), pp. 605-612, ISSN 0141-0229

Pscheidt, B. & Glieder, A.(2008) Yeast cell factories for fine chemical and API production, *Microbial Cell Factories* 7:25 , ISSN: 1475-2859

Ralston, L.; Subramanian, S.; Matsuno, M. & Yu, O. (2005), Partial reconstruction of flavonoid and isoflavonoid biosynthesis in yeast using soybean type I and type II chalcone isomerases, *Plant Physiology*, Vol.137, No.4, (April 2005), pp. 1375-1388, ISSN 0032-0889

Rao, Z.; Ma, Z.; Shen, W.; Fang, H.; Zhuge, J. & Wang, X. Engineered *Saccharomyces cerevisiae* that produces 1,3-propanediol from D-glucose, *Journal of Applied Microbiology*, Vol.105. No.6 (December 2008), pp. 768-776, ISSN 1364-5072

Remize, F.; Roustan, J.L.; Sablayrolles, J.M.; Barre, P. & Dequin, S. (1999), Glycerol overproduction by engineered *Saccharomyces cerevisiae* wine yeast strains leads to substantial changes in by-product formation and to a stimulation of fermentation rate in stationary phase, *Applied and Environmental Microbiology*, Vol.65, No.1, (January 1999), pp. 143–149, ISSN 0099-2240

Ribeiro B.D.; de Castro A.M.; Coelho M.A. & Freire D.M., (2011), Production and use of lipases in bioenergy: a review from the feedstocks to biodiesel production. *Enzyme Res.*, Epub 2011 Jul 7. ISSN: 2090-0406

Ro, D.; Paradise, E.M.; Ouellet, M.; Fisher, K.J.; Newman, K.L.; Ndungu, J.M.; Ho, K.A.; Eachus, R.A.; Ham, T.S.; Kirby, J.; Chang, M.C.Y.; Withers, S.T.; Shiba, Y.; Sarpong, R. & Keasling, J.D. (2006), Production of the antimalarial drug precursor artemisinic acid in engineered yeast, *Nature*, Vol.440, (13 April 2006), pp.940-943, ISSN 0028-0836

Rodriguez, S.; Kayser, M.M. & Stewart, J.D, (1999), Improving the stereoselectivity of bakers' yeast reductions by genetic engineering, *Organic Letters*, Vol.1, No.8, (October 1999), pp. 1153-1155, ISSN 1523-7060

Rodriguez S., Schroeder K.T., Kayser M.M. & Stewart J.D. (2000), Asymmetric synthesis of β-hydroxy esters and α-alkyl-β-hydroxy esters by recombinant *Escherichia coli* expressing enzymes from baker's yeast, *Journal of Organic Chemistry*, Vol.65, No.8, (April 2000), pp. 2586-2587, ISSN 0022-3263

Rodriguez, S.; Kayser, M.M. & Stewart, J.D. (2001), Highly Stereoselective Reagent for β-Keto Ester Reductions by Genetic Engineering of Baker's Yeast, *Journal of the American Chemical Society*, Vol.123, No.8, (February 2001), pp. 1547-1555, ISSN 0002-7863

Roustan, J.L.; Chu, A.R.; Moulin, G. & Bigey, F. (2005), A novel lipase/acyltransferase from the yeast *Candida albicans*: expression and characterisation of the recombinant enzyme, *Applied Microbiology and Biotechnology*, Vol.68, No.2, (August 2005), pp. 203-212, ISSN 0175-7598

Sarvary, I.; Almqvist, F. & Frejd, T. (2001), Asymmetric reduction of ketones with catecholborane using 2,6-BODOL complexes of titanium(IV) as catalysts, *Chemistry - An European Journal*, Vol.7, No.10, (May 2001), pp. 2158–2166, ISSN 0947-6539

Servi, S.; (1990), Baker's Yeast in Organic Synthesis, *Synthesis*, (1990), pp. 1-25, ISSN 1570-2693

Shiba, Y.; Paradise, E.M.; Kirby, J.; Ro, D.K. & Keasling, J.D. Engineering of the pyruvate dehydrogenase bypass in *Saccharomyces cerevisiae* for high-level production of isoprenoids, *Metabolic Engineering*, Vol.9, No.2, (March 2007), pp. 160-168, ISSN 1096-7176

Steen, E.J.; Chan, R.; Prasad, N.; Myers, S.; Petzold, C.J.; Redding, A.; Ouellet, M. & Keasling, J.D. (December 2008), Metabolic engineering of *Saccharomyces cerevisiae* for the production of butanol, In: *Microbial Cell Factories,* 3.12.2008, Available from: http://www.ncbi.nlm.nih.gov/pmc/articles/PMC2621116

Stewart, J.D.; Reed, K.W.; Martiez, C.A.; Zhu, J.; Chen, G. & Kayser, M.M. (1998), Recombinant baker's yeast as a whole-cell catalyst for asymmetric Baeyer−Villiger oxidations, *Journal of American Chemical Society*, Vol.120, No.15, (1998), pp.3541-3548, ISSN 0002-7863

Stewart, J.D. (2000), Organic transformations catalyzed by engineered yeast cells and related systems, *Current Opinion in Biotechnology*, Vol.11, No.4 (August 2000), pp. 363-368, ISSN 0958-1669

Szczebara, F.M.; Chandelier, C.; Villeret, C.; Masurel, A.; Bourot, S.; Duport, C.; Blanchard, S.; Groisillier, A.; Testet, E.; Costaglioli, P.; Cauet, G.; Degryse, E.; Balbuena, D.; Winter, J.; Achstetter, T.; Spagnoli, R.; Pompon, D. & Dumas, B. (2003), Total biosynthesis of hydrocortisone from a simple carbon source in yeast, *Nature Biotechnology,* Vol.21, (6 January 2003), pp. 143 – 149, ISSN 1087-0156

Takahashi, S.; Ueda, M.; Atomi, H.; Beer, H.D.; Bornscheuer, U.T.; Schmid, R.D. & Tanaka, A. (1998), Extracellular production of active *Rhizopus oryzae* lipase by *Saccharomyces cerevisiae, Journal of Fermentation and Bioengineering*, 1998, Vol.86, No.2, (1998), pp. 164-168, ISSN 0922-338X

Takahashi, S.; Ueda, M. & Tanaka, A.; (1999), Independent production of two molecular forms of a recombinant *Rhizopus oryzae* lipase by KEX2-engineered strains of *Saccharomyces cerevisiae, Applied Microbiology and Biotechnology*, Vol.52, No.4, (October 1999), pp. 534-540, ISSN 0175-7598

Tanino, T.; Ohno, T.; Aoki, T.; Fukuda, H. & Kondo, A. (2007), Development of yeast cells displaying *Candida antarctica* lipase B and their application to ester synthesis reaction, *Applied Microbiology and Biotechnology*, Vol.75, No.6, (July 2007), pp. 1319-1325, ISSN 0175-7598

Toivari, M.H.; Ruohonen, L.; Miasnikov, A.N.; Richard, P. & Penttilä, M. (2007) Metabolic engineering of *Saccharomyces cerevisiae* for conversion of D-glucose to xylitol and other five-carbon sugars and sugar alcohols, *Applied Environmental Microbiology*. Vol. 73, No. 17, pp. 5471-5476, ISSN 0099-2240

Tokuhiro, K.; Muramatsu, M.; Ohto, C.; Kawaguchi, T.; Obata, S.; Muramoto, N.; Hirai, M.; Takahashi, H.; Kondo, A.; Sakuradani, E. & Shimizu, S. (2009), Overproduction of geranylgeraniol by metabolically engineered *Saccharomyces cerevisiae, Applied and Environmental Microbiology*, Vol.75, No.17 (September 2009), pp. 5536-5543. ISSN 0099-2240

Ueda, M. & Tanaka, A. (2000), Cell surface bioengineering of yeast: construction of arming yeast with biocatalyst, *Journal of Bioscience and Bioengineering*, Vol.90, No.2 (2000), pp. 125-136, ISSN 1389-1723

Wang, Z.X.; Zhuge, J.; Fang, H. & Prior, B.A. (2001), Glycerol production by microbial fermentation: a review, *Biotechnology Advances*, Vol.19, No.3, (June 2001), pp. 201-223, ISSN 0734-9750

Washida, M.; Takahashi, S.; Ueda, M. & Tanaka, A. (2001), Spacer-mediated display of active lipase on the yeast cell surface, *Applied Microbiology and Biotechnology*, Vol.56, No.5-6, (September 2001), pp. 681-686, ISSN 0175-7598

Veen, M. & Lang, C. (2004), Production of lipid compounds in the yeast *Saccharomyces cerevisiae, Applied Microbiology and Biotechnology*, Vol.63, No. 6, (February 2008), pp. 635-646, ISSN 0175-7598

Verwaal, R.; Wang, J.; Meijnen, J.; Visser, H.; Sandmann, G.; Berg, J.A. & Ooyen A.J.J. (2007), High-Level production of beta-carotene in *Saccharomyces cerevisiae* by successive transformation with carotenogenic genes from *Xanthophyllomyces dendrorhous, Applied and Environmental Microbiology*, Vol.73, No.13, (July 2007), pp. 4342–4350, ISSN 0099-2240

Yan, Y.; Kohli, A. & Koffas, M.A. (2005), Biosynthesis of natural flavanones in *Saccharomyces cerevisiae, Applied and Environmental Microbiology*, Vol.71, No.9, (September 2005), pp. 5610-5613, ISSN 0099-2240

Yazawa, H.; Iwahashi, H.; Kamisaka, Y.; Kimura, K. & Uemura, H. (2010), Improvement of polyunsaturated fatty acids synthesis by the coexpression of CYB5 with desaturase genes in *Saccharomyces cerevisiae, Applied Microbiology and Biotechnology*, Vol.87, No. 6, (August 2010), pp. 2185–2193, ISSN 0175-7598

Young, E.; Lee, S. & Alper H. (November 2010), Optimizing pentose utilization in yeast: the need for novel tools and approaches, In: *Biotechnology for Biofuels* 2010, 16.09.2010, Available from:
http://www.biotechnologyforbiofuels.com/content/3/1/24

12

Animal Models for Hydrodynamic Gene Delivery

Michalis Katsimpoulas[1,*], Dimitris Zacharoulis[2],
Nagy Habib[3] and Alkiviadis Kostakis[1]
[1]*Center for Experimental Surgery, Biomedical Research Foundation of the Academy of Athens*
[2]*Department of Surgery, University Hospital of Larissa, Mezourlo*
[3]*Imperial College London, Faculty of Medicine, London*
[1,2]*Greece*
[3]*UK*

1. Introduction

Hydrodynamic gene delivery (HGD) is an established method of the last decade where macromolecules, non-normally permeable to cell membrane, are delivered intracellular. The basic principle is that a large volume of solution, containing genes or oligonucleotides, RNA, proteins or other compounds, is infused rapidly in circulation to permit the entrance of these substances to parenchymal cells. Excellent review papers are published summarizing the principles, the applications and the progress that has been made in this field[1-10] The aim of this chapter is to describe the basic principles of the hydrodynamic gene delivery, the surgical procedures in all animal models and the reflection of our scope for the future development.

2. Basics of hydrodynamic gene delivery

Hydrodynamic gene delivery is developed based in our knowledge on vasculature, fluid properties and cell membrane permeability. Based on the studies of Zhang et al.[1], Kobayashi et al.[2], Lecocq et al.[3] and Al-Dosari et al.[5], the rapid infusion of large solution containing macromolecules, non-normally permeable to cell membrane, generates high hydrodynamic pressure in the circulation refluxing to the target organ. The enlarged perivenous area, by the extravasation of the infused solution, generates high pressure in the exterior of the cells. When this pressure reaches a certain level, defects (pores) are been created on the cell membrane leading to the insertion of the macromolecules intracellular. After a few seconds, while the pressure declines at post-injection period, the defects are restoring, trapping inside the cytoplasm the infused molecules. Finally, the body adapts the volume overload within time and the homeostasis is again balanced. We believe that the sequence of these events occur in all surgical protocols, from tail vein injection in mice to specific organ infusion in large animals.

---

* Corresponding Author

The vehicle for the molecules is Normal Saline, Ringer's Solution or Phosphate Buffered Saline and the dosage range from 0.1 to 10 mg/kg, depending on the application. The molecules can be DNA, as plasmid, fragment or artificial chromosome, RNA, single or double stranded, genomic or not, polymers, proteins and small compounds. The main application of the hydrodynamic delivery is the therapy studies, especially genes encoding secretory proteins which can be even isolated and purified. A different area of interest is the gene function analysis in a whole animal instead of a cell culture; the transfection is applied with the hydrodynamic technique and the conclusions are withdrawn from the whole animal. The study of DNA sequences can be achieved with HGD, where promoters, introns, enhancers, suppressors can be delivered intracellular in whole animal. Moreover new animal models are possible to be created for study human diseases, mainly viral due to the fact that they are species specific[5,8,10].

3. Animal models

3.1 Mice

The majority of the research protocols of the hydrodynamic delivery in mice are based on tail vein injection, due to the ease of application and the positive results[11,14-53]. In the mouse tail there are four blood vessels, the lateral ones being the veins. The animal is placed in a restrain box or in light sedation and prior to injection the tail vein should be dilated with a tourniquet, immersed in hot water (40 ^{0}C for 1-2 min) (photo 1)or placed in incubator at 37^{0}C for 10-15 min. A 27g-30g needle at a very shallow angle is commonly used, while the injection of small amount of fluid is mandatory for the verification of the correct placement of the needle (photo 2). Then, the solution of a total 1.5-2 ml, equivalent to 8-10% of body weight, is rapidly infused, within 5 to10 seconds[4].

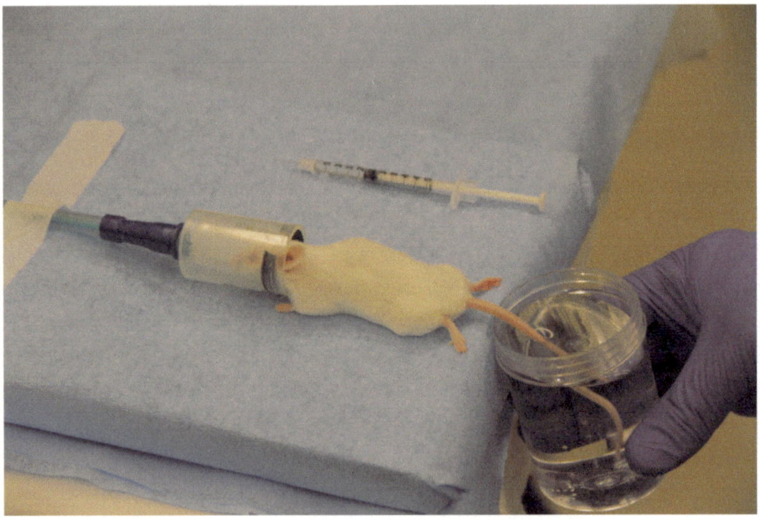

Photo 1. The animal is light sedation with the tail is placed in hot water for dilatation of the veins

A surgical procedure has been established by Zhang et al[15] for delivering macromolecules in liver under general anesthesia. After a midline incision at the abdomen, the portal vein was identified and the solution was injected with a 30g needle while the hepatic vein and the inferior vena cava had been occluded. Modification of this procedure, by the same group, involved the solution being infused in the hepatic vein with occlusion of the portal vein, vena cava and the hepatic artery. At the end of the infusion gentle pressure was applied at the insertion point to minimize hemorrhage. The midline incision was closed in standard technique.

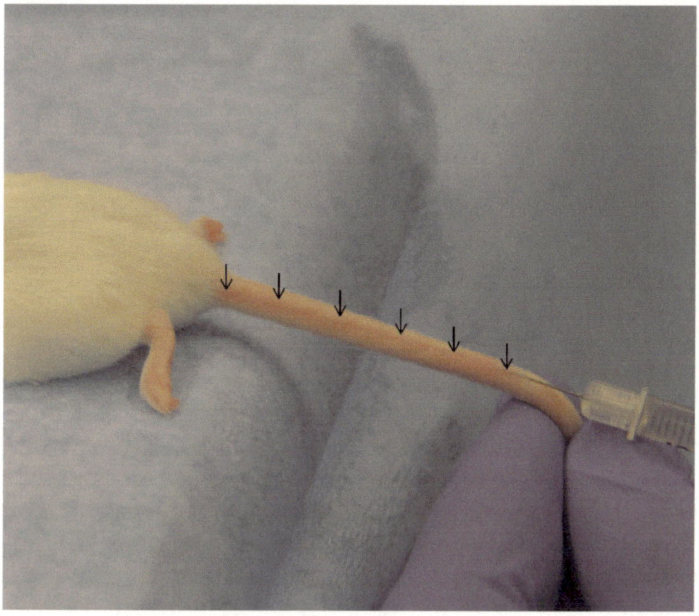

Photo 2. The tail is rotated 90° degrees and the needle is inserted in a shallow angle-arrows are depicting the dilated vein

For hydrodynamic gene delivery in mice muscle, under general anesthesia, a small latex tourniquet is applied above the knee and a small incision is made to expose the great saphenous vein. The infusion of the gene solution precedes the placement of a 30g catheter in the tail vein, and two minutes later the tourniquet is removed[5,6].

3.2 Rat

The tail vein injection is the most common procedure in rats[57-65] as in mice with small modifications. For hydrodynamic infusion a needle or an over-the-needle intravenous catheter 23-27g can be used. The over-the-needle catheter insertion in the vein is followed by needle withdrawal and catheter advancement (photo 3). Blood entrance in the catheter confirms the correct placement of the catheter, secured by taping to the tail and connected with an extension to the infusion pump at a rate of 1-2 ml/sec[54].

For selective infusion in rat's liver, the animal should be in general anesthesia and with the aid of a midline incision, the inferior vena cava (IVC) is isolated with 4-0 suture above and below the liver. At a midpoint, a 21-gauge needle is positioned while around the hole a 4-0 suture is secured to minimize leakage. Two minutes after the injection, the sutures above and below the IVC are removed and midline incision is closed with 4-0 absorbable suture[7,8,9].

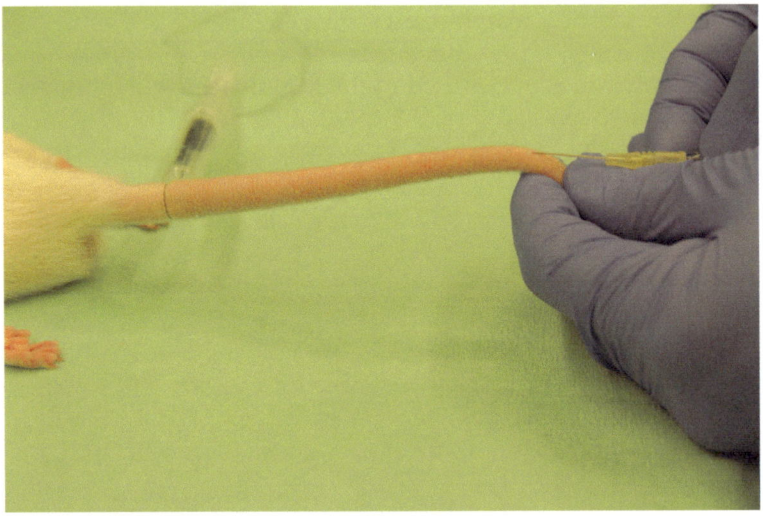

Photo 3. The vein is dilated with the aim of a tourniquet and an over-the-needle catheter is inserted.

A surgical protocol, modifying the above technique, has been developed by Inoue et al.[10] for selective infusion in liver. A 0.5mm silicon tube was advanced to the inferior vena cava (IVC) through the right common iliac vein for rapid injection of the solution, whereas the suprahepatic and the infrahepatic IVC were clamped. At the end of the procedure, clamps were removed and gently pressure was applied in the puncture point to minimize hemorrhage. Sawyer's group[67] verified Inoue results and calculated the optimum volume of DNA solution per g of liver to be 0.6 ml and the flow rate at 15 ml of solution per min per g liver for optimal hydrodynamic gene delivery to the liver.

A different approach for infusion in liver is to insert a 26g elastic cannula at the middle of the hepatic artery, through the gastroduodenal artery, and the celiac artery, the portal vein and the inferior vena cava are temporally clamped. Injection of the solution in the arterial vasculature is followed, fifteen second later, by clamps removal and gastroduodenal artery ligation[11].

For hydrodynamic infusion in kidney, a 26-g catheter is inserted distally to clamped renal vein, at its origin, and blood flow is re-established immediately after the rapid infusion. Pressure should be applied for homeostasis and the abdomen is closed in standard technique [68,12].

Hydrodynamic gene delivery in hind limb muscle is carried out through an inserted 28g needle catheter in the saphenous vein in direction to the knee, while a placed tourniquet, at a proximal part, forces the solution into the small veins ending up in the muscle tissue cells for 1-2 min post-injection[13,14,15,16](photo 4). According to Danialou et al.[17] gene delivery in rat's muscle was applied by a 23 g butterfly needle positioned in the femoral artery, directed distally, and the obstruction of venous circulation was originated by a clamp on the femoral vein. Before the hydrodynamic infusion 2 mg of papaverin and 2 mg of histamine (1 mg/ml) in normal saline were injected and, 20 minutes following the infusion, the vascular clamp was removed.

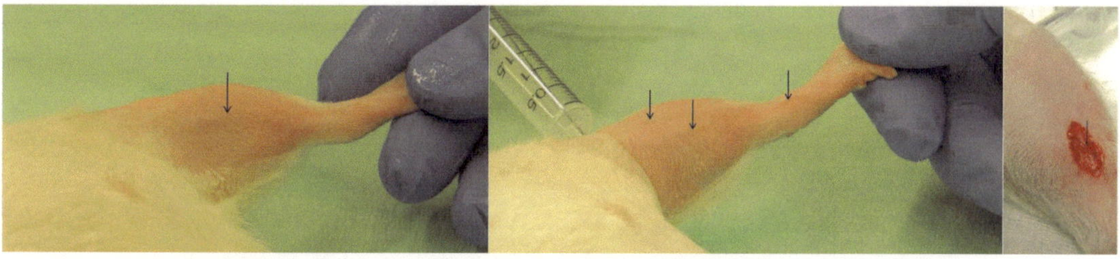

Photo 4. The saphenous vein prior to dilatation(left), after the placement of a tourniquet (middle), exposed after a small skin incision (right).

A protocol for hydrodynamic gene delivery in rat's heart was developed by de Carvalho and associates[18]. An incision at the fourth intercostal space exposed the heart and the aorta was identified, dissected from surrounding tissue and clamped. An injection, of 200 μl PBS solution containing naked DNA in the left ventricle over 3 seconds, was followed, twenty seconds later, by clamp removal. During the procedure an ultrasound probe was attached toward the heart with an intensity of 1 MHz and power of 2 W/cm^2

3.3 Rabbit

Eastman et al.[19], in 2002, applied hydrodynamic gene delivery in larger animal model than rodents, avoiding heart failure that could be encountered when applied due to the large amount of infused solution. In anesthetized rabbit, in dorsal recumbent position, a right paramedian incision at the neck revealed the right jugular for the placement of a 20-g angiographic sheath and the insertion a 0.025 inch hydrophilic guide wire into a hepatic vein, under fluoroscopic control. The wire was replaced by a 5F balloon occlusion catheter in selected hepatic vein and inflated. A small amount of contrast material confirmed the proper position and the medium containing the macromolecule was infused rapidly at a rate 5 ml/sec at an 80 ml total volume. At the end of infusion the catheter was withdrawn and the incision was closed in standard technique.

The researchers modified this technique to block venous outflow during infusion. A balloon catheter was inserted in the hepatic vein, as described above, while from a second skin incision, over the femoral vein, a 20-g angiographic sheath was positioned and a 0.025 hydrophilic guide wire was inserted, through the sheath, into the proximal inferior vena cava (IVC) under fluoroscopic guidance (photo 5). A 7F introducer sheath substituted the angiographic sheath, reaching the distal IVC and enabling the introduction of a 5F XXL angioplasty catheter to the intrahepatic portion of the IVC to inhibit all venous outflow. After the infusion all catheters were removed.

Subsequently, a whole organ isolation protocol was demonstrated with the primary aid of a balloon catheter positioned, as already described through the jugular vein, above the upper hepatic veins. A second balloon catheter was inserted through the femoral vein between the lower hepatic veins and the renal veins, while, a third 4F pediatric pigtail infusion catheter was advanced from the controlateral femoral vein to a point between the two occlusion catheters. Vasovagal response was diminished by glycopyrrolate injection (0.01 mg/Kg) followed by the inflation of the catheters, and injection of a radiopaque solution, prior to the gene delivery, demonstrated the absence of a leakage. At the end of the surgical procedure all catheters were removed and bleeding was controlled.

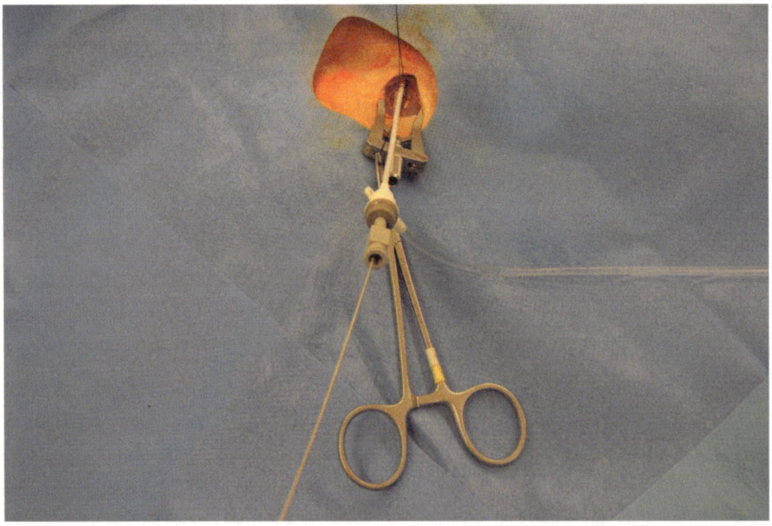

Photo 5. The sheath is being inserted in the femoral vein of a rabbit.

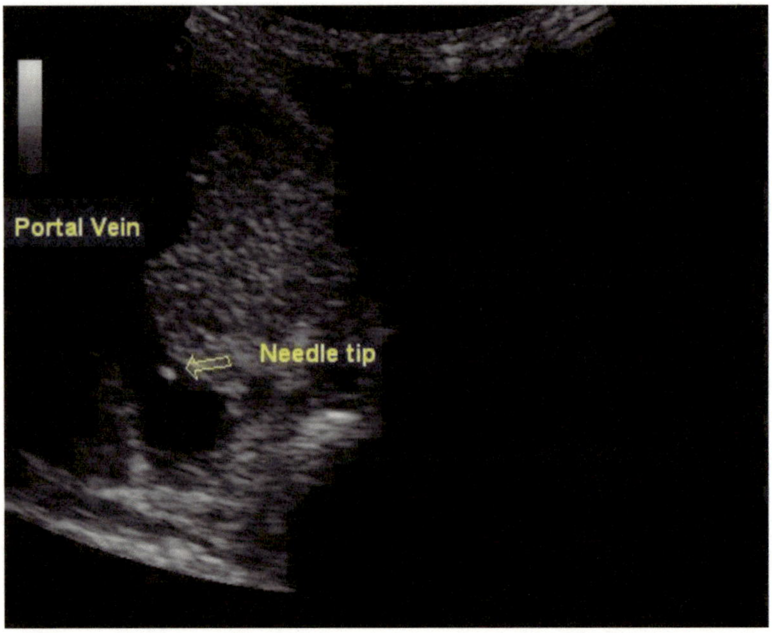

Photo 6. Instertion of a needle catheter in the portal vein, under ultrasound guidance.

The same group revealed the possibility to infuse percutaneously in the portal system, with a 20-g needle under ultrasound guidance, the gene solution (photo 6). A small amount of contrast material was firstly delivered to confirm the correct placement of the needle in order a 0.025 glide wire to be advanced through the needle at liver parenchyma. An angiographic sheath was, then, placed over the wire and the position was verified again fluoroscopically. After the infusion, the catheter was removed and hemostasis was achieved with manual pressure at the insertion point.

3.4 Swine

Swine is the most popular large animal model for hydrodynamic gene delivery among the researcher groups the past years. Yoshino and associates[20] have developed a surgical protocol based on Eastman's research. Upon the insertion of an occlusion balloon catheter, through the right external jugular vein, into the left hepatic vein under fluoroscopic guidance, the hepatic artery and the portal vein were identified and clamped after a midline incision. The injection of 200 ml of saline, to wash out the liver blood through a catheter placed proximal to the clamp in the portal vein, was followed by the occlusion of the hepatic veins. The occlusion catheter was removed thirty seconds after the application of the hydrodynamic delivery, along with the clamps at the portal and hepatic veins, and the hepatic artery. The midline incision was closed in standard technique.

A modification of the technique above was introduced by Suda et al.[68] and Kamimura et al.[21]. Through an 18g catheter placed in the jugular vein, a 0.035 guide wire was inserted into the inferior vena cava (IVC) under fluoroscopic guidance. A 9F balloon catheter was advanced into the right lateral liver lobe over the guide wire, while two occlusions balloons 8F were inserted, one from the femoral vein into the IVC and the other one from the superior mesenteric vein into the portal trunk, to block leakage from the injected solution. Suda's group[68], also, demonstrated a successful gene delivery in kidney with the aid of the balloon catheter inserted to the right renal vein, instead of the right liver lobe, and the IVC occlusion balloon, using the technique as already described.

Habib group[22,23] developed a minimal invasive surgical protocol to deliver gene in liver lobe's segment. After a right paramedian incision at the neck, the right external jugular vein was identified and isolated from the surrounding tissues and an introducer sheath was positioned in place. A hydrophilic catheter 5F was advanced, over a 0.035 guide wire previously positioned under fluoroscopy to the right hepatic vein, at the periphery of the right lateral liver lobe. A custom made 7F rigid balloon catheter, with multiple holes at the tip (photo 7), replaced the hydrophilic catheter remained in place for 10 minutes, after the infusion, to verify fluoroscopically the occlusion of the hepatic vein (photo 8). At the end of the procedure, the catheter was deflated, removed from the external jugular vein and a 3-0 absorbable ligature was used proximal to the puncture site to control bleeding.

Danialou et al.[76], described a surgical protocol for gene delivery to hind limb muscle introducing an 8F catheter in the femoral artery directing distally, while a tourniquet was placed around the limb proximal to the catheter. 10 mg of papaverine was delivered in the arterial line over 30 sec, while femoral vein had been clamped, followed, 5 min later, by the injection of 10 mg of papaverine mixed with 10 mg of histamine. The gradual removal of the tourniquet, within 10 min from the gene infusion, was followed by catheter and clamp removal. The effects of histamine, if present, were counteracted by 50 mg of diphenhydramine and dextran volume expander. A different approach, for hydrodynamic delivery in muscle, has been described by Kamimura and associates[24]. Over a 0.035 guide wire, positioned to the femoral vein under fluoroscopic guidance through an 18g catheter in the jugular vein, an 8F balloon catheter is positioned. The isolated hind limb, by a rubber band proximally sited, remained in place until the infused balloon was slowly deflated 20 min after the infusion.

For hydrodynamic gene transfer to heart tissue, Alino et al.[25] placed a 7F Swan-Ganz catheter and a 6 F multipurpose catheter at the coronary sinus under fluoroscopic guidance, through two 7F sheaths in the right and the left jugular veins, respectively. The coronary sinus vein was sealed by the Swan-Ganz catheter while the plasmid was injected through the multipurpose catheter and three minutes later the catheters were removed.

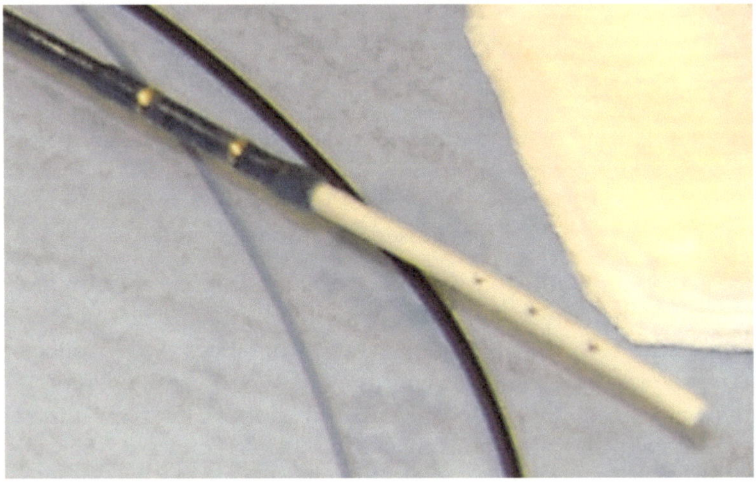

Photo 7. The custom made rigid-catheter with multiple holes at the side and a single at the center, enabling the infusion of large volumes of solution in small period of time without leakage.

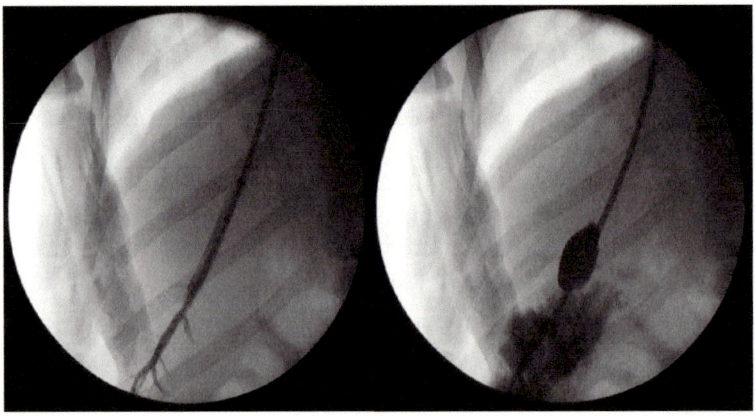

Photo 8. Placement of the catheter under pluoroscopic assistance (left), and the infused liver lobe with the ballon inflated without leakage of the gene solution and a small amount of contrast material (right).

3.5 Other animals - Human

3.5.1 Canine

Canine, as an animal model, is limited, mainly, in muscle gene delivery. After sedation, a 20 g catheter is inserted in the great saphenous vein while a cuff has been positioned just above the stifle and inflated to 300 mm Hg pressure, to block blood flow. Before infusing the gene solution, 25ml of natural saline containing 4.2 mg of papaverine and 150 units of heparin are delivered in 10sec, and five minutes later the solution containing the pDNA is injected at a rate of 2ml/sec[72,75,26].

3.5.2 Primates

In nonhuman primates, Brunetti-Pierri's group[27,28] has established several different surgical protocols for gene delivery in liver. Saline solution (200ml) was firstly infused, followed by the vector injection in the portal vein, while its distal part and the hepatic artery had been occluded by vessel loops. The loops were retained in place for 30 minutes and the portal vein at the injection point was sutured by monofilament suture. In the second surgical protocol, the main strategy was to occlude not only the blood inflow but the outflow, as well. The surgical procedure described above was repeated and two umbilical tapes were passed around the infrahepatic IVC and suprahepatic IVC. 20ml/kg of normal saline were intravenously infused to prevent hypotension during total hepatic occlusion. Saline was again infused, at a volume of 12 to 13% of the total blood volume, through the portal vein catheter followed by the rapid occlusion of the infrahepatic and suprahepatic IVC and the hydrodynamic injection. In case of severe hypotension (mean arterial pressure less than 60 mm Hg) phenylephrine, at a dose of 40 mg in 250 ml of saline, was administrated. In the third protocol, a catheter was placed in the IVC to retrieve the unabsorbed vector modifying the previous protocol, while in the next protocol the retrieved vector was reinjected to the portal vein

A minimal invasive protocol was developed by the same research group. Two sheaths, 4F and 11 F, were positioned in the right and the left femoral vein, respectively, along with a 4F sheath in the left femoral artery. From the right femoral sheath, a custom made occlusion balloon-catheter was advanced in the IVC right arterial junction and deflated 7.5 to 15 min after the infusion from the catheter mounted in the hepatic artery, through the left femoral artery. Mild hypotension resolved with 20 ml/kg saline and phenylephrine to effect[29].

In nonhuman primates, hydrodynamic delivery in muscle has also been well established. Under general anesthesia, after the placement of a 20g intravenous catheter into the distal cephalic vein, for the forearm, or into the small saphenous vein, for the hind limb, a tourniquet, above the elbow or the knee, is inflated to a pressure of 450 mmHg. The gene solution is rapidly infusion at a rate of 2 ml/sec and 2 minutes post-injection the tourniquet is deflated[55,72,73,30,31,32].

3.5.3 Human

There is only one publication about hydrodynamic gene delivery in humans, by Khorsandi et al.[82]. A modified balloon catheter was introduced in the femoral vein, after the infusion of local anesthetic at the puncture site, and reached the middle hepatic vein, under fluoroscopy guidance. Before injecting the gene solution, the correct placement of the catheter and the absence of back flow were verified by injecting contrast material. Upon the completion of the infusion, the catheter remained in place for 10 minutes and then removed; the patients were discharged from hospital the same day.

4. Conclusion and future perspectives

Hydrodynamic delivery of macromolecules, membrane-impermeable, to cytoplasm is a physical method that gains popularity over the last years. The simplicity of the procedures as already described, the lack of sophisticated equipment and the produced results, lead the hydrodynamic delivery to become a clinically feasible procedure in the near future. There is already one clinical trial on humans and we believe that the refinement of the procedure and the delivery in tissues other than liver or muscle, with minimal tissue damage and maximal delivery, will aim to new drug discovery and potential treatment of several human diseases.

5. Acknoledgements

We thank P. Moustarda, C. Dimitriou, E Balafa and N. Koutsogiorgo for their contribution to the chapter.

6. References

[1] Liu D, Knapp JE. Hydrodynamics-based gene delivery. Curr Opin Mol Ther. 2001 Apr;3(2):192-7. [PubMed: 11338933]
[2] Hagstrom JE. Plasmid-based gene delivery to target tissues in vivo: the intravascular approach. Curr Opin Mol Ther. 2003 Aug;5(4):338-44. [PubMed: 14513675]
[3] Hodges BL, Scheule RK. Hydrodynamic delivery of DNA. Expert Opin Biol Ther. 2003 Sep;3(6):911-8. [PubMed: 12943450]
[4] Kobayashi N, Nishikawa M, Takakura Y. The hydrodynamics-based procedure for controlling the pharmacokinetics of gene medicines at whole body, organ and cellular levels. Adv Drug Deliv Rev. 2005 Apr 5;57(5):713-31. [PubMed: 1575775]
[5] Al-Dosari MS, Knapp JE, Liu D. Hydrodynamic delivery. Adv Genet. 2005;54:65-82. [PubMed: 16096008]
[6] Gao X, Kim KS, Liu D. Nonviral gene delivery: what we know and what is next. AAPS J. 2007 Mar 23;9(1):E92-104. [PubMed: 17408239]
[7] Herweijer H, Wolff JA. Gene therapy progress and prospects: hydrodynamic gene delivery. Gene Ther. 2007 Jan;14(2):99-107. [PubMed: 17167496]
[8] Suda T, Liu D.Hydrodynamic gene delivery: its principles and applications. Mol Ther. 2007 Dec;15(12):2063-9. [PubMed: 17912237]
[9] Lewis DL, Wolff JA. Systemic siRNA delivery via hydrodynamic intravascular injection. Adv Drug Deliv Rev. 2007 Mar 30;59(2-3):115-23. [PubMed: 17442446]
[10] Bonamassa B, Hai L, Liu D.Hydrodynamic gene delivery and its applications in pharmaceutical research. Pharm Res. 2011 Apr;28(4):694-701. [PubMed: 21191634]
[11] Zhang G, Gao X, Song YK, Vollmer R, Stolz DB, Gasiorowski JZ, Dean DA, Liu D. Hydroporation as the mechanism of hydrodynamic delivery. Gene Ther. 2004 Apr;11(8):675-82.
[12] Kobayashi N, Kuramoto T, Yamaoka K, Hashida M, Takakura Y.Hepatic uptake and gene expression mechanisms following intravenous administration of plasmid DNA by conventional and hydrodynamics-based procedures. J Pharmacol Exp Ther. 2001 Jun;297(3):853-60.
[13] Lecocq M, Andrianaivo F, Warnier MT, Wattiaux-De Coninck S, Wattiaux R, Jadot M. Uptake by mouse liver and intracellular fate of plasmid DNA after a rapid tail vein injection of a small or a large volume. J Gene Med. 2003 Feb;5(2):142-56.
[14] Liu F, Song Y, Liu D. Hydrodynamics-based transfection in animals by systemic administration of plasmid DNA. Gene Ther. 1999 Jul;6(7):1258-66. [PubMed: 10455434]

[15] Zhang G, Budker V, Wolff JA. High levels of foreign gene expression in hepatocytes after tail vein injections of naked plasmid DNA. Hum Gene Ther. 1999 Jul 1;10(10):1735-7. [PubMed: 10428218]
[16] Jiang J, Yamato E, Miyazaki J. Intravenous delivery of naked plasmid DNA for in vivo cytokine expression. Biochem Biophys Res Commun. 2001 Dec 21;289(5):1088-92. [PubMed: 11741303]
[17] Kitajima M, Tsuyama Y, Miyano-Kurosaki N, Takaku H. Anti-tumor effect of intravenous TNFalpha gene delivery naked plasmid DNA using a hydrodynamics-based procedure. Nucleosides Nucleotides Nucleic Acids. 2005;24(5-7):647-50. [PubMed: 16248005]
[18] Arad U, Zeira E, El-Latif MA, Mukherjee S, Mitchell L, Pappo O, Galun E, Oppenheim A. Liver-targeted gene therapy by SV40-based vectors using the hydrodynamic injection method. Hum Gene Ther. 2005 Mar;16(3):361-71. [PubMed: 15812231]
[19] Al-Dosari M, Zhang G, Knapp JE, Liu D. Direct assessment of promoter activity of human cytochrome p450 genes using optimized transfection in vitro and in vivo. Biosci Rep. 2006 Jun;26(3):217-29. [PubMed: 16850252]
[20] Sebestyén MG, Budker VG, Budker T, Subbotin VM, Zhang G, Monahan SD, Lewis DL, Wong SC, Hagstrom JE, Wolff JA. Mechanism of plasmid delivery by hydrodynamic tail vein injection. I. Hepatocyte uptake of various molecules. J Gene Med. 2006 Jul;8(7):852-73. [PubMed: 16724360]
[21] Budker VG, Subbotin VM, Budker T, Sebestyén MG, Zhang G, Wolff JA. Mechanism of plasmid delivery by hydrodynamic tail vein injection. II.Morphological studies. J Gene Med. 2006 Jul;8(7):874-88. [PubMed: 16718734]
[22] Pergolizzi RG, Jin G, Chan D, Pierre L, Bussel J, Ferris B, Leopold PL, Crystal RG. Correction of a murine model of von Willebrand disease by gene transfer. Blood. 2006 Aug 1;108(3):862-9. [PubMed: 16638935]
[23] Bell JB, Podetz-Pedersen KM, Aronovich EL, Belur LR, McIvor RS, Hackett PB. Preferential delivery of the Sleeping Beauty transposon system to livers of mice by hydrodynamic injection. Nat Protoc. 2007;2(12):3153-65[PubMed: 18079715]
[24] Fukushima M, Hattori Y, Tsukada H, Koga K, Kajiwara E, Kawano K, Kobayashi T, Kamata K, Maitani Y. Adiponectin gene therapy of streptozotocin-induced diabetic mice using hydrodynamic injection. J Gene Med. 2007 Nov;9(11):976-85. [PubMed: 17868184]
[25] Dames P, Laner A, Maucksch C, Aneja MK, Rudolph C. Targeting of the glucocorticoid hormone receptor with plasmid DNA comprising glucocorticoid response elements improves nonviral gene transfer efficiency in the lungs of mice. J Gene Med. 2007 Sep;9(9):820-9. [PubMed: 17668918]
[26] Hibbitt OC, Harbottle RP, Waddington SN, Bursill CA, Coutelle C, Channon KM, Wade-Martins R. Delivery and long-term expression of a 135 kb LDLR genomic DNA locus in vivo by hydrodynamic tail vein injection. J Gene Med. 2007 Jun;9(6):488-97. [PubMed: 17471590]
[27] Aronovich EL, Bell JB, Belur LR, Gunther R, Koniar B, Erickson DC, Schachern PA, Matise I, McIvor RS, Whitley CB, Hackett PB. Prolonged expression of a lysosomal enzyme in mouse liver after Sleeping Beauty transposon-mediated gene delivery: implications for non-viral gene therapy of mucopolysaccharidoses. J Gene Med. 2007 May;9(5):403-15. [PubMed: 17407189]
[28] Rudolph C, Sieverling N, Schillinger U, Lesina E, Plank C, Thünemann AF, Schönberger H, Rosenecker J. Thyroid hormone (T3)-modification of polyethyleneglycol (PEG)-

polyethyleneimine (PEI) graft copolymers for improved gene delivery to hepatocytes. Biomaterials. 2007 Apr;28(10):1900-11. [PubMed: 17196251]

[29] Nguyen AT, Dow AC, Kupiec-Weglinski J, Busuttil RW, Lipshutz GS. Evaluation of gene promoters for liver expression by hydrodynamic gene transfer. J Surg Res. 2008 Jul;148(1):60-6. [PubMed: 18570932]

[30] Jeong YS, Kim EJ, Shim CK, Hou JH, Kim JM, Choi HG, Kim WK, Oh YK. Modulation of biodistribution and expression of plasmid DNA following mesenchymal progenitor cell-based delivery. J Drug Target. 2008 Jun;16(5):405-14. [PubMed: 18569285]

[31] Chilukuri N, Duysen EG, Parikh K, Sun W, Doctor BP, Lockridge O, Saxena A. Adenovirus-mediated gene transfer of human butyrylcholinesterase results in persistent high-level transgene expression in vivo. Chem Biol Interact. 2008 Sep 25;175(1-3):327-31. [PubMed: 18499092]

[32] Li W, Ma N, Ong LL, Kaminski A, Skrabal C, Ugurlucan M, Lorenz P, Gatzen HH, Lützow K, Lendlein A, Pützer BM, Li RK, Steinhoff G. Enhanced thoracic gene delivery by magnetic nanobead-mediated vector. J Gene Med. 2008 Aug;10(8):897-909. [PubMed: 18481827]

[33] Wooddell CI, Reppen T, Wolff JA, Herweijer H. Sustained liver-specific transgene expression from the albumin promoter in mice following hydrodynamic plasmid DNA delivery. J Gene Med. 2008 May;10(5):551-63. [PubMed: 18330848]

[34] Chen Q, Khoury M, Chen J. Expression of human cytokines dramatically improves reconstitution of specific human-blood lineage cells in humanized mice. Proc Natl Acad Sci U S A. 2009 Dec 22;106(51):21783-8. [PubMed: 19966223]

[35] Woodard LE, Keravala A, Jung WE, Wapinski OL, Yang Q, Felsher DW, Calos MP. Impact of hydrodynamic injection and phiC31 integrase on tumor latency in a mouse model of MYC-induced hepatocellular carcinoma. PLoS One. 2010 Jun 29;5(6):e11367. [PubMed: 20614008]

[36] Belcher JD, Vineyard JV, Bruzzone CM, Chen C, Beckman JD, Nguyen J, Steer CJ,Vercellotti GM.Heme oxygenase-1 gene delivery by Sleeping Beauty inhibits vascular stasis in a murine model of sickle cell disease. J Mol Med (Berl). 2010 Jul;88(7):665-75. [PubMed: 20306336]

[37] Chen ZY, Liang K, Qiu RX. Targeted gene delivery in tumor xenografts by the combination of ultrasound-targeted microbubble destruction and polyethylenimine to inhibit surviving gene expression and induce apoptosis. J Exp Clin Cancer Res. 2010 Nov 23;29:152. [PubMed: 21092274]

[38] Keng VW, Tschida BR, Bell JB, Largaespada DA. Modeling hepatitis B virus X-induced hepatocellular carcinoma in mice with the Sleeping Beauty transposon system. Hepatology. 2011 Mar;53(3):781-90. [PubMed: 21374658]

[39] Cheng L, Wang J, Li X, Xing Q, Du P, Su L, Wang S. Interleukin-6 induces Gr-1+CD11b+ myeloid cells to suppress CD8+ T cell-mediated liver injury in mice. PLoS One. 2011 Mar 4;6(3):e17631. [PubMed: 21394214]

[40] Miyakawa N, Nishikawa M, Takahashi Y, Ando M, Misaka M, Watanabe Y, Takakura Y. Prolonged circulation half-life of interferon γ activity by gene delivery of interferon γ-serum albumin fusion protein in mice. J Pharm Sci. 2011 Jun;100(6):2350-7. [PubMed: 21246562]

[41] Schüttrumpf J, Milanov P, Abriss D, Roth S, Tonn T, Seifried E. Transgene loss and changes in the promoter methylation status as determinants for expression duration in nonviral gene transfer for factor IX. Hum Gene Ther. 2011 Jan;22(1):101-6. [PubMed: 20677911]

[42] Kim HS, Kim JC, Lee YK, Kim JS, Park YS. Hepatic control elements promote long-term expression of human coagulation factor IX gene in hydrodynamically transfected mice. J Gene Med. 2011 Jul;13(7-8):365-72. [PubMed: 21710610]
[43] Shahaf G, Moser H, Ozeri E, Mizrahi M, Abecassis A, Lewis EC. α-1-Antitrypsin Gene Delivery Reduces Inflammation, Increases T-Regulatory Cell Population Size and Prevents Islet Allograft Rejection. Mol Med. 2011;17(9-10):1000-11. [PubMed: 21670848]
[44] Doherty JE, Huye L, Yusa K, Zhou L, Craig N, Wilson M. Hyperactive piggyBac gene transfer in human cells and in vivo. Hum Gene Ther. 2011 Oct 12. [PubMed: 21992617]
[45] Bu X, Zhou Y, Zhang H, Qiu W, Chen L, Cao H, Fang L, Wen P, Tan R, Yang J. Systemic administration of naked plasmid encoding HGF attenuates puromycin aminonucleoside-induced damage of murine glomerular podocytes. Am J Physiol Renal Physiol. 2011 Oct;301(4):F784-92. [PubMed: 21775482]
[46] Abedini F, Ismail M, Hosseinkhani H, Ibrahim TA, Omar AR, Chong PP, Bejo MH, Domb AJ. Effects of CXCR4 siRNA/dextran-spermine nanoparticles on CXCR4 expression and serum LDH levels in a mouse model of colorectal cancer metastasis to the liver. Cancer Manag Res. 2011;3:301-9. [PubMed: 21931504]
[47] Tripathi SK, Goyal R, Ansari KM, Ravi Ram K, Shukla Y, Chowdhuri DK, Gupta KC. Polyglutamic acid-based nanocomposites as efficient non-viral gene carriers in vitro and in vivo. Eur J Pharm Biopharm. 2011 Nov;79(3):473-84. [PubMed: 21820510]
[48] Shi B, Keough E, Matter A, Leander K, Young S, Carlini E, Sachs AB, Tao W, Abrams M, Howell B, Sepp-Lorenzino L. Biodistribution of small interfering RNA at the organ and cellular levels after lipid nanoparticle-mediated delivery. J Histochem Cytochem. 2011 Aug;59(8):727-40. [PubMed: 21804077]
[49] Hibbitt O, Wade-Martins R. High capacity extrachromosomal gene expression vectors. Methods Mol Biol. 2011;738:19-40. [PubMed: 21431717]
[50] Keravala A, Chavez CL, Hu G, Woodard LE, Monahan PE, Calos MP. Long-term phenotypic correction in factor IX knockout mice by using phiC31 integrase-mediated gene therapy. Gene Ther. 2011 Aug;18(8):842-8. [PubMed: 21412285]
[51] Parikh K, Duysen EG, Snow B, Jensen NS, Manne V, Lockridge O, Chilukuri N. Gene-delivered butyrylcholinesterase is prophylactic against the toxicity of chemical warfare nerve agents and organophosphorus compounds. J Pharmacol Exp Ther. 2011 Apr;337(1):92-101. [PubMed: 21205915]
[52] Fernandez CA, Baumhover NJ, Duskey JT, Khargharia S, Kizzire K, Ericson MD, Rice KG. Metabolically stabilized long-circulating PEGylated polyacridine peptide polyplexes mediate hydrodynamically stimulated gene expression in liver. Gene Ther. 2011 Jan;18(1):23-37. [PubMed: 20720577]
[53] Cao M, Khan JA, Kang BY, Mehta JL, Hermonat PL. Dual AAV/IL-10 Plus STAT3 Anti-Inflammatory Gene Delivery Lowers Atherosclerosis in LDLR KO Mice, but without Increased Benefit. Int J Vasc Med. 2012;2012:524235. [PubMed: 21915378]
[54] Waynforth HB, Flecknell P. *Specific surgical operations. Experimental and Surgical Technique in the Rat. 1992 Academic Press Ltd, London. 2nd ed, pp 44–48.*
[55] Wooddell CI, Subbotin VM, Sebestyén MG, Griffin JB, Zhang G, Schleef M, Braun S, Huss T, Wolff JA. Muscle damage after delivery of naked plasmid DNA into skeletal muscles is batch dependent. Hum Gene Ther. 2011 Feb;22(2):225-35. [PubMed: 20942645]

[56] Itaka K, Osada K, Morii K, Kim P, Yun SH, Kataoka K. Polyplex nanomicelle promotes hydrodynamic gene introduction to skeletal muscle. J Control Release. 2010 Apr 2;143(1):112-9. [PubMed: 20043959]

[57] Maruyama H, Higuchi N, Nishikawa Y, Kameda S, Iino N, Kazama JJ, Takahashi N, Sugawa M, Hanawa H, Tada N, Miyazaki J, Gejyo F. High-level expression of naked DNA delivered to rat liver via tail vein injection. J Gene Med. 2002 May-Jun;4(3):333-41. [PubMed: 12112650]

[58] Higuchi N, Maruyama H, Kuroda T, Kameda S, Iino N, Kawachi H, Nishikawa Y, Hanaw, H, Tahara H, Miyazaki J, Gejyo F. Hydrodynamics-based delivery of the viral interleukin-10 gene suppresses experimental crescentic glomerulonephritis in Wistar-Kyoto rats. Gene Ther. 2003 Aug;10(16):1297-310. [PubMed: 12883526]

[59] Liu H, Hanawa H, Yoshida T, Elnaggar R, Hayashi M, Watanabe R, Toba K, Yoshida K, Chang H, Okura Y, Kato K, Kodama M, Maruyama H, Miyazaki J, Nakazawa M, Aizawa Y. Effect of hydrodynamics-based gene delivery of plasmid DNA encoding interleukin-1 receptor antagonist-Ig for treatment of rat autoimmune myocarditis: possible mechanism for lymphocytes and noncardiac cells. Circulation. 2005 Apr 5;111(13):1593-600. [PubMed: 15795329]

[60] Elnaggar R, Hanawa H, Liu H, Yoshida T, Hayashi M, Watanabe R, Abe S, Toba K, Yoshida K, Chang H, Minagawa S, Okura Y, Kato K, Kodama M, Maruyama H, Miyazaki J, Aizawa Y. The effect of hydrodynamics-based delivery of an IL-13-Ig fusion gene for experimental autoimmune myocarditis in rats and its possible mechanism. Eur J Immunol. 2005 Jun;35(6):1995-2005. [PubMed: 15902684]

[61] Abe S, Hanawa H, Hayashi M, Yoshida T, Komura S, Watanabe R, Lie H, Chang H, Kato K, Kodama M, Maruyama H, Nakazawa M, Miyazaki J, Aizawa Y. Prevention of experimental autoimmune myocarditis by hydrodynamics-based naked plasmid DNA encoding CTLA4-Ig gene delivery. J Card Fail. 2005 Sep;11(7):557-64. [PubMed: 16198253]

[62] Toietta G, Mane VP, Norona WS, Finegold MJ, Ng P, McDonagh AF, Beaudet AL, Lee B. Lifelong elimination of hyperbilirubinemia in the Gunn rat with a single injection of helper-dependent adenoviral vector. Proc Natl Acad Sci USA. 2005 Mar 15;102(11):3930-5. [PubMed: 15753292]

[63] Tada M, Hatano E, Taura K, Nitta T, Koizumi N, Ikai I, Shimahara Y. High volume hydrodynamic injection of plasmid DNA via the hepatic artery results in a high level of gene expression in rat hepatocellular carcinoma induced by diethylnitrosamine. J Gene Med. 2006 Aug;8(8):1018-26. [PubMed: 16779866]

[64] Takekubo M, Tsuchida M, Haga M, Saitoh M, Hanawa H, Maruyama H, Miyazaki J, Hayashi J. Hydrodynamics-based delivery of plasmid DNA encoding CTLA4-Ig prolonged cardiac allograft survival in rats. J Gene Med. 2008 Mar;10(3):290-7. [PubMed: 18074399]

[65] Dimmock D, Brunetti-Pierri N, Palmer DJ, Beaudet AL, Ng P. Correction of hyperbilirubinemia in gunn rats using clinically relevant low doses of helper-dependent adenoviral vectors. Hum Gene Ther. 2011 Apr;22(4):483-8. [PubMed: 20973621]

[66] Salehi S, Eckley L, Sawyer GJ, Zhang X, Dong X, Freund JN, Fabre JW. Intestinal lactase as an autologous beta-galactosidase reporter gene for in vivo gene expression studies. Hum Gene Ther. 2009 Jan;20(1):21-30. [PubMed: 20377368]

[67] Sawyer GJ, Grehan A, Dong X, Whitehorne M, Seddon M, Shah AM, Zhang X, Salehi S, Fabre JW. Low-volume hydrodynamic gene delivery to the rat liver via an isolated

segment of the inferior vena cava: efficiency, cardiovascular response and intrahepatic vascular dynamics. J Gene Med. 2008 May;10(5):540-50. [PubMed: 18307279]

[68] Suda T, Suda K, Liu D. Computer-assisted hydrodynamic gene delivery. Mol Ther. 2008 Jun;16(6):1098-104. [PubMed: 18398428]

[69] Inoue S, Hakamata Y, Kaneko M, Kobayashi E. Gene therapy for organ grafts using rapid injection of naked DNA: application to the rat liver. Transplantation. 2004 Apr 15;77(7):997-1003. [PubMed: 15087760]

[70] Tada M, Hatano E, Taura K, Nitta T, Koizumi N, Ikai I, Shimahara Y. High volume hydrodynamic injection of plasmid DNA via the hepatic artery results in a high level of gene expression in rat hepatocellular carcinoma induced by diethylnitrosamine. J Gene Med. 2006 Aug;8(8):1018-26. [PubMed: 16779866]

[71] Xing Y, Pua EC, Lu X, Zhong P. Low-amplitude ultrasound enhances hydrodynamic-based gene delivery to rat kidney. Biochem Biophys Res Commun. 2009 Aug 14;386(1):217-22. [PubMed: 19523454]

[72] Hagstrom JE, Hegge J, Zhang G, Noble M, Budker V, Lewis DL, Herweijer H, Wolff JA. A facile nonviral method for delivering genes and siRNAs to skeletal muscle of mammalian limbs. Mol Ther. 2004 Aug;10(2):386-98. [PubMed: 15294185]

[73] Sebestyén MG, Hegge JO, Noble MA, Lewis DL, Herweijer H, Wolff JA. Progress toward a nonviral gene therapy protocol for the treatment of anemia. Hum Gene Ther. 2007 Mar;18(3):269-85. [PubMed: 17376007]

[74] Shi Q, Wang H, Tran C, Qiu X, Winnik FM, Zhang X, Dai K, Benderdour M, Fernandes JC. Hydrodynamic delivery of chitosan-folate-DNA nanoparticles in rats with adjuvant-induced arthritis. J Biomed Biotechnol. 2011;2011:148763. [PubMed: 21274258]

[75] Su LT, Gopal K, Wang Z, Yin X, Nelson A, Kozyak BW, Burkman JM, Mitchell MA, Low DW, Bridges CR, Stedman HH. Uniform scale-independent gene transfer to striated muscle after transvenular extravasation of vector. Circulation. 2005 Sep 20;112(12):1780-8. [PubMed: 16157771]

[76] Danialou G, Comtois AS, Matecki S, Nalbantoglu J, Karpati G, Gilbert R, Geoffroy P, Gilligan S, Tanguay JF, Petrof BJ. Optimization of regional intraarterial naked DNA-mediated transgene delivery to skeletal muscles in a large animal model. Mol Ther. 2005 Feb;11(2):257-66. [PubMed: 15668137]

[77] Pinto de Carvalho L, Takeshita D, Carillo BA, Garcia Lisbôa BC, Molina G, Beutel A, Yasumura EG, Takiya CM, Valero VB, Ribeiro de Campos R Jr, Dohmann HF, Han SW. Hydrodynamics- and ultrasound-based transfection of heart with naked plasmid DNA. Hum Gene Ther. 2007 Dec;18(12):1233-43. [PubMed: 18021018]

[78] Eastman SJ, Baskin KM, Hodges BL, Chu Q, Gates A, Dreusicke R, Anderson S, Scheule RK. Development of catheter-based procedures for transducing the isolated rabbit liver with plasmid DNA. Hum Gene Ther. 2002 Nov 20;13(17):2065-77. [PubMed: 12490001]

[79] Yoshino H, Hashizume K, Kobayashi E. Naked plasmid DNA transfer to the porcine liver using rapid injection with large volume. Gene Ther. 2006 Dec;13(24):1696-702. [PubMed: 16871229]

[80] Kamimura K, Suda T, Xu W, Zhang G, Liu D. Image-guided, lobe-specific hydrodynamic gene delivery to swine liver. Mol Ther. 2009 Mar;17(3):491-9. [PubMed: 19156134]

[81] Katsimpoulas M, Zacharoulis D, Rountas C, Dimitriou C, Mantziaras G, Kostomitsopoulos N, Habib N, Kostakis A. Minimal invasive technique for gene delivery in porcine liver lobe segment. J Invest Surg. 2011;24(1):13-7.

[82] Khorsandi SE, Bachellier P, Weber JC, Greget M, Jaeck D, Zacharoulis D, Rountas C, Helmy S, Helmy A, Al-Waracky M, Salama H, Jiao L, Nicholls J, Davies AJ, Levicar N, Jensen S, Habib N.Minimally invasive and selective hydrodynamic gene therapy of liver segments in the pig and human. Cancer Gene Ther. 2008 Apr;15(4):225-30.

[83] Kamimura K, Zhang G, Liu D. Image-guided, intravascular hydrodynamic gene delivery to skeletal muscle in pigs. Mol Ther. 2010 Jan;18(1):93-100.

[84] Aliño SF, José Herrero M, Bodi V, Noguera I, Mainar L, Dasí F, Sempere A, Sánchez M, Díaz A, Sabater L, Lledó S. Naked DNA delivery to whole pig cardiac tissue by coronary sinus retrograde injection employing non-invasive catheterization. J Gene Med. 2010 Nov;12(11):920-6.

[85] Qiao C, Li J, Zheng H, Bogan J, Li J, Yuan Z, Zhang C, Bogan D, Kornegay J, Xiao X. Hydrodynamic limb vein injection of adeno-associated virus serotype 8 vector carrying canine myostatin propeptide gene into normal dogs enhances muscle growth. Hum Gene Ther. 2009 Jan;20(1):1-10. [PubMed: 18828709]

[86] Brunetti-Pierri N, Ng T, Iannitti DA, Palmer DJ, Beaudet AL, Finegold MJ, Carey KD, Cioffi WG, Ng P. Improved hepatic transduction, reduced systemic vector dissemination, and long-term transgene expression by delivering helper-dependent adenoviral vectors into the surgically isolated liver of nonhuman primates. Hum Gene Ther. 2006 Apr;17(4):391-404. [PubMed: 12490001]

[87] Brunetti-Pierri N, Ng P. Progress towards the clinical application of helper-dependent adenoviral vectors for liver and lung gene therapy. Curr Opin Mol Ther. 2006 Oct;8(5):446-54. [PubMed: 17078387]

[88] Brunetti-Pierri N, Stapleton GE, Law M, Breinholt J, Palmer DJ, Zuo Y, Grove NC, Finegold MJ, Rice K, Beaudet AL, Mullins CE, Ng P. Efficient, long-term hepatic gene transfer using clinically relevant HDAd doses by balloon occlusion catheter delivery in nonhuman primates. Mol Ther. 2009 Feb;17(2):327-33. [PubMed: 19050700]

[89] Zhang G, Budker V, Williams P, Subbotin V, Wolff JA. Efficient expression of naked dna delivered intraarterially to limb muscles of nonhuman primates. Hum Gene Ther. 2001 Mar 1;12(4):427-38. [PubMed: 11242534]

[90] Vigen KK, Hegge JO, Zhang G, Mukherjee R, Braun S, Grist TM, Wolff JA. Magnetic resonance imaging-monitored plasmid DNA delivery in primate limb muscle. Hum Gene Ther. 2007 Mar;18(3):257-68. [PubMed: 17376006]

[91] Toromanoff A, Chérel Y, Guilbaud M, Penaud-Budloo M, Snyder RO, Haskins ME, Deschamps JY, Guigand L, Podevin G, Arruda VR, High KA, Stedman HH, Rolling F, Anegon I, Moullier P, Le Guiner C. Safety and efficacy of regional intravenous (r.i.) versus intramuscular (i.m.) delivery of rAAV1 and rAAV8 to nonhuman primate skeletal muscle. Mol Ther. 2008 Jul;16(7):1291-9. [PubMed: 18461055]

Part 4

Hormones and Receptors

13

Role of β-Arrestins in Endosomal Sorting of G Protein-Coupled Receptors

Rohit Malik[1] and Adriano Marchese[1,2,*]
[1]*Program in Molecular Biology*
[2]*Department of Molecular Pharmacology and Therapeutics, Loyola University Chicago*
USA

1. Introduction

G protein-coupled receptors (GPCRs) are members of a large gene family encoded by approximately 1000 members (Lefkowitz, 2007; Pierce et al., 2002). They are also known as seven transmembrane domain (7TM) receptors as they are characterized by a single polypeptide that has seven membrane spanning helical domains. Members of this family include receptors that bind to biogenic amines, chemokines, opioids, lipids, among many more. They mediate a wide variety of physiological processes ranging from neurotransmission, cardiovascular function, leukocyte chemotaxis and analgesia. GPCRs are important clinically as many have been implicated in many diseases and are targets for approximately 50% of all marketed medicines (Drews, 2000).

Upon binding to their cognate ligands GPCRs mediate down stream signaling via heterotrimeric GTP-binding proteins (G protein). G proteins are comprised of an α-subunit (Gα) and a tightly associated dimer of β and γ-subunits (Gβγ). In the inactive state Gα is bound to GDP and exists in an inactive conformation. Once the GPCR is activated by its cognate ligand, conformational changes in the receptor induce the exchange of GDP for GTP on Gα leading to its activation and dissociation from the βγ subunits. The activated Gα (Gα-GTP) and dissociated βγ subunits activate downstream effector molecules contributing to GPCR signaling. One effector molecule is adenylyl cyclase and upon its activation it leads to the production of cyclic AMP (cAMP), which in turn activates the protein kinase A (PKA), a serine/threonine kinase that phosphorylates many different substrates. Another effector molecule that is activated is phospholipase C, which catalyzes the hydrolysis of phosphatidyl 4,5 bisphosphate to produce inositol 1,4,5-trisphosphate and diacyclglycerol, which in turn leads to calcium release from intracellular stores and activation of protein kinase C, respectively. Recently, it has become apparent that GPCRs may also signal independent of the G protein, and this typically involves proteins known as β-arrestins (DeWire et al., 2007).

To ensure that signals are of the appropriate magnitude and duration GPCR signaling is tightly regulated. This is critical because perturbations in the regulatory processes that control GPCR signaling may contribute to the cause of human pathologies (Hernandez et al., 2003).

---

* Corresponding Author

Regulation of signaling is complex and involves multiple steps. It involves inactivation of the G protein and degradation of second messengers (Nelson et al., 2007; Perry et al., 2002). Regulation also involves a series of complex events that occur at the level of the receptor. Regulation may occur by second-messenger-dependent protein kinases PKA and PKC, which may also phosphorylate GPCRs, in addition to other downstream factors, when they become activated. Additional regulation occurs by another family of serine/threonine kinases known as G protein-coupled receptor kinases (GRKs), which phosphorylate activated receptors resulting in the binding of another class of proteins known as arrestins (Fig 1) (Krupnick and Benovic, 1998). Here we will focus on recent advances that have linked non-visual arrestins to a novel function by regulating the endosomal sorting machinery with a role in trafficking of GPCRs into the degradative pathway.

2. Traditional roles of β-arrestins

Mammalian arrestins comprise a family of four members that can be sub-divided into two groups: visual (arrestin-1 and arrestin-4) and non-visual arrestins (β-arrestin-1 and β-arrestin-2, also known as arrestin-2 and arrestin-3, respectively) (Gurevich and Gurevich, 2006a). Visual arrestins are restricted in distribution to the visual system. Arrestin-1 is found in high abundance in rod cells whereas arrestin-4 is found in cone cells. Non-visual arrestins are ubiquitously expressed and likely regulate the signaling of many GPCRs.

The classical known function of arrestins, in conjunction with GRKs, is to mediate GPCR desensitization, a phenomenon in which responsiveness to chronic stimulation is attenuated (Fig. 1). Arrestins were initially identified in the visual system as a 48-kDa protein capable of blocking light induced signaling of rhodopsin (Wilden et al., 1986). Subsequently, β-arrestins were identified and found to function in non-visual systems in an analogous manner to visual arrestins in that they blocked agonist-induced signaling of the β_2-adrenergic receptor (β_2AR) (Lohse et al., 1990). Arrestins are typically recruited to GPCRs that are activated by their cognate ligands and phosphorylated by GRKs (Krupnick and Benovic, 1998). Arrestin binding uncouples the G protein from the receptor via steric hindrance culminating in attenuated signaling (Lohse et al., 1990). Arrestins may also contribute to signal termination by promoting degradation of second messengers (Perry et al., 2002) (Nelson et al., 2007) and also by promoting GPCR internalization through clathrin-coated pits based on their ability to bind to components of the internalization machinery (Goodman et al., 1996; Laporte et al., 1999).

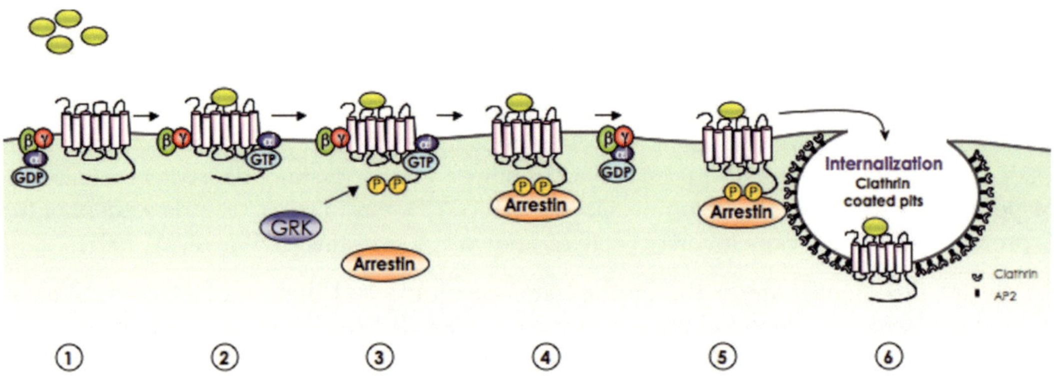

Fig. 1. Schematic depicting β-arrestin mediated desensitization and internalization of GPCRs.

1) Under basal conditions, the heterotrimeric GTP binding protein (G protein), which is comprised of an α-subunit (Gα) and a tightly associated dimer of β and γ-subunits (Gβγ), is bound to GDP and is in an inactive conformation. **2)** Upon binding to its cognate ligand (yellow oval), conformational changes in the GPCR induce the exchange of GDP for GTP on the α-subunit (Gα) leading to its activation and dissociation from the βγ subunits. The activated α-subunit (Gα-GTP) and βγ subunits activate downstream effector molecules contributing to GPCR signaling. **3)** Signaling is rapidly terminated in part by GPCR kinase (GRK) recruitment to the activated receptor, which phosphorylates the receptor on serine and/or threonine residues that are located within the carboxy-terminal tail and/or on the intracellular loops. **4)** Arrestins are rapidly recruited to the phosphorylated receptor, which upon binding uncouple the receptor from the associated Gα subunit via steric hindrance, contributing to signal termination and to the phenomenon known as desensitization. **5)** Arrestins subsequently interact with clathrin and AP2, two important components of the internalization machinery. **6)** This results in the recruitment of activated and phosphorylated receptors for internalization via clathrin coated pits.

The crystal structure of β-arrestins revealed that they consist of two distinct elongated domains (N and C domains) and a carboxy-terminal tail (C-tail) (Han et al., 2001; Hirsch et al., 1999; Milano et al., 2002). The N and C domains are independent folding units composed of anti-parallel beta sheets connected by a short linker or hinge sequence. Under basal conditions or in an inactive state, the N and C domains are compact and the C-tail is anchored to the N domain. Upon binding to phosphorylated GPCRs arrestins undergo significant conformational changes in which a rearrangement of the N and C domains occurs and the C-tail is released (Gurevich and Gurevich, 2006b). The C-tail of β-arrestins contains the binding sites for clathrin and AP2, two important components of the internalization machinery (Goodman et al., 1996; Laporte et al., 1999). The exposure of the C-tail enables clathrin and AP2 binding and subsequent recruitment of receptors to clathrin-coated pits for internalization. The arrestin C-tail also contains sites for post-translational modifications that also regulate the ability of arrestins to promote GPCR internalization. Nitrosylation of β-arrestin-2 on the terminal cysteine residue 409 increases its ability to interact with clathrin and facilitate β_2AR internalization (Ozawa et al., 2008). SUMOylation of β-arrestin-2 on lysine residue 400 has also been shown to have a role in β_2AR internalization, likely by facilitating interactions with AP2 (Wyatt et al., 2011). Detailed molecular insight regarding how these events are coordinated to properly execute β-arrestins roles in GPCR internalization remains to be determined.

3. Role of β-arrestins in GPCR recycling

Once removed from the plasma membrane by internalization receptors are delivered to early endosomes (Hanyaloglu and von Zastrow, 2008; Marchese et al., 2008). Once on endosomes, receptors may enter into the recycling pathway and return to the plasma membrane giving rise to functional resensitization of GPCR responsiveness. Upon agonist activation the β_2AR is rapidly phosphorylated and internalized onto endosomes before it is rapidly recycled to the plasma membrane (Krueger et al., 1997; Pippig et al., 1995; von Zastrow and Kobilka, 1992). Internalization onto endosomes may be a prerequisite for receptor resensitization of signaling. The acidic environment of the endosomal compartment may induce a conformational change in the receptor enabling an endosomal associated phosphatase to dephosphorylate the receptor (Krueger et al., 1997). Dephosphorylation of the receptor may facilitate its entry into the recycling pathway for delivery to the plasma membrane and functional resensitization.

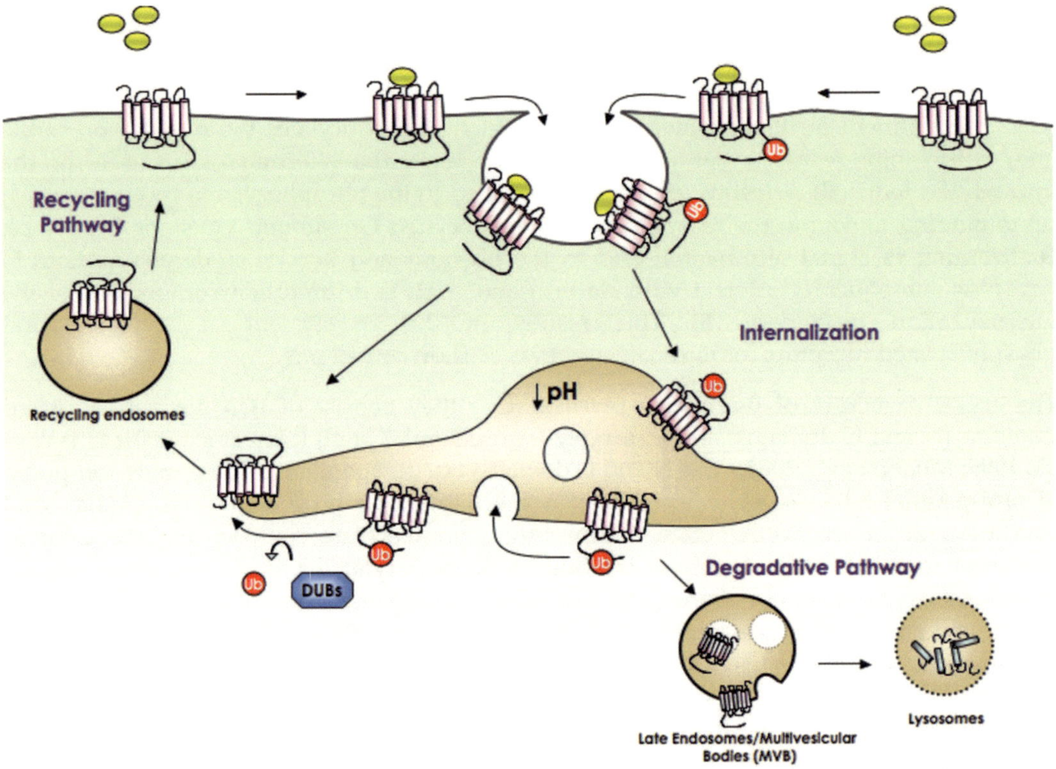

Fig. 2. Trafficking pathways for GPCRs.

Upon activation by its cognate ligand (oval), GPCRs are typically sequestered into specialized microdomains of the plasma membrane that are responsible for endocytosis. These areas of the plasma membrane pinch off, forming vesicles that eventually fuse with early endosomes, where the receptors are delivered. Some receptors may be modified with ubiquitin moieties at the plasma membrane, although the ubiquitin moiety is not required to promote receptor internalization. Once on early endosomes, GPCRs are sorted for either entering the recycling or degradative pathways. GPCRs that enter the degradative pathway are typically modified with ubiquitin, which serves as a signal for sorting into multivesicular bodies (MVBs). MVBs fuse with lysosomes where degradation of the receptor occurs. GPCRs that are not ubiquitinated may enter the recycling pathway and are returned to the plasma membrane via recycling endosomes, giving rise to functional resensitization of signaling. The ubiquitin moieties on some receptors may be removed by the action of deubiquitinating enzymes (DUBs), which may occur on endosomes, and redirect receptors targeted for the degradative pathway into the recycling pathway.

β-arrestins may also mediate receptor recycling at a post-internalization step by playing a direct role on endosomes. For many GPCRs, the interaction with β-arrestins is transient.

Once β-arrestins bind and desensitization and receptor internalization have initiated, the two proteins disassociate while the receptor internalizes onto endosomes (Oakley et al., 1999; Oakley et al., 2001; Oakley et al., 2000). Once on endosomes the receptor rapidly recycles to the plasma membrane. However, for a subset of GPCRs, not only does β-arrestin binding promote desensitization and internalization, β-arrestins may co-internalize and co-localize with the receptor on endosomes. The ability of β-arrestin to co-internalize with GPCRs is dependent upon their ability to form a stable complex with the GPCR. This slows down the recycling of the receptor, likely by preventing the receptor from efficiently entering into the recycling pathway. Receptors that form a stable interaction with β-arrestins tend to have multiple serine/threonine residue clusters near the end of their carboxy-terminal tail (Oakley et al., 1999; Oakley et al., 2001). Phosphorylation of these clusters promotes high affinity β-arrestin binding thereby allowing it to remain bound to the receptor while internalization onto endosomes occurs. The stability of the interaction and the trafficking pattern of β-arrestins also correlates with their ubiquitination status (Shenoy and Lefkowitz, 2003; Shenoy et al., 2009). Upon receptor activation, β-arrestins may also be subject to ubiquitination (Shenoy et al., 2001). β-arrestin-2 interacts with the E3 ubiquitin ligase Mdm2 and the deubiquitinating enzyme USP33 (Shenoy et al., 2001; Shenoy et al., 2009). Upon activation of GPCRs for which the interaction with β-arrestin-2 is transient, such as β_2AR, ubiquitination of β-arrestin-2 is rapid, but it is rapidly deubiquitinated by USP33. In contrast, upon activation of GPCRs for which the interaction with β-arrestin-2 is stable, such as the vasopressin V2 receptor, ubiquitination is also rapid, but it is not deubiquitinated by USP33, resulting in sustained ubiquitination. One possible explanation to account for this difference between receptors is that β-arrestin-2 may undergo differential conformational changes when bound to distinct receptors which may affect its ability to interact with USP33 (Shenoy et al., 2009). Therefore discrete conformational changes regulated by different GPCRs may control β-arrestin deubiquitination and thus their ability to co-traffic with receptors onto endosomes.

β-arrestins may also play a role in GPCR recycling independent of their role in promoting GPCR internalization. Although, β-arrestins are not required for internalization of the N-formyl peptide receptor (FPR), they are required for promoting FPR recycling. Agonist-induced internalization of FPR occurs in a β-arrestin-independent fashion, as FPR internalization is not altered in mouse embryonic fibroblasts (MEFs) isolated from β-arrestin-1 and β-arrestin-2 knockout out mice (Vines et al., 2003). However, recycling of the receptor is impaired in the β-arrestin knock-out MEFs and restored in cells in which β-arrestin-1 and β-arrestin-2 are re-expressed. Although not required for internalization, agonist activation promotes β-arrestin binding to FPR and its co-internalization with the receptor onto endosomes. Once on endosomes β-arrestins promote recycling through a mechanism that remains to be determined.

4. β-arrestins as adaptors for E3 ubiquitin ligases

In addition to entering the recycling pathway, GPCRs may be sorted into the degradative pathway (Fig. 2) for delivery to lysosomes where they are degraded (Hanyaloglu and von Zastrow, 2008; Marchese et al., 2008). The functional consequence of this is long-term attenuation of signaling giving rise to a phenomenon known as downregulation. β-arrestin may serve as an adaptor to recruit E3 ubiquitin ligases directly to activated receptors to mediate their ubiquitination. The ubiquitin moiety serves as a sorting signal on endosomes for targeting the receptor into the degradative pathway. Receptor ubiquitination occurs at the plasma membrane even though the attached ubiquitin serves as a sorting signal on endosomes. Receptor mutants lacking ubiquitination sites internalize normally but are defective in their ability to be sorted or targeted to lysosomes (Marchese and Benovic, 2001; Shenoy et al., 2001). A role for β-arrestins as an E3 ubiquitin ligase adaptor for receptor ubiquitination was first suggested in studies examining the regulation of β_2AR (Shenoy et al., 2001). Agonist-promoted ubiquitination of β_2AR is impaired in MEFs isolated from β-arrestin-2 knock-out mice, suggesting that β-arrestin-2 mediates ubiquitination of β_2AR. In contrast, β-arrestin-1 is not involved in β_2AR ubiquitination. β-arrestin-2 interacts with the HECT-domain E3 ubiquitin ligase Nedd4, which mediates ubiquitination of β_2AR (Shenoy et al., 2008). Although β-arrestin-2 interacts with Mdm2, it does not mediate ubiquitination of β_2AR (Shenoy et al., 2001). Depletion of Nedd4 by siRNA attenuates β_2AR ubiquitination and lysosomal targeting and the interaction between β_2AR and Nedd4 is dependent upon the presence of β-arrestin-2 (Shenoy et al., 2008). This is consistent for a role of β-arrestin-2 serving as an adaptor to recruit Nedd4 to β_2AR.

Alternatively, Nedd4 may be recruited to β_2AR independent of β-arrestin-2 involving the arrestin domain-containing protein ARRDC3 (Nabhan et al., 2010). ARRDs are a family of 6 mammalian proteins related to yeast proteins called arrestin-related trafficking (ART) adaptor proteins that were first characterized in *Saccharomyces cerevisiae* (Lin et al., 2008). Collectively these proteins have been referred to as alpha-arrestins to distinguish them from β-arrestins (Alvarez, 2008). Although sharing very little amino acid sequence identity with mammalian β-arrestins, bioinformatics modeling revealed that alpha-arrestins have an arrestin-fold consisting of N and C domains. This overall architecture may represent a conserved structural design found in seemingly distantly related proteins. The retromer component Vps26, although it does not share primary sequence identity with mammalian arrestins, has structurally similar folded N and C domains (Shi et al., 2006). One key distinguishing feature of ARRDs compared to β-arrestins or visual arrestins is that ARRDs have a long carboxy-terminal region harboring PY motifs (Lin et al., 2008). PY motifs are short stretches of amino acids typically found in the context of PP*x*Y and PPPY, where P is a proline residue, *x* is any amino acid and Y is a tyrosine residue (Einbond and Sudol, 1996). PY motifs are typically recognized by WW domains, which are domains of approximately 30 amino acids containing two highly conserved tryptophan residues (Macias et al., 2002). The Nedd4-like family of HECT-domain E3 ubiquitin ligases, comprised of 9 members in the human genome, is characterized by the presence of 4 tandemly linked WW domains (Ingham et al., 2004). Substrates that have PY motifs interact directly with their cognate E3s via the WW domains, however, many substrates do not have PY motifs and are believed to interact indirectly with these E3s through an adaptor protein that harbors a PY motif (Shearwin-Whyatt et al., 2006). Recently, it was reported that ARRDC3 interacts with β_2AR and serves as an adaptor for Nedd4-dependent ubiquitination of the receptor (Nabhan et al., 2010). Depletion of ARRDC3 attenuates agonist-induced ubiquitination and lysosomal targeting of β_2AR (Nabhan et al., 2010). ARRDC3 is predicted to bind directly to Nedd4 via its PY motif and to β_2AR. Mammalian arrestins do not contain PY motifs and it is unclear how Nedd4 interacts with β-arrestin-2. It remains to be determined how the actions of β-arrestin-2 and ARRDC3 are coordinated to mediate Nedd4-dependent β_2AR ubiquitination.

Interestingly, β-arrestin-1 interacts with another member of the Nedd4-like family of E3 ubiquitin ligases known as AIP4 (Bhandari et al., 2007). AIP4 mediates ubiquitination and endosomal sorting of the GPCR chemokine receptor C-X-C receptor 4 (CXCR4) (Marchese et al., 2003). As with all members of this family, AIP4 has a C2 domain, 4 tandemly-linked WW domains and the catalytic HECT domain. Unlike Nedd4, however, AIP4 has a proline-rich region that may bind to a subset of SH3 domains (Angers et al., 2004). The interaction between β-arrestin-1 and AIP4 is direct and it is mediated by AIP4 WW domains I and II, not III and IV (Bhandari et al., 2007). In addition to interacting with PY motifs, WW domains are also known to interact with phosphorylated serine or threonine residues in the presence of an adjacent proline residue (p(S/T)P) (Lu et al., 1999; Verdecia et al., 2000). Presently, it remains to be determined how the AIP4 WW-domains interact with β-arrestin-1. Interestingly, a non-canonical WW-mediated interaction involving AIP4 has been recently reported (Bhandari et al., 2009). The WW domains of AIP4 have been shown to interact with phosphorylated serine residues in the absence of nearby proline residues. Phosphorylation of serine residues within the C-tail of CXCR4 mediates a direct interaction with AIP4 via WW domains I and II (Bhandari et al., 2009). A receptor mutant in which C-tail serine residues 324 and 325 were changed to alanine residues (S324/5A) attenuated binding to AIP4 (Bhandari et al., 2009; Marchese and Benovic, 2001). Accordingly, the S324/5A CXCR4 receptor mutant shows defective ubiquitination and degradation (Bhandari et al., 2009). Agonist-promoted phosphorylation of these residues occurs at the plasma membrane leading to AIP4 recruitment to the plasma membrane where the receptor is ubiquitinated. Therefore for CXCR4 ubiquitination, an adaptor is not required because the Nedd4-like E3 AIP4 can interact directly with the receptor. Similar to the mechanism of the interaction between AIP4 and CXCR4, the WW domains of Nedd4 may also interact with its substrates via phosphorylated serine residues in the absence of nearby proline residues (Edwin et al., 2010). Therefore, phosphorylated serine residues in the absence of proline residues may represent a general recognition motif for WW domains. To our knowledge, Nedd4 and AIP4 are the only Nedd4-like E3 ubiquitin ligases assigned to the ubiquitination of mammalian GPCRs, but it is tempting to speculate that CXCR4 recognition by AIP4 may represent a general mechanism by which Nedd4-like E3 family members interact with and ubiquitinate GPCRs.

5. β-arrestins as adaptors for endosomal sorting

Although β-arrestin-1 interacts with AIP4 it is not involved in ubiquitination of CXCR4. Depletion of β-arrestin-1 by siRNA does not block CXCR4 ubiquitination and nor does it block CXCR4 internalization (Bhandari et al., 2007). However, β-arrestin-1 depletion blocks CXCR4 degradation by preventing its trafficking from early to late endosomes, suggesting a role for β-arrestin-1 at a sorting step on endosomes (Bhandari et al., 2007). AIP4 also mediates sorting of CXCR4 on endosomes, in addition to its role at the plasma membrane in mediating CXCR4 ubiquitination (Marchese et al., 2003). Both β-arrestin-1 and AIP4 co-localize with CXCR4 on early endosomes where they likely function to sort CXCR4 into the degradative pathway. Therefore β-arrestin-1 acts on early endosomes as an endosomal adaptor molecule to regulate the trafficking of ubiquitinated CXCR4 from early endosomes to lysosomes (Bhandari et al., 2007; Malik and Marchese, 2010).

One function of endosomal β-arrestin-1 is likely to recruit ubiquitinated CXCR4 to the endosomal sorting required for transport (ESCRT) pathway, the degradative pathway that targets ubiquitinated receptors for lysosomal degradation [Fig. 3 and (Malik and Marchese, 2010)]. The ESCRT machinery consists of four discrete multi-protein complexes named ESCRT 0-III (Henne et al., 2011). The ESCRT complexes act in a sequential and a coordinated manner to target ubiquitinated cargo into intraluminal vesicles (ILVs) of multivesicular bodies, which then fuse with lysosomes where degradation occurs. Delivery into the ESCRT pathway is thought to occur when the ubiquitin moiety on ubiquitinated receptors is initially recognized by the ubiquitin binding domains (UBD) found in ESCRT 0 (Raiborg and Stenmark, 2009; Shields et al., 2009). ESCRT-0 subsequently recruits ESCRT-I to the endosomal membrane, followed by recruitment of ESCRT II and III, culminating in sorting of the receptor into ILVs [reviewed in (Raiborg and Stenmark, 2009)]. ESCRT-0 is comprised of two proteins: HRS (hepatocyte growth factor-regulated tyrosine kinase substrate) and STAM (signal transduction adaptor molecule). β-arrestin-1 interacts with ESCRT-0 through a direct interaction with STAM-1 (Malik and Marchese, 2010). Therefore it is possible that β-arrestin-1 facilitates ubiquitinated CXCR4 recruitment to ESCRT-0 for entry into the ESCRT pathway.

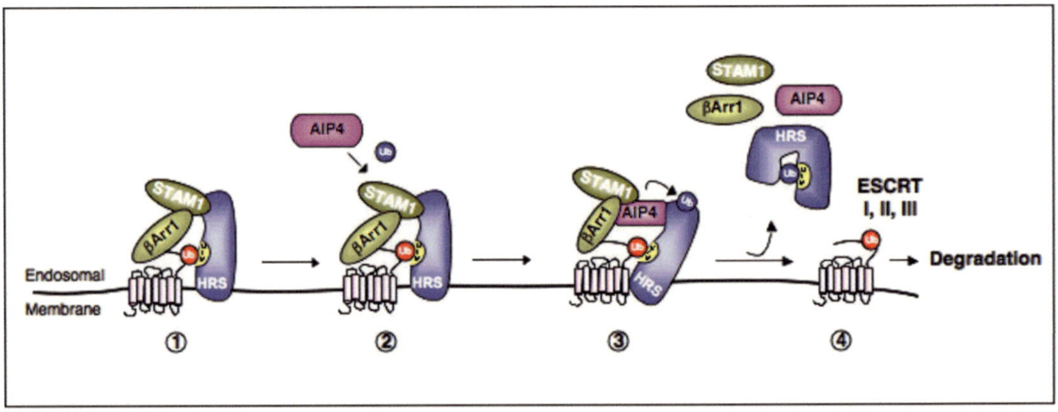

Fig. 3. Proposed model for the roles of β-arrestins in endosomal sorting of CXCR4.

1) β-arrestin-1 (βArr1) interacts with ESCRT-0 through a direct interaction with STAM-1. STAM-1 together with HRS form the core components of ESCRT-0. Although the ubiquitin moiety on CXCR4 may be recognized by the ubiquitin interacting motif (UIM) present in HRS, one role of β-arrestin-1 may be to facilitate the interaction of ubiquitinated CXCR4 with ESCRT-0. **2)** A second role for β-arrestin-1 in conjunction with STAM-1 may serve to recruit the E3 ubiquitin ligase AIP4 for ubiquitination of HRS. **3)** The ubiquitin moiety on HRS may induce an auto-inhibitory conformation through an interaction with its UIM domain to relieve HRS of its sorting function. **4)** The precise molecular details remain unknown, but this somehow leads to the delivery of ubiquitinated CXCR4 to downstream elements of the ESCRT machinery for subsequent entry into ILVs and delivery to lysosomes for degradation.

Surprisingly, depletion of STAM-1 by siRNA accelerates CXCR4 degradation, indicating that STAM-1 negatively regulates CXCR4 entry into the degradative pathway (Malik and Marchese, 2010). This function of STAM-1 is dependent on the interaction with β-arrestin-1 as disruption of this interaction by expressing fragments encoding the minimal binding regions of each respective protein also accelerates CXCR4 degradation. Although detailed mechanistic insight is lacking, the function of the STAM-1/β-arrestin-1 interaction may be to regulate the ubiquitination status of HRS. Disruption of the STAM-1/β-arrestin-1 interaction decreases CXCR4-induced ubiquitination HRS, indicating that STAM-1/β-arrestin-1 function to mediate HRS ubiquitination. As β-arrestin-1 interacts with AIP4, it may be that together with STAM-1 they function as adaptors for recruiting AIP4 to endosomes to mediate ubiquitination of HRS. This suggests that ubiquitination of HRS serves to delay endosomal sorting of CXCR4. One possible explanation is that the ubiquitin moiety on HRS may interact with its own UIM, a type of ubiquitin binding domain, resulting in an auto-inhibitory conformation that inhibits the ability of HRS to perform its sorting function (Hoeller et al., 2006). Alternatively, the ubiquitin moiety on HRS may be required to interact with other ubiquitin binding domains present in other proteins of the ESCRT machinery and possibly other factors. Regardless, this may represent a way in which receptors can regulate their own sorting efficiency (Malik and Marchese, 2010).

Remarkably, linking GPCRs to the ESCRT machinery may be an evolutionary conserved function of β-arrestins (Herrador et al., 2010; Herranz et al., 2005). Rim8, an arrestin-like molecule in *Saccharomyces cerevisiae* related to PalF, an arrestin-like molecule in the fungus *Aspergillus nidulans,* interacts with components of the ESCRT machinery (Herrador et al., 2010). Sequence homology predicts that Rim8 and PalF share a limited amount of sequence identity with mammalian arrestins. In fungi, PalF may interact with a putative seven transmembrane domain (7TM) receptor in an analogous manner to which β-arrestins interact with ligand activated GPCRs (Herranz et al., 2005). Rim8 and PalF are involved in a signaling cascade that senses the pH of the environment. In fungi, pH is recognized in part by the putative 7TM receptor called PalH. Akaline pH is thought to activate PalH and to promote its binding to PalF. The predicted cytoplasmic tail of PalH interacts with PalF and this interaction may be necessary to activate the intracellular signaling pathway necessary for pH sensing. Intriguingly, genetic screens have revealed that components of the ESCRT machinery, including ESCRT-I, ESCRT-II and ESCRT-III subunits Snf7 and Vps20, but not ESCRT-0 and ESCRT-III subunits Vps2 and Vps24, are also necessary for pH sensing (Xu et al., 2004). Rim8, the orthologue of PalF binds to Rim21, a 7TM receptor with pH sensing capabilities (Herrador et al., 2010). Not only does the arrestin-like molecule Rim8 interact with the receptor Rim21 it also interacts with the ESCRT machinery. Rim8 interacts with ESCRT-I subunits Vps23 and Vps28 and as genetic evidence suggests that ESCRT-0 is not involved in pH sensing signaling (Xu et al., 2004), therefore the arrestin-like protein Rim8 may link the 7TM receptor Rim21 directly to ESCRT-I. This raises the intriguing possibility that arrestin recruitment of GPCRs to the ESCRT machinery may represent a conserved function.

6. Conclusion

In addition to their roles in GPCR desensitization, internalization and recycling, β-arrestins may also function in GPCR endosomal sorting. In this capacity they may serve to link ubiquitinated GPCRs to the ESCRT machinery for subsequent targeting to lysosomes. This may represent an evolutionary conserved function of β-arrestins, suggesting they may play a broad role in GPCR endosomal sorting. Further studies are required to gain greater mechanistic insight into the process by which β-arrestins integrate with the ESCRT machinery to target GPCRs for lysosomal degradation.

7. Acknowledgments

Grants from the National Institutes of Health (GM075159 and DA026040 to A.M.) supported this work. R.M. was supported in part by a predoctoral fellowship from the American Heart Association (0910098G).

8. References

Angers, A., Ramjaun, A.R., and McPherson, P.S. (2004). The HECT domain ligase itch ubiquitinates endophilin and localizes to the trans-Golgi network and endosomal system. *J Biol Chem.* 279 (12), pp.11471-11479.

Bhandari, D., Robia, S.L., and Marchese, A. (2009). The E3 ubiquitin ligase atrophin interacting protein 4 binds directly to the chemokine receptor CXCR4 via a novel WW domain-mediated interaction. *Mol Biol Cell.* 20 (5), pp.1324-1339.

Bhandari, D., Trejo, J., Benovic, J.L., and Marchese, A. (2007). Arrestin-2 interacts with the ubiquitin-protein isopeptide ligase atrophin-interacting protein 4 and mediates endosomal sorting of the chemokine receptor CXCR4. *J Biol Chem.* 282 (51), pp.36971-36979.

DeWire, S.M., Ahn, S., Lefkowitz, R.J., and Shenoy, S.K. (2007). Beta-arrestins and cell signaling. *Annu Rev Physiol.* 69 pp.483-510.

Drews, J. (2000). Drug discovery: a historical perspective. *Science.* 287 (5460), pp.1960-1964.

Edwin, F., Anderson, K., and Patel, T.B. (2010). HECT domain-containing E3 ubiquitin ligase Nedd4 interacts with and ubiquitinates Sprouty2. *J Biol Chem.* 285 (1), pp.255-264.

Einbond, A., and Sudol, M. (1996). Towards prediction of cognate complexes between the WW domain and proline-rich ligands. *FEBS Lett.* 384 (1), pp.1-8.

Goodman, O.B., Jr., Krupnick, J.G., Santini, F., Gurevich, V.V., Penn, R.B., Gagnon, A.W., Keen, J.H., and Benovic, J.L. (1996). Beta-arrestin acts as a clathrin adaptor in endocytosis of the beta2-adrenergic receptor. *Nature.* 383 (6599), pp.447-450.

Gurevich, E.V., and Gurevich, V.V. (2006a). Arrestins: ubiquitous regulators of cellular signaling pathways. *Genome Biol.* 7 (9), pp.236.

Gurevich, V.V., and Gurevich, E.V. (2006b). The structural basis of arrestin-mediated regulation of G-protein-coupled receptors. *Pharmacol Ther.* 110 (3), pp.465-502.

Han, M., Gurevich, V.V., Vishnivetskiy, S.A., Sigler, P.B., and Schubert, C. (2001). Crystal structure of beta-arrestin at 1.9 A: possible mechanism of receptor binding and membrane Translocation. *Structure.* 9 (9), pp.869-880.

Hanyaloglu, A.C., and von Zastrow, M. (2008). Regulation of GPCRs by endocytic membrane trafficking and its potential implications. *Annu Rev Pharmacol Toxicol.* 48 pp.537-568.

Henne, W.M., Buchkovich, N.J., and Emr, S.D. (2011). The ESCRT pathway. *Dev Cell.* 21 (1), pp.77-91.

Hernandez, P.A., Gorlin, R.J., Lukens, J.N., Taniuchi, S., Bohinjec, J., Francois, F., Klotman, M.E., and Diaz, G.A. (2003). Mutations in the chemokine receptor gene CXCR4 are associated with WHIM syndrome, a combined immunodeficiency disease. *Nat Genet.* 34 (1), pp.70-74.

Herrador, A., Herranz, S., Lara, D., and Vincent, O. (2010). Recruitment of the ESCRT machinery to a putative seven-transmembrane-domain receptor is mediated by an arrestin-related protein. *Mol Cell Biol.* 30 (4), pp.897-907.

Herranz, S., Rodriguez, J.M., Bussink, H.J., Sanchez-Ferrero, J.C., Arst, H.N., Jr., Penalva, M.A., and Vincent, O. (2005). Arrestin-related proteins mediate pH signaling in fungi. *Proc Natl Acad Sci U S A.* 102 (34), pp.12141-12146.

Hirsch, J.A., Schubert, C., Gurevich, V.V., and Sigler, P.B. (1999). The 2.8 A crystal structure of visual arrestin: a model for arrestin's regulation. *Cell.* 97 (2), pp.257-269.

Hoeller, D., Crosetto, N., Blagoev, B., Raiborg, C., Tikkanen, R., Wagner, S., Kowanetz, K., Breitling, R., Mann, M., Stenmark, H., and Dikic, I. (2006). Regulation of ubiquitin-binding proteins by monoubiquitination. *Nat Cell Biol.* 8 (2), pp.163-169.

Ingham, R.J., Gish, G., and Pawson, T. (2004). The Nedd4 family of E3 ubiquitin ligases: functional diversity within a common modular architecture. *Oncogene.* 23 (11), pp.1972-1984.

Krueger, K.M., Daaka, Y., Pitcher, J.A., and Lefkowitz, R.J. (1997). The role of sequestration in G protein-coupled receptor resensitization. Regulation of beta2-adrenergic receptor dephosphorylation by vesicular acidification. *J Biol Chem.* 272 (1), pp.5-8.

Krupnick, J.G., and Benovic, J.L. (1998). The role of receptor kinases and arrestins in G protein-coupled receptor regulation. *Annu Rev Pharmacol Toxicol.* 38 pp.289-319.

Laporte, S.A., Oakley, R.H., Zhang, J., Holt, J.A., Ferguson, S.S., Caron, M.G., and Barak, L.S. (1999). The beta2-adrenergic receptor/betaarrestin complex recruits the clathrin adaptor AP-2 during endocytosis. *Proc Natl Acad Sci U S A.* 96 (7), pp.3712-3717.

Lefkowitz, R.J. (2007). Seven transmembrane receptors: something old, something new. *Acta Physiol (Oxf).* 190 (1), pp.9-19.

Lin, C.H., MacGurn, J.A., Chu, T., Stefan, C.J., and Emr, S.D. (2008). Arrestin-related ubiquitin-ligase adaptors regulate endocytosis and protein turnover at the cell surface. *Cell.* 135 (4), pp.714-725.

Lohse, M.J., Benovic, J.L., Codina, J., Caron, M.G., and Lefkowitz, R.J. (1990). beta-Arrestin: a protein that regulates beta-adrenergic receptor function. *Science*. 248 (4962), pp.1547-1550.

Lu, P.J., Zhou, X.Z., Shen, M., and Lu, K.P. (1999). Function of WW domains as phosphoserine- or phosphothreonine-binding modules. *Science*. 283 (5406), pp.1325-1328.

Macias, M.J., Wiesner, S., and Sudol, M. (2002). WW and SH3 domains, two different scaffolds to recognize proline-rich ligands. *FEBS Lett*. 513 (1), pp.30-37.

Malik, R., and Marchese, A. (2010). Arrestin-2 interacts with the endosomal sorting complex required for transport machinery to modulate endosomal sorting of CXCR4. *Mol Biol Cell*. 21 (14), pp.2529-2541.

Marchese, A., and Benovic, J.L. (2001). Agonist-promoted ubiquitination of the G protein-coupled receptor CXCR4 mediates lysosomal sorting. *J Biol Chem*. 276 (49), pp.45509-45512.

Marchese, A., Paing, M.M., Temple, B.R., and Trejo, J. (2008). G protein-coupled receptor sorting to endosomes and lysosomes. *Annu Rev Pharmacol Toxicol*. 48 pp.601-629.

Marchese, A., Raiborg, C., Santini, F., Keen, J.H., Stenmark, H., and Benovic, J.L. (2003). The E3 ubiquitin ligase AIP4 mediates ubiquitination and sorting of the G protein-coupled receptor CXCR4. *Dev Cell*. 5 (5), pp.709-722.

Milano, S.K., Pace, H.C., Kim, Y.M., Brenner, C., and Benovic, J.L. (2002). Scaffolding functions of arrestin-2 revealed by crystal structure and mutagenesis. *Biochemistry*. 41 (10), pp.3321-3328.

Nabhan, J.F., Pan, H., and Lu, Q. (2010). Arrestin domain-containing protein 3 recruits the NEDD4 E3 ligase to mediate ubiquitination of the beta2-adrenergic receptor. *EMBO Rep*. 11 (8), pp.605-611.

Nelson, C.D., Perry, S.J., Regier, D.S., Prescott, S.M., Topham, M.K., and Lefkowitz, R.J. (2007). Targeting of diacylglycerol degradation to M1 muscarinic receptors by beta-arrestins. *Science*. 315 (5812), pp.663-666.

Oakley, R.H., Laporte, S.A., Holt, J.A., Barak, L.S., and Caron, M.G. (1999). Association of beta-arrestin with G protein-coupled receptors during clathrin-mediated endocytosis dictates the profile of receptor resensitization. *J Biol Chem*. 274 (45), pp.32248-32257.

Oakley, R.H., Laporte, S.A., Holt, J.A., Barak, L.S., and Caron, M.G. (2001). Molecular determinants underlying the formation of stable intracellular G protein-coupled receptor-beta-arrestin complexes after receptor endocytosis*. *J Biol Chem*. 276 (22), pp.19452-19460.

Oakley, R.H., Laporte, S.A., Holt, J.A., Caron, M.G., and Barak, L.S. (2000). Differential affinities of visual arrestin, beta arrestinl, and beta arrestin2 for G protein-coupled receptors delineate two major classes of receptors. *J Biol Chem*. 275 (22), pp.17201-17210.

Ozawa, K., Whalen, E.J., Nelson, C.D., Mu, Y., Hess, D.T., Lefkowitz, R.J., and Stamler, J.S. (2008). S-nitrosylation of beta-arrestin regulates beta-adrenergic receptor trafficking. *Mol Cell*. 31 (3), pp.395-405.

Perry, S.J., Baillie, G.S., Kohout, T.A., McPhee, I., Magiera, M.M., Ang, K.L., Miller, W.E., McLean, A.J., Conti, M., Houslay, M.D., and Lefkowitz, R.J. (2002). Targeting of cyclic AMP degradation to beta 2-adrenergic receptors by beta-arrestins. *Science.* 298 (5594), pp.834-836.

Pierce, K.L., Premont, R.T., and Lefkowitz, R.J. (2002). Seven-transmembrane receptors. *Nat Rev Mol Cell Biol.* 3 (9), pp.639-650.

Pippig, S., Andexinger, S., and Lohse, M.J. (1995). Sequestration and recycling of beta 2-adrenergic receptors permit receptor resensitization. *Mol Pharmacol.* 47 (4), pp.666-676.

Raiborg, C., and Stenmark, H. (2009). The ESCRT machinery in endosomal sorting of ubiquitylated membrane proteins. *Nature.* 458 (7237), pp.445-452.

Shearwin-Whyatt, L., Dalton, H.E., Foot, N., and Kumar, S. (2006). Regulation of functional diversity within the Nedd4 family by accessory and adaptor proteins. *Bioessays.* 28 (6), pp.617-628.

Shenoy, S.K., and Lefkowitz, R.J. (2003). Trafficking patterns of beta-arrestin and G protein-coupled receptors determined by the kinetics of beta-arrestin deubiquitination. *J Biol Chem.* 278 (16), pp.14498-14506.

Shenoy, S.K., McDonald, P.H., Kohout, T.A., and Lefkowitz, R.J. (2001). Regulation of receptor fate by ubiquitination of activated beta 2-adrenergic receptor and beta-arrestin. *Science.* 294 (5545), pp.1307-1313.

Shenoy, S.K., Modi, A.S., Shukla, A.K., Xiao, K., Berthouze, M., Ahn, S., Wilkinson, K.D., Miller, W.E., and Lefkowitz, R.J. (2009). Beta-arrestin-dependent signaling and trafficking of 7-transmembrane receptors is reciprocally regulated by the deubiquitinase USP33 and the E3 ligase Mdm2. *Proc Natl Acad Sci U S A.* 106 (16), pp.6650-6655.

Shenoy, S.K., Xiao, K., Venkataramanan, V., Snyder, P.M., Freedman, N.J., and Weissman, A.M. (2008). Nedd4 mediates agonist-dependent ubiquitination, lysosomal targeting, and degradation of the beta2-adrenergic receptor. *J Biol Chem.* 283 (32), pp.22166-22176.

Shi, H., Rojas, R., Bonifacino, J.S., and Hurley, J.H. (2006). The retromer subunit Vps26 has an arrestin fold and binds Vps35 through its C-terminal domain. *Nat Struct Mol Biol.* 13 (6), pp.540-548.

Shields, S.B., Oestreich, A.J., Winistorfer, S., Nguyen, D., Payne, J.A., Katzmann, D.J., and Piper, R. (2009). ESCRT ubiquitin-binding domains function cooperatively during MVB cargo sorting. *J Cell Biol.* 185 (2), pp.213-224.

Verdecia, M.A., Bowman, M.E., Lu, K.P., Hunter, T., and Noel, J.P. (2000). Structural basis for phosphoserine-proline recognition by group IV WW domains. *Nat Struct Biol.* 7 (8), pp.639-643.

Vines, C.M., Revankar, C.M., Maestas, D.C., LaRusch, L.L., Cimino, D.F., Kohout, T.A., Lefkowitz, R.J., and Prossnitz, E.R. (2003). N-formyl peptide receptors internalize but do not recycle in the absence of arrestins. *J Biol Chem.* 278 (43), pp.41581-41584.

von Zastrow, M., and Kobilka, B.K. (1992). Ligand-regulated internalization and recycling of human beta 2-adrenergic receptors between the plasma membrane and endosomes containing transferrin receptors. *J Biol Chem*. 267 (5), pp.3530-3538.

Wilden, U., Hall, S.W., and Kuhn, H. (1986). Phosphodiesterase activation by photoexcited rhodopsin is quenched when rhodopsin is phosphorylated and binds the intrinsic 48-kDa protein of rod outer segments. *Proc Natl Acad Sci U S A*. 83 (5), pp.1174-1178.

Wyatt, D., Malik, R., Vesecky, A.C., and Marchese, A. (2011). Small ubiquitin-like modifier modification of arrestin-3 regulates receptor trafficking. *J Biol Chem*. 286 (5), pp.3884-3893.

Xu, W., Smith, F.J., Jr., Subaran, R., and Mitchell, A.P. (2004). Multivesicular body-ESCRT components function in pH response regulation in Saccharomyces cerevisiae and Candida albicans. *Mol Biol Cell*. 15 (12), pp.5528-5537.

14

Structure and Function of the Lipolysis Stimulated Lipoprotein Receptor

Christophe Stenger, Catherine Corbier and Frances T. Yen
EA4422 Lipidomix and UR AFPA, ENSAIA, INPL, University of Lorraine
France

1. Introduction

1.1 Role of lipoproteins

Lipoproteins provide the means of transport of hydrophobic lipids in the circulation (Havel & Kane, 1995). Composed of a single monolayer of phospholipids surrounding a neutral lipid core, the primary purpose of these spherical lipid particles is to deliver two major classes of lipids - cholesterol and fatty acids (FA) - to the different peripheral tissues. Cholesterol is carried in the form of free unesterified form in the phospholipid monolayer, or as cholesteryl esters (CE) in the nucleus of the lipoprotein. FA are transported in the form of triglycerides (TGs), also found in the hydrophobic core of the lipid particles along with the CE. Proteins that are amphipathic in nature are found associated with the lipoprotein and are referred to as apolipoproteins (apo). They serve as cofactors for lipid-modifying enzymes and proteins, including lipoprotein lipase, hepatic lipase, lecithin-cholesterol acyltransferase and cholesteryl-ester transfer protein. Apolipoproteins also serve as ligands that bind to specific sites of lipoprotein receptors, providing a means by which the lipoproteins are bound and then internalized through endocytosis via these receptors located on cell surface membranes.

1.2 Triglyceride-rich lipoproteins

TGs transported by lipoproteins can be either of endogenous or exogenous origin (Figure 1). Upon absorption through the intestinal wall, dietary TGs are repackaged by the enterocyte into very large TG-rich chylomicrons containing apoB48 which are then released into the circulation (Havel & Kane, 1995). These lipoproteins distribute their TG load to the periphery via interaction with the lipoprotein lipase (LpL) anchored by heparan sulfate proteoglycans (HSPG) on the capillary endothelium. This enzyme catalyzes the hydrolysis of TG of the chylomicrons to free FA (FFA) which are then taken up by the peripheral tissues including adipose tissue and skeletal muscle. This process leads to a depletion of the chylomicron's neutral lipid core, resulting in the formation of smaller particles called chylomicron remnants. These particles have lower TG content due to the lipolytic action, and are enriched in CE as compared to the original chylomicrons. During the transformation, these smaller particles acquire apoE and exchange other apolipoproteins with other lipoprotein classes. The half-life of both chylomicrons and remnants is very short, reported to be less than 15 min in studies using radioactively-labelled retinyl palmitate (Berr & Kern, 1984; Cortner *et al.*, 1987). Numerous studies indicate that the principal site of removal of these residues from the circulation is the liver (Attie *et al.*, 1982), and involves a series of complex processes in the space of Disse that include binding to HSPG, interacting with hepatic and lipoprotein lipases, and acquiring additional apoE (Mahley & Huang, 2007). The final step culminates in the delivery to the cell through receptor-mediated endocytosis, leading to internalization and degradation of the lipid particles in the lysosomal compartment of the hepatocyte.

Very-low density lipoproteins (VLDL) are produced by the liver and transport endogenously-derived lipids (Havel & Kane, 1995). Similar to chylomicrons, these TG-rich lipid particles deliver lipids to the peripheral tissues through the LpL system, and acquire apoE and exchange other apolipoproteins with other lipoproteins. However, in humans, it is distinguished from intestinally-produced chylomicrons in that it contains apoB-100. Upon hydrolysis of TG by LpL, VLDL is converted to intermediate-density lipoproteins (IDL), and eventually to LDL. IDL is removed quickly from the circulation, and rarely detected in plasma. LDL which contains only one apoB100 per particle displays a half-life of 3-4 days, and represents the major carrier of CE in the circulation. LDL is ultimately removed from the circulation by the liver through receptor-mediated endocytosis through the LDL-receptor (LDL-R).

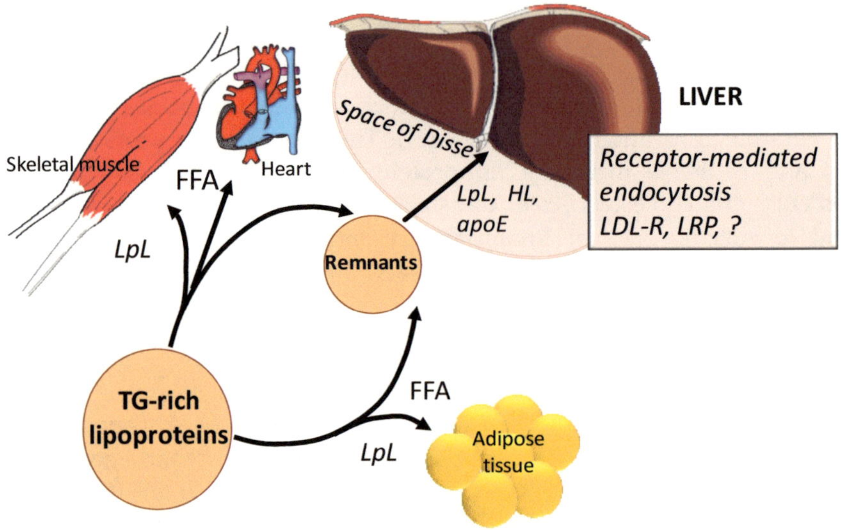

Fig. 1. General diagram for TG-rich lipoprotein processing in the circulation. Lipoproteins carrying TG of exogenous (chylomicrons) or endogenous (hepatic-derived VLDL) origins circulate in the plasma to deliver their lipids to the peripheral tissues (adipose tissue, cardiac and skeletal muscle). The lipase system consisting of lipoprotein lipase (LpL) bound to the endothelial wall hydrolyzes the TG of these particles, and the lipolytic products (FFA) are delivered to the different tissues. The resulting particles reduced in TG content (chylomicron remnants, IDL, LDL) are then taken up for catabolism by the liver through receptor-mediated endocytosis. Although LDL-R and LRP1 have both been implicated, evidence suggests the presence of other pathway(s) involved.

1.2.1 Receptor-mediated clearance of TG-rich lipoproteins

1.2.1.1 LDL-R

A major advance was made in the lipoprotein field by the discovery of the LDL-R by MS Brown and JL Goldstein (Brown & Goldstein, 1986; Goldstein & Brown, 2009). Their Nobel-prize winning research clearly showed the pivotal role of this receptor in body cholesterol homeostasis (Brown & Goldstein, 1997). The LDL-R that is defective in subjects with familial hypercholesterolemia (FH) (Goldstein *et al.*, 1995) accounts for most of LDL removal. Although this receptor binds to both apoB100 and apoE, it does not seem to be the principal pathway for the removal of chylomicrons and their remnants. Indeed, clearance of chylomicron remnants proceeds at normal rates in humans and animals with genetic lesions impairing LDL-R function (Kita *et al.*, 1982; Rubinsztein *et al.*, 1990). Transgenic studies using LDL-R$^{-/-}$ mice nevertheless indicate that the LDL-R does participate, albeit partially to the clearance of chylomicron remnants (de Faria *et al.*, 1996; Ishibashi *et al.*, 1996).

The identification of the receptor(s) involved with the LDL-R in the liver-specific capture of chylomicrons has remained a difficult task, and has been a subject of many lively debates. Initial efforts resulted in the identification of apoE-binding proteins not directly involved in receptor-mediated endocytosis of lipoproteins (Beisiegel *et al.*, 1988). Other potential candidates proposed were later found not to be expressed in the liver such as the VLDL-receptor and macrophage TG-rich lipoprotein receptor (Ramprasad *et al.*, 1995; Sakai *et al.*, 1994). A candidate receptor that has received the greatest attention is the LDL-receptor related protein, LRP1, which belongs to the LDL-R family (Herz *et al.*, 1988).

1.2.1.2 LRP1

LRP1 was originally discovered by cloning based on homologous recombination (Herz *et al.*, 1988). Because of its shared domains to the LDL-R, its primary role was thought to be in the removal of lipoproteins from the circulation. Studies revealed that while this protein cannot bind LDL, it could recognize and bind β-VLDL isolated from cholesterol-fed rabbits, but only when these lipid particles are enriched with added exogenous recombinant apoE (Kowal *et al.*, 1989). It was later revealed that LRP was in fact the alpha$_2$-macroglobulin receptor (Strickland *et al.*, 1990), with multiple ligands including among others, activated macroglobulin, tissue-type (tPA) and urokinase (uPA) plasminogen activators, coagulation factors IXa, VIIIa, VIIa, TFPI, complement C3, and thrombospondin-1 (Herz & Strickland, 2001). LRP plays an important role in physiology since the absence of both alleles is embryonic lethal in mice (Herz *et al.*, 1992). However, no remnant accumulation was observed in hepatic LRP1-inactive mice, and only a modest increase in apoB48-containing lipoproteins was observed when both hepatic LRP1 and LDL-R were inactive, suggesting that LRP plays a back-up role in the uptake of lipoproteins in the absence of the LDL-R (Rohlmann *et al.*, 1998). Furthermore, LRP1 binds both apoE2/2 and apoE3/3 isoforms equally well (Beisiegel *et al.*, 1989), which could not completely explain the mechanisms underlying type III hyperlipidemia which is often associated with the apoE2/2 phenotype (Soutar, 1989).

Recent evidence has shown that LRP1 selective inactivation in adipose tissue increases LpL activity, but prevents uptake of TG in adipose tissue, thereby preventing weight gain and insulin resistance (Hofmann *et al.*, 2007). The authors propose that LRP1 mediates uptake of chylomicrons by adipocytes, and that this accounts for a large part of dietary TG clearance. However, it is difficult to imagine chylomicrons gaining direct access to adipocytes since unlike the liver with its space of Disse, the capillary endothelium is not fenestrated, therefore limiting access of these large bulky particles. The data nevertheless clearly show that LRP1 is necessary for proper processing of functional LpL. In absence of LRP1, the enzyme is present, but fails to deliver lipid to the underlying adipose tissue. If this is indeed the function of LRP1 in adipose tissue, this might also be the case for LRP1 in the liver. Thus, the modest hyperlipidemic effect in hepatic LRP and LDL-R inactive mice could be explained by mechanisms other than impaired receptor-mediated endocytosis, but rather by those involving LpL function. Other pathways as yet unidentified were clearly involved in the removal of TG-rich lipoproteins.

This chapter focuses on the biochemical characterization, identification and function of the lipoprotein receptor which plays an important role in the removal of TG-rich lipoproteins during the postprandial phase.

2. Biochemical characterization of LSR

2.1 Original identification and characterization of an LDL-receptor independent pathway in LDL-receptor negative fibroblasts from a subject homozygous for familial hypercholesterolemia

A study by Bihain *et al* revealed that FFA could regulate lipoprotein receptor activity, by showing that FFA inhibit binding of LDL to its receptor (Bihain *et al.*, 1989). During these studies, it was discovered that in the presence of oleate, fibroblasts from a patient homozygous for familial hypercholesterolemia (FH fibroblasts) and therefore deficient in LDL-R were able to internalize a significant amount of LDL. A series of experiments were conducted to biochemically characterize this pathway apparently independent of the LDL-R (Bihain & Yen, 1992). The uptake of ^{125}I-radiolabelled LDL increased in a dose-dependent manner with the concentration of oleate used. The first question was to ask if cell viability was modified by the presence of FFA, which was not the case. Kinetic studies revealed that the internalization of LDL was time and dose-dependent, where saturation was achieved at 50 µg LDL protein/ml. Degradation products, measured as TCA-soluble ^{125}I released into the media appeared approximately 30 min after addition of oleate, which corresponds closely to the time required for delivery of ligands to the lysosome after internalization. Chloroquine, which is an inhibitor of lysosomal enzymes, prevented the appearance of ^{125}I-products, confirming that the ^{125}I-LDL internalized was delivered to the lysosome for degradation. It therefore appeared that the uptake of LDL into the cell through this LDL-receptor independent pathway was achieved through endocytosis, implying the presence of a receptor binding site of a specific nature on the cell surface plasma membrane (Bihain & Yen, 1992).

Different FFAs were used to ascertain the nature of the potential binding site on the plasma membranes of the FH fibroblasts. It was found that the addition of a double bond rendered the FFA more efficient in activating this LDL-R independent pathway, as compared to the corresponding saturated analogs (Bihain & Yen, 1992). Interestingly, oleate, one of the more abundant FFA in plasma was found to be the most efficient activator of this pathway. FFA analogs with uncharged residues such as oleyl alcohol or oleyl acetate also demonstrated a similar effect on LDL internalization, although slightly less efficient as compared to oleate (Bihain & Yen, 1992).

Binding studies at 4°C revealed that in the presence of oleate, ^{125}I-LDL bound with high affinity to a saturable and specific binding site on FH fibroblasts (Bihain & Yen, 1992). Bound LDL could be released by suramin, a polysulfated aromatic sodium salt, much in the same manner that heparin is used to release LDL bound to the LDL-R (Bihain & Yen, 1992). By comparing the amount of LDL bound to this oleate-induced binding site at 4°C to that released by suramin, it was determined that suramin released lipoprotein that was specifically bound to the cell surface. Treatment of cells with suramin during incubations with oleate and ^{125}I-LDL simultaneously prevented the accumulation of cellular ^{125}I-LDL as well as TCA soluble products, demonstrating that this binding site, once occupied by the ligand LDL, mediates internalization and subsequent degradation of the lipoprotein in the lysosome. Scatchard analysis of binding curves revealed characteristics of a single binding site with a Kd of 12.3 μg/ml and Bmax of 78.4 ng of LDL protein.

Competition studies indicated that this binding site displayed a higher affinity for the TG-rich VLDL as compared to LDL (Bihain & Yen, 1992). This affinity for VLDL was directly related to the size of the particle and therefore TG content of the VLDL fraction, with the highest affinity being for the larger $VLDL_1$ fraction, as compared to that for $VLDL_2$ and $VLDL_3$. Further studies revealed that chylomicrons or lipid emulsions containing recombinant apoE could compete with ^{125}I-LDL for binding in the presence of FFA (Yen *et al.*, 1994). Digestion of apoB100 on the LDL with mild pronase treatment renders the LDL unable to bind this oleate-induced binding site, thus clearly demonstrating a direct protein-protein interaction between the lipoprotein ligand and the cell surface binding site. Derivatization of LDL with cyclohexanedione (CHD) modifies arginine residues of apoB resulting in a CHD-LDL unable to bind the LDL-R. Interestingly, CHD-LDL was able to compete for binding with ^{125}I-LDL as well as normal unlabeled LDL. This modified LDL has been used to demonstrate the presence of a LDL-receptor independent pathway in human subjects (Simons *et al.*, 1975), suggesting that perhaps this binding site may be involved in LDL-R-independent clearance. Finally, removal of cell surface proteins by mild trypsin treatment of the cell surface eliminated the cell's capacity to exhibit oleate-induced binding, uptake and degradation of ^{125}I-LDL (Bihain & Yen, 1992), clearly showing the protein nature of this cell binding site.

Taken together, the initial characterization of this LDL-receptor independent pathway revealed that the FFA-induced binding, internalization and degradation of LDL was mediated by protein-protein interaction with a cell surface protein that recognized either apoB100 or apoE. Binding of the lipoprotein to this binding site in the presence of FFA leads to endocytosis, ultimately leading to lysosomal degradation of the lipoprotein particle.

2.2 Identity of the lipolysis stimulated lipoprotein receptor, LSR distinct from LDL-R and LRP

In view of the cell type used in these studies, it was clear that this binding site was not the LDL-R itself. Indeed, in FH fibroblasts, due to a significant deletion of the promoter region, these cells are unable to synthesize the LDL-R (Hobbs *et al.*, 1987). Furthermore, it was previously demonstrated that FFA inhibit binding of LDL to the LDL-R. Other data that characterize LDL binding outside of the cell using liver membrane assays described later will further confirm the distinct identity from the LDL-R of this FFA binding site. LDL was nevertheless used in these assays for technical reasons. First of all, LDL is easily isolated from human plasma in sufficient quantities for the type of studies that were performed. Secondly, LDL contains only one molecule of the large 500 kDa apoB100 that does not dissociate from the lipoprotein particle. Therefore, the ^{125}I-radiolabel on the apoB100 of LDL provides an accurate measurement of the amount of lipoprotein particle itself bound, internalized and degraded by the cells. This is unlike the TG-rich lipoproteins, chylomicrons and VLDL, which contain a number of different apolipoproteins including apoE, as well as the apoA- and the apoC- classes of apolipoproteins. These apolipoproteins readily exchange between different lipoprotein particles, rendering it difficult to precisely measure the kinetics of the endocytosis. It was therefore deliberately chosen to continue using LDL as a ligand to study this pathway.

At the time, the LRP1 had been cloned and then identified as being 100% homologous to the α_2-macroglobulin receptor. Questions arose as to whether this oleate-induced binding site could be LRP1. A number of evidence indicated that such was not the case. While it was necessary to supplement β-VLDL with exogenous recombinant apoE before binding by LRP, this was not necessary for β-VLDL binding in the presence of oleate in FH fibroblasts (Bihain & Yen, 1992). Experiments using activated α_2-macroglobulin as a known LRP ligand demonstrated that there was no effect of oleate on LRP-mediated internalization and degradation of this ligand (Yen *et al.*, 1994). Nor was the oleate-induced receptor inhibited by the 39 kDa receptor-associated protein (RAP) at concentrations shown to affect LRP activity (Yen *et al.*, 1994). Furthermore, it had been shown that LRP was able to bind and recognize apoE2/2 (Beisiegel *et al.*, 1989), while this oleate-induced pathway for lipoproteins was unable to bind to VLDL isolated from a type III hypertriglyceridemic patient with the apoE2/2 phenotype (Yen *et al.*, 1994).

Biochemical characterization of both receptors therefore revealed two distinct identities. It is interesting to note the two different approaches used to identify these two receptors. LRP was cloned based on its homology to specific domains of the LDL-R. The biochemical characterization of LRP revealed a multiligand receptor, for which lipoproteins were not necessarily the most important of ligands. Indeed, there are now numerous members of the LDL-R family that have been identified, all exhibiting diverse functions in physiology, at the level of both periphery and central nervous system (Herz & Strickland, 2001). While the protein and gene had been identified, it remained to determine the actual physiological role in lipoprotein metabolism of this receptor related to LDL-R. On the other hand, LSR was first identified and characterized functionally, as a receptor on the cell surface able to bind, internalize and degrade lipoproteins in the presence of FFA. It remained to identify the protein and gene responsible, in order to validate the physiological role of this receptor.

By virtue of its activation by FFA, it was thought that this receptor would be active primarily during times when lipolysis is active, in other words during postprandial lipemia when there is increased levels of chylomicrons in the circulation. Indeed, lipases on the endothelial wall hydrolyze TG of chylomicrons and VLDL, two classes of lipoproteins that display the highest affinity for this LDL-receptor independent pathway. Because of this lipolysis-dependent step, the as yet unidentified receptor responsible for this pathway was named the lipolysis stimulated lipoprotein receptor, or LSR.

2.3 LSR in the hepatocyte

2.3.1 Biochemical characterization of hepatic LSR

The clearance of chylomicrons has previously been shown to take place in the liver through a receptor-mediated process (Sherrill & Dietschy, 1978). The LSR model had been biochemically characterized in FH fibroblasts, much in the same way as the LDL-R had been identified in normal human fibroblasts. Studies revealed that LSR activity was indeed found in primary cultured hepatocytes (Yen *et al.*, 1994). However, although LDL cannot bind LDL-R in the presence of FFA, LDL-R activity in liver cells in the absence of oleate yielded interpretation of results difficult at times. Assays were developed using cell-free liver membranes in order to directly measure binding of lipoproteins to LSR. Binding studies using isolated liver total or plasma membranes revealed that LSR binding to ^{125}I-LDL in the presence of oleate was Ca^{2+}-independent (Mann *et al.*, 1995). Indeed, LDL-R being a Ca^{2+}-dependent receptor (Goldstein *et al.*, 1983), this provided additional evidence for the distinct identity between LSR and LDL-R.

Further biochemical characterization using these assays revealed that the binding characteristics were similar to those observed for FH fibroblasts (Yen *et al.*, 1994). Saturated FFA were less efficient in the activation of LSR as compared to oleate. Scatchard plots revealed again a single binding site with half maximum binding occurring at 23 µg LDL protein/ml. Heparan sulfate proteoglycans had been previously speculated to play a role in endocytosis of TG-rich lipoproteins (Eisenberg *et al.*, 1992; Mulder *et al.*, 1993; Williams *et al.*, 1992). However, pretreatment of isolated liver membranes with heparinase/heparitinase and chondroitanase had no significant effect on oleate-induced binding of ^{125}I-LDL to LSR, unlike trypsin, which diminished LSR activity in a time-dependent manner leading to the disappearance of receptor binding after 60 minute treatment (Mann *et al.*, 1995). Perfusion of livers with trypsin to degrade cell surface proteins before preparation of membranes also significantly diminished LSR binding activity by 80%, demonstrating that the majority of LSR activity measured occurs on the cell surface exposed on the extracellular side (Mann *et al.*, 1995).

2.3.2 Reversibility of LSR activation by FFA

The development of these liver membrane assays allowed closer examination of the activation of LSR by FFA. Experiments revealed that the activation of LSR by FFA is reversible, but only if the binding site remains unoccupied (Mann *et al.*, 1995). Albumin provided a means to remove FFA from the membranes after the activation step. Indeed, if membranes were washed with buffer containing albumin after incubation with oleate, binding of the lipoprotein to LSR did not occur. If after one series of activation by oleate and deactivation using albumin were performed, it was still possible to wash the membranes to remove the albumin, and then re-incubate the membranes in the presence of oleate with ^{125}I- LDL. Binding of the lipoprotein to LSR was detected, indicating that FFA binding is reversible. However, if liver membranes are incubated with oleate, followed by the radiolabelled lipoprotein, the LSR-LDL binding complex remains stable. Washing with even very high concentrations of albumin after formation of the LDL-LSR complex is unable to dissociate the ligand from the receptor. These results suggested a reversible conformational change of LSR in the presence of FFA that reveals the apoB, apoE binding site. Once the ligand-receptor complex is formed, this appears to stabilize the active LSR conformation. It is also possible that the FFA associated with the receptor or the surrounding phospholipid environment may be trapped within the complex, and inaccessible to the albumin in the washing buffer. These potential mechanisms remain to be tested in isolated systems using purified receptor.

2.4 *In vitro* evidence for this receptor's role in the hepatic clearance of triglyceride-rich lipoproteins

A number of circumstantial evidence for LSR's potential role in hepatic clearance of lipoproteins was brought to light using these cell and membrane assays for LSR activity in the liver. Indeed, the measure of oleate-induced ^{125}I-LDL binding under conditions in which maximal binding capacity of LSR is achieved represents an estimation of the apparent number of LSR on the hepatocyte membrane. Rats were sacrificed under non-fasted and fasted conditions, and plasma hepatocyte membranes isolated to measure maximal LSR binding activity. A strong negative correlation ($r = -0.828$, $p < 0.001$) was observed between the apparent number of LSR on the cell surface and the plasma TG levels measured at the time of the sacrifice of the animals (Mann *et al.*, 1995). This implied that the higher LSR binding activity was, the lower the plasma TG levels. On the other hand, under fasting conditions, there was no significant correlation between these 2 parameters. Therefore, even in a normal sample population of laboratory rats, the variation in LSR corresponds to their ability to remove TG from the circulation, but only during the fed or postprandial state. This was the first *in vitro* circumstantial evidence pointing towards a potential role of LSR as a rate limiting factor for the removal of TG-rich lipoproteins during the postprandial phase.

2.4.1 Lactoferrin

Studies in the literature on the regulation of postprandial lipemia demonstrated that the milk protein, lactoferrin, when injected *iv* in rats led to increased TG during the postprandial phase by inhibiting clearance of chylomicrons (Huettinger *et al.*, 1988). Indeed, lactoferrin contains a cluster of arginine residues, which is also found in the apoE binding site for the LDL-R. Further investigation showed that while lactoferrin delays chylomicron remnant clearance, it has no effect on α_2-macroglobulin uptake in the liver, which is one of the principal ligands of LRP as discussed in an earlier section. Lactoferrin was found to inhibit LSR activity as a lipoprotein receptor in FH fibroblasts (Yen *et al.*, 1994) and LSR binding activity in liver membranes (Mann *et al.*, 1995). It was also observed that lactoferrin's inhibitory effect was present only if LSR was in its FFA-activated form. Indeed, the presence of lactoferrin before addition of oleate had no effect on LSR's ability to bind the lipoprotein ligand. Levels of membrane-associated oleate was not altered in the presence of lactoferrin, demonstrating that lactoferrin did not inhibit LSR activity by binding or removing FFA from the membranes (Mann *et al.*, 1995). This supports the earlier data described above suggesting two distinct conformations of LSR, one conformation of the receptor in the absence of oleate, and a different conformation in the presence of oleate.

2.4.2 ApoCIII

ApoCIII is an apolipoprotein associated with TG-rich lipoproteins that plays a role in the modulation of plasma TG levels as a lipase inhibitor. Delayed clearance of TG-rich lipoproteins has been observed in mice overexpressing apoCIII (Aalto-Setala *et al.*, 1992; Ito *et al.*, 1990). In addition, targeted disruption of apoCIII gene in mice leads to increased removal rates of chylomicron remnants associated with reduced TG levels during the postprandial phase (Maeda *et al.*, 1994). Supplementation of VLDL with apoCIII led to decreased binding and uptake through the LSR pathway in primary cultures of rat hepatocytes (Mann *et al.*, 1997). This was in contrast to apoCII, which displayed no such impact on VLDL binding to LSR. Furthermore, the degree of sialyation of apoCIII appeared to influence its ability to inhibit VLDL binding to this receptor (Mann *et al.*, 1997). Therefore, the hypertriglyceridemic effect reported for apoCIII may be due in part to its inhibition of binding of VLDL to the LSR receptor, resulting in reduced hepatic capacity for the removal of TG-rich lipoproteins from the circulation.

2.4.3 39 kDa receptor-associated protein

Finally, the 39 kDa receptor-associated protein (RAP) is a protein that co-purifies with LRP1, and which was shown to inhibit binding of LRP ligands to their receptor (Herz *et al.*, 1991; Moestrup & Gliemann, 1991; Williams *et al.*, 1992). RAP overexpression using adenovirus vector in wild-type and LDL-R$^{-/-}$ mice led to an accumulation of plasma cholesterol and TG, as well as apoB-48 and apoE particles (Willnow *et al.*, 1994). In this model, RAP levels were increased to a large extent in the animals. It was found that RAP, at concentrations similar to those used in this animal study inhibited oleate-induced binding, uptake and degradation of ^{125}I-LDL through the LSR pathway (Troussard *et al.*, 1995). Lineweaver-Burk analysis revealed that this was due to a change in maximal binding capacity, rather than to a change in affinity (Troussard *et al.*, 1995). Other studies also reported that at these levels, RAP could also affect LDL-R activity (Medh *et al.*, 1995; Mokuno *et al.*, 1994), therefore showing that the modifications in plasma lipids in RAP-overexpressed mice were most likely to be due to inhibition of multiple lipoprotein receptors, rather than just LRP1 alone.

The use of the different proteins that modulate TG and postprandial lipemia revealed a number of *in vitro* circumstantial evidence for the potential physiological role of LSR. Indeed, each protein shown to demonstrate a hypertriglyceridemic effect was determined to influence LSR activity as a lipoprotein receptor, whether directly on the receptor or indirectly through ligand binding to LSR. With this circumstantial evidence pointing towards a role of LSR in the removal of TG-rich lipoproteins during the postprandial phase, it remained to actually identify the protein or proteins responsible for LSR activity.

3. Molecular characterization of LSR

3.1 Purification of LSR

Ligand blots were performed using human FH fibroblast lysates, in which cell lysate proteins were separated by SDS-polyacrylamide gel electrophoresis under non-reducing conditions, and then transferred to nitrocellulose (Yen *et al.*, 1994). Interestingly, binding of ^{125}I-LDL in the presence of oleate was observed for 2 protein bands migrating at apparent molecular mass of 115 and 85 kDa. Mild trypsin treatment of cells before preparation of cell lysates demonstrated that these 2 proteins were located on the cell surface. These ligand blots were performed in the absence of Ca^{2+}, which therefore eliminated the possibility that LDL was binding to the Ca^{2+}-dependent LDL-R.

Upon identification of LSR activity in rat hepatocytes, liver membranes were prepared followed by treatment with *n*-octylglucoside (Mann *et al.*, 1995), a mild non-ionic detergent often used to solubilize integral membrane proteins. An LSR-enriched fraction was prepared following further purification of the solubilized protein fraction by anion exchange chromatography, and was tested in ligand blots. These binding studies again revealed 2 bands at 115 and 90 kDa. A third band was also observed at around 230 kDa, which could represent either a homodimeric form of the 115 kDa band, or a heterodimeric form of the two lower molecular mass bands (Mann *et al.*, 1995). The results of these ligand blots did reveal that this receptor, even when immobilized on nitrocellulose could be activated by oleate in absence of the phospholipid environment. This would therefore be consistent with the previous data suggesting a conformational change in the protein occurring following direct interaction of the FFA with the LSR proteins rather than with the surrounding membrane bilayer.

The ligand blots allowed the study of LSR protein interaction with other proteins, including the 39kDa RAP (Troussard *et al.*, 1995). Using a recombinant 39 kDa RAP fusion protein, it was found that this protein could bind directly to the same bands identified as displaying LSR binding activity. Interestingly, this binding did not require oleate. Furthermore, although preincubation with high concentrations of RAP fusion protein inhibited binding of the LDL ligand to oleate-activated LSR, LDL was unable to compete for RAP fusion binding to LDL. This suggested that the LSR binding site for RAP was distinct from that for the apoB component of LDL.

3.2 Identification and cloning of the gene candidate

Even after having identified LSR protein bands, it proved to be very difficult to isolate the LSR receptor. Indeed, ligand blots led to the speculation that this was indeed a complex composed of multiple subunits. It was discovered that the protein migrating around 240 kDa rapidly degraded into multiple bands of different molecular masses once removed from the membrane phospholipid environment (Mann *et al.*, 1995; Yen *et al.*, 1999). Because of the lability of this complex, the purification of sufficient quantities in reduced form with adequate purity for microsequencing ended up being a major obstacle. Indeed, it was difficult to isolate the different bands from other contaminating proteins of similar molecular masses. Antibodies were prepared against the large 240 kDa complex by splicing out the band directly from the SDS gel and injecting into rabbits (Yen *et al.*, 1999). These antibodies were validated in Western blots, identifying the same proteins also shown to bind radiolabelled LDL in the presence of oleate in corresponding ligand blots. This same antibody was also able to inhibit LSR binding activity in liver membranes, as well as LSR receptor activity in primary cultures of rat hepatocytes. Immunoprecipitation studies were performed with these antibodies using ^{35}S-metabolically-labelled rat hepatocytes. Separation of the proteins under non-reduced conditions revealed 3 predominant bands at 240, 180 kDa, and a third band migrating at 70 kDa previously unidentified in ligand blots, while the 115 and 90 kDa proteins were also present, but only as weak bands (Yen *et al.*, 1999). Under reduced conditions, all of these bands were detected, but only weakly, and 2 principal subunit bands with the molecular mass of 68 and 56 kDa were observed, with the 68 kDa band appearing sometimes as a doublet. The other bands originally identified in human FH fibroblasts and liver membranes were therefore most likely heterotrimer or tetramers of these 2 subunits, hereafter referred to as α (68 kDa) and β (56 kDa), respectively. The lower molecular mass band of the doublet was referred to as α′. Mild trypsin treatment of the hepatocytes before preparation of the cell lysates provided evidence that these two subunits were located on the cell surface. Two-dimensional electrophoresis in which the 240 kDa complex was separated under non-reduced conditions, then isolated and then separated under reduced conditions revealed that these 2 bands were directly derived from the larger protein complex (Yen *et al.*, 1999).

This validated antibody was therefore used to screen a phage expression library of rat liver cDNA, from which a potential candidate gene was identified with an open reading frame within a Kozak consensus sequence (Yen *et al.*, 1999).

3.3 Bioinformatic analysis of the candidate gene

Bioinformatic analysis of the predicted protein sequence of the candidate cDNA revealed a number of domains potentially indicative of a receptor (Bihain & Yen, 1998; Yen *et al.*, 1999) (Figure 2).

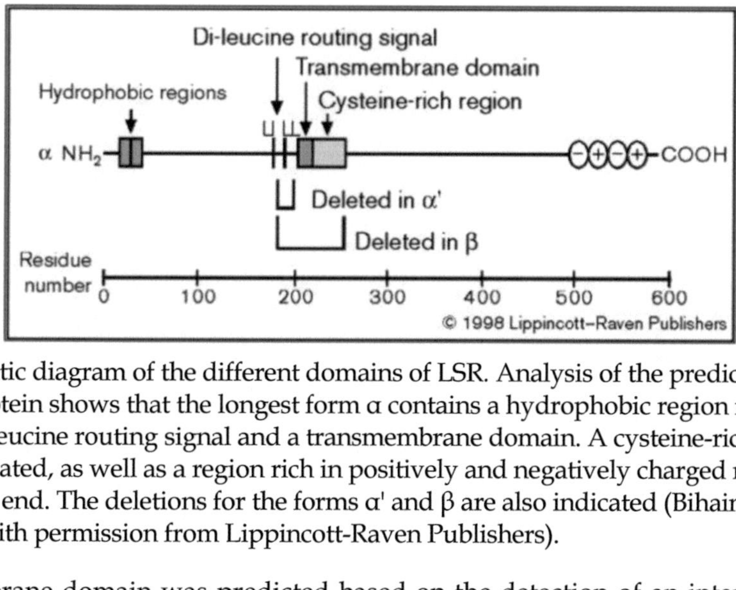

Fig. 2. Schematic diagram of the different domains of LSR. Analysis of the predicted sequence of the LSR protein shows that the longest form α contains a hydrophobic region near the NH_2 terminal, a dileucine routing signal and a transmembrane domain. A cysteine-rich region is found as indicated, as well as a region rich in positively and negatively charged residues near the carboxylic end. The deletions for the forms α' and β are also indicated (Bihain & Yen, 1998, reproduced with permission from Lippincott-Raven Publishers).

A transmembrane domain was predicted based on the detection of an internal cluster of hydrophobic residues. Several motifs related to endocytosis were detected, including a phosphorylation site NPGY that potentially represents a clathrin-binding site (Chen *et al.*, 1990), as well as a dileucine lysosomal targeting signal (Dietrich *et al.*, 1994; Shin *et al.*, 1991). Both of these domains were located on the N-terminal side with respect to the putative transmembrane domain, suggesting that the N-terminal was exposed intracellularly. On the N-terminal side of the protein, a smaller group of hydrophobic residues separated by a proline residue was located, which could be associated with the membrane phospholipids, providing a potential site of interaction with FFA. On the C-terminal side of the transmembrane domain, a cysteine-rich region was found that is often observed in the family of cytokine receptor proteins (Bazan, 1990). On the C-terminal end of the protein was located a group of alternating negatively and positively charged amino acids, which could represent the lipoprotein binding site. LSR is a phosphorylated receptor, with six phosphopeptides recently identified (Villen *et al.*, 2007), for which the function remains to be determined.

Northern blots revealed that mRNA of this candidate was detected in abundance in the liver, and in lesser levels in the kidney and lung, with none being detected in the muscle, spleen brain or heart (Yen *et al.*, 1999). Multiple mRNAs detected in the liver led to a closer examination of the mRNA derived from this potential LSR gene. Indeed, RT-PCR using different primer sets with overlapping sequences revealed in actual fact 3 distinct mRNAs with base numbers corresponding to LSR 2097, 2040 and 1893. The predicted molecular mass for each corresponded to 65.8, 63.8 and 58.3 kDa, respectively, matching rather closely to the 3 bands, α, α′ and β, identified in the previously described immuno-precipitation studies (Yen *et al.*, 1999). The predicted sequence of α corresponds to the full length protein containing all domains described. In α′, the sequence is practically identical with the exception of the loss of the dileucine repeat, suggesting that the intracellular routing of this subunit may differ from that of α. The β subunit no longer contains the sequences for endocytosis and lysosomal targeting, the transmembrane domain and the cysteine-rich domain. This subunit however still contains both N-terminal hydrophobic domain for potential interaction with FFA or the cell membrane, as well as the C-terminal potential binding site for lipoproteins. Therefore, while the subunits α and α′ are associated with the phospholipid bilayer, the β subunit, although associated with the LSR complex, may be located either extracellularly or intracellularly. The role of each 3 subunit either as individual polypeptides or as complexes associated on the cell membrane surface remains to be clearly defined.

The human *lsr* gene is located on chromosome 19 (19q13.12) (Bihain & Yen, 2005)(Genbank Gene ID: 51599). Interestingly, the LDL-R is located on chromosome 19 as well, but on 19p13.2. Furthermore, the *lsr* gene is found upstream of a number of genes involved in lipid and lipoprotein metabolism including *LIPE*, which codes for the hormone sensitive lipase of adipose tissue involved in the insulin-controlled FFA release from adipose tissue. Further downstream is the apoE/apoCI,CII,CIV gene cluster, in which apoE is one of the potential ligands for LSR. The *lsr* gene structure contains 10 exons, from which are derived 3 mRNA products, 2 of which are alternatively-spliced products. Sequence analysis revealed that the mRNA LSR 2097 (α) cited above represents the full-length sequence of all 10 exons, while the mRNA LSR 2040 (α′) and 1893 (β) represent alternatively spliced products in which either exon 4 (α′) or exon 4 and 5 (β) are deleted.

4. Functional validation of LSR's function

4.1 *In vitro* studies

Studies were conducted on CHO-K1 cells transiently transfected with the subunits α and β (Yen *et al.*, 1999). The transfection of each subunit individually led to increased binding and internalization of ^{125}I-LDL in the presence of oleate. Only the co-transfection of both subunits completely restored LSR activity as a lipoprotein receptor leading to degradation of the lipoprotein particle. These results, along with the immunoprecipitation studies described earlier suggested that LSR present as a multimeric complex composed of at least α and β serves as a receptor for lipoproteins leading to delivery of the particle to lysosomes for degradation via endocytosis. Each subunit appears to retain the capacity to bind the lipoprotein in the presence of oleate, consistent with the domains that are present in both polypeptides. However, the β polypeptide does not contain the transmembrane domain, and it remains to be determined how it is expressed on the cell surface in the absence of the other LSR subunits. These functional *in vitro* data nevertheless supported the notion that the protein products derived from this candidate gene were responsible for LSR activity (Figure 3).

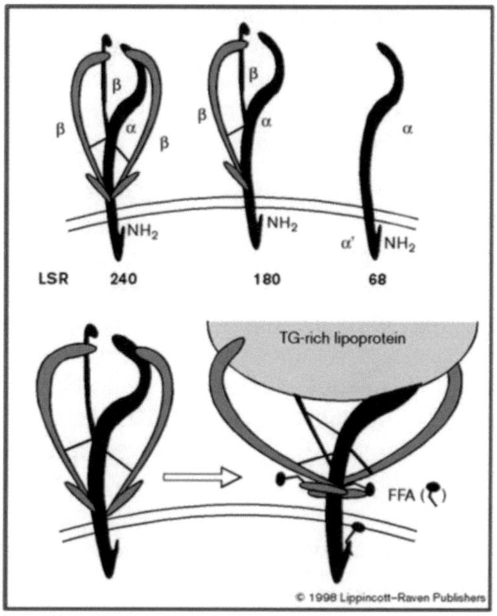

Fig. 3. Hypothetical model of lipolysis stimulated receptor and its activation by FFA. (Top panel) Analysis of both biochemical data and that from Figure 2 allows us to propose that LSR could exist on the cell surface as a multimeric complex of one α or α' (solid) subunit containing the transmembrane domain associated by disulfide bridges with two or three β (gray) subunits, which would be exposed completely on the extracellular side. The predicted molecular masses would be 240 and 180 kDa. It is also possible that α or α' subunits would be present alone on the plasma membrane surface. (Bottom panel) When FFAs are present, they bind to the hydrophobic regions, causing a conformational change of the LSR multimeric complex. This, in turn, exposes a binding site that, in this case, recognizes the apoE of a TG-rich lipoprotein, leading to its binding, internalization and degradation. (Bihain & Yen, 1998, reproduced with permission from Lippincott-Raven Publishers)

4.2 *In vivo* studies

The next step was then to determine LSR physiological function in an animal model. The *lsr* gene was inactivated in 129/Ola ES cells by the deletion of a gene segment containing exons 2-5, and then injected into mouse embryos. Complete suppression of this gene proved to be lethal at the embryonic stage (Mesli *et al.*, 2004). Even though 3 LSR-/- males were produced in the initial reproduction, they proved to be weak and sterile. Mortality of the homozygote embryos occurred between days 12.5 and 15.3 of the gestation period. Interestingly, the livers were abnormally small in these mice, suggesting that LSR was important for normal liver and embryonic development. Heterozygote $LSR^{+/-}$ mice proved however to be viable. Both immunoblots and quantitative PCR analyses demonstrated that hepatic LSR protein and mRNA were reduced by at least 50% in the $LSR^{+/-}$ mice (Yen *et al.*, 2008b).

Phenotypic analysis revealed that these $LSR^{+/-}$ mice with reduced expression of LSR displayed both increased postprandial lipemia and reduced clearance of TG-rich lipid particles (Yen *et al.*, 2008b). These results are striking in that even in heterozygote mice expressing only one allele, the lower expression of LSR has a considerable impact on their ability to clear lipoproteins from the circulation during the postprandial state. Interestingly, if LDL-R was also absent in $LSR^{+/-}$ mice, postprandial lipemia was increased almost 2-fold as compared to $LSR^{+/-}$ mice with normal LDL-R activity, suggesting a cooperativity between these two receptors in the removal of apoB-containing lipoproteins.

Both plasma cholesterol and TG levels increased in these mice when placed on a Western-type diet containing high fat and cholesterol, primarily due to an increase in both TG-rich lipoproteins and LDL (Yen *et al.*, 2008b). This was accompanied by increased lipid deposits in the aorta of the $LSR^{+/-}$ mice as compared to controls on the same diet, consistent with the atherogenic nature of these lipid particles. Interestingly, body weight gain of the $LSR^{+/-}$ but not $LSR^{+/+}$ mice on the Western-type diet, was directly correlated with the increased plasma lipid levels (Yen *et al.*, 2008b). This provided clear *in vivo* evidence for the physiological role of LSR in lipoprotein clearance during the postprandial phase, as well as a potential link provided by this receptor that could explain mixed hyperlipidemia (hypercholesterolemia and hypertriglyceridemia), weight gain and atherosclerosis. A recent study confirmed this in which adenovirus-mediated expression of siRNA specifically targeting hepatic LSR led to a significant increase in postprandial triglyceridemia, accompanied by increased levels of both apoB and apoE (Narvekar *et al.*, 2009).

Continued monitoring of $LSR^{+/-}$ mice on standard laboratory chow diet revealed significant weight gain and increased leptin with age as compared to control mice (Stenger *et al.*, 2010). This was most marked in female $LSR^{+/-}$ mice, where body mass was increased 1.5 fold due to increased body fat mass, accompanied by a disproportionate 3-fold increase in plasma leptin, a satiety hormone produced by the adipose tissue. Therefore, even under normal dietary conditions, a deficit in this receptor associated with elevated postprandial lipemia can lead to anomalies in both weight and plasma leptin. Indeed, hyperleptinemia is often observed in obesity, normally due to increased fat mass. It can be concluded therefore that by virtue of its role in the removal of apoB and apoE-containing lipoproteins from the circulation, LSR also contributes to the regulation of lipid distribution between the liver and the peripheral tissues.

4.3 LSR's physiological role

4.3.1 Clearance of TG-rich lipoproteins

The *in vivo* data described above clearly point towards LSR playing an important role in the removal of TG-rich lipoproteins during the postprandial phase, a time in which there is considerable influx of chylomicrons containing dietary-derived lipids. As these particles enter the portal system, they are acted upon by the hepatic and lipoprotein lipases anchored to the endothelium by HSPG. The resulting hydrolysis of the TG core leads to the release of FFA in the highly fenestrated space of Disse. This increased influx of FFA results in optimal activity of LSR, thus allowing the rapid removal of TG-rich particles and their remnants through this pathway. LSR therefore acts downstream of the lipolytic process as a hepatic receptor for the removal of TG-rich particles from the circulation (Figure 4). This conclusion is supported by a recent paper which reported that the removal of apoB-containing lipoproteins was inhibited in animals with reduced sulfation of heparan sulfate proteoglycans in the liver (MacArthur *et al.*, 2007). Indeed, it was suggested that the removal of lipoproteins was mediated by a downstream pathway independent of the LDL-R or LRP1 (Mahley & Huang, 2007).

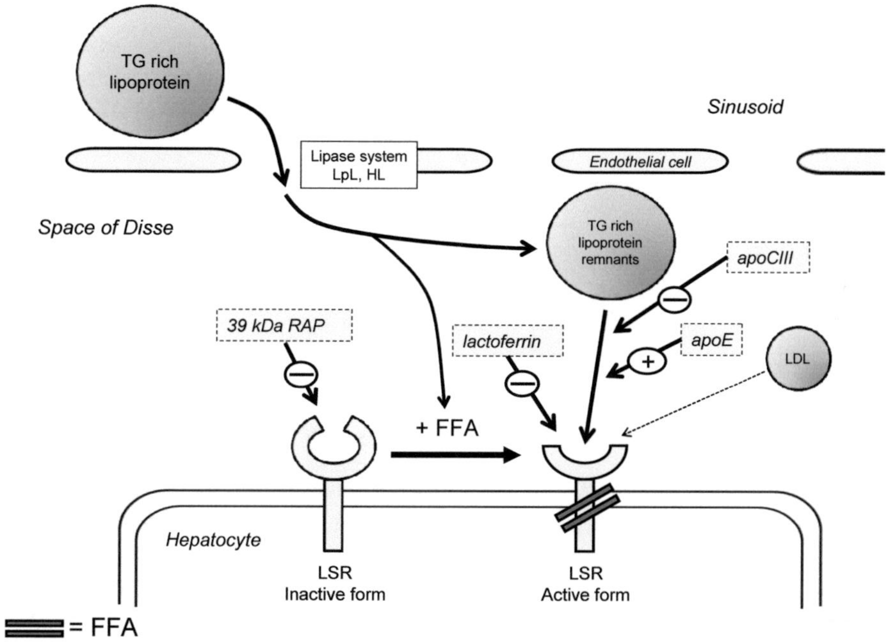

Fig. 4. Diagram for LSR physiological function. During the postprandial phase, there is increased influx of chylomicrons (TG-rich lipoprotein) into the space of Disse, providing substrate for the lipase system (LpL and hepatic lipase, HL). This produces high levels of FFA, which activate LSR, leading to a conformational change exposing an apoB,E binding site for the TG rich particles and/or their remnants. A similar mechanism could occur for TG-rich VLDL. Evidence shows that different proteins (39 kDa RAP, lactoferrin or apoCIII) that can lead to increased plasma TG levels can also affect LSR activity. LSR can also bind apoB of LDL, and may serve as an LDL-receptor independent pathway, most particularly in conditions in which LDL-R is deficient or inactive.

4.3.2 LDL-receptor independent pathway

Although acting principally as a receptor for TG-rich lipoproteins, LSR can also bind and endocytose LDL. Its ability to recognize CHD-treated LDL clearly distinguishes this receptor from the LDL-R. If LSR participates in the removal of LDL, it can occur but only under certain conditions. Indeed, LDL is a cholesterol-rich particle and contains very low levels of TG, and thus is not a preferred substrate for lipases. It therefore would not be able to generate sufficient lipolytic products to activate LSR, and its removal would occur primarily via the LDL-R. During the postprandial phase, the high influx of FFA in the space of Disse due to lipolytic activity on chylomicrons would inhibit binding of LDL to the LDL-R, while at the same time activating LSR. LDL could therefore be internalized by LSR during the postprandial phase, but only if concentrations are sufficiently high to be able to compete with the higher affinity TG-rich lipoproteins for binding to LSR (Figure 4). Indeed, this hypothesis is supported by the increase in LDL and plasma cholesterol observed in $LSR^{+/-}$ mice on a high-fat cholesterol-containing diet (Yen *et al.*, 2008b), as well as the increased plasma cholesterol following specific hepatic LSR knockout (Narvekar *et al.*, 2009).

This could provide an explanation for the large amounts of LDL that are cleared in FH subjects lacking the LDL receptor. Indeed, J Shepherd and colleagues postulated that this occurred via an LDL-R independent pathway based on their kinetic studies using CHD-LDL (Simons *et al.*, 1975). However, this liver-specific receptor-mediated process that was able to remove large quantities of LDL in both humans and animal models remained unidentified at the time.

The LSR model therefore addresses two issues concerning 1) the removal of apoB,E-containing TG-rich lipoproteins and 2) the clearance of LDL through an LDL-receptor independent pathway with the liver as the final destination.

4.4 Regulation of LSR

4.4.1 Leptin regulation of postprandial lipemia through its effect of LSR

In the *in vivo* studies described above, $LSR^{+/-}$ mice demonstrated increased weight gain and a disproportionate increase in leptin levels with age, suggesting a potential connection between leptin and LSR. Leptin is an adipokine that plays a key role in the regulation of food intake and energy storage. Leptin exerts its anorectic effect at the level of the central nervous system by acting on hypothalamic neurons through its interaction with the long form of its receptor, ObRb. This binding leads to increased expression of α-melanocyte stimulating hormone (α-MSH), a potent appetite inhibitor. However, leptin has also been described as a potent stimulator of synaptic transmission, by increasing N-methyl-D-aspartate (NMDA) receptor at the cell surface (Moult *et al.*, 2011; Oomura *et al.*, 2006), neuroprotection and neurogenesis (Garza *et al.*, 2008). Hence, leptin could represent much more than a simple on-off switch system controlling food intake and appetite behavior.

The leptin receptor family is widely expressed in many tissue types, including the liver, and leptin itself has been shown to exert other peripheral effects including insulin production by the pancreas (Kahn *et al.*, 2005; Morioka *et al.*, 2007). The distention of the stomach following a meal can lead to increased leptin secretion by adipocytes, thus controlling appetite and energy storage, consistent with its effect as a satiety factor. It was recently demonstrated that leptin can regulate postprandial lipemia by increasing hepatic LSR protein levels (Stenger *et al.*, 2010). Indeed, physiological concentrations of leptin (1 to 10 ng/mL) significantly and rapidly (within 1 h), increased LSR protein levels *in vitro* in Hepa1-6 cells. The leptin effect on LSR was confirmed *in vivo* in wild-type mice injected intra-peritoneally 8 days with leptin. Evidence indicated that leptin mediated this effect by promoting *lsr* gene transcription through the canonical MAPK/ERK pathway that is activated following leptin interaction with its receptor. Interestingly, these leptin-treated mice displayed lower body mass, despite no significant change in food intake. Furthermore, endogenous liver TG output in the form of VLDL was higher, most likely to deliver energy substrate to the peripheral tissues. Thus, during the postprandial phase, leptin directly affects dietary lipid metabolism and storage through its action on LSR at the liver. This mechanism represents a lever for the regulation of dietary lipid uptake, degradation and distribution, thus contributing to maintaining an appropriate lipid homeostasis (Stenger *et al.*, 2010) (Figure 5A).

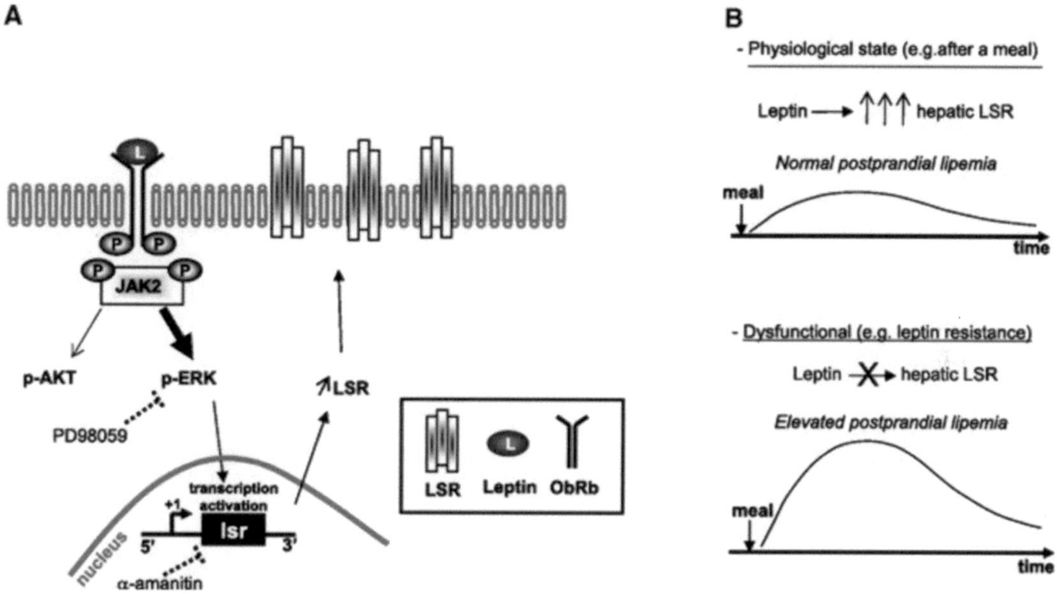

Fig. 5. Schematic diagram for leptin-mediated up-regulation of LSR and potential consequences on postprandial lipemia. A) We propose that the canonical leptin-induced signaling pathway involving phosphorylation of ERK leads to increased transcription of the lsr gene and increased protein levels of LSR at the surface of hepatocyte membranes. B) Under normal conditions, physiological levels of leptin are sufficient to maintain an optimal amount of LSR at the cell surface, permitting clearance of TG-rich lipoproteins during the postprandial phase. However, when leptin interaction with its receptor in the liver is impaired, such as in leptin resistance often observed in obesity, LSR protein levels at the cell surface may no longer be optimal, leading to decreased efficiency in removing lipids from the cell surface, resulting in elevated postprandial lipemia (Stenger *et al*, 2010, reproduced with permission from Faseb J).

4.5 Other roles of LSR

Other studies have recently appeared reporting other biological activities associated with LSR. LSR has been shown to be critical for the formation of tight tricellular contacts in epithelial cells, implying that this receptor is important for the maintenance of epithelial barrier function (Masuda *et al.*, 2011). In view of the multimericity of LSR, it will prove interesting to determine the role of each of the LSR polypeptide subunits in the formation of these cellular junctions. A function has been identified for the full-length LSR α subunit as host receptor for the binary toxin, *Clostridium difficile* transferase (CDT) (Papatheodorou *et al.*, 2011). By using a genome-wide haploid genetic screen on a derivative cell line of the human leukemia cell line KBM7, HAP1, the authors identified LSR as the potential host receptor, and then demonstrated increased reactivity to the toxin in cells expressing LSR. This would therefore point towards a specific role for one of the polypeptides of LSR. It remains however to determine if the other LSR subunits also demonstrate this activity.

LSR expression was found to be inducible by p53 (Jazag *et al.*, 2005; Kannan *et al.*, 2001), and is specifically expressed in certain cancers, including bladder and colon (Herbsleb *et al.*, 2008), which would suggest a role during cancer development. siRNA studies used to inhibit LSR expression in bladder cancer cells led towards increased invasion capacity and cell motility of the cells, suggesting that p53-induced LSR expression may be aimed towards inhibiting the cancerous properties of cells. Additional studies remain to determine LSR's exact role in such circumstances.

Finally, a homolog to LSR was identified by positional cloning with similar alternatively-sliced mRNA products, that could potentially relate to the susceptibility of mice to type 2 diabetes (Dokmanovic-Chouinard *et al.*, 2008). Indeed, LSR does not demonstrate any distinct homology to other known receptors, which is why little is known about the relationship between its structure and function. The embryonic lethality of LSR when both alleles are absent does show that this gene and its protein product(s) are essential for development and viability of the embryo and newborn. While this may be due in part to its function as a lipoprotein receptor, it may also be related to the different functions identified here involving the integrity of tricellular junctions, as well as a role in the cell cycle.

5. LSR as molecular link between hyperlipidemia, obesity and atherosclerosis

Currently, all data on the physiological role of LSR are based on cell and animal models. The role of leptin in the regulation of LSR as well as the change in distribution of tissue lipids led towards the hypothesis that LSR could provide a molecular link between hyperlipidemia and obesity. Indeed, it was found that LSR expression was significantly diminished in obese mouse models, including the *ob/ob*, *db/db* and diet-induced obese (DIO) mouse (Narvekar *et al.*, 2009; Yen *et al.*, 2008a). If leptin either by its absence (*ob/ob*) or due to a leptin resistance (*db/db* or DIO) is unable to increase LSR levels during the postprandial phase, this could lead to suboptimal levels of this receptor. This in turn could explain the abnormalities found in plasma lipid profile often associated with obesity, most notably in the form of increased plasma TG and elevated postprandial lipemia (Figure 5B).

The $LSR^{+/-}$ animal model demonstrates that even with 50% of LSR protein levels, this can lead eventually to increased weight gain and body fat mass. These changes are relatively small during early adulthood, but can develop with time into overweight issues leading to obesity, as was observed in aging $LSR^{+/-}$ female mice (Stenger *et al.*, 2010). In addition, if dietary fat content is increased under conditions in which LSR is sub-optimal, this leads to accelerated accumulation of fat mass. Interestingly, $LSR^{+/-}$ mice on a 10-wk high fat diet exhibited both hyperleptinemia and impaired phosphorylation of the ERK kinase pathway, associated with a decrease in hepatic leptin receptor expression, as compared to age-matched $LSR^{+/+}$ littermates on the same diet. These results suggested that the low levels of LSR are directly associated with the appearance of a high-fat diet-induced peripheral leptin resistance, for which the mechanisms remain to be elucidated. Peripheral leptin resistance combined with suboptimal levels of LSR during the critical moment of the postprandial phase leads to decreased efficiency in the removal of these atherogenic TG-rich lipoproteins from the circulation, as witnessed by the increased lipid deposits in the aorta of $LSR^{+/-}$ mice versus controls on a high fat and cholesterol-containing diet. (Yen *et al.*, 2008). It remains to characterize the physiopathology of these lipid deposits and the molecular mechanisms involved. Nevertheless, together these data show significant modifications in lipid status directly related to LSR that may explain the link between hyperlipidemia and obesity, two significant risk factors for atherosclerosis.

6. Conclusions and perspectives

This chapter describes the biochemical characterization and identification of a receptor activated by FFA that plays an important role in lipid homeostasis as receptor for TG-rich lipoproteins during the postprandial phase. Its role in the removal of dietary lipids in the form of chylomicrons from the circulation directly impacts on lipid distribution in the peripheral tissues.

It remains to be determined the importance of LSR in man. Numerous loci have been associated with obesity, body mass index and adiposity, including that located on chromosome 19q13, the loci on which the *lsr* gene is located. Indeed, linkage studies have identified this region as being associated with TG levels and obesity (Feitosa *et al.*, 2006; Long *et al.*, 2003). This locus has also been identified to contribute to serum lipid levels and dyslipidemia in genome wide association studies (Aulchenko *et al.*, 2009; Kathiresan *et al.*, 2009). It now remains to identify patients with deficiencies in LSR. It is unlikely that patients completely lacking LSR will be identified, since absence of this gene is lethal. Instead, it is possible that patients with genetic polymorphisms of the LSR gene leading to dysfunctional LSR may be discovered that are associated with hypertriglyceridemia or mixed hyperlipidemia.

A majority of research has focused on elevated LDL cholesterol levels, which is an established risk factor for cardiovascular disease. The delineation of the role of LDL-R, as well as the discovery of the statins led to a better understanding in the control of cholesterol homeostasis (Goldstein & Brown, 2009). This, combined with an heightened public awareness of the increased risk of cardiovascular disease associated with high cholesterol, led to a steady decline in average plasma cholesterol levels in recent years. Nevertheless, cardiovascular disease remains top in the list of principal causes for mortality. Other therapeutic strategies have not proven to be conclusive (Couzin, 2008), demonstrating that other potential targets are needed for the development of other therapeutic strategies. Elevated postprandial lipemia is often observed in both obesity and patients with coronary heart disease. Indeed, TGs are an independent risk factor for cardiovascular disease (Goldberg *et al.*, 2011; Lopez-Miranda *et al.*, 2006). Such being the case, it is also important to find new ways to treat hypertriglyceridemia or mixed hyperlipidemia.

The LSR model provides new insight in the molecular mechanisms involved in the development of hyperlipidemia and represents a molecular between hyperlipidemia, weight gain, and atherosclerosis. Further investigation is required to understand the underlying regulatory pathways involved. Nevertheless, it is clear that LSR is a new and promising therapeutic target for the treatment of mixed hyperlipidemia. Optimal activation of LSR by preventive or therapeutic means can be envisaged to normalize clearance of lipoproteins during the postprandial lipemia, thereby preventing the accumulation of these high-risk lipid particles in the circulation.

7. Acknowledgements

The work described here represents the efforts of many, but most particularly that of Bernard E. Bihain, who discovered this receptor with F.T. Yen. Although he is pursuing other avenues of research, he will not be forgotten for his work and dedication in bringing to light a novel lipoprotein receptor, this despite enormous skepticism from the scientific community. The authors also dedicate this chapter to Patrice André, Nelly Clossais-Besnard, and Vincent Lotteau of INSERM U391 in Rennes where the gene was identified, and proven not until much later to be the right one.

8. References

Aalto-Setala, K., Fisher, E.A., Chen, X., Chajek-Shaul, T., Hayek, T., Zechner, R., Walsh, A., Ramakrishnan, R., Ginsberg, H.N. & Breslow, J.L. (1992). Mechanism of hypertriglyceridemia in human apolipoprotein (apo) CIII transgenic mice. Diminished very low density lipoprotein fractional catabolic rate associated with increased apo CIII and reduced apo E on the particles. *J Clin Invest*, Vol. 90, No. 5, pp. 1889-1900

Attie, A.D., Pittman, R.C. & Steinberg, D. (1982). Hepatic catabolism of low density lipoprotein: mechanisms and metabolic consequences. *Hepatology*, Vol. 2, No. 2, pp. 269-281

Aulchenko, Y.S., Ripatti, S., Lindqvist, I., Boomsma, D., Heid, I.M., Pramstaller, P.P., Penninx, B.W., Janssens, A.C., Wilson, J.F., Spector, T., Martin, N.G., Pedersen, N.L., Kyvik, K.O., Kaprio, J., Hofman, A., Freimer, N.B., Jarvelin, M.R., Gyllensten, U., Campbell, H., Rudan, I., Johansson, A., Marroni, F., Hayward, C., Vitart, V., Jonasson, I., Pattaro, C., Wright, A., Hastie, N., Pichler, I., Hicks, A.A., Falchi, M., Willemsen, G., Hottenga, J.J., de Geus, E.J., Montgomery, G.W., Whitfield, J., Magnusson, P., Saharinen, J., Perola, M., Silander, K., Isaacs, A., Sijbrands, E.J., Uitterlinden, A.G., Witteman, J.C., Oostra, B.A., Elliott, P., Ruokonen, A., Sabatti, C., Gieger, C., Meitinger, T., Kronenberg, F., Doring, A., Wichmann, H.E., Smit, J.H., McCarthy, M.I., van Duijn, C.M. & Peltonen, L. (2009). Loci influencing lipid levels and coronary heart disease risk in 16 European population cohorts. *Nat Genet*, Vol. 41, No. 1, pp. 47-55

Bazan, J.F. (1990). Structural design and molecular evolution of a cytokine receptor superfamily. *Proc Natl Acad Sci U S A*, Vol. 87, No. 18, pp. 6934-6938

Beisiegel, U., Weber, W., Havinga, J.R., Ihrke, G., Hui, D.Y., Wernette-Hammond, M.E., Turck, C.W., Innerarity, T.L. & Mahley, R.W. (1988). Apolipoprotein E-binding proteins isolated from dog and human liver. *Arteriosclerosis*, Vol. 8, No. 3, pp. 288-297

Beisiegel, U., Weber, W., Ihrke, G., Herz, J. & Stanley, K.K. (1989). The LDL-receptor-related protein, LRP, is an apolipoprotein E-binding protein. *Nature*, Vol. 341, No. 6238, pp. 162-164

Berr, F. & Kern, F., Jr. (1984). Plasma clearance of chylomicrons labeled with retinyl palmitate in healthy human subjects. *J Lipid Res*, Vol. 25, No. 8, pp. 805-812

Bihain, B.E., Deckelbaum, R.J., Yen, F.T., Gleeson, A.M., Carpentier, Y.A. & Witte, L.D. (1989). Unesterified fatty acids inhibit the binding of low density lipoproteins to the human fibroblast low density lipoprotein receptor. *J Biol Chem*, Vol. 264, No. 29, pp. 17316-17321

Bihain, B.E. & Yen, F.T. (1992). Free fatty acids activate a high-affinity saturable pathway for degradation of low-density lipoproteins in fibroblasts from a subject homozygous for familial hypercholesterolemia. *Biochemistry*, Vol. 31, No. 19, pp. 4628-4636

Bihain, B.E. & Yen, F.T. (1998). The lipolysis stimulated receptor: a gene at last. *Curr Opin Lipidol*, Vol. 9, No. 3, pp. 221-224

Bihain, B.E. & Yen, F.T. (2005). Receptors for circulating lipoproteins (Article in French). *Médecine Clinique endocrinologie & diabète*, Vol. Special edition, No. pp. 2-6

Brown, M.S. & Goldstein, J.L. (1986). A receptor-mediated pathway for cholesterol homeostasis. *Science*, Vol. 232, No. 4746, pp. 34-47

Brown, M.S. & Goldstein, J.L. (1997). The SREBP pathway: regulation of cholesterol metabolism by proteolysis of a membrane-bound transcription factor. *Cell*, Vol. 89, No. 3, pp. 331-340

Chen, W.J., Goldstein, J.L. & Brown, M.S. (1990). NPXY, a sequence often found in cytoplasmic tails, is required for coated pit-mediated internalization of the low density lipoprotein receptor. *J Biol Chem*, Vol. 265, No. 6, pp. 3116-3123

Cortner, J.A., Coates, P.M., Le, N.A., Cryer, D.R., Ragni, M.C., Faulkner, A. & Langer, T. (1987). Kinetics of chylomicron remnant clearance in normal and in hyperlipoproteinemic subjects. *J Lipid Res*, Vol. 28, No. 2, pp. 195-206

Couzin, J. (2008). Clinical trials and tribulations. Cholesterol veers off script. *Science*, Vol. 322, No. 5899, pp. 220-223

de Faria, E., Fong, L.G., Komaromy, M. & Cooper, A.D. (1996). Relative roles of the LDL receptor, the LDL receptor-like protein, and hepatic lipase in chylomicron remnant removal by the liver. *J Lipid Res*, Vol. 37, No. 1, pp. 197-209

Dietrich, J., Hou, X., Wegener, A.M. & Geisler, C. (1994). CD3 gamma contains a phosphoserine-dependent di-leucine motif involved in down-regulation of the T cell receptor. *Embo J*, Vol. 13, No. 9, pp. 2156-2166

Dokmanovic-Chouinard, M., Chung, W.K., Chevre, J.C., Watson, E., Yonan, J., Wiegand, B., Bromberg, Y., Wakae, N., Wright, C.V., Overton, J., Ghosh, S., Sathe, G.M., Ammala, C.E., Brown, K.K., Ito, R., LeDuc, C., Solomon, K., Fischer, S.G. & Leibel, R.L. (2008). Positional cloning of "Lisch-Like", a candidate modifier of susceptibility to type 2 diabetes in mice. *PLoS Genet*, Vol. 4, No. 7, pp. e1000137

Eisenberg, S., Sehayek, E., Olivecrona, T. & Vlodavsky, I. (1992). Lipoprotein lipase enhances binding of lipoproteins to heparan sulfate on cell surfaces and extracellular matrix. *J Clin Invest*, Vol. 90, No. 5, pp. 2013-2021

Feitosa, M.F., Rice, T., North, K.E., Kraja, A., Rankinen, T., Leon, A.S., Skinner, J.S., Blangero, J., Bouchard, C. & Rao, D.C. (2006). Pleiotropic QTL on chromosome 19q13 for triglycerides and adiposity: the HERITAGE Family Study. *Atherosclerosis*, Vol. 185, No. 2, pp. 426-432

Garza, J.C., Guo, M., Zhang, W. & Lu, X.Y. (2008). Leptin increases adult hippocampal neurogenesis in vivo and in vitro. *J Biol Chem*, Vol. 283, No. 26, pp. 18238-18247

Goldberg, I.J., Eckel, R.H. & McPherson, R. (2011). Triglycerides and heart disease: still a hypothesis? *Arterioscler Thromb Vasc Biol*, Vol. 31, No. 8, pp. 1716-1725

Goldstein, J.L., Basu, S.K. & Brown, M.S. (1983). Receptor-mediated endocytosis of low-density lipoprotein in cultured cells. *Methods Enzymol*, Vol. 98, No. pp. 241-260

Goldstein, J.L. & Brown, M.S. (2009). The LDL receptor. *Arterioscler Thromb Vasc Biol*, Vol. 29, No. 4, pp. 431-438

Goldstein, J.L., Hobbs, H.H. & Brown, M.S. (1995). Familial hypercholesterolemia, In: *The Metabolic and Molecular Bases of Inherited Disease*, Scriver, C. R., Beaudet, A. L., Sly, W. S. and Valle, D., pp. 1981-2030, McGraw-Hill, Inc., New York

Havel, R.J. & Kane, J.P. (1995). Structure and Metabolism of plasma lipoproteins, In: *The Metabolic and Molecular Bases of Inherited Disease*, Scriver, C. R., Beaudet, A. L., Sly, W. S. and Valle, D., pp. 1841-1851, McGraw-Hill, Inc., New York

Herbsleb, M., Birkenkamp-Demtroder, K., Thykjaer, T., Wiuf, C., Hein, A.M., Orntoft, T.F. & Dyrskjot, L. (2008). Increased cell motility and invasion upon knockdown of lipolysis stimulated lipoprotein receptor (LSR) in SW780 bladder cancer cells. *BMC Med Genomics*, Vol. 1, No. pp. 31

Herz, J., Clouthier, D.E. & Hammer, R.E. (1992). LDL receptor-related protein internalizes and degrades uPA-PAI-1 complexes and is essential for embryo implantation. *Cell*, Vol. 71, No. 3, pp. 411-421

Herz, J., Goldstein, J.L., Strickland, D.K., Ho, Y.K. & Brown, M.S. (1991). 39-kDa protein modulates binding of ligands to low density lipoprotein receptor-related protein/alpha 2-macroglobulin receptor. *J Biol Chem*, Vol. 266, No. 31, pp. 21232-21238

Herz, J., Hamann, U., Rogne, S., Myklebost, O., Gausepohl, H. & Stanley, K.K. (1988). Surface location and high affinity for calcium of a 500-kd liver membrane protein closely related to the LDL-receptor suggest a physiological role as lipoprotein receptor. *Embo J*, Vol. 7, No. 13, pp. 4119-4127

Herz, J. & Strickland, D.K. (2001). LRP: a multifunctional scavenger and signaling receptor. *J Clin Invest*, Vol. 108, No. 6, pp. 779-784

Hobbs, H.H., Brown, M.S., Russell, D.W., Davignon, J. & Goldstein, J.L. (1987). Deletion in the gene for the low-density-lipoprotein receptor in a majority of French Canadians with familial hypercholesterolemia. *N Engl J Med*, Vol. 317, No. 12, pp. 734-737

Hofmann, S.M., Zhou, L., Perez-Tilve, D., Greer, T., Grant, E., Wancata, L., Thomas, A., Pfluger, P.T., Basford, J.E., Gilham, D., Herz, J., Tschop, M.H. & Hui, D.Y. (2007). Adipocyte LDL receptor-related protein-1 expression modulates postprandial lipid transport and glucose homeostasis in mice. *J Clin Invest*, Vol. 117, No. 11, pp. 3271-3282

Huettinger, M., Retzek, H., Eder, M. & Goldenberg, H. (1988). Characteristics of chylomicron remnant uptake into rat liver. *Clin Biochem*, Vol. 21, No. 2, pp. 87-92

Ishibashi, S., Perrey, S., Chen, Z., Osuga, J., Shimada, M., Ohashi, K., Harada, K., Yazaki, Y. & Yamada, N. (1996). Role of the low density lipoprotein (LDL) receptor pathway in the metabolism of chylomicron remnants. A quantitative study in knockout mice lacking the LDL receptor, apolipoprotein E, or both. *J Biol Chem*, Vol. 271, No. 37, pp. 22422-22427

Ito, Y., Azrolan, N., O'Connell, A., Walsh, A. & Breslow, J.L. (1990). Hypertriglyceridemia as a result of human apo CIII gene expression in transgenic mice. *Science*, Vol. 249, No. 4970, pp. 790-793

Jazag, A., Ijichi, H., Kanai, F., Imamura, T., Guleng, B., Ohta, M., Imamura, J., Tanaka, Y., Tateishi, K., Ikenoue, T., Kawakami, T., Arakawa, Y., Miyagishi, M., Taira, K., Kawabe, T. & Omata, M. (2005). Smad4 silencing in pancreatic cancer cell lines using stable RNA interference and gene expression profiles induced by transforming growth factor-beta. *Oncogene*, Vol. 24, No. 4, pp. 662-671

Kahn, B.B., Alquier, T., Carling, D. & Hardie, D.G. (2005). AMP-activated protein kinase: ancient energy gauge provides clues to modern understanding of metabolism. *Cell Metab*, Vol. 1, No. 1, pp. 15-25

Kannan, K., Amariglio, N., Rechavi, G., Jakob-Hirsch, J., Kela, I., Kaminski, N., Getz, G., Domany, E. & Givol, D. (2001). DNA microarrays identification of primary and secondary target genes regulated by p53. *Oncogene*, Vol. 20, No. 18, pp. 2225-2234

Kathiresan, S., Willer, C.J., Peloso, G.M., Demissie, S., Musunuru, K., Schadt, E.E., Kaplan, L., Bennett, D., Li, Y., Tanaka, T., Voight, B.F., Bonnycastle, L.L., Jackson, A.U., Crawford, G., Surti, A., Guiducci, C., Burtt, N.P., Parish, S., Clarke, R., Zelenika, D., Kubalanza, K.A., Morken, M.A., Scott, L.J., Stringham, H.M., Galan, P., Swift, A.J., Kuusisto, J., Bergman, R.N., Sundvall, J., Laakso, M., Ferrucci, L., Scheet, P., Sanna, S., Uda, M., Yang, Q., Lunetta, K.L., Dupuis, J., de Bakker, P.I., O'Donnell, C.J., Chambers, J.C., Kooner, J.S., Hercberg, S., Meneton, P., Lakatta, E.G., Scuteri, A., Schlessinger, D., Tuomilehto, J., Collins, F.S., Groop, L., Altshuler, D., Collins, R., Lathrop, G.M., Melander, O., Salomaa, V., Peltonen, L., Orho-Melander, M., Ordovas, J.M., Boehnke, M., Abecasis, G.R., Mohlke, K.L. & Cupples, L.A. (2009). Common variants at 30 loci contribute to polygenic dyslipidemia. *Nat Genet*, Vol. 41, No. 1, pp. 56-65

Kita, T., Goldstein, J.L., Brown, M.S., Watanabe, Y., Hornick, C.A. & Havel, R.J. (1982). Hepatic uptake of chylomicron remnants in WHHL rabbits: a mechanism genetically distinct from the low density lipoprotein receptor. *Proc Natl Acad Sci U S A*, Vol. 79, No. 11, pp. 3623-3627

Kowal, R.C., Herz, J., Goldstein, J.L., Esser, V. & Brown, M.S. (1989). Low density lipoprotein receptor-related protein mediates uptake of cholesteryl esters derived from apoprotein E-enriched lipoproteins. *Proc Natl Acad Sci U S A*, Vol. 86, No. 15, pp. 5810-5814

Long, J.R., Liu, P.Y., Liu, Y.J., Lu, Y., Xiong, D.H., Elze, L., Recker, R.R. & Deng, H.W. (2003). APOE and TGF-beta1 genes are associated with obesity phenotypes. *J Med Genet*, Vol. 40, No. 12, pp. 918-924

Lopez-Miranda, J., Perez-Martinez, P., Marin, C., Moreno, J.A., Gomez, P. & Perez-Jimenez, F. (2006). Postprandial lipoprotein metabolism, genes and risk of cardiovascular disease. *Curr Opin Lipidol*, Vol. 17, No. 2, pp. 132-138

MacArthur, J.M., Bishop, J.R., Stanford, K.I., Wang, L., Bensadoun, A., Witztum, J.L. & Esko, J.D. (2007). Liver heparan sulfate proteoglycans mediate clearance of triglyceride-rich lipoproteins independently of LDL receptor family members. *J Clin Invest*, Vol. 117, No. 1, pp. 153-164

Maeda, N., Li, H., Lee, D., Oliver, P., Quarfordt, S.H. & Osada, J. (1994). Targeted disruption of the apolipoprotein C-III gene in mice results in hypotriglyceridemia and protection from postprandial hypertriglyceridemia. *J Biol Chem*, Vol. 269, No. 38, pp. 23610-23616

Mahley, R.W. & Huang, Y. (2007). Atherogenic remnant lipoproteins: role for proteoglycans in trapping, transferring, and internalizing. *J Clin Invest*, Vol. 117, No. 1, pp. 94-98

Mann, C.J., Khallou, J., Chevreuil, O., Troussard, A.A., Guermani, L.M., Launay, K., Delplanque, B., Yen, F.T. & Bihain, B.E. (1995). Mechanism of activation and functional significance of the lipolysis-stimulated receptor. Evidence for a role as chylomicron remnant receptor. *Biochemistry*, Vol. 34, No. 33, pp. 10421-10431

Mann, C.J., Troussard, A.A., Yen, F.T., Hannouche, N., Najib, J., Fruchart, J.C., Lotteau, V., Andre, P. & Bihain, B.E. (1997). Inhibitory effects of specific apolipoprotein C-III isoforms on the binding of triglyceride-rich lipoproteins to the lipolysis-stimulated receptor. *J Biol Chem*, Vol. 272, No. 50, pp. 31348-31354

Masuda, S., Oda, Y., Sasaki, H., Ikenouchi, J., Higashi, T., Akashi, M., Nishi, E. & Furuse, M. (2011). LSR defines cell corners for tricellular tight junction formation in epithelial cells. *J Cell Sci*, Vol. 124, No. Pt 4, pp. 548-555

Medh, J.D., Fry, G.L., Bowen, S.L., Pladet, M.W., Strickland, D.K. & Chappell, D.A. (1995). The 39-kDa receptor-associated protein modulates lipoprotein catabolism by binding to LDL receptors. *J Biol Chem*, Vol. 270, No. 2, pp. 536-540

Mesli, S., Javorschi, S., Berard, A.M., Landry, M., Priddle, H., Kivlichan, D., Smith, A.J., Yen, F.T., Bihain, B.E. & Darmon, M. (2004). Distribution of the lipolysis stimulated receptor in adult and embryonic murine tissues and lethality of LSR-/- embryos at 12.5 to 14.5 days of gestation. *Eur J Biochem*, Vol. 271, No. 15, pp. 3103-3114

Moestrup, S.K. & Gliemann, J. (1991). Analysis of ligand recognition by the purified alpha 2-macroglobulin receptor (low density lipoprotein receptor-related protein). Evidence that high affinity of alpha 2-macroglobulin-proteinase complex is achieved by binding to adjacent receptors. *J Biol Chem*, Vol. 266, No. 21, pp. 14011-14017

Mokuno, H., Brady, S., Kotite, L., Herz, J. & Havel, R.J. (1994). Effect of the 39-kDa receptor-associated protein on the hepatic uptake and endocytosis of chylomicron remnants and low density lipoproteins in the rat. *J Biol Chem*, Vol. 269, No. 18, pp. 13238-13243

Morioka, T., Asilmaz, E., Hu, J., Dishinger, J.F., Kurpad, A.J., Elias, C.F., Li, H., Elmquist, J.K., Kennedy, R.T. & Kulkarni, R.N. (2007). Disruption of leptin receptor expression in the pancreas directly affects beta cell growth and function in mice. *J Clin Invest*, Vol. 117, No. 10, pp. 2860-2868

Moult, P.R., Cross, A., Santos, S.D., Carvalho, A.L., Lindsay, Y., Connolly, C.N., Irving, A.J., Leslie, N.R. & Harvey, J. (2011). Leptin regulates AMPA receptor trafficking via PTEN inhibition. *J Neurosci*, Vol. 30, No. 11, pp. 4088-4101

Mulder, M., Lombardi, P., Jansen, H., van Berkel, T.J., Frants, R.R. & Havekes, L.M. (1993). Low density lipoprotein receptor internalizes low density and very low density lipoproteins that are bound to heparan sulfate proteoglycans via lipoprotein lipase. *J Biol Chem*, Vol. 268, No. 13, pp. 9369-9375

Narvekar, P., Berriel Diaz, M., Krones-Herzig, A., Hardeland, U., Strzoda, D., Stohr, S., Frohme, M. & Herzig, S. (2009). Liver-specific loss of lipolysis-stimulated lipoprotein receptor triggers systemic hyperlipidemia in mice. *Diabetes*, Vol. 58, No. 5, pp. 1040-1049

Oomura, Y., Hori, N., Shiraishi, T., Fukunaga, K., Takeda, H., Tsuji, M., Matsumiya, T., Ishibashi, M., Aou, S., Li, X.L., Kohno, D., Uramura, K., Sougawa, H., Yada, T., Wayner, M.J. & Sasaki, K. (2006). Leptin facilitates learning and memory performance and enhances hippocampal CA1 long-term potentiation and CaMK II phosphorylation in rats. *Peptides*, Vol. 27, No. 11, pp. 2738-2749

Papatheodorou, P., Carette, J.E., Bell, G.W., Schwan, C., Guttenberg, G., Brummelkamp, T.R. & Aktories, K. (2011). Lipolysis-stimulated lipoprotein receptor (LSR) is the host receptor for the binary toxin Clostridium difficile transferase (CDT). *Proc Natl Acad Sci U S A*, Vol. 108, No. 39, pp. 16422-16427

Ramprasad, M.P., Li, R., Gianturco, S.H. & Bradley, W.A. (1995). Purification of the human THP-1 monocyte-macrophage triglyceride-rich lipoprotein receptor. *Biochem Biophys Res Commun*, Vol. 210, No. 2, pp. 491-497

Rohlmann, A., Gotthardt, M., Hammer, R.E. & Herz, J. (1998). Inducible inactivation of hepatic LRP gene by cre-mediated recombination confirms role of LRP in clearance of chylomicron remnants. *J Clin Invest*, Vol. 101, No. 3, pp. 689-695

Rubinsztein, D.C., Cohen, J.C., Berger, G.M., van der Westhuyzen, D.R., Coetzee, G.A. & Gevers, W. (1990). Chylomicron remnant clearance from the plasma is normal in familial hypercholesterolemic homozygotes with defined receptor defects. *J Clin Invest*, Vol. 86, No. 4, pp. 1306-1312

Sakai, J., Hoshino, A., Takahashi, S., Miura, Y., Ishii, H., Suzuki, H., Kawarabayasi, Y. & Yamamoto, T. (1994). Structure, chromosome location, and expression of the human very low density lipoprotein receptor gene. *J Biol Chem*, Vol. 269, No. 3, pp. 2173-2182

Sherrill, B.C. & Dietschy, J.M. (1978). Characterization of the sinusoidal transport process responsible for uptake of chylomicrons by the liver. *J Biol Chem*, Vol. 253, No. 6, pp. 1859-1867

Shin, J., Dunbrack, R.L., Jr., Lee, S. & Strominger, J.L. (1991). Phosphorylation-dependent down-modulation of CD4 requires a specific structure within the cytoplasmic domain of CD4. *J Biol Chem*, Vol. 266, No. 16, pp. 10658-10665

Simons, L.A., Reichl, D., Myant, N.B. & Mancini, M. (1975). The metabolism of the apoprotein of plasma low density lipoprotein in familial hyperbetalipoproteinaemia in the homozygous form. *Atherosclerosis*, Vol. 21, No. 2, pp. 283-298

Soutar, A.K. (1989). Lipoprotein-remnant receptors: second receptor verified? *Nature*, Vol. 341, No. 6238, pp. 106-107

Stenger, C., Hanse, M., Pratte, D., Mbala, M.L., Akbar, S., Koziel, V., Escanye, M.C., Kriem, B., Malaplate-Armand, C., Olivier, J.L., Oster, T., Pillot, T. & Yen, F.T. (2010). Up-regulation of hepatic lipolysis stimulated lipoprotein receptor by leptin: a potential lever for controlling lipid clearance during the postprandial phase. *Faseb J*, Vol. 24, No. 11, pp. 4218-4228

Strickland, D.K., Ashcom, J.D., Williams, S., Burgess, W.H., Migliorini, M. & Argraves, W.S. (1990). Sequence identity between the alpha 2-macroglobulin receptor and low density lipoprotein receptor-related protein suggests that this molecule is a multifunctional receptor. *J Biol Chem*, Vol. 265, No. 29, pp. 17401-17404

Troussard, A.A., Khallou, J., Mann, C.J., Andre, P., Strickland, D.K., Bihain, B.E. & Yen, F.T. (1995). Inhibitory effect on the lipolysis-stimulated receptor of the 39-kDa receptor-associated protein. *J Biol Chem*, Vol. 270, No. 29, pp. 17068-17071

Villen, J., Beausoleil, S.A., Gerber, S.A. & Gygi, S.P. (2007). Large-scale phosphorylation analysis of mouse liver. *Proc Natl Acad Sci U S A*, Vol. 104, No. 5, pp. 1488-1493

Williams, K.J., Fless, G.M., Petrie, K.A., Snyder, M.L., Brocia, R.W. & Swenson, T.L. (1992). Mechanisms by which lipoprotein lipase alters cellular metabolism of lipoprotein(a), low density lipoprotein, and nascent lipoproteins. Roles for low density lipoprotein receptors and heparan sulfate proteoglycans. *J Biol Chem*, Vol. 267, No. 19, pp. 13284-13292

Williams, S.E., Ashcom, J.D., Argraves, W.S. & Strickland, D.K. (1992). A novel mechanism for controlling the activity of alpha 2-macroglobulin receptor/low density lipoprotein receptor-related protein. Multiple regulatory sites for 39-kDa receptor-associated protein. *J Biol Chem*, Vol. 267, No. 13, pp. 9035-9040

Willnow, T.E., Sheng, Z., Ishibashi, S. & Herz, J. (1994). Inhibition of hepatic chylomicron remnant uptake by gene transfer of a receptor antagonist. *Science*, Vol. 264, No. 5164, pp. 1471-1474

Yen, F.T., Mann, C.J., Guermani, L.M., Hannouche, N.F., Hubert, N., Hornick, C.A., Bordeau, V.N., Agnani, G. & Bihain, B.E. (1994). Identification of a lipolysis-stimulated receptor that is distinct from the LDL receptor and the LDL receptor-related protein. *Biochemistry*, Vol. 33, No. 5, pp. 1172-1180

Yen, F.T., Masson, M., Clossais-Besnard, N., Andre, P., Grosset, J.M., Bougueleret, L., Dumas, J.B., Guerassimenko, O. & Bihain, B.E. (1999). Molecular cloning of a lipolysis-stimulated remnant receptor expressed in the liver. *J Biol Chem*, Vol. 274, No. 19, pp. 13390-13398

Yen, F.T., Masson, M., Roitel, O., Bonnard, L., Notet, V., Pratte, D., Magueur, E. & Bihain, B.E. (2008a). Elevated posptrandial lipemia and increased apoB-containing lipoproteins in mice with single allele inactivation of hepatic lipolysis stimulated receptor (LSR): Evidence for link between hyperlipidemia, obesity, and atherosclerosis. *Arterioscler Thromb Vasc Biol*, Vol. 28, No. 6, pp. e36

Yen, F.T., Roitel, O., Bonnard, L., Notet, V., Pratte, D., Stenger, C., Magueur, E. & Bihain, B.E. (2008b). Lipolysis stimulated lipoprotein receptor: A novel molecular link between hyperlipidemia, weight gain, and atherosclerosis in mice. *J Biol Chem*, Vol. 283, No. 37, pp. 25650-25659

15

Ghrelin, a Gastric Hormone with Diverse Functions

Ziru Li[1], Jie Luo[1], Yin Li[1] and Weizhen Zhang[1,2]
[1]*Department of Physiology and Pathophysiology*
Peking University Health Science Center, Beijing
[2]*Department of Surgery, University of Michigan Medical Center, Ann Arbor, MI*
[1]*China*
[2]*USA*

1. Introduction

Ghrelin is a 28 amino acid peptide hormone derived from the X/A like endocrine cells located in the gastric mucosa. Since its discovery more than a decade ago, about 5000 papers have been reported. The scope of these reports is broad and ranges from the gene structure, transcriptional regulation, posttranslation modification, ligand-receptor binding activities, intracellular signaling, to various biological functions such as food intake, energy balance, glucose and lipid metabolism, cardiovascular disease and immunomodulation. In this chapter, we will summarize the advance of our knowledge on this fascinating hormone, ghrelin.

2. The history of ghrelin discovery

2.1 Development of GHS (growth hormone secretagogues)

Growth hormone (GH) was first identified for its effect on longitudinal growth. Recombinant growth hormone is therefore used as the treatment for individuals with short stature in various conditions (Chipman, 1993; Jorgensen & Christiansen, 1993). However, GH replacement therapy encounters several drawbacks such as low bioavailability and side effects upon its administration. Moreover, most growth hormone-deficient individuals exhibit a secretory defect rather than a primary deficiency in the production of growth hormone.

A synthetic hexapeptide, His-D-Trp-Ala-Trp-D-Phe-Lys-NH2 (GHRP-6), was identified as a potent stimulator on GH release *in vitro* and *in vivo* by an unknown mechanism (Bowers et al., 1984). Because of its poor oral bioavailability and short half-life in human serum, GHRP-6 was selected only as a structural model to design non-peptide mimetics (Bowers et al., 1992). However, as a peptide, GHRP-6 did not readily lend itself to optimization of pharmacokinetic properties by medicinal chemistry. The benzodiazepine like template containing aromatic substitution was therefore discovered as potential non-peptide structure to model the GHRP-6 structure (Bowers et al., 1980). Based on the structure-activity relation, a non-peptidyl GH secretagogue L-692, 429, was identified. This non-peptidyl GH secretagogue synergizes with GHRP-6 to stimulate GH release and cAMP production (Smith et al., 1993).

In parallel with developing structure activity relationships for the benzolactams, alternative approach was proposed. It has been suggested that the derivatives of frequently occurring units was used as a useful approach to design receptor agonists and antagonists (Evans et al., 1988). These recurring structural units were termed "privileged structures" and had been recognized (Patchett et al., 1995) as hydrophobic double ring systems contributing to receptor binding of many antagonists of biogenic amines. By replacing d-Trp by *O*-benzyl-d-serine and incorporating a methanesulfonyl amido group, a new drug with excellent oral bioavailability and specificity in its release of GH without significant effect on plasma levels of other hormones was discovered and named L-163,191 (MK-0677) (Patchett et al., 1995).

2.2 Discovery of GHS-R

Extensive researches had been focused on the regulation of GH secretion in the 90s, due to its potential in the therapeutics of GH deficient related diseases. GH secretion is mainly regulated by GH-releasing hormone (GHRH) (Frohman & Jansson, 1986) and somatostatin (Katakami et al., 1986, which are stimulatory and inhibitory hormones respectively. Studies on GHS synthetic peptides and non-peptides which stimulate GH release indicated a distinct third pathway in addition to GHRH and somatostatin and the presence of a unique receptor (Bowers et al., 1984). Their activity was not blocked by the opiate receptor antagonist naloxone. Furthermore, these molecules were neither GHRH receptor agonists nor somatostatin receptor antagonists (Cheng et al., 1989; Blake & Smith, 1991; Cheng et al., 1991). This novel receptor was identified in porcine and rat anterior pituitary membranes using MK-0677 as a ligand. The binding of this receptor is Mg^{2+}-dependent and could be inhibited by GTP-gamma-S, indicating that it is a G-protein-coupled receptor (GPCR) (Pong et al., 1996). Activation of this GPCR by GHS results in the stimulation and amplified pulsatile release of growth hormone, and it was therefore named as GHS-R. This GHS-R defines a novel neuroendocrine pathway for the control of pulsatile GH release. The full-length sequence of this receptor was identified by using the expression-cloning strategy in Xenopus oocytes. Activation of GHS-R induces a transient increase in the concentration of intracellular calcium in somatotrophs (Bresson-Bepoldin & Dufy-Barbe, 1994), the increase in the level of inositol trisphosphate (IP3) (Mau et al., 1995) and activity of protein kinase-C (PKC) (Cheng et al., 1991).

2.3 Identification of ghrelin

Cloning of the GHS-R cDNA allows the engineering of cell lines stably expressing the GHS-R, which are essential for identification of endogenous GHS-R ligands. Through a process of reverse pharmacology, the endogenous nature ligand for GHS-R1a was identified from the gastric extracts and named as ghrelin. By detecting the intracellular calcium concentration in CHO cells expressing GHS-R1a as a functional readout, Kojima *et al.* screened several tissue extracts from brain, lung, heart, kidney, stomach and intestine, and unexpectedly discovered the active agonist for the GHS-R1a in the stomach rather than in the brain (Kojima et al., 1999). The active component was purified by successive chromatography including gel filtration, ion-exchange high-performance liquid chromatography (HPLC) and reverse-phase HPLC (RP-HPLC) (Kojima et al., 1999). Its amino acid sequence in rat was finally determined as GSSFLSPEHQKAQQRKESKKPPAKLQPR in which the serine-3 (Ser^3) is n-octanoylated. This octanoylation is necessary for its binding and activation of the GHS-

R. This 28 amino acid peptide was named ghrelin based on the word root "ghre" in proto-indo-european language for "growth" to depict its ability of stimulating the release of growth hormone (Kojima et al., 1999). Ghrelin is the only known peptide modified by a fatty acid, and it shares no structure homology with growth hormone secretagogues. Human ghrelin was cloned using the primers based on rat ghrelin cDNA under low stringency condition. Only two amino acids are different between human and rat preproghrelin (Kojima et al., 1999), indicating that ghrelin is highly conserved between species.

3. The gene structure of ghrelin

3.1 Ghrelin gene

Localized on the chromosome 3p25–26, the human ghrelin gene comprises five exons, same as the mouse gene (Tanaka et al., 2001a; Kanamoto et al., 2004). The short first exon contains only 20 base pairs, which encode part of the 5′-untranslated region. There are two different transcriptional initiation sites in the ghrelin gene; one occurs at -80 and the other at -555 relative to the ATG initiation codon, resulting in two distinct mRNA transcripts (transcript-A and transcript-B) (Kanamoto et al., 2004).

The 5′-flanking region of the human ghrelin gene contains a TATA box-like sequence (TATATAA; -585 to -579), as well as putative binding sites for several transcription factors such as AP2, basic helix-loop-helix (bHLH), hepatocyte nuclear factor-5, NF-κB, and half-sites for estrogen and glucocorticoid response elements (Tanaka et al., 2001a; Kishimoto et al., 2003; Kanamoto et al., 2004). Neither a typical GC nor a CAAT box was found.

3.2 Ghrelin gene-derived transcripts

The 28 amino acids of the functional ghrelin peptide are encoded by ghrelin gene which consists of 4 introns and 5 exons, including a non-coded exon 1. In rat, mouse and pig, the codon for Gln^{14} (CAG) is used as an alternative splicing signal to create two different ghrelin mRNAs (Hosoda et al., 2000b). One encodes the ghrelin precursor, and another encodes des-Gln^{14}-ghrelin precursor: a spliced variant without the codon for glutamine in position 14 (pre-pro-des-Gln^{14}- ghrelin mRNA) (Hosoda et al., 2003). Des-Gln^{14}-ghrelin is identical to ghrelin, except for the deletion of Gln^{14}. Nearly all of the cDNA clones isolated from human stomach encodes the pre-pro-ghrelin mRNA, only a few cDNA clones encodes the pre-pro-des-Gln^{14}-ghrelin mRNA (Hosoda et al., 2003).

Another splicing variant was detected in the mouse testis (Tanaka et al., 2001b). This variant, a ghrelin gene-derived transcript (GGDT), comprises the 68-bp 5′-unique sequence located between exons 3 and 4 of the ghrelin gene, and the remaining 252 bp sequence identical to the exons 4 and 5 of mouse ghrelin gene. GGDT encodes a 54 amino acid peptide consisting of 12 amino acid residues which is unrelated to the ghrelin gene and the COOH-terminal 42-amino acid sequence of mouse ghrelin precursor. The function of this variant is not clear.

In addition, a human exon 3-deleted mRNA transcript has been described in breast and prostate cancer (Yeh et al., 2005). The murine homologue (exon 4-deleted mRNA transcript) has also been identified in comprehensive murine tissues (Jeffery et al., 2005). Theses variants are likely products of the ghrelin gene attributed to alternative splicing mechanisms.

4. Posttranslational modification of ghrelin

In vitro and *in vivo* studies have demonstrated that the prohormone convertase 1/3 (PC1/3) is the only identified enzyme responsible for processing of proghrelin into ghrelin (Zhu et al., 2006). This endoprotease co-localizes with ghrelin in gastric endocrine cells and processes the 94 amino acids human ghrelin precursor into the 28 amino acids mature ghrelin through limited proteolytic cleavage (Ozawa et al., 2007).

Octanoylation of the third amino acid Serine is the most unique posttranslational modification of ghrelin. Ghrelin is the only protein currently known to be octanoylated. This octanoylation is essential for binding of ghrelin with its receptor, GHS-R1a. The enzyme that catalyzes the octanoylation of ghrelin was identified by two individual study groups and designated as ghrelin O-acyltransferase (GOAT) in 2008 (Gutierrez et al., 2008; Yang et al., 2008a). GOAT is a member of the family membrane-bound O-acyltransferases, with its structure conserved among different species. Transcripts for both GOAT and ghrelin are present predominantly in stomach and pancreas. Genetic disruption of the GOAT gene in mice leads to complete absence of octanoylated ghrelin in circulation. The design of GOAT inhibitors may therefore provide a new approach for the treatment of obesity and diabetes. *In vitro* study has demonstrated that GOAT activity could be significantly inhibited by an octanoylated ghrelin pentapeptide and other end-products (Yang et al., 2008b), suggesting the existence of a negative feedback regulation. In addition to the octanoylation, different acylation by other fatty acid has been reported. Studies have revealed some naturally occurring variants of ghrelin such as decanoyl ghrelin based on the different acylation on the serine-3 residue, which exhibit physiological function similar to octanoylated ghrelin (Ghigo et al., 2005).

In addition, ghrelin has been reported to be phosphorylated by protein kinase C α, β, and δ at serine-18 residue which affects the secondary structure and membrane binding property of ghrelin (Dehlin et al., 2008). The intracellular function of phosphorylated ghrelin remains unknown, but the phosphorylation may associate with subcellular localization of ghrelin, especially des-acyl ghrelin.

5. Ghrelin family members

The preproghrelin, encoded by ghrelin mRNA transcript A, contains a 23 amino-acid signal peptide and a 94 amino-acid proghrelin (1–94). The latter includes the 28 amino acid mature ghrelin (1–28) and a 66 amino acid tail (29–94) (Kojima et al., 1999; Jeffery et al., 2005). This proghrelin can be cleaved by prohormone convertase 1/3 (PC1/3) in mouse gastric mucosa to generate bona fide ghrelin *in vivo* (Zhu et al., 2006). Different products of ghrelin gene have been generated by alternative splicing and post-translational modification.

5.1 N-octanoyl ghrelin

Ghrelin was purified from the rat stomach through a series of steps of chromatography: gel filtration, two ion-exchange HPLC steps, and a final reverse-phase HPLC (RP-HPLC) procedure. The second ion-exchange HPLC yielded two active peaks, from which ghrelin and des-Gln^{14}-ghrelin were purified, respectively (Hosoda et al., 2000b). Analysis of ghrelin revealed the unique structure of a 28-amino acid peptide with its third amino acid Serine *n*-octanoylated. This modification is essential for ghrelin's activity for binding with GHSR1a. It is the first known case of a peptide hormone modified by a fatty acid. The sequence of ghrelin is highly conserved between species. Only two amino acid residues are different between rat and human (Kojima et al., 1999). There is no structural homology between ghrelin and peptide GHSs such as GHRP-6 or hexarelin.

5.2 Des-Gln^{14}-ghrelin

A second type of ghrelin peptide has been purified from rat stomach and identified as des-Gln^{14}-ghrelin (Hosoda et al., 2000b). Except for the deficiency of Gln^{14}, des-Gln^{14}-ghrelin is identical to ghrelin, with the same *n*-octanoic acid modification and potency of activities. The deletion of Gln^{14} attributes to the CAG codon encoding Gln, which is a splicing signal. Analysis of ghrelin gene structure reveals that an intron exists between Gln^{13} and Gln^{14} of the ghrelin sequence. The 3'-end of the intron has two tandem CAG sequences. The AGs of these sequences may serve as the splicing signals (McKeown, 1992). When the first AG is used as the splicing signal, prepro-ghrelin mRNA is produced and the second CAG is translated into Gln^{14}. On the other hand, when the second AG is used, prepro-des-Gln^{14}-ghrelin mRNA is generated to produce des-Gln^{14}-ghrelin missing Gln^{14} (Hosoda et al., 2000b). These findings confirm that des-Gln^{14}-ghrelin is processed from the ghrelin gene by alternative splicing rather than a distinct genomic product from ghrelin. Des-Gln^{14}-ghrelin is present in low amounts in the stomach, indicating that ghrelin is the major functional form.

5.3 Des-acyl ghrelin

Des-acyl ghrelin, the nonacylated form of ghrelin, also exists at significant levels in both stomach and blood (Hosoda et al., 2000a). In blood, des-acyl ghrelin circulates in amount far greater than acylated ghrelin. The clearance rates of inactive forms of peptide hormones are often reduced compared with their respective active forms. Ghrelin in the plasma binds to high-density lipoproteins (HDLs) that contain a plasma esterase, paraoxonase, and clusterin (Beaumont et al., 2003). Because a fatty acid is attached to the third amino acid Serine of ghrelin via an ester bond, paraoxonase may be involved in de-acylation of acyl-modified ghrelin. Thus parts of des-acyl ghrelin may originate from the de-acylation of acyl ghrelin.

Des-acyl ghrelin does not compete the binding sites of GHS-R1a with acyl ghrelin in hypothalamus and pituitary. Consistent with this finding, des-acyl ghrelin shows no GH-releasing and other endocrine activities associated with activation of GHS-R1a. However, des-acyl ghrelin shares with active acyl-ghrelin some non-endocrine actions, including the modulation of cell proliferation and adipogenesis (Cassoni et al., 2004). Emerging evidences suggest the existence of a distinct receptor other than GHS-R1a, which des-acyl ghrelin may bind with. Des-acyl ghrelin has been reported to either stimulate (Toshinai et al., 2006) or inhibit (Asakawa et al., 2005; Matsuda et al., 2006) food intake in a GHS-R1a independent pathway. It has been showed that ghrelin and des-acyl ghrelin both recognize common high-affinity binding sites on H9c2 cardiomyocytes (Baldanzi et al., 2002), which do not express the classical ghrelin receptor GHS-R1a, suggesting the existence of a novel ghrelin receptor subtype in the cardiovascular system.

5.4 Obestatin

Recent evidence suggests that the 66 amino acid tail of proghrelin (29–94) can either circulate as a full-length peptide (C-ghrelin) or be processed to smaller peptides, mainly obestatin (11-23). Obestatin was proposed as a novel peptide derived from C terminal of proghrelin (Zhang et al., 2005). This new peptide was named 'obestatin' based on its appetite-suppressing potential. The structure of obestatin was deduced as a 23 amino acid sequence of proghrelin 53–75 according to mass spectrometric analysis. Because of the C-terminal Gly–Lys motif, amidation of obestatin was assumed and confirmed as a prerequisite for its biological activity (Zhang et al., 2005). Detail mechanism on how this peptide is processed to a 23 amino acid peptide remains to be explored. Obestatin was originally extracted from rat stomach and has subsequently been shown to be an active circulating peptide (Zizzari et al., 2007; Harada et al., 2008). However, it had been found that no evidence for obestatin peptide circulating as distinct entity in the human and rat blood (Bang et al., 2007). Specific immunoassays directed to the N-terminus, C-terminus, and mid-region of proghrelin (29–94) all detect the same molecule of Mr ~7kDa, suggesting that the only form present in circulation is close to the full length 66- amino acid C-ghrelin (Bang et al., 2007). Further studies are necessary to clarify this issue.

5.5 Other ghrelin forms

Several other minor forms of the ghrelin peptides were isolated in the purification of human ghrelin from the stomach (Hosoda et al., 2003). These peptides could be classified into four groups by the type of acylation at the third amino acid Serine: nonacylated, octanoylated (C8:0), decanoylated (C10:0), and decenoylated (C10:1). Two kinds of different length peptides, either 27 or 28 amino acids, were found. The 27 amino acid peptide lacks the C terminal Arg28 and derives from the same ghrelin precursor. Both synthetic octanoylated and decanoylated ghrelins increase the concentration of intracellular Ca^{2+} in GHS-R-expressing cells and stimulate GH release in rats to a similar degree.

The expression of a novel human mRNA transcript designated as exon 3-deleted mRNA transcript has been reported in breast and prostate cancer (Yeh et al., 2005). A murine homologue, exon 4-deleted mRNA transcript, has also been identified in a widely range of tissues (Jeffery et al., 2005). Deletion of exon 3 from the preproghrelin transcript results in a truncated C-peptide with a unique 16-amino-acid peptide tail, COOH-terminal D3 peptide, which begins with a potential peptide cleavage site [Arg–Arg]. Previous studies indicate that this unique COOH-terminal D3 peptide (RPQPTSDRPQALLTSL) does not affect cell proliferation or cell apoptosis in prostate cancer cell lines. However, the comprehensive expression of this C-ghrelin suggests that it may possess some unidentified functions.

6. Tissue distribution of ghrelin

6.1 Stomach and other gastrointestinal organs

In all vertebrate species, stomach is the primary organ producing ghrelin (Ariyasu et al., 2001). Ghrelin immune-reactive cells are more abundant in the fundus than in the pylorus (Date et al., 2000; Tomasetto et al., 2001; Yabuki et al., 2004). In situ hybridization and immunohistochemical analysis indicate that ghrelin-containing cells are a distinct endocrine cell type found in the gastric mucosa (Date et al., 2000; Rindi et al., 2002). Four types of endocrine cells have been identified in the oxyntic mucosa: ECL cells, D cells, enterochromaffin (EC) cells, and X/A-like cells (Capella et al., 1969; Solcia et al., 1975; Grube & Forssmann, 1979). The granule contents of X/A-like cells had not been identified until the discovery of ghrelin. The X/A-like cells contain round, compact, electron dense granules that are filled with ghrelin (Date et al., 2000; Dornonville de la Cour et al., 2001; Yabuki et al., 2004). The X/A-like cells account for about 20% of the endocrine cell population in adult oxyntic glands and can be stained by an antibody specific to the NH2-terminal, acylated portion of ghrelin, indicating that ghrelin in the secretory granules of X/A-like cells has already been acylated. However, the number of X/A-like cells and ghrelin concentration in the fetal stomach is very low and gradually increases after birth (Hayashida et al., 2002).

Ghrelin-immunoreactive cells are also found in the duodenum, jejunum, ileum, and colon (Date et al., 2000; Hosoda et al., 2000a; Sakata et al., 2002), among which ghrelin concentration gradually decreases from the duodenum to the colon. The pancreas is another ghrelin-producing organ especially during the embryo development and the production of ghrelin decreases rapidly after birth. The cell type that produces ghrelin in the pancreatic islets remains controversial, whether it might be the α cells, β cells, newly identified islet epsilon (ε) cells, or an unknown novel type of islet cells (Date et al., 2002; Wierup et al., 2002; Prado et al., 2004; Wierup et al., 2004).

6.2 Central nervous system – Hypothalamus

Ghrelin is the endogenous ligand for GHS-R which is mainly expressed in the hypothalamus and pituitary (Howard et al., 1996; Guan et al., 1997). Despite that the content of ghrelin is very low (Kojima et al., 1999; Hosoda et al., 2000a), ghrelin immnuo-reactivity has been detected in the hypothalamic arcuate nucleus, an important region for controlling appetite (Kojima et al., 1999; Lu et al., 2002). In addition, ghrelin immune-reactive cell bodies in the hypothalamus are distributed in a continuum filling the inter-nuclear space between the lateral hypothalamus (LH), arcuate (ARC), ventromedial (VMH), dorsomedial (DMH), and paraventricular hypothalamic nuclei (PVH) and the ependymal layer of the third ventricle (Cowley et al., 2003). These ghrelin-containing axons innervate neurons that contain neuropeptide Y (NPY) and agouti-related protein (AgRP) and may stimulate the release of these orexigenic peptides. These histological findings are consistent with the functional studies in which injection of ghrelin into the cerebral ventricles of rats potently stimulates food intake.

Ghrelin has also been found in the pituitary gland (Korbonits et al., 2001a; Korbonits et al., 2001b), where it may influence the release of GH in an autocrine or paracrine manner. Stimulation of primary pituitary cells with ghrelin increases their intracellular Ca^{2+} concentration (Bennett et al., 1997; Guan et al., 1997; McKee et al., 1997a). The expression level of ghrelin in the pituitary is high in the neonates and declines with puberty. Pituitary tumors, such as adenomas, corticotroph tumors, and gonadotroph tumors contain ghrelin peptides as well.

6.3 Other tissues

Two major forms of ghrelin: *n*-octanoyl and des-acyl ghrelin, are found in circulation (Hosoda et al., 2000a). The concentrations of *n*-octanoyl ghrelin and total ghrelin in human plasma are 10–20 fmol/ml and 100–150 fmol/ml respectively. Plasma ghrelin concentration increases in fasting condition and reduces after habitual feeding (Cummings et al., 2001; Tschop et al., 2001a). In addition, plasma ghrelin level is lower in obese subjects than the age matched lean controls (Tschop et al., 2001b; Hansen et al., 2002; Shiiya et al., 2002).

Ghrelin mRNA is also detected in the kidney, especially in the glomeruli (Mori et al., 2000; Gnanapavan et al., 2002). Moreover, peptide extracts from mouse kidney contain both *n*-octanoyl and des-acyl ghrelin. Significant relationship between plasma ghrelin concentration and the serum creatinine level has been demonstrated. In patients with end-stage renal disease, plasma ghrelin level increases 2.8-fold compared with those with normal renal function (Yoshimoto et al., 2002). This finding suggests that the kidney is an important organ for clearance and/or degradation of ghrelin.

In the reproductive system, ghrelin immune-reactive cells have been identified in interstitial Leydig cells and at a low level in Sertoli cells (Barreiro et al., 2002; Tena-Sempere et al., 2002). The ghrelin receptor has also been detected in germ cells, mainly in pachytene spermatocytes, as well as in somatic Sertoli and Leydig cells (Gaytan et al., 2004). During embryo development, ghrelin-immunoreactive cells are detectable in cytotrophoblast cells in first-trimester human placenta but disappear in third-trimester placenta (Gualillo et al., 2001). Ghrelin-containing cells are also detected in syncytiotrophoblast cells of the human placenta and in the cytoplasm of labyrinth trophoblasts of the rat placenta.

6.4 Ghrelin producing cell lines

Ghrelin can be detected in several cultured cell lines. One of which is TT cells, a human thyroid medullary carcinoma cell line (Kanamoto et al., 2001). TT cells express ghrelin mRNA. Conditioned medium and cellular extracts of TT cells contain both *n*-octanoyl ghrelin and des-acyl ghrelin. Other cultured cells such as the kidney-derived cell line NRK-49F (Mori et al., 2000), gastric carcinoid ECC10 cells (Kishimoto et al., 2003), and the cardiomyocyte cell line HL-1 (Iglesias et al., 2004) express ghrelin as well.

Some patients with neuroendocrine tumors also accompany with high level of ghrelin in plasma. It had been reported that a patient with a malignant neuroendocrine pancreatic tumor with ghrelin immunoreactivity and a high circulating ghrelin level (Corbetta et al., 2003). A patient with a metastasizing gastric neuroendocrine tumor has also been reported to have extremely high circulating level of ghrelin (Tsolakis et al., 2004). In the latter case, the patient developed diabetes mellitus and hypothyroidism. In both cases, GH and IGF-I levels were within the normal range, and the patients had no clinical features of acromegaly.

7. Regulation of ghrelin biosynthesis

In the 5′-flanking region of the ghrelin gene, several E-box consensus sequences exist (Kanamoto et al., 2004). Destruction or site-directed mutagenesis of these sites decreases the promoter activity in TT cells. Upstream stimulatory factors (USF), members of the bHLH-LZ family of transcription factors, bind to these E-box elements and may thus modulate the expression of ghrelin gene. Ghrelin promoter activity in ECC10 cells is up-regulated by glucagon and its second messenger cAMP (Kishimoto et al., 2003), suggesting that a high level of ghrelin production may be related to increased glucagon in fasting condition.

The biosynthesis of ghrelin is tightly linked with the energy status and regulated by different nutritional ingredients. Fasting increases ghrelin, des-acyl ghrelin and C-ghrelin, while demonstrates no effect on obestatin (Zhang et al., 2005; Bang et al., 2007). Conversely, feeding decreases ghrelin, des-acyl ghrelin and C-ghrelin (Zhang et al., 2005; Bang et al., 2007). Postprandial acylated ghrelin level falls more quickly than total ghrelin. The postprandial decrease of ghrelin can be attributed mainly to the increase of the serum glucose concentration, because the total ghrelin, acyl ghrelin, and des-acyl ghrelin all decrease significantly after a high-carbohydrate meal (Sedlackova et al., 2008). Other nutrients, such as high-fat meal, induces minor (Monteleone et al., 2003) but more persistent (Erdmann et al., 2003; Foster-Schubert et al., 2008) post-prandial suppression in circulating total ghrelin than high-carbohydrate meal in humans. Ingestion of either medium-chain fatty acids (MCFAs, such as n-octanoic acid) or medium-chain triglycerides (MCTs, such as glyceryl trioctanoate) increases the stomach contents of acyl ghrelin without changing the total ghrelin in mouse (Nishi et al., 2005). This finding suggests that medium-chain fatty acids can be utilized directly for the acyl modification of ghrelin. Oral ingestion of a physiological dose of essential amino acids leads to a continuous rise in serum ghrelin level in humans (Groschl et al., 2003; Knerr et al., 2003). Insoluble dietary fiber ingestion may influence ghrelin level as well (Gruendel et al., 2007).

Chronic energy imbalance such as obesity or anorexia also alters ghrelin production. Body mass index in human is negatively correlated with the production of ghrelin. Plasma ghrelin concentration is low and post-prandial decrease in ghrelin is attenuated in the obese individuals (Erdmann et al., 2005). Patients with anorexia nervosa have significant increase in serum total and acyl ghrelin levels (Harada et al., 2008), which return to normal when the disease is cured and the body weight restored (Otto et al., 2001). Furthermore, total ghrelin level is inversely associated with fat cell volume (Purnell et al., 2003) and specifically in women with total fat mass and fat mass/lean mass ratio, whereas in men with abdominal fat mass and fat distribution index (Makovey et al., 2007).

There may also be a developmental regulation of ghrelin gene-derived peptides expression. It has been showed that plasma total ghrelin concentration decreases abruptly after birth, contrasting with a 3-fold increase in the concentration of acylated ghrelin (Chanoine & Wong, 2004).

Little has been known on the intracellular mechanism by which ghrelin expression is regulated. Our studies suggest that gastric mammalian target of rapamycin (mTOR) activity is critical for the modulation of ghrelin production (Xu et al., 2009; Xu et al., 2010).

8. Ghrelin receptors and their distribution

The GHS receptor gene is located on chromosome 3q26.2 and encodes for two transcripts, 1a and 1b which encode a full-length receptor GHS-R1a and a shortened version GHS-R1b respectively (McKee et al., 1997a). Type 1a encodes a 366-amino acid polypeptide containing all seven TM (transmembrane) domains, while type 1b encodes a truncated 289-amino acid polypeptide with only five TM domains. Both sequences are identical from the Met translation site to Leu at the 265th amino acid position (Howard et al., 1996).

The GHS-R1a is comprehensively distributed in a variety of tissues and has high affinity with ghrelin and GHSs. GHS-R1b also has a widespread expression but does not compete with GHS-R1a to bind ghrelin or synthetic GHS (Gnanapavan et al., 2002) and its function remains controversial. GHS-R1a is highly expressed in the hypothalamus and pituitary, consistent with ghrelin's functional role in the control of appetite and growth hormone release.

8.1 Type 1a GHS receptor

The GHS-R1a gene contains a single, ~2kb intron separating two exons and encodes a GPCR with seven transmembrane (TM) domains (McKee et al., 1997b). Its ligand activation domains are in the second TM domain (TM2) and the third TM domain (TM3). These key amino acid residues are essential for binding and activation by different ligands, and have been evolutionarily conserved for 400 million years, highlighting its importance in fundamental physiological processes (Palyha et al., 2000). GHS-R1a mRNA is detected at 18 and 31 weeks of gestation, indicating that ghrelin might be active early in fetal development (Hayashida et al., 2002; Nakahara et al., 2006). Comparison of the predicted human rat, pig and sheep GHS-R1a amino acid sequences reveals 91.8–95.6% sequence homology (Palyha et al., 2000).

8.1.1 GHS-R1a structure

The human GHS-R1a consists of 366 amino acids with a molecular mass of approximately 41 kDa (Howard et al., 1996; Schwartz et al., 2006) and belongs to family A of G-protein-coupled receptors (GPCRs) (Bockaert & Pin, 1999). It spans the membrane with seven α-helix hydrophobic domains which joint each other by three intra- and extracellular domains, beginning with an extracellular N-terminal domain and ending with an intracellular C-terminal domain (Bockaert & Pin, 1999). The N-terminal domain forms a β-hairpin structure and the TM domains a round calyx-like structure which is attributed to Pro residues in the center of the TM helices. TM3 occupies the central position among the TM segments and TM5 is the most peripheral (Pedretti et al., 2006).

GHS-R1a possesses three conserved residues Glu-Arg-Tyr at the intracellular end of transmembrane 3 (TM3) domain in position 140–142 (ERY/DRY motif), which are important for the isomerization between the active and inactive conformation, and two cysteine (Cys116 and Cys198) residues on the extracellular loop 1 and 2 respectively forming a disulfide bond (Bockaert & Pin, 1999; Petersenn, 2002). Mutagenesis of TM sites reveals much of the functional domains within the GHS-R1a. Amino acids like E124, D99, and R102 of TM2 and TM3 contribute to the ligand activation, while E124, F119, and Q120 of TM3 mediate binding activity (Palyha et al., 2000). Consistent with this finding, investigation using antibody-binding studies directed at different regions of the ghrelin receptor, has also identified the amino acid residue E124 as an important binding site for GHSs, mediating a salt bridge interaction with basic moieties that appear to act independent of receptor activation (Feighner et al., 1998). Meanwhile, the importance of the hydrophobic octanoyl moiety attached to ghrelin's serine-3 residue to GHS-R1a binding has been demonstrated (Bednarek et al., 2000), and an 'active core' sequence of Gly-Ser-Ser (octanoyl)-Phe essential for receptor activation has been identified.

8.1.2 Ligand binding domains

GHS-R1a transduces information provided by both ghrelin and GHSs. This concept has been explained based on the existence of a common binding domain as demonstrated by molecular modeling and site-directed mutagenesis studies (Feighner et al., 1998). The binding domain should allow a conserved structure of agonists to recognize a complementary binding pocket that accommodates the variable part of the ligand overlapping in the agonist binding site (Bondensgaard et al., 2004). This domain is imposed by the orientation of transmembrane segments that determine the stereo- and geometric specificity of the ligand's entry and binding to the transmembrane core.

The binding domain of GHS-R1a involves six amino acids located in TM3, TM6 or TM7 (Holst et al., 2006). In addition, binding of the natural ligand ghrelin with GHS-R1a requires the ligand to interact with one pocket formed by polar amino acids in TM2/TM3 and one formed by non-polar amino acids in TM5/TM6, respectively (Pedretti et al., 2006). The synthetic peptidyl and non-peptidyl GHSs share a common binding pocket in the TM3 region of the GHS-R1a, although there are other distinct binding sites in the receptor that seem to be selective for particular classes of agonists (Feighner et al., 1998). While the main binding pocket for small amines is deep in the pocket created by the TM domains, small peptides bind to extracellular epitopes as well. The basic amine common to peptidyl (GHRP-6) and non-peptidyl (MK-0677) GHS establishes an electrostatic interaction with Glu^{124} in the TM3 domain (Feighner et al., 1998). Substitution of glutamine for glutamic acid [Glu124Gln mutant] in human GHS-R1a inactivates its function. Furthermore, mutation of Arg^{283} in TM6, which interacts with Glu^{124}, abolishes both agonist-induced activation and constitutive signaling (Feighner et al., 1998; Holst et al., 2003). In addition, disruption of the disulfide bond between Cys^{116} and Cys^{198} in the extracellular portion of the receptor by mutating Cys^{116} into alanine [Cys116Ala mutant] completely abolishes the activity of all agonists (Feighner et al., 1998; Palyha et al., 2000).

8.1.3 Constitutive activity

Evidences on the physiological relevance of GHS-R1a constitutive activity are steadily emerging. The molecular mechanism of such constitutive activity relates to three aromatic residues located in TM6 and TM7, namely PheVI:16, PheVII:06 and PheVII:09. These residues promote the formation of a hydrophobic core between helices 6 and 7 to ensure proper docking of the extracellular end of TM7 into TM6 (Holst et al., 2004). Phe279Leu and Ala204Glu polymorphisms have been reported to be associated with short stature and obesity respectively (Wang et al., 2004), indicating that constitutive activity of GHS-R1a *in vivo* might be imperative for normal growth and development of the human body (Holst & Schwartz, 2006). The Phe279 mutation is a conservative amino acid exchange in the sixth transmembrane domain of GHS-R1a, and the Ala204 is located in the second extracellular loop. Both amino acid residues are highly conserved (Wang et al., 2004).

When transfected to HEK-293 cells, the derived GHS-R1a receptor displays reduced basal activity and lower expression at the plasma membrane (Pantel et al., 2006), although responsiveness to ghrelin is intact (Holst & Schwartz, 2006; Pantel et al., 2006). In addition, GHS-R1a transfection leads to constitutively stimulated CRE (cAMP-responsive element) activity in HEK-293 cells (Holst et al., 2003). When overexpressed in COS-7 cells, GHS-R1a possesses a constitutive activity in the turnover of IP3, which is approximately 50% of the maximal agonist induced activity (Holst et al., 2003; Holst et al., 2006). However, GHS-R1a does not show any constitutive activity in the pituitary cell line RC-4B/ C40, indicating that GHS-R1a constitutive activity may depend on the cellular context (Falls et al., 2006). The only known ligand blocking GHS-R1a constitutive activity is D-Arg^{1}-D-Phe^{5}-D-Trp^{7,9}-Leu ^{11}-substance P (Holst et al., 2003; Holst et al., 2006).

8.1.4 Endocytosis of GHS-R1a

Kinetic studies of GHS-R1a internalization based on radioligand binding experiments and confocal microscopy have demonstrated that GHSR-1a is internalized in a time dependent manner, which peaks at approximately 20 min after ghrelin stimulation. The ghrelin- GHSR-1a complex is internalized principally by a clathrin-mediated mechanism and through the endosomal trafficking pathway (Camina et al., 2004). Once the ligand-receptor complex is internalized into vesicles, GHSR-1a is sorted into endosomes to be recycled back to membrane. About 360 min after agonist removal, the level of GHS-R1a receptors on the cell surface rises close to 100% of its original level. The process is not affected by cycloheximide, suggesting that GHS-R1a originates from endosomes rather than de novo receptor synthesis. This notion is further demonstrated by the observation that fluorescence associated with GHS-R1a-EGFP in CHO cells reappears on the membrane in the similar kinetics after a ghrelin pulse. Moreover, fluorescence emitted by the EGFP-labeled GHS-R1a in cells co-localizes with the early endosomal protein 1 but not with cathepsin (lysosomal marker), indicating that GHS-R1a is not targeted to lysosomal compartments. Thus, most of GHS-R1a appearing at the cell surface originates from endocytosed receptors, leading to the complete restoration of binding capacity and functionality.

Function experiments also demonstrate the theory of GHS-R1a recycling. GH response to twice consecutive pulses of ghrelin is blunted when pulses are separated by short interval (60 min), but it restores the initial amplitude when the second pulse is administered after 180, 240 or 360 min, which fits well with the kinetics of receptor recycling. The slower recycling (3–6 h) of this receptor compared to other GPCRs is determined by the stability of the complex GHSR-1a/β-arrestin during clathrin-mediated endocytosis because this complex appears to dictate the profile of the receptor re-sensitization (Luttrell & Lefkowitz, 2002).

In the normal healthy organism, ghrelin receptor desensitization and endocytosis govern the ability of cells to respond to ghrelin and thereby regulate intracellular signaling to avoid a permanent stimulation of target cells. Moreover, receptor re-sensitization determines the frequency of the response to ghrelin. Deficiencies in this attenuation system may lead to an uncontrolled or defective stimulation of target cells, which causes alteration in its intracellular signaling and subsequent pathological changes.

8.2 Non-type 1a GHS receptors

Remarkable differences in the binding profile among ghrelin, synthetic peptidyl (hexarelin) and non-peptidyl (MK-0677) GHSs have been reported (Ong et al., 1998a; Ong et al., 1998b; Papotti et al., 2000; Holst et al., 2004) , suggesting the presence of a novel unidentified receptor subtype in tissues that do not express GHS-R1a or express the receptor at a low level. The existence of various GHS-R1a homologues, the lack of a definite phenotype in GHS-R1a knockout mice as well as the presence of multiple endogenous ghrelin-like ligands, indicate the existence of multiple receptors for ghrelin and GHSs.

8.2.1 Other GPCR homologues of GHS-R1a

Based on its peptide sequence and intracellular signaling, GHS-R1a is often included in a small family of receptors for small polypeptides such as the receptor for motilin (52% homology), neurotensin receptor-1 (NTS-R1) and NTS-R2 subtype (33–35% homology), neuromedin U receptor-1 (NMU-R1) and NMU-R2 subtype (~ 30% homology), and the orphan receptor GPR39 (27–32% homology) (McKee et al., 1997b; Tan et al., 1998).

GHS-R1a shares the highest homology with GPR38, the receptor for motilin (McKee et al., 1997b; Petersenn, 2002). GRP38 is mostly distributed in the thyroid gland, bone marrow, stomach and gastrointestinal smooth muscles, and less widely expressed in the neuroendocrine tissues than GHS-R1a. Both GRP38 and GHS-R1a receptors mediate the pulsatile biological effects upon continuous stimulation and increase gastrointestinal motility. In response to motilin, GRP38 couples with Gαq/Gα13, increases cytosolic free Ca^{2+} by an IP_3-dependent Ca^{2+} release mechanism (Huang et al., 2005).

Another group of GPCR homologues of GHS-R1a are NTS receptor, namely NTS-R1 and NTS-R2. NTS is a 13-amino-acid polypeptide, which has high sequence homology with neuromedin N. In the CNS, NTS-Rs have been identified in the hypothalamus, amygdala and nucleus accumbens. NTS-Rs involve in modulation of the dopaminergic system in the CNS, while act on the small intestine endocrine cells to increases acid secretion and regulate smooth muscle contraction in the peripheral tissues (Vincent et al., 1999). NTS-R1 mainly functions through Gq/11-PLC/IP3, but it also activates cGMP/cAMP signaling and ERK1/2 phosphorylation (Amar et al., 1985; Bozou et al., 1989; Poinot-Chazel et al., 1996; Vincent et al., 1999). NTS-R2 induces intracellular Ca^{2+} signaling and ERK phosphorylation (Botto et al., 1997; Sarret et al., 2002).

Neuromedin U is a 23~25 amino acids polypeptide which is the ligand for other members of this receptor family, the NMU-R1/R2 receptors. NMU is highly conserved among species and widely distributed in the body especially at high levels in the brain where it mediates effects on food intake opposite to ghrelin (Jethwa et al., 2006), likely by cross-talking with the anorectic leptin system (Jethwa et al., 2006; Nogueiras et al., 2006). The NMU-R1 subtype is mainly expressed in the periphery, while NMU-R2 is mostly in the brain (Brighton et al., 2004). Both receptors are involved in the regulation of smooth muscle contraction, gastric acid secretion, insulin secretion, ion transport in the gut, feeding behavior and stress (Brighton et al., 2004). Both NMU-R1 and NMU-R2 signal through Gq/11, PLC and Ca^{2+} (Brighton et al., 2004).

The other GHS-R1a homologue is GRP39, which remains controversial about whether or not obestatin binds it (Zhang et al., 2005; Dun et al., 2006; Lauwers et al., 2006; Tremblay et al., 2007). GRP39 is distributed in a variety of tissues, but mostly in brain regions (McKee et al., 1997b). Compared with GHS-R1a and GRP38, GRP39 has a very long C-terminal and two potential palmitoylation sites, which creates a 4th intracellular loop (Morello & Bouvier, 1996; McKee et al., 1997b). Activation of GRP39 by Zn^{2+} induces PLC signaling, CRE- and SRE-dependent transcriptional activity, and cAMP production (Holst et al., 2007). Except for the sequence homology with GHS-R1a, the relationship between GPR39 and ghrelin system remains indefinable.

8.2.2 Type 1b GHS receptor

Type 1b GHS receptor (GHS-R1b) contains 298 amino acids corresponding to the first five TM domains encoded by exon 1, plus a unique 24 amino acid tail encoded by an alternatively spliced intron sequence. Neither ghrelin nor GHS binds or causes any response to GHS-R1b receptor (Smith et al., 1997) in cells transfected with GHS-R1b. Since GHS-R1b is comprehensively expressed in various tissues (Gnanapavan et al., 2002), it is reasonable to assume that this receptor possesses some unidentified biological functions. A report in 2007 has revealed that GHS-R1b decreases the cell surface expression of GHS-R1a and acts as a repressor of the constitutive activity of GHS-R1a when overexpressed in HEK-293 cells (Chu et al., 2007). This finding suggests that GHS-R1b may act as an endogenous modulator for GHS-R1a constitutive activity. This fascinating action may attribute to its capability of forming heterodimers with full length GPCR receptors, causing altered biological properties compared to the original receptor.

8.2.3 CD36 receptor

CD36, a membrane glycoprotein of Mr ~84kDa belonging to the scavenger receptor type-B family (Bodart et al., 1999), distributes in a wide range of tissues including adipose tissue, platelets, monocytes/macrophages, dendritic cells, and microvascular endothelium (Febbraio et al., 2001). CD36 is implicated in multiple physiological functions such as fatty acid/lipid transportation, insulin resistance, antigen presentation and regulation of angiogenesis, as well as pathophysiological processes related to the formation of macrophage foam cell and atherosclerotic lesions (Febbraio et al., 2001). In a perfused heart preparation, hexarelin elicits vasoconstriction. This effect is likely mediated via CD36 because no similar effect occurs in CD36 null mice and rats (Bodart et al., 2002). The signal transduction pathways mediating the vasoconstrictive effect of hexarelin involve both L-type calcium channels and protein kinase C (Bodart et al., 1999).

The effect of hexarelin on the inhibition of cholesterol accumulation might be also mediated by CD36. Hexarelin reduces internalization of oxLDL by interfering with CD36 on the same interaction site as that of oxLDL (Avallone et al., 2006). In addition, hexarelin increases cholesterol efflux in macrophages through activation of CD36 and GHS-R1a which enhances expression of the ABCA1 and ABCG1 transporters and therefore improves cholesterol efflux from macrophages (Avallone et al., 2006).

8.3 Tissue distribution of ghrelin receptors

The ghrelin receptor is ubiquitously distributed in both the CNS and peripheral tissues. In the rat brain, GHS-R1a signals are detected in hypothalamic nuclei including anteroventral preoptic nucleus, anterior hypothalamic area, suprachiasmatic nucleus, lateroanterior hypothalamic nucleus, supraoptic nucleus, ventromedial hypothalamic nucleus, arcuate nucleus, paraventricular nucleus and tuberomammillary nucleus as well as in the pituitary gland (Guan et al., 1997). This distribution is consistent with its physiological function associated with energy metabolism and GH release. In addition to the hypothalamus, GHS-R1a mRNA is also expressed in several other discrete regions of the rat brain such as dentate gyrus, CA2 and CA3 regions of the hippocampal formation, thalamic regions, and several nuclei within the brain stem including pars compacta of the substantia nigra, ventral tegmental area, median and dorsal raphe nuclei and laterodorsal tegmental nucleus (Guan et al., 1997). GHS-R1a may therefore play a role in the control of other functions such as memory, anxiety and some endocrine physiology.

In the peripheral tissues, GHS-R1a mRNA is detected in the stomach and intestine (Date et al., 2000), pancreas (Guan et al., 1997), kidney (Mori et al., 2000), heart and aorta (Nagaya et al., 2001c), adipose tissues (Gnanapavan et al., 2002), as well as in various endocrine neoplasms (de Keyzer et al., 1997; Korbonits et al., 2001a; Papotti et al., 2001). Such a wide distribution of GHS-R1a is consistent with the reported broad functions beyond the control of GH release and food intake. Existence of a novel ghrelin receptor subtype with a binding profile different from the GHS-R1a in the peripheral tissues, in human thyroid and breast tumors, as well as in related cancer cell lines has been reported but requires further confirmation (Cassoni et al., 2001; Volante et al., 2002).

9. Intracellular signaling pathways of ghrelin receptors

Binding of ghrelin to GHS-R1a induces a profound change in the transmembrane α helices, which alters the conformation of intracellular loops and uncovers masked G protein binding sites interacting with G proteins. This interaction promotes the release of guanosine diphosphate (GDP) bound to the G protein α subunit and exchange for guanosine triphosphate (GTP), followed by activation of G protein subunits which initiates intracellular signaling responses via a series of intracellular molecules.

Intracellular calcium is the well-characterized signaling mode used by GHS-1a receptor related studies, as its endogenous ligand ghrelin was discovered by monitoring intracellular calcium flux in recombinant cells expressing the GHSR-1a (Howard et al., 1996; Kojima et al., 1999). Ghrelin activates at least two signal transduction pathways associated with calcium regulation. In neuropeptide Y (NPY)-containing neurons of the hypothalamus, ghrelin induced increase in intracellular calcium concentration is dependent on the calcium influx through N-type calcium channels activated by the AC-cAMP-PKA signaling pathway following the binding of the Gs protein to GHS-R1a (Kohno et al., 2003). In addition to this cAMP/ PKA pathway, ghrelin also evokes the intracellular calcium signaling by an alternative pathway through Gq/PLC/IP3 (Howard et al., 1996). Coupling of ghrelin to the GHS-R1a activates phospholipase C to generate IP3 and diacylglycerol (DAG). IP3 then triggers the release of calcium from IP3 sensitive intracellular calcium stores (Deghenghi et al., 2001), whereas DAG is responsible for the activation of PKC. The initial calcium current is followed by sustained calcium entry through L- and T-type calcium channels via membrane depolarization (Chen et al., 1990; Smith et al., 1993). Different binding sites on the GHSR-1a have been proposed to potentially account for these two distinct components of ghrelin receptor signaling (Cassoni et al., 2001).

In addition to the well-characterized activation of calcium currents, ghrelin also activates the MAP kinase (MAPK) signaling pathway. In 3T3-L1 preadipocytes, ghrelin induces a rapid activation of ERK1/2, while inhibition of ERK signaling by PD98059 significantly attenuates the proliferative activities of ghrelin (Kim et al., 2004). In human and rat adrenal zona glomerulosa cells, the proliferative effect of ghrelin involves the activation of tyrosine kinase-dependent MAPK p42/p43 mechanism and appears to be independent of PKA and

PKC (Carreira et al., 2004; Mazzocchi et al., 2004). Pretreatment of cells with *Gi/o* inhibitor (pertussis toxin), PKC inhibitors (staurosporin and GF109203X), or PI3K inhibitor (wortmannin) significantly attenuates ghrelin-induced ERK1/2 phosphorylation, suggesting that multiple signaling pathways are involved in activation of MAPK induced by ghrelin (Mousseaux et al., 2006).

AMP-activated protein kinase (AMPK) plays a pivotal role in the regulation of energy metabolism. Ghrelin increases food intake through the activation of AMPK and acetyl-CoA carboxylase (ACC), and the inhibition of fatty acid synthase (FAS) in the ventral medial hypothalamus (Lopez et al., 2008). The molecular mechanism by which ghrelin regulates AMPK is still unknown, although tumor suppressor LKB1 may be an upstream regulator of AMPK kinase (Hardie, 2004). It is worth noting that AMPK's metabolic activity induced by ghrelin receptor is tissue-specific. In the rat liver, ghrelin inhibits AMPK activity to evoke a series of lipogenic and glucogenic gene expression and an increase in triglyceride content without changing the mitochondrial oxidative enzyme activities (Barazzoni et al., 2005). Ghrelin can also inhibit vascular inflammation through the activation of the calmodulin-dependent kinase kinase (CaMKK), AMPK and endothelial nitric oxide synthase (eNOS) (Xu et al., 2008).

Insulin signaling pathway is also regulated by ghrelin through the insulin receptor substrate (IRS-1) associated PI3K activity and Akt phosphorylation. In hepatoma cells, ghrelin increases IRS-1 associated PI3K activity (Murata et al., 2002) and inhibits Akt kinase activity. In addition, ghrelin partially reverses the down-regulation of insulin on phosphoenol pyruvate carboxykinase (PEPCK) mRNA expression, a rate-limiting enzyme of gluconeogenesis.

10. Ghrelin's functions

Few gastrointestinal peptide hormones attract more attention than ghrelin, not only because it is the only identified orexigenic hormone circulating in the blood, but also it possesses the unique structure of octanoyl acid modification. In addition to its classical effect on GH release, ghrelin exercises a wide range of physiological functions ranging from regulation of food intake and energy metabolism, control of glucose and lipid homeostasis, modulation of cardiovascular function, to immunomodulation.

10.1 Ghrelin and energy balance

10.1.1 Ghrelin and obesity

Hyperphagia, weight gain, and increased adiposity occurs after chronic administration of ghrelin either systemically or intracerebroventricularly (Tschop et al., 2000; Nakazato et al., 2001). Chronic central administration of ghrelin increases adipose deposition independently of food intake (Nakazato et al., 2001) in pair-fed animals. Further studies demonstrate that central ghrelin stimulates triglyceride (TG) uptake and lipogenesis, while inhibits lipid oxidation in white adipocytes (Theander-Carrillo et al., 2006). Peripheral daily administration of ghrelin for two weeks causes a significant increase in fat mass as measured by dual energy X-ray absorptiometry (DXA) (Tschop et al., 2000). Consistent with these observations, the ghrelin receptor null mice are protected against the full development of diet-induced obesity (Zigman et al., 2005).

10.1.2 Ghrelin and cachexia/anorexia

Cachexia is characterized by involuntary weight loss due to persistently negative nitrogen balance resulting from a diverse series of pathological conditions such as malnutrition, chronic infectious diseases, immunodeficiency and malignancy. Cachexia is always accompanied by anorexia, while the reduction in food intake alone cannot explain the metabolic changes and wasting associated with cachexia. Ghrelin and ghrelin receptor agonists are fascinating candidates for the treatment of cachexia, due to their unique anabolic effects such as stimulating GH secretion, food intake, gastric motility and adiposity. High levels of circulating acylated ghrelin are detected in cachexia patients associated with lung cancer (Shimizu et al., 2003), chronic heart failure (Nagaya et al., 2001b), renal failure (Yoshimoto et al., 2002), chronic liver disease (Tacke et al., 2003) and anorexia nervosa (Otto et al., 2001), which may represent a compensatory response to an organism's wasting. Both ghrelin and acylated to total ghrelin ratio are elevated in cancer induced cachexia and inversely associated with body mass index (BMI). Whether increased circulating ghrelin levels indicate a ghrelin resistance in cachexia remains to be determined.

10.2 Ghrelin and food intake

Both animal experiments and clinical studies demonstrate that ghrelin induces a rapid increase in food intake in rodents and humans (Muccioli et al., 2002). Originating from the stomach, ghrelin activates arcuate NPY/AgRP neurons to increase the secretion of NPY, AgRP and GABA. NPY subsequently modulates the activity of postsynaptic secondary order neurons in the paraventricular nucleus, dorsomedial nucleus and lateral hypothalamic area to stimulate food intake, while GABA inactivates the proopiomelanocortin neurons and inhibits the anorectic melanocortin signaling pathway.

The intracellular mechanisms that mediate NPY/AgRP neuronal activation in response to ghrelin in appetite regulation have been demonstrated to be associated with the lipid metabolism in the hypothalamus (Lopez et al., 2008). After binding of ghrelin, the ghrelin receptor activates hypothalamic AMPK. AMPK then phosphorylates the acetyl-CoA carboxylase, decreases malonyl-CoA level, and suppresses lipid synthesis (Kola et al., 2005), leading to the activation of carnitine palmitoyltransferase 1 (CPT1). Activated CPT1 then accelerates lipid transport into mitochondria to catabolize lipid. Ghrelin regulates mitochondrial oxidation in the hypothalamic cells mainly through uncoupling protein 2 (UCP2) (Andrews et al., 2008), ninety percent of which co-expresses the ghrelin receptor. Experiments on isolated hypothalamic synaptosomes demonstrate that ghrelin increases oxygen consumption and reactive oxygen species (ROS) (Yamagishi et al., 2001), which are always associated with increased transcription and activity of UCP2 (Echtay et al., 2002). The produced ROS are de-gradated by UCP2 (Brand et al., 2004) in order to allow continuous CPT1-promoted fatty acid β oxidation that supports the bioenergetic needs to maintain firing of NPY/AgRP neurons and stimulate food intake. Compound C, an AMPK inhibitor, suppresses appetite stimulated by ghrelin. In UCP2-deficient mice, compound C fails to affect appetite. Taken together, these results suggest that lipid metabolism through AMPK/CPT1/UCP2 in the hypothalamic neurons is a key modulator for the regulation of appetite by ghrelin.

10.3 Ghrelin and glucose homeostasis

It has been demonstrated that ghrelin contributes to the modulation of glucose homeostasis. Blocking the effect of endogenous ghrelin by D-Lys3-GHRP-6 significantly improves the intraperitoneal glucose tolerance test (IGTT), indicating that endogenous ghrelin involves in the regulation of insulin release and blood glucose homeostasis (Dezaki et al., 2004). Although ghrelin has been demonstrated to inhibit the activity of the glucosensing neurons in the dorsal vagal complex of rats (Penicaud et al., 2006; Wang et al., 2008), most available data validate that ghrelin modulates insulin secretion and sensitivity, and hepatic glucose production.

The inhibitory effect of ghrelin on insulin secretion has been profoundly demonstrated. In the dissected and perfused rat pancreas, ghrelin significantly inhibits the insulin release in response to increasing glucose concentrations, while demonstrates no significant effect on basal insulin release (Egido et al., 2002). Moreover, the level of insulin released from the perfused pancreas is markedly increased by either blocking the GHS-R1a or immuno-neutralizating the endogenous ghrelin. Glucose stimulated insulin release is also enhanced in the pancreas islets isolated from ghrelin-null mice (Dezaki et al., 2006).

In vitro experiments show that ghrelin attenuates the inhibitory effects of insulin on expression of PEPCK and up-regulates the gluconeogenesis in cultured rat hepatoma cells (Murata et al., 2002). It also decreases the phosphorylation levels of protein kinase B (PKB) and glycogen synthase kinase (pGSK). All of these results provide evidences that ghrelin has a direct effect on hepatic glucose metabolism (Murata et al., 2002). However, GHS-R1a expression in the liver has not yet been demonstrated. Whether the effect of ghrelin on hepatic glucose metabolism is exerted via a novel ghrelin receptor subtype remains to be explored.

10.4 Ghrelin and lipid metabolism

Chronic central administration of ghrelin increases adipose deposition independently of food intake (Nakazato et al., 2001; Theander-Carrillo et al., 2006). Furthermore, various fat storage promoting enzymes such as lipoprotein lipase, acetyl-CoA carboxylase (ACC), fatty acid synthase(FAS), and stearoyl-CoA desaturase–1(SCD1) are markedly increased, while the fat oxidation rate limiting enzymes such as carnitine palmitoyl transferase–1α (CPT1) and uncoupling proteins (UCPs) are significantly decreased (Theander-Carrillo et al., 2006).

Peripheral administration of ghrelin has also been implicated on the regulation of lipid metabolism, with effects on liver, skeletal muscle and adipose tissue. In the liver, ghrelin increases lipogenic genes expression and triglyceride content, while reduces AMPK activity leading to lower fatty acid oxidation (Barazzoni et al., 2005). In the gastrocnemius muscle, ghrelin increases mitochondrial oxidative enzyme activities and reduces triglyceride content (Barazzoni et al., 2005). Ghrelin selectively increases peroxisome proliferator activated receptor γ to reduce muscle fat content in skeletal muscle (Barazzoni et al., 2005). In adipocytes, ghrelin stimulates lipogenesis partly by the insulin-induced glucose uptake (Patel et al., 2006), antagonizes lipolysis induced by isoproterenol (Larhammar, 1996), and stimulates the proliferation and differentiation of preadipocytes (Kim et al., 2004).

10.5 Ghrelin and gastrointestinal motility

Ghrelin is a strong gastrokinetic agent, having the motilin-like ability to stimulate motor activity in the gastrointestinal tract (Trudel et al., 2002). Ghrelin triggers the migrating motor complex in the fasted state (Fujino et al., 2003; De Winter et al., 2004; Tack et al., 2006) and accelerates gastric emptying in the postprandial state in animals and humans (Asakawa et al., 2001b; De Winter et al., 2004; Binn et al., 2006; Levin et al., 2006). Ghrelin also accelerates the transit of the small intestine but has no effect on the colon (Trudel et al., 2002). The contractile response of the stomach to the intravenous administration of ghrelin has also been reported (Masuda et al., 2000). The prokinetic effect of ghrelin on motility is mediated through activation of vagal afferents because atropine or vagotomy blocks the contractions induced by ghrelin in these urethane-anesthetized rats (Fujino et al., 2003; Fukuda et al., 2004).

Post-operative ileus is a major cause of complications and prolonged hospitalization. Ghrelin reverses the delay in gastric emptying in post-operative patients (Trudel et al., 2002). Treatment with TZP-101, a synthetic ghrelin receptor agonist or RC-1139, a ghrelin analogue, shows promising results of acceleration of gastric emptying in rats with ileus induced by morphine or surgery (Poitras et al., 2005; Venkova et al., 2007). In addition, intravenous administration of ghrelin has been shown to accelerate gastric emptying in patients with diabetic, idiopathic and post-vagotomy gastroparesis (Murray et al., 2005; Tack et al., 2005; Binn et al., 2006). However, controversy still exists on the use of ghrelin for treatment of gastroparesis because that the study number of patients was small and the methods employed to measure gastric emptying varied from study to study (Tack et al., 2005; Peeters, 2006). In addition, the unwanted effects of ghrelin on glucose and lipid metabolism should be considered. The future of ghrelin application in gastrointestinal motility disorder is largely dependent on our understanding of the mechanism by which ghrelin stimulates gastrointestingal motility (Peeters, 2006).

10.6 Ghrelin and memory, depression and anxiety

It was firstly demonstrated that ghrelin can increase memory retention when injected i.c.v. in rats (Carlini et al., 2002). Latter studies showed similar results when injected into the hippocampus, amygdala and dorsal raphe nucleus in rats (Carlini et al., 2004), as well as injected i.c.v. in chronically food-restricted mice (Carlini et al., 2008). Results from step-down tests suggest that ghrelin might modulate specific molecular intermediates involved in the memory acquisition/consolidation processes (Carlini et al., 2010a). The precise mechanism by which ghrelin affects memory is currently unknown. It has been revealed that circulating ghrelin can reach the hippocampas, increase spine synapse density in CA1, and induce long-term potentiation (LTP), which are paralleled by enhanced spatial learning and memory (Diano et al., 2006). Further studies suggest that ghrelin enhances spatial memory by activating the PI3K signaling pathway. The NOS/NO pathway might also involve in the effects of ghrelin on memory consolidation. Intra-hippocampal administration of ghrelin increases the NOS activity dose-dependently and reduces the threshold for LTP generation in dentate gyrus of rats (Carlini et al., 2010b). Additional studies have showed that ghrelin's effects on memory may depend on the availability of 5-HT (Carlini et al., 2007). In contrast to the enhanced effects of ghrelin on memory in rodents, ghrelin could decrease memory retention in neonatal chicks (Carvajal et al., 2009) .

The connection between ghrelin and depression was firstly demonstrated by administrating ghrelin antisense DNA which induces an anti-depressant response (Kanehisa et al., 2006). This observation is in agreement with an increase in depression-like behavior in rats with central administration of ghrelin (Schanze et al., 2008). In contrast to these findings, Lutter and colleagues have found that increasing ghrelin levels produces antidepressant-like responses which may be induced by the activation of orexin neurons in the lateral hypothalamus (Lutter et al., 2008). It has also been revealed that chronic social defeat stress, a rodent model of depression, persistently increases ghrelin levels, whereas growth hormone secretagogue receptor (GHSR) null mice show increased deleterious effects of chronic defeat (Kluge et al., 2009). Clinical studies also reveal conflicting results. One study has reported that the levels of acylated and des-acylated ghrelin are lower in depressive patients before and after citalopram treatment than in the control group (Olszanecka-Glinianowicz et al., 2010). Other studies have shown no significant difference in patients with major depression (Asakawa et al., 2001a; Nakashima et al., 2008; Ogaya et al., 2011). Despite of these controversial findings, emerging evidences suggest that ghrelin plays an important role in depression. Ghrelin gene polymorphism has been found to be associated with depression (Carlini et al., 2002). In the brain regions critical for the regulation of cognitive behavior, ghrelin has been reported to increase expression of glutamate receptor metabotropic 5 (Grm5) mRNA, GABAA-3α(Gabra3) and GABAA-5α receptor (Gabra5) subunit mRNA in the amygdala, to influence the central serotonin system (Schanze et al., 2008), and to inhibit 5-HT release (Carlini et al., 2004). Since it has been reported that decreased 5-HT activity can elicit depressive like behavior (Temel et al., 2007), it is reasonable to believe that ghrelin might be associated with depression.

Animal and clinical studies indicate that ghrelin is also associated with anxiety. Both central and peripheral administration of ghrelin induces anxiogenesis (Asakawa et al., 2001a; Carlini et al., 2002; Carlini et al., 2004; Carvajal et al., 2009). Ghrelin gene expression is increased by stresses in mice (Asakawa et al., 2001a; Hansson et al., 2011) and psychological stress may increase plasma ghrelin levels in humans (Rouach et al., 2007). Administration of anti-sense DNA for ghrelin in rats has been reported to induce an antidepressant response, while blockade of ghrelin decrease anxiety-like behavior (Kanehisa et al., 2006). The enhanced effects of ghrelin on anxiety may be mediated by corticotropin-releasing hormone (CRH) (Asakawa et al., 2001a) and inhibition of serotonin release (Carlini et al., 2004). In contrast to these observations, increasing circulating ghrelin has been reported to produce anxiolytic-like responses, while no longer were these anxiolytic-like behavioral responses observed when GHSR-null mice were calorie restricted (Lutter et al., 2008). These findings are consistent with a report showing low ghrelin cell activity in high-anxiety Wistar Kyoto rats (Kristensson et al., 2007). The reasons for the conflicting results are currently unknown (Andrews, 2011).

The finding of ghrelin's effects on brain cognitive functions will provide new therapies for mental disorders.

10.7 Ghrelin and cardiovascular disease

Numerous studies suggest that ghrelin exercises a wide array of cardiovascular activities including the vasodilation, improvement of myocardial function (Chang et al., 2004; Li et al., 2006) and endothelium protection (Li et al., 2004; Tesauro et al., 2005; Rossi et al., 2007).

These effects may involve both peripheral and central mechanisms (Lin et al., 2004). Microinjection of ghrelin into the nucleus of the solitary tract significantly decreases the mean arterial pressure and heart rate through its action on this nucleus (Matsumura et al., 2002; Lin et al., 2004). The direct effects of ghrelin on cardiovascular function are supported by the mRNA expression of both ghrelin and its receptor in the heart and aortas (Nagaya et al., 2001a; Gnanapavan et al., 2002). In addition, [^{125}I-His9] ghrelin, a radio-labeled ghrelin, has been shown to bind to the heart and to peripheral vascular tissue (Katugampola et al., 2001). *In vitro* studies demonstrated that ghrelin inhibits apoptosis of cardiomyocytes (Baldanzi et al., 2002), improves myocardial function during ischemia/reperfusion (Chang et al., 2004) and reduces infarct size (Frascarelli et al., 2003). Moreover, the radiolabeled ghrelin signal increases in atherosclerotic regions (Katugampola et al., 2001), suggesting that ghrelin receptor expression is up-regulated and ghrelin may participate in the development of atherosclerosis.

Congestive heart failure (CHF) is an often-fatal condition in which the heart muscles become weakened and lack the strength to adequately pump blood throughout the body. In patients with CHF, ghrelin decreases mean arterial pressure without increasing heart rate. Chronic treatment with ghrelin improves left-ventricular (LV) function and exercise capacity, as well as attenuates the development of LV remodeling and cachexia in a rat model of chronic heart failure (Nagaya et al., 2001c; Nagaya et al., 2004). Treatment with synthetic GHS such as GHRP-6 or hexarelin improves LV function, prevents cardiac damage after ischemia, and attenuates fibroblast proliferation (Locatelli et al., 1999; Iwase et al., 2004; Xu et al., 2007). In summary, emerging data supports the role of ghrelin and its analogs as novel therapeutic candidates for CHF.

10.8 Ghrelin and immunomodulation

The expression of ghrelin and GHS-R in cells of the immune system has been detected in several lymphoid organs (Gnanapavan et al., 2002) and leukocyte subsets including T and B cells, monocytes (Dixit et al., 2004). Such widespread expression of ghrelin receptor in the immune system supports a role for ghrelin in the regulation of immune-related functions. This notion is further supported by previous studies demonstrating that ghrelin regulates immune cell proliferation, activation and secretion of proinflamatory cytokines (Dixit et al., 2004; Taub, 2008). Chronic administration of a ghrelin analogue to old mice for 3 weeks has been demonstrated to stimulate growth, cellularity and differentiation of the thymus, and to increase T-cell production (Koo et al., 2001) which enhances resistance to the initiation of neoplasms and subsequent metastasis in animals inoculated with a transplantable lymphoma cell line, EL4. In addition, GHSs promotes thymic engraftment in bone marrow transplant of SCID mice (Koo et al., 2001).

Upon binding with ghrelin, GHS-R elicits a potent intracellular calcium release in primary and cultured human T cells. In addition, GHS-R is preferentially associated with GM1$^+$ lipid rafts upon cellular activation (Dixit et al., 2004). These findings suggest that GHS-R is actually expressed on the surface of T cells and functionally active. Activation of T cells by ghrelin forms the lamellipodia and remodeling of actin cytoskeleton, leading to polarization and directional migration (Dixit et al., 2004).

Monocytes are important sources of proinflammatory cytokines. Initial studies show that ghrelin acts to inhibit the production of IL-1β and IL-6 via the GHS-R because this inhibition of cytokines is blunted by GHS-R antagonists (Dixit et al., 2004). All these data suggest a novel role for ghrelin in immune cell function as a negative regulator of inflammatory cytokine.

10.9 Ghrelin and cell differentiation

Dependent on the type of cells, ghrelin can either stimulate or inhibit the cell differentiation. In 3T3-L1 cell lines, over-expression of ghrelin significantly inhibits differentiation of adipocytes (Zhang et al., 2004) and markedly decreases mRNA and protein levels of PPARγ, a marker of adipocyte differentiation. Over-expression of ghrelin in adipose tissue under the control of FABP4 promoter (Zhang et al., 2008) significant decreases the amount of adipose tissue and renders the mice resistant to diet induced obesity, which indicate that ghrelin may impair the development of adipose tissue (Zhang et al., 2008).

In C2C12 cells lines, ghrelin significantly increases the differentiation of premyocytes into myocytes. Expression of both Myo D and myosin heavy chain protein are elevated in cells overexpressing ghrelin (Zhang et al., 2007), indicating the differentiation of myocyte. Exogenous ghrelin stimulates the proliferation of C2C12 myoblasts and promotes these cells to differentiate and fuse into multinucleated myotubes by activating p38 (Filigheddu et al., 2007).

In both osteoblast cell lines and primary cultured osteoblasts, ghrelin stimulates an increase in cell proliferation and differentiation (Fukushima et al., 2005; Delhanty et al., 2006). The expression of GHS-R1a mRNA and immunoreactivity in osteoblast cells has been demonstrated (Fukushima et al., 2005). The proliferative effect of ghrelin is suppressed by inhibitors of extracellular-signal-regulated kinase (ERK) and phosphoinositide-3 kinase (PI3K), indicating that ghrelin stimulates human osteoblast growth via MAPK/PI3K pathways.

11. Cross talk between ghrelin and other hormones

Normal functions of the organism rely on the precise coordination of various hormones. As expected, ghrelin also interacts with other hormones to exercise its biological functions.

11.1 Growth hormone/insulin-like growth factor-1 (IGF-1) axis

Ghrelin is known to evoke a specific, dose-dependent release of GH either *in vitro* or *in vivo* (Kojima et al., 1999). This effect is mediated by GHS-R because GHS-R-null mice fail to show the ghrelin-induced GH secretion (Sun et al., 2004). Although ablation of the ghrelin receptor does not reduce food intake or fat mass, GHS-R-/- mice do demonstrate modest reductions in body weight and exhibit biochemical alterations in IGF-1 levels (Sun et al., 2004). IGF-1 accounts for most of the perceptible, anabolic effects of GH such as the linear growth and increase in skeletal muscle mass (Root, A.W. & Root, M.J., 2002). These results indicate that ghrelin may play a more modulatory role in GH release (Sun et al., 2004).

Administration of growth hormone significantly decreases the serum concentration of acyl ghrelin (Eden Engstrom et al., 2003; Seoane et al., 2007), the total ghrelin secretion from rat stomach (Seoane et al., 2007), while demonstrates no effect on gastric ghrelin level (Qi et al., 2003). These findings indicate that growth hormone exerts a negative feedback action on ghrelin production and secretion. Administration of recombinant human IGF-1 in severe malnutrition patients elevates plasma total ghrelin concentration (Grinspoon et al., 2004). Because IGF-1 always inhibits growth hormone secretion, it is reasonable to assume that IGF-1 may induce ghrelin secretion through the reduction of growth hormone.

11.2 Ghrelin and somatostatin

Study has showed that ghrelin is a functional antagonist of somatostatin (Tannenbaum et al., 2003). This finding is in conformity with early *in vitro* studies of the GHSs demonstrating that this effect is at the level of the pituitary gland (Conley et al., 1995). However, ghrelin effectively stimulates GH secretion in the absence of somatostatin, indicating that its GH-releasing activity is not dependent on inhibiting endogenous somatostatin release (Tannenbaum et al., 2003).

Infusion of somatostatin or somatostatin analog octreotide significantly decreases the plasma acyl and total ghrelin levels (Shimada et al., 2003), indicating that somatostatin probably inhibits ghrelin synthesis by directly acting on the somatostatin receptor present in rat stomach (Silva et al., 2005). Furthermore, somatostatin null mice display an increased ghrelin mRNA in stomach and serum total ghrelin without any alteration in hypothalamic and pituitary ghrelin mRNA and serum acyl ghrelin concentration (Luque et al., 2006). Since serum acylated ghrelin remains unchanged in the somatostatin knockout mice, somatostatin may only affect the transcription and translation of ghrelin, but have no effect on its post-translational modification.

11.3 Ghrelin and NPY/AgRP

Ghrelin receptor is expressed predominantly in NPY/AgRP neurons in the arcuate nucleus of the hypothalamus (Hahn et al., 1998; Willesen et al., 1999). The arcuate NPY/AgRP neurons have been shown to be essential in the control of food intake and body weight (Gropp et al., 2005; Luquet et al., 2005). Functional activation of these neurons by ghrelin increases expression of the orexigenic neuropeptides themselves, NPY and AgRP (Kamegai et al., 2001; Nakazato et al., 2001). NPY receptor antagonists decrease ghrelin induced increase in food intake (Asakawa et al., 2001b; Shintani et al., 2001; Bagnasco et al., 2003), while disruption of NPY and AgRP via targeted mutagenesis abolishes virtually all ghrelin-induced effects (Chen et al., 2004; Luquet et al., 2005). All these results suggest that the orexigenic effect of ghrelin is fully dependent on NPY/AgRP. However, effects of ghrelin on appetite and body weight are not compromised in mice with selective disruption of the NPY gene, suggesting that AgRP may be a sufficient mediator (Tschop et al., 2002). Additional studies have also demonstrated that ghrelin induces more AgRP mRNA expression than NPY mRNA (Kamegai et al., 2001; Tschop et al., 2002).

11.4 Ghrelin and melanocortin

Ghrelin, either originated from blood or from local ghrelin expressing neurons, inhibits melanocortin signaling both directly and indirectly, resulting in an increase of food intake (Cowley et al., 2003). The orexigenic effect of ghrelin has been demonstrated to be attenuated in the Mc3r/Mc4r double knockout mice. This finding suggests that ghrelin stimulates energy intake partly by suppressing hypothalamic melanocortin tone.

Immunohistochemistry and electrophysiology studies have shown that ghrelin acts on NPY neurons, which synapse on and inhibit POMC neurons directly (Cowley et al., 2001) or activates inhibitory GABAergic interneurons innervating POMC and MC4r neurons (Cowley et al., 1999), thereby inhibits melanocortin signaling indirectly.

11.5 Ghrelin and leptin

Leptin is commonly considered as an inhibitor of ghrelin synthesis. Reciprocal relationship has been found between serum concentrations of ghrelin and leptin. Leptin concentration in obese is significantly higher than lean control, whereas ghrelin is lower (Tschop et al., 2001b). Moreover, ghrelin mRNA increases in stomach during fasting whereas leptin and leptin mRNAs decrease (Zhao et al., 2008). Leptin dose-dependently inhibits ghrelin transcription *in vitro* (Zhao et al., 2008) and decreases ghrelin release from isolated rat stomach (Kamegai et al., 2004). Central leptin gene therapy decreases plasma leptin level, whereas increases ghrelin level significantly in the mouse fed with high-fat diet (Dube et al., 2002), indicating that leptin inhibits ghrelin secretion only in peripheral tissues. Thus, peripheral, especially gastric leptin, might repress ghrelin expression through its receptor in gastric mucosa cells.

In the CNS, 57% of neurons activated by peripheral ghrelin express the Ob-R (Traebert et al., 2002), suggesting the co-expression of GHS-R and the Ob-R in majority of neurons. It is therefore proposed that ghrelin and leptin exert their opposite effect on food intake by acting largely on the same population of hypothalamic neurons. Ghrelin-induced increase in food intake in the light phase has been demonstrated to be suppressed by ICV administration of leptin (Nakazato et al., 2001), or pretreatment with anti-NPY immunoglobulin (Nakazato et al., 2001), suggesting that ghrelin and leptin may act via the same cellular pathway.

Leptin has been long considered to cause satiety by depolarizing the POMC neurons, while hyperpolarizing NPY cells (Cowley et al., 2001). Ghrelin substantially blocks this reduction of feeding in rats pretreated with leptin (Nakazato et al., 2001). These results indicate that ghrelin may antagonize leptin action in the regulation of the NPY system.

11.6 Ghrelin and dopamine

When delivered directly into the VTA, ghrelin binds to the VTA neurons and produces a marked increase in food intake that resembles rebound feeding after fasting (Naleid et al., 2005; Abizaid et al., 2006). GHS-R has been detected in about 50–60% VTA dopamine cells. In addition, VTA dopamine cells are innervated by lateral hypothalamic hypocretin/orexin neurones, which are also sensitive to ghrelin (Toshinai et al., 2003). These results indicate that ghrelin might potentially influence the release of dopamine from these cells (Abizaid et al., 2006; Zigman et al., 2006). This concept is supported by numerous studies in which extracellular dopamine content in the nucleus accumbens of rats has been shown to be elevated by peripheral, i.c.v. and intra-VTA injections of ghrelin (Jerlhag et al., 2006; Jerlhag et al., 2007; Quarta et al., 2009). Further study suggests that ghrelin increases dopamine release by improving the dopamine cells excitability (Abizaid et al., 2006). In the presence of ghrelin, dopamine cells in the VTA increase their frequency of action potentials, which appears to be mediated by glutamatergic neurotransmission. These changes are undetectable in dopamine cells from the VTA of GHS-R knockout mice (Abizaid et al., 2006). As in the hypothalamus, ghrelin lowers the threshold of activation of dopamine neurons through mechanisms that involve remodeling the ratio of excitatory versus inhibitory inputs onto these cells (Abizaid et al., 2006).

Peripheral injections of ghrelin increase dopamine turnover in the ventral striatum of mice and rats (Abizaid et al., 2006). Considering that the VTA is protected by the blood–brain barrier and less permeable than the arcuate nuclei to blood-borne substances, ghrelin might be transported into VTA through a saturable mechanism remained to be fully determined (Banks et al., 2002). Another possibility is the indirect effect of ghrelin on afferent neurons which innervate the VTA such as lateral hypothalamic neurons and laterodorsal tegmental nucleus (Guan et al., 1997; Geisler & Zahm, 2005).

11.7 Ghrelin and insulin

In human pancreatic islets, ghrelin receptor immunoreactivity partially overlaps with insulin-positive β-cells (Granata et al., 2007), indicating that human β-cells might also be responsive to ghrelin stimulation. In cultured islet cells, acyl-ghrelin suppresses both basal insulin secretion (Dezaki et al., 2004) and glucose-induced insulin release (Dezaki et al., 2006). The level of insulin released from the perfused pancreas is significantly increased by either blocking the GHS-R1a or immunoneutralizating the endogenous ghrelin. Furthermore, glucose-induced insulin release is greater in islets isolated from ghrelin-null mice than wild type littermates. All these data suggest that ghrelin regulates insulin secretion from the islet cells. The molecular mechanism by which ghrelin suppresses glucose-induced insulin release has been reported to be the attenuation of Ca^{2+} signaling in β-cells via $G_{\alpha i2}$ and Kv channel (Dezaki et al., 2007).

Gastric artery perfusion of insulin significantly inhibits ghrelin release from isolated stomach tissue in rats (Kamegai et al., 2004). Central administration of insulin reduces serum total ghrelin concentration (Ueno et al., 2006), while maintaining euglycemia (Saad et al., 2002; Flanagan et al., 2003). Several clinical observations in humans also indicate that insulin may inhibit ghrelin secretion.

11.8 Ghrelin and glucagon

Both ghrelin and GHS-R1a have been identified in either human or rat pancreatic islets α cells (Date et al., 2002). Ghrelin induces significant increase in glucagon secretion from the pancreas of diabetic rats rather than in normal rats (Adeghate & Parvez, 2002). A possible reason for this difference is that signal transduction involving the calcium pathway is impaired in diabetic rat pancreas.

Glucagon may stimulate the gene transcription of ghrelin as well. Glucagon might elevate the activity of ghrelin gene promoter by the mediation of the second messenger cAMP (Wei et al., 2005). Several studies suggest that glucagon may contribute to the pre-prandial surge of ghrelin. Glucagon receptor is present in endocrine cells in gastric mucosa (Katayama et al., 2007). Glucagon concentration increases during fasting, and plasma acyl ghrelin concentration rises after administration of glucagon in rats. In addition, ghrelin released from the rat stomach is augmented by glucagon (Kamegai et al., 2004).

11.9 Ghrelin and estrogen

Numerous studies report that estrogen up-regulates ghrelin level. The effects of estrogen to stimulate food intake and growth hormone secretion might therefore be partially mediated through ghrelin. Plasma total ghrelin concentration in female patients with anorexia nervosa is significantly elevated (Grinspoon et al., 2004), while ghrelin mRNA level rises dramatically in cultured stomach cells (Sakata et al., 2006) after estrogen administration. Discrepant results on the effect of estrogen on ghrelin have been reported. Precise mechanism for the discrepancy remains unknown, but may relate to the distinction of age (Matsubara et al., 2004), physiological status (Chu et al., 2006) and variation in methods used for estrogen application (Kellokoski et al., 2005; Chu et al., 2006). Estrogen replacement therapy in post-menopausal women has been reported to induce increase (Kellokoski et al., 2005; Lambrinoudaki et al., 2008), no significant change, or even decreases (Chu et al., 2006) in serum total and acyl ghrelin secretion. In female rats, ovariectomy induces a transient augment in plasma acyl ghrelin, ghrelin expressing cells and ghrelin mRNA in stomach (Matsubara et al., 2004).

12. Conclusion

Since its discovery in 1999, ghrelin has attracted a tremendous interest from both academy and industry. It has become one of the most extensively studied fields. This is due to its highly conserved sequence between species, its unique molecular structure, and the ubiquity of ghrelin and receptors which implicates its important physiological function during the development. The multiplicity of physiological functions of ghrelin are revealing gradually. Current evidences show that ghrelin affects GH release, food intake, energy and glucose homeostasis, gastrointestinal, cardiovascular and immune functions, cell proliferation and differentiation, and cognitive behavior. Ghrelin is therefore a critical hormone linking the gastrointestinal activities with organism functions.

13. Acknowledgement

This work was supported by grants from the National Natural Science Foundation of China (81030012, 81170795, 30890043, 30971434, 30871194, 30821001, and 30971085), and the Major National Basic Research Program of China (2010CB912504)

14. References

Abizaid, A., Liu, Z.W., Andrews, Z.B., Shanabrough, M., Borok, E., Elsworth, J.D., Roth, R.H., Sleeman, M.W., Picciotto, M.R., Tschop, M.H., Gao, X.B. & Horvath, T.L. (2006) Ghrelin modulates the activity and synaptic input organization of midbrain dopamine neurons while promoting appetite. *J Clin Invest*, 116,pp 3229-3239.

Adeghate, E. & Parvez, H. (2002) Mechanism of ghrelin-evoked glucagon secretion from the pancreas of diabetic rats. *Neuro Endocrinol Lett*, 23,pp 432-436.

Amar, S., Mazella, J., Checler, F., Kitabgi, P. & Vincent, J.P. (1985) Regulation of cyclic GMP levels by neurotensin in neuroblastoma clone N1E115. *Biochem Biophys Res Commun*, 129,pp 117-125.

Andrews, Z.B. (2011) The extra-hypothalamic actions of ghrelin on neuronal function. *Trends Neurosci*, 34,pp 31-40.

Andrews, Z.B., Liu, Z.W., Walllingford, N., Erion, D.M., Borok, E., Friedman, J.M., Tschop, M.H., Shanabrough, M., Cline, G., Shulman, G.I., Coppola, A., Gao, X.B., Horvath, T.L. & Diano, S. (2008) UCP2 mediates ghrelin's action on NPY/AgRP neurons by lowering free radicals. *Nature*, 454,pp 846-851.

Ariyasu, H., Takaya, K., Tagami, T., Ogawa, Y., Hosoda, K., Akamizu, T., Suda, M., Koh, T., Natsui, K., Toyooka, S., Shirakami, G., Usui, T., Shimatsu, A., Doi, K., Hosoda, H., Kojima, M., Kangawa, K. & Nakao, K. (2001) Stomach is a major source of circulating ghrelin, and feeding state determines plasma ghrelin-like immunoreactivity levels in humans. *J Clin Endocrinol Metab*, 86,pp 4753-4758.

Asakawa, A., Inui, A., Fujimiya, M., Sakamaki, R., Shinfuku, N., Ueta, Y., Meguid, M.M. & Kasuga, M. (2005) Stomach regulates energy balance via acylated ghrelin and desacyl ghrelin. *Gut*, 54,pp 18-24.

Asakawa, A., Inui, A., Kaga, T., Yuzuriha, H., Nagata, T., Fujimiya, M., Katsuura, G., Makino, S., Fujino, M.A. & Kasuga, M. (2001a) A role of ghrelin in neuroendocrine and behavioral responses to stress in mice. *Neuroendocrinology*, 74,pp 143-147.

Asakawa, A., Inui, A., Kaga, T., Yuzuriha, H., Nagata, T., Ueno, N., Makino, S., Fujimiya, M., Niijima, A., Fujino, M.A. & Kasuga, M. (2001b) Ghrelin is an appetite-stimulatory signal from stomach with structural resemblance to motilin. *Gastroenterology*, 120,pp 337-345.

Avallone, R., Demers, A., Rodrigue-Way, A., Bujold, K., Harb, D., Anghel, S., Wahli, W., Marleau, S., Ong, H. & Tremblay, A. (2006) A growth hormone-releasing peptide that binds scavenger receptor CD36 and ghrelin receptor up-regulates sterol transporters and cholesterol efflux in macrophages through a peroxisome proliferator-activated receptor gamma-dependent pathway. *Mol Endocrinol*, 20,pp 3165-3178.

Bagnasco, M., Tulipano, G., Melis, M.R., Argiolas, A., Cocchi, D. & Muller, E.E. (2003) Endogenous ghrelin is an orexigenic peptide acting in the arcuate nucleus in response to fasting. *Regul Pept*, 111,pp 161-167.

Baldanzi, G., Filigheddu, N., Cutrupi, S., Catapano, F., Bonissoni, S., Fubini, A., Malan, D., Baj, G., Granata, R., Broglio, F., Papotti, M., Surico, N., Bussolino, F., Isgaard, J., Deghenghi, R., Sinigaglia, F., Prat, M., Muccioli, G., Ghigo, E. & Graziani, A. (2002) Ghrelin and des-acyl ghrelin inhibit cell death in cardiomyocytes and endothelial cells through ERK1/2 and PI 3-kinase/AKT. *J Cell Biol*, 159,pp 1029-1037.

Bang, A.S., Soule, S.G., Yandle, T.G., Richards, A.M. & Pemberton, C.J. (2007) Characterisation of proghrelin peptides in mammalian tissue and plasma. *J Endocrinol*, 192,pp 313-323.

Banks, W.A., Tschop, M., Robinson, S.M. & Heiman, M.L. (2002) Extent and direction of ghrelin transport across the blood-brain barrier is determined by its unique primary structure. *J Pharmacol Exp Ther*, 302,pp 822-827.

Barazzoni, R., Bosutti, A., Stebel, M., Cattin, M.R., Roder, E., Visintin, L., Cattin, L., Biolo, G., Zanetti, M. & Guarnieri, G. (2005) Ghrelin regulates mitochondrial-lipid metabolism gene expression and tissue fat distribution in liver and skeletal muscle. *Am J Physiol Endocrinol Metab*, 288,pp E228-235.

Barim, A.O., Aydin, S., Colak, R., Dag, E., Deniz, O. & Sahin, I. (2009) Ghrelin, paraoxonase and arylesterase levels in depressive patients before and after citalopram treatment. *Clin Biochem*, 42,pp 1076-1081.

Barreiro, M.L., Gaytan, F., Caminos, J.E., Pinilla, L., Casanueva, F.F., Aguilar, E., Dieguez, C. & Tena-Sempere, M. (2002) Cellular location and hormonal regulation of ghrelin expression in rat testis. *Biol Reprod*, 67,pp 1768-1776.

Beaumont, N.J., Skinner, V.O., Tan, T.M., Ramesh, B.S., Byrne, D.J., MacColl, G.S., Keen, J.N., Bouloux, P.M., Mikhailidis, D.P., Bruckdorfer, K.R., Vanderpump, M.P. & Srai, K.S. (2003) Ghrelin can bind to a species of high density lipoprotein associated with paraoxonase. *J Biol Chem*, 278,pp 8877-8880.

Bednarek, M.A., Feighner, S.D., Pong, S.S., McKee, K.K., Hreniuk, D.L., Silva, M.V., Warren, V.A., Howard, A.D., Van Der Ploeg, L.H. & Heck, J.V. (2000) Structure-function studies on the new growth hormone-releasing peptide, ghrelin: minimal sequence of ghrelin necessary for activation of growth hormone secretagogue receptor 1a. *J Med Chem*, 43,pp 4370-4376.

Bennett, P.A., Thomas, G.B., Howard, A.D., Feighner, S.D., van der Ploeg, L.H., Smith, R.G. & Robinson, I.C. (1997) Hypothalamic growth hormone secretagogue-receptor (GHS-R) expression is regulated by growth hormone in the rat. *Endocrinology*, 138,pp 4552-4557.

Binn, M., Albert, C., Gougeon, A., Maerki, H., Coulie, B., Lemoyne, M., Rabasa Lhoret, R., Tomasetto, C. & Poitras, P. (2006) Ghrelin gastrokinetic action in patients with neurogenic gastroparesis. *Peptides*, 27,pp 1603-1606.

Blake, A.D. & Smith, R.G. (1991) Desensitization studies using perifused rat pituitary cells show that growth hormone-releasing hormone and His-D-Trp-Ala-Trp-D-Phe-Lys-NH2 stimulate growth hormone release through distinct receptor sites. *J Endocrinol*, 129,pp 11-19.

Bockaert, J. & Pin, J.P. (1999) Molecular tinkering of G protein-coupled receptors: an evolutionary success. *EMBO J*, 18,pp 1723-1729.

Bodart, V., Bouchard, J.F., McNicoll, N., Escher, E., Carriere, P., Ghigo, E., Sejlitz, T., Sirois, M.G., Lamontagne, D. & Ong, H. (1999) Identification and characterization of a new growth hormone-releasing peptide receptor in the heart. *Circ Res*, 85,pp 796-802.

Bodart, V., Febbraio, M., Demers, A., McNicoll, N., Pohankova, P., Perreault, A., Sejlitz, T., Escher, E., Silverstein, R.L., Lamontagne, D. & Ong, H. (2002) CD36 mediates the cardiovascular action of growth hormone-releasing peptides in the heart. *Circ Res*, 90,pp 844-849.

Bondensgaard, K., Ankersen, M., Thogersen, H., Hansen, B.S., Wulff, B.S. & Bywater, R.P. (2004) Recognition of privileged structures by G-protein coupled receptors. *J Med Chem*, 47,pp 888-899.

Botto, J.M., Guillemare, E., Vincent, J.P. & Mazella, J. (1997) Effects of SR 48692 on neurotensin-induced calcium-activated chloride currents in the Xenopus oocyte expression system: agonist-like activity on the levocabastine-sensitive neurotensin receptor and absence of antagonist effect on the levocabastine insensitive neurotensin receptor. *Neurosci Lett*, 223,pp 193-196.

Bowers, C.Y., Alster, D.K. & Frentz, J.M. (1992) The growth hormone-releasing activity of a synthetic hexapeptide in normal men and short statured children after oral administration. *J Clin Endocrinol Metab*, 74,pp 292-298.

Bowers, C.Y., Momany, F., Reynolds, G.A., Chang, D., Hong, A. & Chang, K. (1980) Structure-activity relationships of a synthetic pentapeptide that specifically releases growth hormone in vitro. *Endocrinology*, 106,pp 663-667.

Bowers, C.Y., Momany, F.A., Reynolds, G.A. & Hong, A. (1984) On the *in vitro* and *in vivo* activity of a new synthetic hexapeptide that acts on the pituitary to specifically release growth hormone. *Endocrinology*, 114,pp 1537-1545.

Bozou, J.C., Rochet, N., Magnaldo, I., Vincent, J.P. & Kitabgi, P. (1989) Neurotensin stimulates inositol trisphosphate-mediated calcium mobilization but not protein kinase C activation in HT29 cells. Involvement of a G-protein. *Biochem J*, 264,pp 871-878.

Brand, M.D., Affourtit, C., Esteves, T.C., Green, K., Lambert, A.J., Miwa, S., Pakay, J.L. & Parker, N. (2004) Mitochondrial superoxide: production, biological effects, and activation of uncoupling proteins. *Free Radic Biol Med*, 37,pp 755-767.

Bresson-Bepoldin, L. & Dufy-Barbe, L. (1994) GHRP-6 induces a biphasic calcium response in rat pituitary somatotrophs. *Cell Calcium*, 15,pp 247-258.

Brighton, P.J., Szekeres, P.G. & Willars, G.B. (2004) Neuromedin U and its receptors: structure, function, and physiological roles. *Pharmacol Rev*, 56,pp 231-248.

Camina, J.P., Carreira, M.C., El Messari, S., Llorens-Cortes, C., Smith, R.G. & Casanueva, F.F. (2004) Desensitization and endocytosis mechanisms of ghrelin-activated growth hormone secretagogue receptor 1a. *Endocrinology*, 145,pp 930-940.

Capella, C., Solcia, E. & Vassallo, G. (1969) Identification of six types of endocrine cells in the gastrointestinal mucosa of the rabbit. *Arch Histol Jpn*, 30,pp 479-495.

Carlini, V.P., Gaydou, R.C., Schioth, H.B. & de Barioglio, S.R. (2007) Selective serotonin reuptake inhibitor (fluoxetine) decreases the effects of ghrelin on memory retention and food intake. *Regul Pept*, 140,pp 65-73.

Carlini, V.P., Ghersi, M., Schioth, H.B. & de Barioglio, S.R. (2010a) Ghrelin and memory: differential effects on acquisition and retrieval. *Peptides*, 31,pp 1190-1193.

Carlini, V.P., Martini, A.C., Schioth, H.B., Ruiz, R.D., Fiol de Cuneo, M. & de Barioglio, S.R. (2008) Decreased memory for novel object recognition in chronically food-restricted mice is reversed by acute ghrelin administration. *Neuroscience*, 153,pp 929-934.

Carlini, V.P., Monzon, M.E., Varas, M.M., Cragnolini, A.B., Schioth, H.B., Scimonelli, T.N. & de Barioglio, S.R. (2002) Ghrelin increases anxiety-like behavior and memory retention in rats. *Biochem Biophys Res Commun*, 299,pp 739-743.

Carlini, V.P., Perez, M.F., Salde, E., Schioth, H.B., Ramirez, O.A. & de Barioglio, S.R. (2010b) Ghrelin induced memory facilitation implicates nitric oxide synthase activation and decrease in the threshold to promote LTP in hippocampal dentate gyrus. *Physiol Behav*, 101,pp 117-123.

Carlini, V.P., Varas, M.M., Cragnolini, A.B., Schioth, H.B., Scimonelli, T.N. & de Barioglio, S.R. (2004) Differential role of the hippocampus, amygdala, and dorsal raphe nucleus in regulating feeding, memory, and anxiety-like behavioral responses to ghrelin. *Biochem Biophys Res Commun*, 313,pp 635-641.

Carreira, M.C., Camina, J.P., Smith, R.G. & Casanueva, F.F. (2004) Agonist-specific coupling of growth hormone secretagogue receptor type 1a to different intracellular signaling systems. Role of adenosine. *Neuroendocrinology*, 79,pp 13-25.

Carvajal, P., Carlini, V.P., Schioth, H.B., de Barioglio, S.R. & Salvatierra, N.A. (2009) Central ghrelin increases anxiety in the Open Field test and impairs retention memory in a passive avoidance task in neonatal chicks. *Neurobiol Learn Mem*, 91,pp 402-407.

Cassoni, P., Ghe, C., Marrocco, T., Tarabra, E., Allia, E., Catapano, F., Deghenghi, R., Ghigo, E., Papotti, M. & Muccioli, G. (2004) Expression of ghrelin and biological activity of specific receptors for ghrelin and des-acyl ghrelin in human prostate neoplasms and related cell lines. *Eur J Endocrinol*, 150,pp 173-184.

Cassoni, P., Papotti, M., Ghe, C., Catapano, F., Sapino, A., Graziani, A., Deghenghi, R., Reissmann, T., Ghigo, E. & Muccioli, G. (2001) Identification, characterization, and biological activity of specific receptors for natural (ghrelin) and synthetic growth hormone secretagogues and analogs in human breast carcinomas and cell lines. *J Clin Endocrinol Metab*, 86,pp 1738-1745.

Chang, L., Ren, Y., Liu, X., Li, W.G., Yang, J., Geng, B., Weintraub, N.L. & Tang, C. (2004) Protective effects of ghrelin on ischemia/reperfusion injury in the isolated rat heart. *J Cardiovasc Pharmacol*, 43,pp 165-170.

Chanoine, J.P. & Wong, A.C. (2004) Ghrelin gene expression is markedly higher in fetal pancreas compared with fetal stomach: effect of maternal fasting. *Endocrinology*, 145,pp 3813-3820.

Chen, C., Zhang, J., Vincent, J.D. & Israel, J.M. (1990) Two types of voltage-dependent calcium current in rat somatotrophs are reduced by somatostatin. *J Physiol*, 425,pp 29-42.

Chen, H.Y., Trumbauer, M.E., Chen, A.S., Weingarth, D.T., Adams, J.R., Frazier, E.G., Shen, Z., Marsh, D.J., Feighner, S.D., Guan, X.M., Ye, Z., Nargund, R.P., Smith, R.G., Van der Ploeg, L.H., Howard, A.D., MacNeil, D.J. & Qian, S. (2004) Orexigenic action of peripheral ghrelin is mediated by neuropeptide Y and agouti-related protein. *Endocrinology*, 145,pp 2607-2612.

Cheng, K., Chan, W.W., Barreto, A., Jr., Convey, E.M. & Smith, R.G. (1989) The synergistic effects of His-D-Trp-Ala-Trp-D-Phe-Lys-NH2 on growth hormone (GH)-releasing factor-stimulated GH release and intracellular adenosine 3',5'-monophosphate accumulation in rat primary pituitary cell culture. *Endocrinology*, 124,pp 2791-2798.

Cheng, K., Chan, W.W., Butler, B., Barreto, A., Jr. & Smith, R.G. (1991) Evidence for a role of protein kinase-C in His-D-Trp-Ala-Trp-D-Phe-Lys-NH2-induced growth hormone release from rat primary pituitary cells. *Endocrinology*, 129,pp 3337-3342.

Chipman, J.J. (1993) Recent advances in hGH clinical research. *J Pediatr Endocrinol*, 6,pp 325-328.

Chu, K.M., Chow, K.B., Leung, P.K., Lau, P.N., Chan, C.B., Cheng, C.H. & Wise, H. (2007) Over-expression of the truncated ghrelin receptor polypeptide attenuates the constitutive activation of phosphatidylinositol-specific phospholipase C by ghrelin receptors but has no effect on ghrelin-stimulated extracellular signal-regulated kinase 1/2 activity. *Int J Biochem Cell Biol*, 39,pp 752-764.

Chu, M.C., Cosper, P., Nakhuda, G.S. & Lobo, R.A. (2006) A comparison of oral and transdermal short-term estrogen therapy in postmenopausal women with metabolic syndrome. *Fertil Steril*, 86,pp 1669-1675.

Conley, L.K., Teik, J.A., Deghenghi, R., Imbimbo, B.P., Giustina, A., Locatelli, V. & Wehrenberg, W.B. (1995) Mechanism of action of hexarelin and GHRP-6: analysis of the involvement of GHRH and somatostatin in the rat. *Neuroendocrinology*, 61,pp 44-50.

Corbetta, S., Peracchi, M., Cappiello, V., Lania, A., Lauri, E., Vago, L., Beck-Peccoz, P. & Spada, A. (2003) Circulating ghrelin levels in patients with pancreatic and gastrointestinal neuroendocrine tumors: identification of one pancreatic ghrelinoma. *J Clin Endocrinol Metab*, 88,pp 3117-3120.

Cowley, M.A., Pronchuk, N., Fan, W., Dinulescu, D.M., Colmers, W.F. & Cone, R.D. (1999) Integration of NPY, AGRP, and melanocortin signals in the hypothalamic paraventricular nucleus: evidence of a cellular basis for the adipostat. *Neuron*, 24,pp 155-163.

Cowley, M.A., Smart, J.L., Rubinstein, M., Cerdan, M.G., Diano, S., Horvath, T.L., Cone, R.D. & Low, M.J. (2001) Leptin activates anorexigenic POMC neurons through a neural network in the arcuate nucleus. *Nature*, 411,pp 480-484.

Cowley, M.A., Smith, R.G., Diano, S., Tschop, M., Pronchuk, N., Grove, K.L., Strasburger, C.J., Bidlingmaier, M., Esterman, M., Heiman, M.L., Garcia-Segura, L.M., Nillni, E.A., Mendez, P., Low, M.J., Sotonyi, P., Friedman, J.M., Liu, H., Pinto, S., Colmers, W.F., Cone, R.D. & Horvath, T.L. (2003) The distribution and mechanism of action of ghrelin in the CNS demonstrates a novel hypothalamic circuit regulating energy homeostasis. *Neuron*, 37,pp 649-661.

Cummings, D.E., Purnell, J.Q., Frayo, R.S., Schmidova, K., Wisse, B.E. & Weigle, D.S. (2001) A preprandial rise in plasma ghrelin levels suggests a role in meal initiation in humans. *Diabetes*, 50,pp 1714-1719.

Date, Y., Kojima, M., Hosoda, H., Sawaguchi, A., Mondal, M.S., Suganuma, T., Matsukura, S., Kangawa, K. & Nakazato, M. (2000) Ghrelin, a novel growth hormone-releasing acylated peptide, is synthesized in a distinct endocrine cell type in the gastrointestinal tracts of rats and humans. *Endocrinology*, 141,pp 4255-4261.

Date, Y., Nakazato, M., Hashiguchi, S., Dezaki, K., Mondal, M.S., Hosoda, H., Kojima, M., Kangawa, K., Arima, T., Matsuo, H., Yada, T. & Matsukura, S. (2002) Ghrelin is present in pancreatic alpha-cells of humans and rats and stimulates insulin secretion. *Diabetes*, 51,pp 124-129.

de Keyzer, Y., Lenne, F. & Bertagna, X. (1997) Widespread transcription of the growth hormone-releasing peptide receptor gene in neuroendocrine human tumors. *Eur J Endocrinol*, 137,pp 715-718.

De Winter, B.Y., De Man, J.G., Seerden, T.C., Depoortere, I., Herman, A.G., Peeters, T.L. & Pelckmans, P.A. (2004) Effect of ghrelin and growth hormone-releasing peptide 6 on septic ileus in mice. *Neurogastroenterol Motil*, 16,pp 439-446.

Deghenghi, R., Papotti, M., Ghigo, E. & Muccioli, G. (2001) Cortistatin, but not somatostatin, binds to growth hormone secretagogue (GHS) receptors of human pituitary gland. *J Endocrinol Invest*, 24,pp RC1-3.

Dehlin, E., Liu, J., Yun, S.H., Fox, E., Snyder, S., Gineste, C., Willingham, L., Geysen, M., Gaylinn, B.D. & Sando, J.J. (2008) Regulation of ghrelin structure and membrane binding by phosphorylation. *Peptides*, 29,pp 904-911.

Delhanty, P.J., van der Eerden, B.C., van der Velde, M., Gauna, C., Pols, H.A., Jahr, H., Chiba, H., van der Lely, A.J. & van Leeuwen, J.P. (2006) Ghrelin and unacylated ghrelin stimulate human osteoblast growth via mitogen-activated protein kinase (MAPK)/phosphoinositide 3-kinase (PI3K) pathways in the absence of GHS-R1a. *J Endocrinol*, 188,pp 37-47.

Dezaki, K., Hosoda, H., Kakei, M., Hashiguchi, S., Watanabe, M., Kangawa, K. & Yada, T. (2004) Endogenous ghrelin in pancreatic islets restricts insulin release by attenuating Ca2+ signaling in beta-cells: implication in the glycemic control in rodents. *Diabetes*, 53,pp 3142-3151.

Dezaki, K., Kakei, M. & Yada, T. (2007) Ghrelin uses Galphai2 and activates voltage-dependent K+ channels to attenuate glucose-induced Ca2+ signaling and insulin release in islet beta-cells: novel signal transduction of ghrelin. *Diabetes*, 56,pp 2319-2327.

Dezaki, K., Sone, H., Koizumi, M., Nakata, M., Kakei, M., Nagai, H., Hosoda, H., Kangawa, K. & Yada, T. (2006) Blockade of pancreatic islet-derived ghrelin enhances insulin secretion to prevent high-fat diet-induced glucose intolerance. *Diabetes*, 55,pp 3486-3493.

Diano, S., Farr, S.A., Benoit, S.C., McNay, E.C., da Silva, I., Horvath, B., Gaskin, F.S., Nonaka, N., Jaeger, L.B., Banks, W.A., Morley, J.E., Pinto, S., Sherwin, R.S., Xu, L., Yamada, K.A., Sleeman, M.W., Tschop, M.H. & Horvath, T.L. (2006) Ghrelin controls hippocampal spine synapse density and memory performance. *Nat Neurosci*, 9,pp 381-388.

Dixit, V.D., Schaffer, E.M., Pyle, R.S., Collins, G.D., Sakthivel, S.K., Palaniappan, R., Lillard, J.W., Jr. & Taub, D.D. (2004) Ghrelin inhibits leptin- and activation-induced proinflammatory cytokine expression by human monocytes and T cells. *J Clin Invest*, 114,pp 57-66.

Dornonville de la Cour, C., Bjorkqvist, M., Sandvik, A.K., Bakke, I., Zhao, C.M., Chen, D. & Hakanson, R. (2001) A-like cells in the rat stomach contain ghrelin and do not operate under gastrin control. *Regul Pept*, 99,pp 141-150.

Dube, M.G., Beretta, E., Dhillon, H., Ueno, N., Kalra, P.S. & Kalra, S.P. (2002) Central leptin gene therapy blocks high-fat diet-induced weight gain, hyperleptinemia, and hyperinsulinemia: increase in serum ghrelin levels. *Diabetes*, 51,pp 1729-1736.

Dun, S.L., Brailoiu, G.C., Brailoiu, E., Yang, J., Chang, J.K. & Dun, N.J. (2006) Distribution and biological activity of obestatin in the rat. *J Endocrinol*, 191,pp 481-489.

Echtay, K.S., Roussel, D., St-Pierre, J., Jekabsons, M.B., Cadenas, S., Stuart, J.A., Harper, J.A., Roebuck, S.J., Morrison, A., Pickering, S., Clapham, J.C. & Brand, M.D. (2002) Superoxide activates mitochondrial uncoupling proteins. *Nature*, 415,pp 96-99.

Eden Engstrom, B., Burman, P., Holdstock, C. & Karlsson, F.A. (2003) Effects of growth hormone (GH) on ghrelin, leptin, and adiponectin in GH-deficient patients. *J Clin Endocrinol Metab*, 88,pp 5193-5198.

Egido, E.M., Rodriguez-Gallardo, J., Silvestre, R.A. & Marco, J. (2002) Inhibitory effect of ghrelin on insulin and pancreatic somatostatin secretion. *Eur J Endocrinol*, 146,pp 241-244.

Erdmann, J., Lippl, F. & Schusdziarra, V. (2003) Differential effect of protein and fat on plasma ghrelin levels in man. *Regul Pept*, 116,pp 101-107.

Erdmann, J., Lippl, F., Wagenpfeil, S. & Schusdziarra, V. (2005) Differential association of basal and postprandial plasma ghrelin with leptin, insulin, and type 2 diabetes. *Diabetes*, 54,pp 1371-1378.

Evans, B.E., Rittle, K.E., Bock, M.G., DiPardo, R.M., Freidinger, R.M., Whitter, W.L., Lundell, G.F., Veber, D.F., Anderson, P.S., Chang, R.S. & et al. (1988) Methods for drug discovery: development of potent, selective, orally effective cholecystokinin antagonists. *J Med Chem*, 31,pp 2235-2246.

Falls, H.D., Dayton, B.D., Fry, D.G., Ogiela, C.A., Schaefer, V.G., Brodjian, S., Reilly, R.M., Collins, C.A. & Kaszubska, W. (2006) Characterization of ghrelin receptor activity in a rat pituitary cell line RC-4B/C. *J Mol Endocrinol*, 37,pp 51-62.

Febbraio, M., Hajjar, D.P. & Silverstein, R.L. (2001) CD36: a class B scavenger receptor involved in angiogenesis, atherosclerosis, inflammation, and lipid metabolism. *J Clin Invest*, 108,pp 785-791.

Feighner, S.D., Howard, A.D., Prendergast, K., Palyha, O.C., Hreniuk, D.L., Nargund, R., Underwood, D., Tata, J.R., Dean, D.C., Tan, C.P., McKee, K.K., Woods, J.W., Patchett, A.A., Smith, R.G. & Van der Ploeg, L.H. (1998) Structural requirements for the activation of the human growth hormone secretagogue receptor by peptide and nonpeptide secretagogues. *Mol Endocrinol*, 12,pp 137-145.

Filigheddu, N., Gnocchi, V.F., Coscia, M., Cappelli, M., Porporato, P.E., Taulli, R., Traini, S., Baldanzi, G., Chianale, F., Cutrupi, S., Arnoletti, E., Ghe, C., Fubini, A., Surico, N., Sinigaglia, F., Ponzetto, C., Muccioli, G., Crepaldi, T. & Graziani, A. (2007) Ghrelin and des-acyl ghrelin promote differentiation and fusion of C2C12 skeletal muscle cells. *Mol Biol Cell*, 18,pp 986-994.

Flanagan, D.E., Evans, M.L., Monsod, T.P., Rife, F., Heptulla, R.A., Tamborlane, W.V. & Sherwin, R.S. (2003) The influence of insulin on circulating ghrelin. *Am J Physiol Endocrinol Metab*, 284,pp E313-316.

Foster-Schubert, K.E., Overduin, J., Prudom, C.E., Liu, J., Callahan, H.S., Gaylinn, B.D., Thorner, M.O. & Cummings, D.E. (2008) Acyl and total ghrelin are suppressed strongly by ingested proteins, weakly by lipids, and biphasically by carbohydrates. *J Clin Endocrinol Metab*, 93,pp 1971-1979.

Frascarelli, S., Ghelardoni, S., Ronca-Testoni, S. & Zucchi, R. (2003) Effect of ghrelin and synthetic growth hormone secretagogues in normal and ischemic rat heart. *Basic Res Cardiol*, 98,pp 401-405.

Frohman, L.A. & Jansson, J.O. (1986) Growth hormone-releasing hormone. *Endocr Rev*, 7,pp 223-253.

Fujino, K., Inui, A., Asakawa, A., Kihara, N., Fujimura, M. & Fujimiya, M. (2003) Ghrelin induces fasted motor activity of the gastrointestinal tract in conscious fed rats. *J Physiol*, 550,pp 227-240.

Fukuda, H., Mizuta, Y., Isomoto, H., Takeshima, F., Ohnita, K., Ohba, K., Omagari, K., Taniyama, K. & Kohno, S. (2004) Ghrelin enhances gastric motility through direct stimulation of intrinsic neural pathways and capsaicin-sensitive afferent neurones in rats. *Scand J Gastroenterol*, 39,pp 1209-1214.

Fukushima, N., Hanada, R., Teranishi, H., Fukue, Y., Tachibana, T., Ishikawa, H., Takeda, S., Takeuchi, Y., Fukumoto, S., Kangawa, K., Nagata, K. & Kojima, M. (2005) Ghrelin directly regulates bone formation. *J Bone Miner Res*, 20,pp 790-798.

Gaytan, F., Barreiro, M.L., Caminos, J.E., Chopin, L.K., Herington, A.C., Morales, C., Pinilla, L., Paniagua, R., Nistal, M., Casanueva, F.F., Aguilar, E., Dieguez, C. & Tena-Sempere, M. (2004) Expression of ghrelin and its functional receptor, the type 1a growth hormone secretagogue receptor, in normal human testis and testicular tumors. *J Clin Endocrinol Metab*, 89,pp 400-409.

Geisler, S. & Zahm, D.S. (2005) Afferents of the ventral tegmental area in the rat-anatomical substratum for integrative functions. *J Comp Neurol*, 490,pp 270-294.

Ghigo, E., Broglio, F., Arvat, E., Maccario, M., Papotti, M. & Muccioli, G. (2005) Ghrelin: more than a natural GH secretagogue and/or an orexigenic factor. *Clin Endocrinol (Oxf)*, 62,pp 1-17.

Gnanapavan, S., Kola, B., Bustin, S.A., Morris, D.G., McGee, P., Fairclough, P., Bhattacharya, S., Carpenter, R., Grossman, A.B. & Korbonits, M. (2002) The tissue distribution of the mRNA of ghrelin and subtypes of its receptor, GHS-R, in humans. *J Clin Endocrinol Metab*, 87,pp 2988.

Granata, R., Settanni, F., Biancone, L., Trovato, L., Nano, R., Bertuzzi, F., Destefanis, S., Annunziata, M., Martinetti, M., Catapano, F., Ghe, C., Isgaard, J., Papotti, M., Ghigo, E. & Muccioli, G. (2007) Acylated and unacylated ghrelin promote proliferation and inhibit apoptosis of pancreatic beta-cells and human islets: involvement of 3',5'-cyclic adenosine monophosphate/protein kinase A, extracellular signal-regulated kinase 1/2, and phosphatidyl inositol 3-Kinase/Akt signaling. *Endocrinology*, 148,pp 512-529.

Grinspoon, S., Miller, K.K., Herzog, D.B., Grieco, K.A. & Klibanski, A. (2004) Effects of estrogen and recombinant human insulin-like growth factor-I on ghrelin secretion in severe undernutrition. *J Clin Endocrinol Metab*, 89,pp 3988-3993.

Gropp, E., Shanabrough, M., Borok, E., Xu, A.W., Janoschek, R., Buch, T., Plum, L., Balthasar, N., Hampel, B., Waisman, A., Barsh, G.S., Horvath, T.L. & Bruning, J.C. (2005) Agouti-related peptide-expressing neurons are mandatory for feeding. *Nat Neurosci*, 8,pp 1289-1291.

Groschl, M., Knerr, I., Topf, H.G., Schmid, P., Rascher, W. & Rauh, M. (2003) Endocrine responses to the oral ingestion of a physiological dose of essential amino acids in humans. *J Endocrinol*, 179,pp 237-244.

Grube, D. & Forssmann, W.G. (1979) Morphology and function of the entero-endocrine cells. *Horm Metab Res*, 11,pp 589-606.

Gruendel, S., Otto, B., Garcia, A.L., Wagner, K., Mueller, C., Weickert, M.O., Heldwein, W. & Koebnick, C. (2007) Carob pulp preparation rich in insoluble dietary fibre and polyphenols increases plasma glucose and serum insulin responses in combination with a glucose load in humans. *Br J Nutr*, 98,pp 101-105.

Gualillo, O., Caminos, J., Blanco, M., Garcia-Caballero, T., Kojima, M., Kangawa, K., Dieguez, C. & Casanueva, F. (2001) Ghrelin, a novel placental-derived hormone. *Endocrinology*, 142,pp 788-794.

Guan, X.M., Yu, H., Palyha, O.C., McKee, K.K., Feighner, S.D., Sirinathsinghji, D.J., Smith, R.G., Van der Ploeg, L.H. & Howard, A.D. (1997) Distribution of mRNA encoding the growth hormone secretagogue receptor in brain and peripheral tissues. *Brain Res Mol Brain Res*, 48,pp 23-29.

Gutierrez, J.A., Solenberg, P.J., Perkins, D.R., Willency, J.A., Knierman, M.D., Jin, Z., Witcher, D.R., Luo, S., Onyia, J.E. & Hale, J.E. (2008) Ghrelin octanoylation mediated by an orphan lipid transferase. *Proc Natl Acad Sci U S A*, 105,pp 6320-6325.

Hahn, T.M., Breininger, J.F., Baskin, D.G. & Schwartz, M.W. (1998) Coexpression of Agrp and NPY in fasting-activated hypothalamic neurons. *Nat Neurosci*, 1,pp 271-272.

Hansen, T.K., Dall, R., Hosoda, H., Kojima, M., Kangawa, K., Christiansen, J.S. & Jorgensen, J.O. (2002) Weight loss increases circulating levels of ghrelin in human obesity. *Clin Endocrinol (Oxf)*, 56,pp 203-206.

Hansson, C., Haage, D., Taube, M., Egecioglu, E., Salome, N. & Dickson, S.L. (2011) Central administration of ghrelin alters emotional responses in rats: behavioural, electrophysiological and molecular evidence. *Neuroscience*, 180,pp 201-211.

Harada, T., Nakahara, T., Yasuhara, D., Kojima, S., Sagiyama, K., Amitani, H., Laviano, A., Naruo, T. & Inui, A. (2008) Obestatin, acyl ghrelin, and des-acyl ghrelin responses to an oral glucose tolerance test in the restricting type of anorexia nervosa. *Biol Psychiatry*, 63,pp 245-247.

Hardie, D.G. (2004) The AMP-activated protein kinase pathway--new players upstream and downstream. *J Cell Sci*, 117,pp 5479-5487.

Hayashida, T., Nakahara, K., Mondal, M.S., Date, Y., Nakazato, M., Kojima, M., Kangawa, K. & Murakami, N. (2002) Ghrelin in neonatal rats: distribution in stomach and its possible role. *J Endocrinol*, 173,pp 239-245.

Holst, B., Cygankiewicz, A., Jensen, T.H., Ankersen, M. & Schwartz, T.W. (2003) High constitutive signaling of the ghrelin receptor--identification of a potent inverse agonist. *Mol Endocrinol*, 17,pp 2201-2210.

Holst, B., Egerod, K.L., Schild, E., Vickers, S.P., Cheetham, S., Gerlach, L.O., Storjohann, L., Stidsen, C.E., Jones, R., Beck-Sickinger, A.G. & Schwartz, T.W. (2007) GPR39 signaling is stimulated by zinc ions but not by obestatin. *Endocrinology*, 148,pp 13-20.

Holst, B., Holliday, N.D., Bach, A., Elling, C.E., Cox, H.M. & Schwartz, T.W. (2004) Common structural basis for constitutive activity of the ghrelin receptor family. *J Biol Chem*, 279,pp 53806-53817.

Holst, B., Lang, M., Brandt, E., Bach, A., Howard, A., Frimurer, T.M., Beck-Sickinger, A. & Schwartz, T.W. (2006) Ghrelin receptor inverse agonists: identification of an active peptide core and its interaction epitopes on the receptor. *Mol Pharmacol*, 70,pp 936-946.

Holst, B. & Schwartz, T.W. (2006) Ghrelin receptor mutations--too little height and too much hunger. *J Clin Invest*, 116,pp 637-641.

Hosoda, H., Kojima, M., Matsuo, H. & Kangawa, K. (2000a) Ghrelin and des-acyl ghrelin: two major forms of rat ghrelin peptide in gastrointestinal tissue. *Biochem Biophys Res Commun*, 279,pp 909-913.

Hosoda, H., Kojima, M., Matsuo, H. & Kangawa, K. (2000b) Purification and characterization of rat des-Gln14-Ghrelin, a second endogenous ligand for the growth hormone secretagogue receptor. *J Biol Chem*, 275,pp 21995-22000.

Hosoda, H., Kojima, M., Mizushima, T., Shimizu, S. & Kangawa, K. (2003) Structural divergence of human ghrelin. Identification of multiple ghrelin-derived molecules produced by post-translational processing. *J Biol Chem*, 278,pp 64-70.

Howard, A.D., Feighner, S.D., Cully, D.F., Arena, J.P., Liberator, P.A., Rosenblum, C.I., Hamelin, M., Hreniuk, D.L., Palyha, O.C., Anderson, J., Paress, P.S., Diaz, C., Chou, M., Liu, K.K., McKee, K.K., Pong, S.S., Chaung, L.Y., Elbrecht, A., Dashkevicz, M., Heavens, R., Rigby, M., Sirinathsinghji, D.J., Dean, D.C., Melillo, D.G., Patchett, A.A., Nargund, R., Griffin, P.R., DeMartino, J.A., Gupta, S.K., Schaeffer, J.M., Smith, R.G. & Van der Ploeg, L.H. (1996) A receptor in pituitary and hypothalamus that functions in growth hormone release. *Science*, 273,pp 974-977.

Huang, J., Zhou, H., Mahavadi, S., Sriwai, W., Lyall, V. & Murthy, K.S. (2005) Signaling pathways mediating gastrointestinal smooth muscle contraction and MLC20 phosphorylation by motilin receptors. *Am J Physiol Gastrointest Liver Physiol*, 288,pp G23-31.

Iglesias, M.J., Pineiro, R., Blanco, M., Gallego, R., Dieguez, C., Gualillo, O., Gonzalez-Juanatey, J.R. & Lago, F. (2004) Growth hormone releasing peptide (ghrelin) is synthesized and secreted by cardiomyocytes. *Cardiovasc Res*, 62,pp 481-488.

Iwase, M., Kanazawa, H., Kato, Y., Nishizawa, T., Somura, F., Ishiki, R., Nagata, K., Hashimoto, K., Takagi, K., Izawa, H. & Yokota, M. (2004) Growth hormone-releasing peptide can improve left ventricular dysfunction and attenuate dilation in dilated cardiomyopathic hamsters. *Cardiovasc Res*, 61,pp 30-38.

Jeffery, P.L., Duncan, R.P., Yeh, A.H., Jaskolski, R.A., Hammond, D.S., Herington, A.C. & Chopin, L.K. (2005) Expression of the ghrelin axis in the mouse: an exon 4-deleted mouse proghrelin variant encodes a novel C terminal peptide. *Endocrinology*, 146,pp 432-440.

Jerlhag, E., Egecioglu, E., Dickson, S.L., Andersson, M., Svensson, L. & Engel, J.A. (2006) Ghrelin stimulates locomotor activity and accumbal dopamine-overflow via central cholinergic systems in mice: implications for its involvement in brain reward. *Addict Biol*, 11,pp 45-54.

Jerlhag, E., Egecioglu, E., Dickson, S.L., Douhan, A., Svensson, L. & Engel, J.A. (2007) Ghrelin administration into tegmental areas stimulates locomotor activity and increases extracellular concentration of dopamine in the nucleus accumbens. *Addict Biol*, 12,pp 6-16.

Jethwa, P.H., Smith, K.L., Small, C.J., Abbott, C.R., Darch, S.J., Murphy, K.G., Seth, A., Semjonous, N.M., Patel, S.R., Todd, J.F., Ghatei, M.A. & Bloom, S.R. (2006) Neuromedin U partially mediates leptin-induced hypothalamo-pituitary adrenal (HPA) stimulation and has a physiological role in the regulation of the HPA axis in the rat. *Endocrinology*, 147,pp 2886-2892.

Jorgensen, J.O. & Christiansen, J.S. (1993) Brave new senescence: GH in adults. *Lancet*, 341,pp 1247-1248.

Kamegai, J., Tamura, H., Shimizu, T., Ishii, S., Sugihara, H. & Oikawa, S. (2004) Effects of insulin, leptin, and glucagon on ghrelin secretion from isolated perfused rat stomach. *Regul Pept*, 119,pp 77-81.

Kamegai, J., Tamura, H., Shimizu, T., Ishii, S., Sugihara, H. & Wakabayashi, I. (2001) Chronic central infusion of ghrelin increases hypothalamic neuropeptide Y and Agouti-related protein mRNA levels and body weight in rats. *Diabetes*, 50,pp 2438-2443.

Kanamoto, N., Akamizu, T., Hosoda, H., Hataya, Y., Ariyasu, H., Takaya, K., Hosoda, K., Saijo, M., Moriyama, K., Shimatsu, A., Kojima, M., Kangawa, K. & Nakao, K. (2001) Substantial production of ghrelin by a human medullary thyroid carcinoma cell line. *J Clin Endocrinol Metab*, 86,pp 4984-4990.

Kanamoto, N., Akamizu, T., Tagami, T., Hataya, Y., Moriyama, K., Takaya, K., Hosoda, H., Kojima, M., Kangawa, K. & Nakao, K. (2004) Genomic structure and characterization of the 5'-flanking region of the human ghrelin gene. *Endocrinology*, 145,pp 4144-4153.

Kanehisa, M., Akiyoshi, J., Kitaichi, T., Matsushita, H., Tanaka, E., Kodama, K., Hanada, H. & Isogawa, K. (2006) Administration of antisense DNA for ghrelin causes an antidepressant and anxiolytic response in rats. *Prog Neuropsychopharmacol Biol Psychiatry*, 30,pp 1403-1407.

Katakami, H., Arimura, A. & Frohman, L.A. (1986) Growth hormone (GH)-releasing factor stimulates hypothalamic somatostatin release: an inhibitory feedback effect on GH secretion. *Endocrinology*, 118,pp 1872-1877.

Katayama, T., Shimamoto, S., Oda, H., Nakahara, K., Kangawa, K. & Murakami, N. (2007) Glucagon receptor expression and glucagon stimulation of ghrelin secretion in rat stomach. *Biochem Biophys Res Commun*, 357,pp 865-870.

Katugampola, S.D., Pallikaros, Z. & Davenport, A.P. (2001) [125I-His(9)]-ghrelin, a novel radioligand for localizing GHS orphan receptors in human and rat tissue: up-regulation of receptors with athersclerosis. *Br J Pharmacol*, 134,pp 143-149.

Kellokoski, E., Poykko, S.M., Karjalainen, A.H., Ukkola, O., Heikkinen, J., Kesaniemi, Y.A. & Horkko, S. (2005) Estrogen replacement therapy increases plasma ghrelin levels. *J Clin Endocrinol Metab*, 90,pp 2954-2963.

Kim, M.S., Yoon, C.Y., Jang, P.G., Park, Y.J., Shin, C.S., Park, H.S., Ryu, J.W., Pak, Y.K., Park, J.Y., Lee, K.U., Kim, S.Y., Lee, H.K., Kim, Y.B. & Park, K.S. (2004) The mitogenic and antiapoptotic actions of ghrelin in 3T3-L1 adipocytes. *Mol Endocrinol*, 18,pp 2291-2301.

Kishimoto, M., Okimura, Y., Nakata, H., Kudo, T., Iguchi, G., Takahashi, Y., Kaji, H. & Chihara, K. (2003) Cloning and characterization of the 5(')-flanking region of the human ghrelin gene. *Biochem Biophys Res Commun*, 305,pp 186-192.

Kluge, M., Schussler, P., Schmid, D., Uhr, M., Kleyer, S., Yassouridis, A. & Steiger, A. (2009) Ghrelin plasma levels are not altered in major depression. *Neuropsychobiology*, 59,pp 199-204.

Knerr, I., Groschl, M., Rascher, W. & Rauh, M. (2003) Endocrine effects of food intake: insulin, ghrelin, and leptin responses to a single bolus of essential amino acids in humans. *Ann Nutr Metab*, 47,pp 312-318.

Kohno, D., Gao, H.Z., Muroya, S., Kikuyama, S. & Yada, T. (2003) Ghrelin directly interacts with neuropeptide-Y-containing neurons in the rat arcuate nucleus: Ca2+ signaling via protein kinase A and N-type channel-dependent mechanisms and cross-talk with leptin and orexin. *Diabetes*, 52,pp 948-956.

Kojima, M., Hosoda, H., Date, Y., Nakazato, M., Matsuo, H. & Kangawa, K. (1999) Ghrelin is a growth-hormone-releasing acylated peptide from stomach. *Nature*, 402,pp 656-660.

Kola, B., Hubina, E., Tucci, S.A., Kirkham, T.C., Garcia, E.A., Mitchell, S.E., Williams, L.M., Hawley, S.A., Hardie, D.G., Grossman, A.B. & Korbonits, M. (2005) Cannabinoids and ghrelin have both central and peripheral metabolic and cardiac effects via AMP-activated protein kinase. *J Biol Chem*, 280,pp 25196-25201.

Koo, G.C., Huang, C., Camacho, R., Trainor, C., Blake, J.T., Sirotina-Meisher, A., Schleim, K.D., Wu, T.J., Cheng, K., Nargund, R. & McKissick, G. (2001) Immune enhancing effect of a growth hormone secretagogue. *J Immunol*, 166,pp 4195-4201.

Korbonits, M., Bustin, S.A., Kojima, M., Jordan, S., Adams, E.F., Lowe, D.G., Kangawa, K. & Grossman, A.B. (2001a) The expression of the growth hormone secretagogue receptor ligand ghrelin in normal and abnormal human pituitary and other neuroendocrine tumors. *J Clin Endocrinol Metab*, 86,pp 881-887.

Korbonits, M., Kojima, M., Kangawa, K. & Grossman, A.B. (2001b) Presence of ghrelin in normal and adenomatous human pituitary. *Endocrine*, 14,pp 101-104.

Kristensson, E., Sundqvist, M., Hakanson, R. & Lindstrom, E. (2007) High gastrin cell activity and low ghrelin cell activity in high-anxiety Wistar Kyoto rats. *J Endocrinol*, 193,pp 245-250.

Lambrinoudaki, I.V., Christodoulakos, G.E., Economou, E.V., Vlachou, S.A., Panoulis, C.P., Alexandrou, A.P., Kouskouni, E.E. & Creatsas, G.C. (2008) Circulating leptin and ghrelin are differentially influenced by estrogen/progestin therapy and raloxifene. *Maturitas*, 59,pp 62-71.

Larhammar, D. (1996) Evolution of neuropeptide Y, peptide YY and pancreatic polypeptide. *Regul Pept*, 62,pp 1-11.

Lauwers, E., Landuyt, B., Arckens, L., Schoofs, L. & Luyten, W. (2006) Obestatin does not activate orphan G protein-coupled receptor GPR39. *Biochem Biophys Res Commun*, 351,pp 21-25.

Levin, F., Edholm, T., Schmidt, P.T., Gryback, P., Jacobsson, H., Degerblad, M., Hoybye, C., Holst, J.J., Rehfeld, J.F., Hellstrom, P.M. & Naslund, E. (2006) Ghrelin stimulates gastric emptying and hunger in normal-weight humans. *J Clin Endocrinol Metab*, 91,pp 3296-3302.

Li, L., Zhang, L.K., Pang, Y.Z., Pan, C.S., Qi, Y.F., Chen, L., Wang, X., Tang, C.S. & Zhang, J. (2006) Cardioprotective effects of ghrelin and des-octanoyl ghrelin on myocardial injury induced by isoproterenol in rats. *Acta Pharmacol Sin*, 27,pp 527-535.

Li, W.G., Gavrila, D., Liu, X., Wang, L., Gunnlaugsson, S., Stoll, L.L., McCormick, M.L., Sigmund, C.D., Tang, C. & Weintraub, N.L. (2004) Ghrelin inhibits proinflammatory responses and nuclear factor-kappaB activation in human endothelial cells. *Circulation*, 109,pp 2221-2226.

Lin, Y., Matsumura, K., Fukuhara, M., Kagiyama, S., Fujii, K. & Iida, M. (2004) Ghrelin acts at the nucleus of the solitary tract to decrease arterial pressure in rats. *Hypertension*, 43,pp 977-982.

Locatelli, V., Rossoni, G., Schweiger, F., Torsello, A., De Gennaro Colonna, V., Bernareggi, M., Deghenghi, R., Muller, E.E. & Berti, F. (1999) Growth hormone-independent cardioprotective effects of hexarelin in the rat. *Endocrinology*, 140,pp 4024-4031.

Lopez, M., Lage, R., Saha, A.K., Perez-Tilve, D., Vazquez, M.J., Varela, L., Sangiao-Alvarellos, S., Tovar, S., Raghay, K., Rodriguez-Cuenca, S., Deoliveira, R.M., Castaneda, T., Datta, R., Dong, J.Z., Culler, M., Sleeman, M.W., Alvarez, C.V., Gallego, R., Lelliott, C.J., Carling, D., Tschop, M.H., Dieguez, C. & Vidal-Puig, A. (2008) Hypothalamic fatty acid metabolism mediates the orexigenic action of ghrelin. *Cell Metab*, 7,pp 389-399.

Lu, S., Guan, J.L., Wang, Q.P., Uehara, K., Yamada, S., Goto, N., Date, Y., Nakazato, M., Kojima, M., Kangawa, K. & Shioda, S. (2002) Immunocytochemical observation of ghrelin-containing neurons in the rat arcuate nucleus. *Neurosci Lett*, 321,pp 157-160.

Luque, R.M., Gahete, M.D., Hochgeschwender, U. & Kineman, R.D. (2006) Evidence that endogenous SST inhibits ACTH and ghrelin expression by independent pathways. *Am J Physiol Endocrinol Metab*, 291,pp E395-403.

Luquet, S., Perez, F.A., Hnasko, T.S. & Palmiter, R.D. (2005) NPY/AgRP neurons are essential for feeding in adult mice but can be ablated in neonates. *Science*, 310,pp 683-685.

Lutter, M., Sakata, I., Osborne-Lawrence, S., Rovinsky, S.A., Anderson, J.G., Jung, S., Birnbaum, S., Yanagisawa, M., Elmquist, J.K., Nestler, E.J. & Zigman, J.M. (2008) The orexigenic hormone ghrelin defends against depressive symptoms of chronic stress. *Nat Neurosci*, 11,pp 752-753.

Luttrell, L.M. & Lefkowitz, R.J. (2002) The role of beta-arrestins in the termination and transduction of G-protein-coupled receptor signals. *J Cell Sci*, 115,pp 455-465.

Makovey, J., Naganathan, V., Seibel, M. & Sambrook, P. (2007) Gender differences in plasma ghrelin and its relations to body composition and bone - an opposite-sex twin study. *Clin Endocrinol (Oxf)*, 66,pp 530-537.

Masuda, Y., Tanaka, T., Inomata, N., Ohnuma, N., Tanaka, S., Itoh, Z., Hosoda, H., Kojima, M. & Kangawa, K. (2000) Ghrelin stimulates gastric acid secretion and motility in rats. *Biochem Biophys Res Commun*, 276,pp 905-908.

Matsubara, M., Sakata, I., Wada, R., Yamazaki, M., Inoue, K. & Sakai, T. (2004) Estrogen modulates ghrelin expression in the female rat stomach. *Peptides*, 25,pp 289-297.

Matsuda, K., Miura, T., Kaiya, H., Maruyama, K., Shimakura, S., Uchiyama, M., Kangawa, K. & Shioda, S. (2006) Regulation of food intake by acyl and des-acyl ghrelins in the goldfish. *Peptides*, 27,pp 2321-2325.

Matsumura, K., Tsuchihashi, T., Fujii, K., Abe, I. & Iida, M. (2002) Central ghrelin modulates sympathetic activity in conscious rabbits. *Hypertension*, 40,pp 694-699.

Mau, S.E., Witt, M.R., Bjerrum, O.J., Saermark, T. & Vilhardt, H. (1995) Growth hormone releasing hexapeptide (GHRP-6) activates the inositol (1,4,5)-trisphosphate/diacylglycerol pathway in rat anterior pituitary cells. *J Recept Signal Transduct Res*, 15,pp 311-323.

Mazzocchi, G., Neri, G., Rucinski, M., Rebuffat, P., Spinazzi, R., Malendowicz, L.K. & Nussdorfer, G.G. (2004) Ghrelin enhances the growth of cultured human adrenal zona glomerulosa cells by exerting MAPK-mediated proliferogenic and antiapoptotic effects. *Peptides*, 25,pp 1269-1277.

McKee, K.K., Palyha, O.C., Feighner, S.D., Hreniuk, D.L., Tan, C.P., Phillips, M.S., Smith, R.G., Van der Ploeg, L.H. & Howard, A.D. (1997a) Molecular analysis of rat pituitary and hypothalamic growth hormone secretagogue receptors. *Mol Endocrinol*, 11,pp 415-423.

McKee, K.K., Tan, C.P., Palyha, O.C., Liu, J., Feighner, S.D., Hreniuk, D.L., Smith, R.G., Howard, A.D. & Van der Ploeg, L.H. (1997b) Cloning and characterization of two human G protein-coupled receptor genes (GPR38 and GPR39) related to the growth hormone secretagogue and neurotensin receptors. *Genomics*, 46,pp 426-434.

McKeown, M. (1992) Alternative mRNA splicing. *Annu Rev Cell Biol*, 8,pp 133-155.

Monteleone, P., Bencivenga, R., Longobardi, N., Serritella, C. & Maj, M. (2003) Differential responses of circulating ghrelin to high-fat or high-carbohydrate meal in healthy women. *J Clin Endocrinol Metab*, 88,pp 5510-5514.

Morello, J.P. & Bouvier, M. (1996) Palmitoylation: a post-translational modification that regulates signalling from G-protein coupled receptors. *Biochem Cell Biol*, 74,pp 449-457.

Mori, K., Yoshimoto, A., Takaya, K., Hosoda, K., Ariyasu, H., Yahata, K., Mukoyama, M., Sugawara, A., Hosoda, H., Kojima, M., Kangawa, K. & Nakao, K. (2000) Kidney produces a novel acylated peptide, ghrelin. *FEBS Lett*, 486,pp 213-216.

Mousseaux, D., Le Gallic, L., Ryan, J., Oiry, C., Gagne, D., Fehrentz, J.A., Galleyrand, J.C. & Martinez, J. (2006) Regulation of ERK1/2 activity by ghrelin-activated growth hormone secretagogue receptor 1A involves a PLC/PKCvarepsilon pathway. *Br J Pharmacol*, 148,pp 350-365.

Muccioli, G., Tschop, M., Papotti, M., Deghenghi, R., Heiman, M. & Ghigo, E. (2002) Neuroendocrine and peripheral activities of ghrelin: implications in metabolism and obesity. *Eur J Pharmacol*, 440,pp 235-254.

Murata, M., Okimura, Y., Iida, K., Matsumoto, M., Sowa, H., Kaji, H., Kojima, M., Kangawa, K. & Chihara, K. (2002) Ghrelin modulates the downstream molecules of insulin signaling in hepatoma cells. *J Biol Chem*, 277,pp 5667-5674.

Murray, C.D., Martin, N.M., Patterson, M., Taylor, S.A., Ghatei, M.A., Kamm, M.A., Johnston, C., Bloom, S.R. & Emmanuel, A.V. (2005) Ghrelin enhances gastric emptying in diabetic gastroparesis: a double blind, placebo controlled, crossover study. *Gut*, 54,pp 1693-1698.

Nagaya, N., Kojima, M., Uematsu, M., Yamagishi, M., Hosoda, H., Oya, H., Hayashi, Y. & Kangawa, K. (2001a) Hemodynamic and hormonal effects of human ghrelin in healthy volunteers. *Am J Physiol Regul Integr Comp Physiol*, 280,pp R1483-1487.

Nagaya, N., Moriya, J., Yasumura, Y., Uematsu, M., Ono, F., Shimizu, W., Ueno, K., Kitakaze, M., Miyatake, K. & Kangawa, K. (2004) Effects of ghrelin administration on left ventricular function, exercise capacity, and muscle wasting in patients with chronic heart failure. *Circulation*, 110,pp 3674-3679.

Nagaya, N., Uematsu, M., Kojima, M., Date, Y., Nakazato, M., Okumura, H., Hosoda, H., Shimizu, W., Yamagishi, M., Oya, H., Koh, H., Yutani, C. & Kangawa, K. (2001b)

Elevated circulating level of ghrelin in cachexia associated with chronic heart failure: relationships between ghrelin and anabolic/catabolic factors. *Circulation*, 104,pp 2034-2038.

Nagaya, N., Uematsu, M., Kojima, M., Ikeda, Y., Yoshihara, F., Shimizu, W., Hosoda, H., Hirota, Y., Ishida, H., Mori, H. & Kangawa, K. (2001c) Chronic administration of ghrelin improves left ventricular dysfunction and attenuates development of cardiac cachexia in rats with heart failure. *Circulation*, 104,pp 1430-1435.

Nakahara, K., Nakagawa, M., Baba, Y., Sato, M., Toshinai, K., Date, Y., Nakazato, M., Kojima, M., Miyazato, M., Kaiya, H., Hosoda, H., Kangawa, K. & Murakami, N. (2006) Maternal ghrelin plays an important role in rat fetal development during pregnancy. *Endocrinology*, 147,pp 1333-1342.

Nakashima, K., Akiyoshi, J., Hatano, K., Hanada, H., Tanaka, Y., Tsuru, J., Matsushita, H., Kodama, K. & Isogawa, K. (2008) Ghrelin gene polymorphism is associated with depression, but not panic disorder. *Psychiatr Genet*, 18,pp 257.

Nakazato, M., Murakami, N., Date, Y., Kojima, M., Matsuo, H., Kangawa, K. & Matsukura, S. (2001) A role for ghrelin in the central regulation of feeding. *Nature*, 409,pp 194-198.

Naleid, A.M., Grace, M.K., Cummings, D.E. & Levine, A.S. (2005) Ghrelin induces feeding in the mesolimbic reward pathway between the ventral tegmental area and the nucleus accumbens. *Peptides*, 26,pp 2274-2279.

Nishi, Y., Hiejima, H., Hosoda, H., Kaiya, H., Mori, K., Fukue, Y., Yanase, T., Nawata, H., Kangawa, K. & Kojima, M. (2005) Ingested medium-chain fatty acids are directly utilized for the acyl modification of ghrelin. *Endocrinology*, 146,pp 2255-2264.

Nogueiras, R., Tovar, S., Mitchell, S.E., Barrett, P., Rayner, D.V., Dieguez, C. & Williams, L.M. (2006) Negative energy balance and leptin regulate neuromedin-U expression in the rat pars tuberalis. *J Endocrinol*, 190,pp 545-553.

Ogaya, M., Kim, J. & Sasaki, K. (2011) Ghrelin postsynaptically depolarizes dorsal raphe neurons in rats in vitro. *Peptides*, 32,pp 1606-1616.

Olszanecka-Glinianowicz, M., Kocelak, P., Wikarek, T., Gruszka, W., Dabrowski, P., Chudek, J. & Zahorska-Markiewicz, B. (2010) Are plasma ghrelin and PYY concentrations associated with obesity-related depression? *Endokrynol Pol*, 61,pp 174-177.

Ong, H., Bodart, V., McNicoll, N., Lamontagne, D. & Bouchard, J.F. (1998a) Binding sites for growth hormone-releasing peptide. *Growth Horm IGF Res*, 8 Suppl B,pp 137-140.

Ong, H., McNicoll, N., Escher, E., Collu, R., Deghenghi, R., Locatelli, V., Ghigo, E., Muccioli, G., Boghen, M. & Nilsson, M. (1998b) Identification of a pituitary growth hormone-releasing peptide (GHRP) receptor subtype by photoaffinity labeling. *Endocrinology*, 139,pp 432-435.

Otto, B., Cuntz, U., Fruehauf, E., Wawarta, R., Folwaczny, C., Riepl, R.L., Heiman, M.L., Lehnert, P., Fichter, M. & Tschop, M. (2001) Weight gain decreases elevated plasma ghrelin concentrations of patients with anorexia nervosa. *Eur J Endocrinol*, 145,pp 669-673.

Ozawa, A., Cai, Y. & Lindberg, I. (2007) Production of bioactive peptides in an in vitro system. *Anal Biochem*, 366,pp 182-189.

Palyha, O.C., Feighner, S.D., Tan, C.P., McKee, K.K., Hreniuk, D.L., Gao, Y.D., Schleim, K.D., Yang, L., Morriello, G.J., Nargund, R., Patchett, A.A., Howard, A.D. & Smith, R.G. (2000) Ligand activation domain of human orphan growth hormone (GH) secretagogue receptor (GHS-R) conserved from Pufferfish to humans. *Mol Endocrinol*, 14,pp 160-169.

Pantel, J., Legendre, M., Cabrol, S., Hilal, L., Hajaji, Y., Morisset, S., Nivot, S., Vie-Luton, M.P., Grouselle, D., de Kerdanet, M., Kadiri, A., Epelbaum, J., Le Bouc, Y. & Amselem, S. (2006) Loss of constitutive activity of the growth hormone secretagogue receptor in familial short stature. *J Clin Invest*, 116,pp 760-768.

Papotti, M., Cassoni, P., Volante, M., Deghenghi, R., Muccioli, G. & Ghigo, E. (2001) Ghrelin-producing endocrine tumors of the stomach and intestine. *J Clin Endocrinol Metab*, 86,pp 5052-5059.

Papotti, M., Ghe, C., Cassoni, P., Catapano, F., Deghenghi, R., Ghigo, E. & Muccioli, G. (2000) Growth hormone secretagogue binding sites in peripheral human tissues. *J Clin Endocrinol Metab*, 85,pp 3803-3807.

Patchett, A.A., Nargund, R.P., Tata, J.R., Chen, M.H., Barakat, K.J., Johnston, D.B., Cheng, K., Chan, W.W., Butler, B., Hickey, G. & et al. (1995) Design and biological activities of L-163,191 (MK-0677): a potent, orally active growth hormone secretagogue. *Proc Natl Acad Sci U S A*, 92,pp 7001-7005.

Patel, A.D., Stanley, S.A., Murphy, K.G., Frost, G.S., Gardiner, J.V., Kent, A.S., White, N.E., Ghatei, M.A. & Bloom, S.R. (2006) Ghrelin stimulates insulin-induced glucose uptake in adipocytes. *Regul Pept*, 134,pp 17-22.

Pedretti, A., Villa, M., Pallavicini, M., Valoti, E. & Vistoli, G. (2006) Construction of human ghrelin receptor (hGHS-R1a) model using a fragmental prediction approach and validation through docking analysis. *J Med Chem*, 49,pp 3077-3085.

Peeters, T.L. (2006) Potential of ghrelin as a therapeutic approach for gastrointestinal motility disorders. *Curr Opin Pharmacol*, 6,pp 553-558.

Penicaud, L., Leloup, C., Fioramonti, X., Lorsignol, A. & Benani, A. (2006) Brain glucose sensing: a subtle mechanism. *Curr Opin Clin Nutr Metab Care*, 9,pp 458-462.

Petersenn, S. (2002) Structure and regulation of the growth hormone secretagogue receptor. *Minerva Endocrinol*, 27,pp 243-256.

Poinot-Chazel, C., Portier, M., Bouaboula, M., Vita, N., Pecceu, F., Gully, D., Monroe, J.G., Maffrand, J.P., Le Fur, G. & Casellas, P. (1996) Activation of mitogen-activated protein kinase couples neurotensin receptor stimulation to induction of the primary response gene Krox-24. *Biochem J*, 320 (Pt 1),pp 145-151.

Poitras, P., Polvino, W.J. & Rocheleau, B. (2005) Gastrokinetic effect of ghrelin analog RC-1139 in the rat. Effect on post-operative and on morphine induced ileus. *Peptides*, 26,pp 1598-1601.

Pong, S.S., Chaung, L.Y., Dean, D.C., Nargund, R.P., Patchett, A.A. & Smith, R.G. (1996) Identification of a new G-protein-linked receptor for growth hormone secretagogues. *Mol Endocrinol*, 10,pp 57-61.

Prado, C.L., Pugh-Bernard, A.E., Elghazi, L., Sosa-Pineda, B. & Sussel, L. (2004) Ghrelin cells replace insulin-producing beta cells in two mouse models of pancreas development. *Proc Natl Acad Sci U S A*, 101,pp 2924-2929.

Purnell, J.Q., Weigle, D.S., Breen, P. & Cummings, D.E. (2003) Ghrelin levels correlate with insulin levels, insulin resistance, and high-density lipoprotein cholesterol, but not with gender, menopausal status, or cortisol levels in humans. *J Clin Endocrinol Metab*, 88,pp 5747-5752.

Qi, X., Reed, J., Englander, E.W., Chandrashekar, V., Bartke, A. & Greeley, G.H., Jr. (2003) Evidence that growth hormone exerts a feedback effect on stomach ghrelin production and secretion. *Exp Biol Med (Maywood)*, 228,pp 1028-1032.

Quarta, D., Di Francesco, C., Melotto, S., Mangiarini, L., Heidbreder, C. & Hedou, G. (2009) Systemic administration of ghrelin increases extracellular dopamine in the shell but not the core subdivision of the nucleus accumbens. *Neurochem Int*, 54,pp 89-94.

Rindi, G., Necchi, V., Savio, A., Torsello, A., Zoli, M., Locatelli, V., Raimondo, F., Cocchi, D. & Solcia, E. (2002) Characterisation of gastric ghrelin cells in man and other mammals: studies in adult and fetal tissues. *Histochem Cell Biol*, 117,pp 511-519.

Root, A.W. & Root, M.J. (2002) Clinical pharmacology of human growth hormone and its secretagogues. *Curr Drug Targets Immune Endocr Metabol Disord*, 2,pp 27-52.

Rossi, F., Bertone, C., Petricca, S. & Santiemma, V. (2007) Ghrelin inhibits angiotensin II-induced migration of human aortic endothelial cells. *Atherosclerosis*, 192,pp 291-297.

Rouach, V., Bloch, M., Rosenberg, N., Gilad, S., Limor, R., Stern, N. & Greenman, Y. (2007) The acute ghrelin response to a psychological stress challenge does not predict the post-stress urge to eat. *Psychoneuroendocrinology*, 32,pp 693-702.

Saad, M.F., Bernaba, B., Hwu, C.M., Jinagouda, S., Fahmi, S., Kogosov, E. & Boyadjian, R. (2002) Insulin regulates plasma ghrelin concentration. *J Clin Endocrinol Metab*, 87,pp 3997-4000.

Sakata, I., Nakamura, K., Yamazaki, M., Matsubara, M., Hayashi, Y., Kangawa, K. & Sakai, T. (2002) Ghrelin-producing cells exist as two types of cells, closed- and opened-type cells, in the rat gastrointestinal tract. *Peptides*, 23,pp 531-536.

Sakata, I., Tanaka, T., Yamazaki, M., Tanizaki, T., Zheng, Z. & Sakai, T. (2006) Gastric estrogen directly induces ghrelin expression and production in the rat stomach. *J Endocrinol*, 190,pp 749-757.

Sarret, P., Gendron, L., Kilian, P., Nguyen, H.M., Gallo-Payet, N., Payet, M.D. & Beaudet, A. (2002) Pharmacology and functional properties of NTS2 neurotensin receptors in cerebellar granule cells. *J Biol Chem*, 277,pp 36233-36243.

Schanze, A., Reulbach, U., Scheuchenzuber, M., Groschl, M., Kornhuber, J. & Kraus, T. (2008) Ghrelin and eating disturbances in psychiatric disorders. *Neuropsychobiology*, 57,pp 126-130.

Schwartz, T.W., Frimurer, T.M., Holst, B., Rosenkilde, M.M. & Elling, C.E. (2006) Molecular mechanism of 7TM receptor activation--a global toggle switch model. *Annu Rev Pharmacol Toxicol*, 46,pp 481-519.

Sedlackova, D., Dostalova, I., Hainer, V., Beranova, L., Kvasnickova, H., Hill, M., Haluzik, M. & Nedvidkova, J. (2008) Simultaneous decrease of plasma obestatin and ghrelin levels after a high-carbohydrate breakfast in healthy women. *Physiol Res*, 57 Suppl 1,pp S29-37.

Seoane, L.M., Al-Massadi, O., Barreiro, F., Dieguez, C. & Casanueva, F.F. (2007) Growth hormone and somatostatin directly inhibit gastric ghrelin secretion. An in vitro organ culture system. *J Endocrinol Invest*, 30,pp RC22-25.

Shiiya, T., Nakazato, M., Mizuta, M., Date, Y., Mondal, M.S., Tanaka, M., Nozoe, S., Hosoda, H., Kangawa, K. & Matsukura, S. (2002) Plasma ghrelin levels in lean and obese humans and the effect of glucose on ghrelin secretion. *J Clin Endocrinol Metab*, 87,pp 240-244.

Shimada, M., Date, Y., Mondal, M.S., Toshinai, K., Shimbara, T., Fukunaga, K., Murakami, N., Miyazato, M., Kangawa, K., Yoshimatsu, H., Matsuo, H. & Nakazato, M. (2003) Somatostatin suppresses ghrelin secretion from the rat stomach. *Biochem Biophys Res Commun*, 302,pp 520-525.

Shimizu, Y., Nagaya, N., Isobe, T., Imazu, M., Okumura, H., Hosoda, H., Kojima, M., Kangawa, K. & Kohno, N. (2003) Increased plasma ghrelin level in lung cancer cachexia. *Clin Cancer Res*, 9,pp 774-778.

Shintani, M., Ogawa, Y., Ebihara, K., Aizawa-Abe, M., Miyanaga, F., Takaya, K., Hayashi, T., Inoue, G., Hosoda, K., Kojima, M., Kangawa, K. & Nakao, K. (2001) Ghrelin, an endogenous growth hormone secretagogue, is a novel orexigenic peptide that antagonizes leptin action through the activation of hypothalamic neuropeptide Y/Y1 receptor pathway. *Diabetes*, 50,pp 227-232.

Silva, A.P., Bethmann, K., Raulf, F. & Schmid, H.A. (2005) Regulation of ghrelin secretion by somatostatin analogs in rats. *Eur J Endocrinol*, 152,pp 887-894.

Smith, R.G., Cheng, K., Schoen, W.R., Pong, S.S., Hickey, G., Jacks, T., Butler, B., Chan, W.W., Chaung, L.Y., Judith, F. & et al. (1993) A nonpeptidyl growth hormone secretagogue. *Science*, 260,pp 1640-1643.

Smith, R.G., Van der Ploeg, L.H., Howard, A.D., Feighner, S.D., Cheng, K., Hickey, G.J., Wyvratt, M.J., Jr., Fisher, M.H., Nargund, R.P. & Patchett, A.A. (1997) Peptidomimetic regulation of growth hormone secretion. *Endocr Rev*, 18,pp 621-645.

Solcia, E., Capella, C., Vassallo, G. & Buffa, R. (1975) Endocrine cells of the gastric mucosa. *Int Rev Cytol*, 42,pp 223-286.

Sun, Y., Wang, P., Zheng, H. & Smith, R.G. (2004) Ghrelin stimulation of growth hormone release and appetite is mediated through the growth hormone secretagogue receptor. *Proc Natl Acad Sci U S A*, 101,pp 4679-4684.

Tack, J., Depoortere, I., Bisschops, R., Delporte, C., Coulie, B., Meulemans, A., Janssens, J. & Peeters, T. (2006) Influence of ghrelin on interdigestive gastrointestinal motility in humans. *Gut*, 55,pp 327-333.

Tack, J., Depoortere, I., Bisschops, R., Verbeke, K., Janssens, J. & Peeters, T. (2005) Influence of ghrelin on gastric emptying and meal-related symptoms in idiopathic gastroparesis. *Aliment Pharmacol Ther*, 22,pp 847-853.

Tacke, F., Brabant, G., Kruck, E., Horn, R., Schoffski, P., Hecker, H., Manns, M.P. & Trautwein, C. (2003) Ghrelin in chronic liver disease. *J Hepatol*, 38,pp 447-454.

Tan, C.P., McKee, K.K., Liu, Q., Palyha, O.C., Feighner, S.D., Hreniuk, D.L., Smith, R.G. & Howard, A.D. (1998) Cloning and characterization of a human and murine T-cell orphan G-protein-coupled receptor similar to the growth hormone secretagogue and neurotensin receptors. *Genomics*, 52,pp 223-229.

Tanaka, M., Hayashida, Y., Iguchi, T., Nakao, N., Nakai, N. & Nakashima, K. (2001a) Organization of the mouse ghrelin gene and promoter: occurrence of a short noncoding first exon. *Endocrinology*, 142,pp 3697-3700.

Tanaka, M., Hayashida, Y., Nakao, N., Nakai, N. & Nakashima, K. (2001b) Testis-specific and developmentally induced expression of a ghrelin gene-derived transcript that encodes a novel polypeptide in the mouse. *Biochim Biophys Acta*, 1522,pp 62-65.

Tannenbaum, G.S., Epelbaum, J. & Bowers, C.Y. (2003) Interrelationship between the novel peptide ghrelin and somatostatin/growth hormone-releasing hormone in regulation of pulsatile growth hormone secretion. *Endocrinology*, 144,pp 967-974.

Taub, D.D. (2008) Novel connections between the neuroendocrine and immune systems: the ghrelin immunoregulatory network. *Vitam Horm*, 77,pp 325-346.

Temel, Y., Boothman, L.J., Blokland, A., Magill, P.J., Steinbusch, H.W., Visser-Vandewalle, V. & Sharp, T. (2007) Inhibition of 5-HT neuron activity and induction of depressive-like behavior by high-frequency stimulation of the subthalamic nucleus. *Proc Natl Acad Sci U S A*, 104,pp 17087-17092.

Tena-Sempere, M., Barreiro, M.L., Gonzalez, L.C., Gaytan, F., Zhang, F.P., Caminos, J.E., Pinilla, L., Casanueva, F.F., Dieguez, C. & Aguilar, E. (2002) Novel expression and functional role of ghrelin in rat testis. *Endocrinology*, 143,pp 717-725.

Tesauro, M., Schinzari, F., Iantorno, M., Rizza, S., Melina, D., Lauro, D. & Cardillo, C. (2005) Ghrelin improves endothelial function in patients with metabolic syndrome. *Circulation*, 112,pp 2986-2992.

Theander-Carrillo, C., Wiedmer, P., Cettour-Rose, P., Nogueiras, R., Perez-Tilve, D., Pfluger, P., Castaneda, T.R., Muzzin, P., Schurmann, A., Szanto, I., Tschop, M.H. & Rohner-Jeanrenaud, F. (2006) Ghrelin action in the brain controls adipocyte metabolism. *J Clin Invest*, 116,pp 1983-1993.

Tomasetto, C., Wendling, C., Rio, M.C. & Poitras, P. (2001) Identification of cDNA encoding motilin related peptide/ghrelin precursor from dog fundus. *Peptides*, 22,pp 2055-2059.

Toshinai, K., Date, Y., Murakami, N., Shimada, M., Mondal, M.S., Shimbara, T., Guan, J.L., Wang, Q.P., Funahashi, H., Sakurai, T., Shioda, S., Matsukura, S., Kangawa, K. & Nakazato, M. (2003) Ghrelin-induced food intake is mediated via the orexin pathway. *Endocrinology*, 144,pp 1506-1512.

Toshinai, K., Yamaguchi, H., Sun, Y., Smith, R.G., Yamanaka, A., Sakurai, T., Date, Y., Mondal, M.S., Shimbara, T., Kawagoe, T., Murakami, N., Miyazato, M., Kangawa, K. & Nakazato, M. (2006) Des-acyl ghrelin induces food intake by a mechanism independent of the growth hormone secretagogue receptor. *Endocrinology*, 147,pp 2306-2314.

Traebert, M., Riediger, T., Whitebread, S., Scharrer, E. & Schmid, H.A. (2002) Ghrelin acts on leptin-responsive neurones in the rat arcuate nucleus. *J Neuroendocrinol*, 14,pp 580-586.

Tremblay, F., Perreault, M., Klaman, L.D., Tobin, J.F., Smith, E. & Gimeno, R.E. (2007) Normal food intake and body weight in mice lacking the G protein-coupled receptor GPR39. *Endocrinology*, 148,pp 501-506.

Trudel, L., Tomasetto, C., Rio, M.C., Bouin, M., Plourde, V., Eberling, P. & Poitras, P. (2002) Ghrelin/motilin-related peptide is a potent prokinetic to reverse gastric postoperative ileus in rat. *Am J Physiol Gastrointest Liver Physiol*, 282,pp G948-952.

Tschop, M., Smiley, D.L. & Heiman, M.L. (2000) Ghrelin induces adiposity in rodents. *Nature*, 407,pp 908-913.

Tschop, M., Statnick, M.A., Suter, T.M. & Heiman, M.L. (2002) GH-releasing peptide-2 increases fat mass in mice lacking NPY: indication for a crucial mediating role of hypothalamic agouti-related protein. *Endocrinology*, 143,pp 558-568.

Tschop, M., Wawarta, R., Riepl, R.L., Friedrich, S., Bidlingmaier, M., Landgraf, R. & Folwaczny, C. (2001a) Post-prandial decrease of circulating human ghrelin levels. *J Endocrinol Invest*, 24,pp RC19-21.

Tschop, M., Weyer, C., Tataranni, P.A., Devanarayan, V., Ravussin, E. & Heiman, M.L. (2001b) Circulating ghrelin levels are decreased in human obesity. *Diabetes*, 50,pp 707-709.

Tsolakis, A.V., Portela-Gomes, G.M., Stridsberg, M., Grimelius, L., Sundin, A., Eriksson, B.K., Oberg, K.E. & Janson, E.T. (2004) Malignant gastric ghrelinoma with hyperghrelinemia. *J Clin Endocrinol Metab*, 89,pp 3739-3744.

Ueno, M., Carvalheira, J.B., Oliveira, R.L., Velloso, L.A. & Saad, M.J. (2006) Circulating ghrelin concentrations are lowered by intracerebroventricular insulin. *Diabetologia*, 49,pp 2449-2452.

Venkova, K., Fraser, G., Hoveyda, H.R. & Greenwood-Van Meerveld, B. (2007) Prokinetic effects of a new ghrelin receptor agonist TZP-101 in a rat model of postoperative ileus. *Dig Dis Sci*, 52,pp 2241-2248.

Vincent, J.P., Mazella, J. & Kitabgi, P. (1999) Neurotensin and neurotensin receptors. *Trends Pharmacol Sci*, 20,pp 302-309.

Volante, M., Allia, E., Gugliotta, P., Funaro, A., Broglio, F., Deghenghi, R., Muccioli, G., Ghigo, E. & Papotti, M. (2002) Expression of ghrelin and of the GH secretagogue receptor by pancreatic islet cells and related endocrine tumors. *J Clin Endocrinol Metab*, 87,pp 1300-1308.

Wang, H.J., Geller, F., Dempfle, A., Schauble, N., Friedel, S., Lichtner, P., Fontenla-Horro, F., Wudy, S., Hagemann, S., Gortner, L., Huse, K., Remschmidt, H., Bettecken, T., Meitinger, T., Schafer, H., Hebebrand, J. & Hinney, A. (2004) Ghrelin receptor gene: identification of several sequence variants in extremely obese children and adolescents, healthy normal-weight and underweight students, and children with short normal stature. *J Clin Endocrinol Metab*, 89,pp 157-162.

Wang, W.G., Chen, X., Jiang, H. & Jiang, Z.Y. (2008) Effects of ghrelin on glucose-sensing and gastric distension sensitive neurons in rat dorsal vagal complex. *Regul Pept*, 146,pp 169-175.

Wei, W., Wang, G., Qi, X., Englander, E.W. & Greeley, G.H., Jr. (2005) Characterization and regulation of the rat and human ghrelin promoters. *Endocrinology*, 146,pp 1611-1625.

Wierup, N., Svensson, H., Mulder, H. & Sundler, F. (2002) The ghrelin cell: a novel developmentally regulated islet cell in the human pancreas. *Regul Pept*, 107,pp 63-69.

Wierup, N., Yang, S., McEvilly, R.J., Mulder, H. & Sundler, F. (2004) Ghrelin is expressed in a novel endocrine cell type in developing rat islets and inhibits insulin secretion from INS-1 (832/13) cells. *J Histochem Cytochem*, 52,pp 301-310.

Willesen, M.G., Kristensen, P. & Romer, J. (1999) Co-localization of growth hormone secretagogue receptor and NPY mRNA in the arcuate nucleus of the rat. *Neuroendocrinology*, 70,pp 306-316.

Xu, G., Li, Y., An, W., Li, S., Guan, Y., Wang, N., Tang, C., Wang, X., Zhu, Y., Li, X., Mulholland, M.W. & Zhang, W. (2009) Gastric mammalian target of rapamycin signaling regulates ghrelin production and food intake. *Endocrinology*, 150,pp 3637-3644.

Xu, G., Li, Y., An, W., Zhao, J., Xiang, X., Ding, L., Li, Z., Guan, Y., Wang, X., Tang, C., Zhu, Y., Wang, N., Li, X., Mulholland, M. & Zhang, W. (2010) Regulation of gastric hormones by systemic rapamycin. *Peptides*, 31,pp 2185-2192.

Xu, X., Jhun, B.S., Ha, C.H. & Jin, Z.G. (2008) Molecular mechanisms of ghrelin-mediated endothelial nitric oxide synthase activation. *Endocrinology*, 149,pp 4183-4192.

Xu, X., Pang, J., Yin, H., Li, M., Hao, W., Chen, C. & Cao, J.M. (2007) Hexarelin suppresses cardiac fibroblast proliferation and collagen synthesis in rat. *Am J Physiol Heart Circ Physiol*, 293,pp H2952-2958.

Yabuki, A., Ojima, T., Kojima, M., Nishi, Y., Mifune, H., Matsumoto, M., Kamimura, R., Masuyama, T. & Suzuki, S. (2004) Characterization and species differences in gastric ghrelin cells from mice, rats and hamsters. *J Anat*, 205,pp 239-246.

Yamagishi, S.I., Edelstein, D., Du, X.L., Kaneda, Y., Guzman, M. & Brownlee, M. (2001) Leptin induces mitochondrial superoxide production and monocyte chemoattractant protein-1 expression in aortic endothelial cells by increasing fatty acid oxidation via protein kinase A. *J Biol Chem*, 276,pp 25096-25100.

Yang, J., Brown, M.S., Liang, G., Grishin, N.V. & Goldstein, J.L. (2008a) Identification of the acyltransferase that octanoylates ghrelin, an appetite-stimulating peptide hormone. *Cell*, 132,pp 387-396.

Yang, J., Zhao, T.J., Goldstein, J.L. & Brown, M.S. (2008b) Inhibition of ghrelin O-acyltransferase (GOAT) by octanoylated pentapeptides. *Proc Natl Acad Sci U S A*, 105,pp 10750-10755.

Yeh, A.H., Jeffery, P.L., Duncan, R.P., Herington, A.C. & Chopin, L.K. (2005) Ghrelin and a novel preproghrelin isoform are highly expressed in prostate cancer and ghrelin activates mitogen-activated protein kinase in prostate cancer. *Clin Cancer Res*, 11,pp 8295-8303.

Yoshimoto, A., Mori, K., Sugawara, A., Mukoyama, M., Yahata, K., Suganami, T., Takaya, K., Hosoda, H., Kojima, M., Kangawa, K. & Nakao, K. (2002) Plasma ghrelin and desacyl ghrelin concentrations in renal failure. *J Am Soc Nephrol*, 13,pp 2748-2752.

Zhang, J.V., Ren, P.G., Avsian-Kretchmer, O., Luo, C.W., Rauch, R., Klein, C. & Hsueh, A.J. (2005) Obestatin, a peptide encoded by the ghrelin gene, opposes ghrelin's effects on food intake. *Science*, 310,pp 996-999.

Zhang, W., Chai, B., Li, J.Y., Wang, H. & Mulholland, M.W. (2008) Effect of des-acyl ghrelin on adiposity and glucose metabolism. *Endocrinology*, 149,pp 4710-4716.

Zhang, W., Zhao, L., Lin, T.R., Chai, B., Fan, Y., Gantz, I. & Mulholland, M.W. (2004) Inhibition of adipogenesis by ghrelin. *Mol Biol Cell*, 15,pp 2484-2491.

Zhang, W., Zhao, L. & Mulholland, M.W. (2007) Ghrelin stimulates myocyte development. *Cell Physiol Biochem*, 20,pp 659-664.

Zhao, Z., Sakata, I., Okubo, Y., Koike, K., Kangawa, K. & Sakai, T. (2008) Gastric leptin, but not estrogen and somatostatin, contributes to the elevation of ghrelin mRNA expression level in fasted rats. *J Endocrinol*, 196,pp 529-538.

Zhu, X., Cao, Y., Voogd, K. & Steiner, D.F. (2006) On the processing of proghrelin to ghrelin. *J Biol Chem*, 281,pp 38867-38870.

Zigman, J.M., Jones, J.E., Lee, C.E., Saper, C.B. & Elmquist, J.K. (2006) Expression of ghrelin receptor mRNA in the rat and the mouse brain. *J Comp Neurol*, 494,pp 528-548.

Zigman, J.M., Nakano, Y., Coppari, R., Balthasar, N., Marcus, J.N., Lee, C.E., Jones, J.E., Deysher, A.E., Waxman, A.R., White, R.D., Williams, T.D., Lachey, J.L., Seeley, R.J., Lowell, B.B. & Elmquist, J.K. (2005) Mice lacking ghrelin receptors resist the development of diet-induced obesity. *J Clin Invest*, 115,pp 3564-3572.

Zizzari, P., Longchamps, R., Epelbaum, J. & Bluet-Pajot, M.T. (2007) Obestatin partially affects ghrelin stimulation of food intake and growth hormone secretion in rodents. *Endocrinology*, 148,pp 1648-1653.

Part 5

Organization & Regulation

16

Polysialylation of the Neural Cell Adhesion Molecule: Setting the Stage for Plasticity Across Scales of Biological Organization

Marie-Claude Amoureux[1,2,3,*]
[1]*Developmental Biology Institute of Marseilles - Luminy, IBDML, Marseille*
[2]*Pharmaxon, Marseille Cedex 9*
[3]*Eurobio-Abcys, Courtaboeuf Cedex B*
France

1. Introduction

In vertebrates, the neuronal cell adhesion molecule (NCAM/CD56) has 3 isoforms resulting from alternative splicing that differ by their size (120, 140 and 180 kDa) and their anchoring at the membrane. Whereas NCAM120 is glycophosphatidyl inositol anchored, NCAM140 and NCAM180 are transmembrane molecules. NCAM180 has an additional intracellular 267 amino acids insert, that differentiate it from NCAM140, but its role remains to be fully elucidated. There are differences regarding the specificity and the level of expression of these isoforms in different cells of the nervous system. Whereas NCAM120 and NCAM140 are preferentially expressed in glial cells NCAM180 seems to be prevalent on neurons. Although differences in expression and function of the NCAM isoforms exist, one common denominator is that they can be post-translationally modified by the addition of long, linear chains of α2,8-linked *N*-acetylneuraminic acid (Neu5Ac) residues. In vertebrates, NCAM is the major acceptor of this unique carbohydrate. This modification occurs on the fifth immunoglobulin domain of NCAM located on the extracellular part of the membrane and common to all 3 isoforms. These polysialylated isoforms have emerged as particularly attractive candidates for promoting plasticity in the central nervous system (CNS). The large negatively charged polysialic acid (PSA) chain of NCAM is postulated to be a spacer that reduces adhesion forces between cells allowing dynamic changes in membrane contacts. However, recent studies indicate that a crucial function of PSA resides in controlling interactions mediated by NCAM. Accumulating evidence also suggests that PSA-NCAM-mediated interactions lead to activation of intracellular signals fundamental to biological functions. An important role of PSA-NCAM appears to be during development, when its expression level is high and where it contributes to the regulation of cell shape, growth or migration. However, PSA-NCAM does persist in adult brain structures such as the hippocampus that display a high degree of plasticity where

---

* Annelise Viallat[2], Coralie Giribone[2], Marion Benezech[2], Philippe Marino[2], Gaelle Millet[2], Isabelle Boquet[2], Jean-Chrétien Norreel[2] and Geneviève Rougon[1]
[1]*Developmental Biology Institute of Marseilles - Luminy, IBDML, Marseille, France*
[2]*Pharmaxon, Marseille cedex 9, France*

it is involved in activity-induced synaptic plasticity. Recent advances in the field of PSA-NCAM research have not only consolidated the importance of this molecule in plasticity processes but also suggested a role for PSA-NCAM in the regulation of higher cognitive functions and psychiatric disorders, metastasis and tissue repair. The presence of PSA-NCAM also outside the nervous system in specific cell types or in pathological conditions, such as cancer, suggests that it is an important feature of NCAM to be investigated.

An underlying theme in this review will be the potential role that polysialylation, a post-translational modification of NCAM, can play in translating plasticity across scales of biological organization. Indeed, the formation of any complex tissue like the CNS involves adaptive changes such as fate determination, differentiation, proliferation, happening via cell dynamic autonomous processes or dynamic interactions with adjacent cells and tissues. In this chapter, we will update interpretations of some earlier findings and highlight how PSA exerts effects on plastic events at the molecular, cellular and tissue levels.

2. Polysialylation controls plasticity at the molecular scale by regulating the molecular behavior of its own carrier, NCAM

Insight into the function of NCAM at the molecular level came from observations based upon NCAM gene knock out, PSA removal/modulation by enzymatic digestion of PSA by endoneuraminidase (EndoN), knock out of the polysialyltransferase coding genes responsible for addition of PSA to NCAM, or the use of mimotope peptides of PSA. Altogether, they converge to indicate that glycosylation of NCAM may be more critical to account for the biological functions than the presence or absence of the core protein itself. The length of the PSA chains isolated from brain NCAM varies widely with development and has been estimated to exceed 50 sialic acid residues (Galuska et al. 2006). Two enzymes, the polysialyltransferases ST8SiaII and ST8SiaIV, are involved in the biosynthesis of PSA chains in mammals. These enzymes share <60% similarity at the amino acid sequence level and are, during postnatal development, differentially expressed in a tissue- and cell type-specific manner with overlapping expression patterns. In vitro studies showed that each enzyme is independently capable of synthesizing PSA on NCAM although ST8SiaII was found to synthesize shorter PSA polymers than ST8SiaIV. Additional studies involving fine structure analysis suggest a comparable quality of polysialylation by ST8SiaII and ST8SiaIV and a distinct synergistic action of the two enzymes in the synthesis of long PSA chains at N-glycosylation site 5 in vivo (Galuska et al., 2008). By analyzing defects in mice with selected combinations of mutant NCAM and polysialyltransferase alleles, Hildebrandt et al. (2009) revealed that the extent of the fiber tract deficiencies was not linked to the total amount of PSA or NCAM, but correlated strictly with the level of NCAM erroneously devoid of PSA during brain development. Hence, PSA is the key regulator of central NCAM functions.

At the molecular level, modelling of light scattering data, with the assumption that the proteins are spherical, led to the conclusion that PSA increased the hydrodynamic radius of NCAM. Moreover, intercellular space was increased by 10 to 15 nm (Yang et al, 1992; Rutishauser, 1996). In this vein, it was accepted that polysialylation was a mechanism controlling the range and magnitude of intermembrane repulsion and thereby cell-cell interactions. Nevertheless, early studies gave no indication on the forces of interactions mediated by NCAM, whether polysialylated or not. The dependence on ionic strength for adhesion of PSA-NCAM expressing cells however suggested that this was due to change in hydrodynamic volume and repulsive forces (Pincus, 1991; Yang et al., 1994). The link between repulsive forces and adhesion suggested that this concept could be generalized to other adhesion molecules (Fujimoto et al., 2001).

Johnson et al. (2005) later showed, by introducing physical forces measurements, that PSA decreased cell-cell interactions by playing a repulsive role on trans-homophilic NCAM-NCAM interactions, as well as on trans-interactions between other membrane-bound proteins, at the contact between cell membranes. Surface force apparatus can be used to quantify the distance dependence of the force between extended surfaces such as membranes (Shi et al., 2010). This technique revealed that PSA was able to overcome NCAM-NCAM homophilic and NCAM-cadherin heterophilic attraction. Physical forces measured to demonstrate such role of PSA were dependent on the amount of PSA and ionic strength. The magnitude of the repulsive effect of PSA is significant since it overwhelms a 3-fold excess of cadherin and at 3 binding sites (Johnson et al., 2005). Furthermore, the properties of the polymer of PSA are independent of other proteins on the membrane. We also demonstrated using two different molecular imaging techniques, Fluorescence Correlation Spectroscopy (FCS) and spot Fluorescence Recovery after Photobleaching (FRAP), that when two live cells were in contact, PSA increased the lateral mobility of NCAM, suggesting more fluidity at cell-cell contacts (Conchonaud et al., 2007). Since the polymer of PSA occupies a tridemensional volume, it was likely that the steric influence of PSA would be at play not only in trans-interactions but also in cis-interactions, involving NCAM or other membrane proteins (Figure 1A). We therefore asked whether PSA affected the molecular mobility of NCAM locally within cell membrane microdomains in isolated cells. Indeed, this could also be a reflection of increased molecular repulsion between adjacent proteins in the same membrane, as suggested by Johnson et al. (2005). By using FRAP and FCS to study NCAM dynamics, we brought the evidence that PSA increased the mobility of NCAM itself at the cell membrane without effect on its trafficking or confinement in small domains. The later depends on the actin cytoskeleton as when the cytoskeleton was disrupted, NCAM like PSA-NCAM were no longer confined in microdomains and acquired a Brownian type mobility with a lateral diffusion 60% of that with intact cytoskeleton. Moreover, without intact cytoskeleton, PSA no longer facilitated NCAM mobility (Figure 1B). Polysialylation effect on NCAM lateral diffusion thus requires the integrity of the link of NCAM with the actin cytoskeleton and/or the confinement of NCAM by the cytoskeleton. Furthermore, PSA effect on NCAM lateral mobility was maintained even when NCAM diffused less rapidly due to cell-cell contacts. Importantly, PSA increased NCAM lateral diffusion when cells were activated by an extracellular factor such as GDNF (Conchonaud et al., 2007). Altogether, these data showed that polysialylation conveys to NCAM an intrinsic capacity to increase its lateral diffusion within the membrane even when the molecules are engaged in interactions or signaling. This leads to postulate that not only molecular interactions, but also the duration of NCAM interactions, are reduced by PSA.

Thus, these data place the addition of a carbohydrate by post-translational modification to a cell surface receptor as an efficient way to control its lateral diffusion in a cell autonomous fashion, and most likely cis-interactions with other proteins in the membrane. Figure 1 recapitulates the modulation of cis and trans-homophilic and heterophilic interactions at cell membranes by PSA via repulsive forces and enhanced mobility. Ultimately, as the resolution of fluorescent imaging improves, it will be exciting to see the imaging of macromolecular assemblies. One of the compelling prospects of this integration is that it should provide molecular models for new mechanistic insights into the signaling processes and the resulting cell behavior.

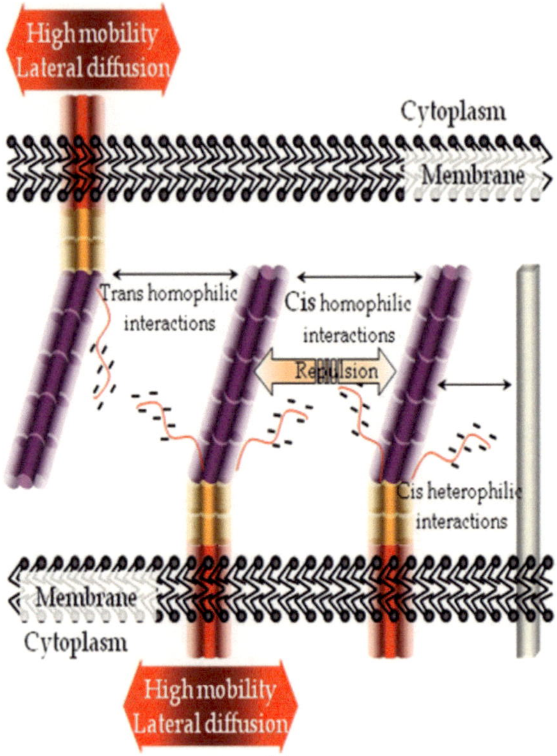

(a)

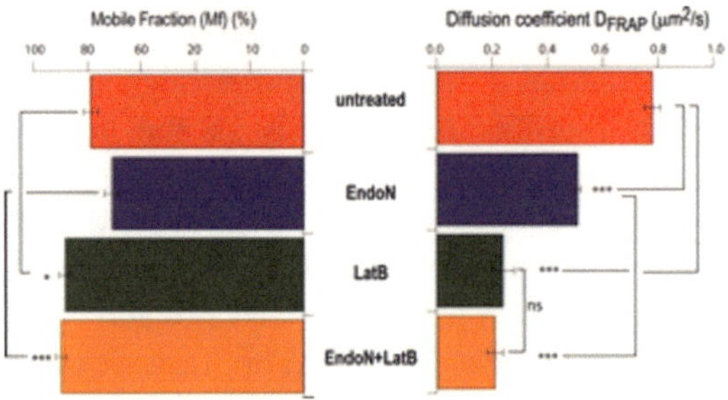

(b)

Fig. 1. (a) Modulation of NCAM lateral diffusion at the membrane, and homophilic and heterophilic trans and cis molecular interactions by PSA. (b) Tracking of GFP-tagged PSA-NCAM140 molecules by FCS without or after treatment of the expressing cell by EndoN allows to calculate the mobile fraction and by FRAP its diffusion. PSA-NCAM or NCAM molecules were as mobile (80%) but PSA increased NCAM diffusion. Disrupting actin organization by Latrunculin B (Lat B) increases the mobile fraction independently of PSA, but suppresses the enhancing effect of PSA on NCAM lateral diffusion.

3. Polysialylation results in plasticity at the cellular scale

There is still a big gap to uncover before getting a full understanding on how the effect of polysialylation of NCAM at the molecular level translates into cellular plasticity. For the purpose of this chapter, only studies describing changes in cellular behavior that occur in a cell autonomous manner or in which the cell carrying PSA is considered as the system of reference will be reviewed. PSA effect during development or on axon regeneration, or in diseased conditions will be addressed in part 4 in an integrated tissue system undergoing changes as a whole. As defined, plasticity at the cellular scale can encompass cell fate decisions, cell division, cell death, cell shape and cell migration.

PSA-NCAM is most conspicuously involved in the change of shape or the migration of different cell types (Decker et al., 2000; Hu et al., 1996). PSA-NCAM is required for the migration of group of cells as chains in the rostral migratory stream (RMS) over millimeters (Durbec & Rougon, 2001). In vivo, ectopic expression of PSA in Schwann cells increases their migration (Bachelin et al., 2010; Papastefanaki et al., 2007). PSA-NCAM can also be considered a cell-mobility enhancing/permissive factor. Specifically, several experiments revealed that PSA–NCAM is required for directional migration in response to concentration gradients of chemoattractants such as PDGF, BDNF, CNTF or GDNF. For example, removal of PSA from purified oligodendrocyte progenitor cell (OPC) significantly reduced lamellipodia formation in response to low concentrations of PDGF, raising the possibility that under these conditions OPC are less responsive to PDGF (Zhang et al., 2004). These observations raised the intriguing possibility that PSA–NCAM could modify the ability of cells to sense accurately growth factor gradients and thereby play a role in the guidance process, as also initially suggested for retinal ganglion axons in vivo (Monnier et al., 2001). We examined the requirement of polysialylation using transfilter migration assays of TE671 cells. We observed that removal of PSA decreased by half the ratio between migrating and non-migrating cells (Conchonaud et al., 2007). Furthermore, like for OPC removal of PSA dramatically reduced the enhancing effect of GDNF on migration. Thus, polysialylation of NCAM plays a role in promoting cell migration but also in efficiently potentiating the chemotactic effect of GDNF in TE671 cells. PSA potentiation of GDNF or PDGF effect illustrates that it may affect NCAM relationship either directly with other molecular partners such as growth factors or their receptors or by priming cells at the level of their cytoskeleton to respond more efficiently. In this cellular system, we found an inter-dependence between migration, cytoskeletal changes, and NCAM lateral diffusion in membranes (Part 2). Polysialylation of NCAM increased actin stress fibers formation, showing that PSA had consequences on the actin cytoskeleton involved in plastic cellular events. Moreover, showing that NCAM lateral diffusion and a directional cell migration both require the cytoskeleton, makes the latter appear as a physical link to NCAM that is required for plasticity at the molecular and cellular level.

Cell differentiation is also characterized by a differential expression of PSA-NCAM (Figure 2). Although being preferentially committed to a restricted either glial or neuronal fate, several studies revealed that cultured PSA-NCAM+ neural progenitors isolated from brain do preserve a relative degree of multipotentiality. With this regard, it is noteworthy that growth factors and neurotransmitters, which belong to the micro-environment of neural cells in vivo, regulate morphogenetic events preceding synaptogenesis such as cell differentiation and death. PSA-NCAM seems to play an active role in the differentiation program of neural cells. For instance, PSA limits the differentiation of progenitors from the subventricular zone (SVZ) (Petridis et al., 2004; Rockle et al., 2008). In this specific system, the effects of PSA removal on cell differentiation and migration seem to be uncoupled as differentiated SVZ neuroblasts due to PSA removal is associated with dispersion into surrounding CNS tissues (Battista & Rutishauser, 2010). It is likely that other factors are also involved in controlling these processes, much like it is the case for neuroblasts reaching the olfactory bulb where signals like reelin influence migration in the OB layers (Courtes et al., 2011; Hack et al., 2002). PSA down regulation in OPC seems to be critical for proper maturation of OPC into myelinating mature oligodendrocytes (Charles et al., 2000, 2002; Koutsoudaki et al., 2010). Polysialylation of NCAM appears to be regulated at the post-translational level since overexpression of polysialyltransferases by ectopic expression is not sufficient to maintain PSA expression and only delays myelination (Coman et al., 2005). These experiments however did not elucidate whether this occurs in a cell-autonomous manner or following interactions with axons to be remyelinated.

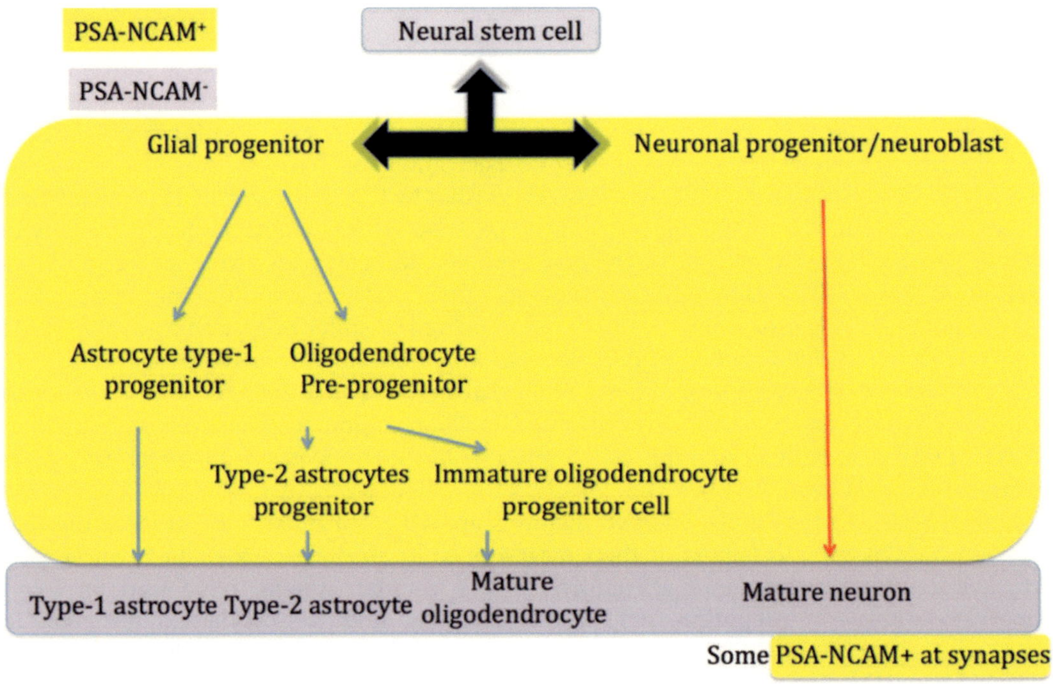

Fig. 2. Schematic view of the progression of CNS cell specification, showing relationships between intermediary phenotypes. Cells expressing PSA-NCAM were specifically shaded on a yellow background and PSA-NCAM- cells shaded in purple. The level of PSA-NCAM expression decreases with the progression towards more differentiated phenotypes.

In vivo, PSA downregulation seems to be associated with cell fate and or functional changes. A most convincing example is given by the experiments showing that the developmental and activity-dependent decline of PSA expression regulates the timing of the maturation of GABAergic inhibition and the onset of ocular dominance plasticity (Di Cristo et al., 2007). Concentrations of PSA significantly decline shortly after eye opening in the adolescent mouse visual cortex; this decline is hindered by visual deprivation. The developmental and activity-dependent regulation of PSA expression is inversely correlated with the maturation of GABAergic innervation. Premature removal of PSA in visual cortex results in precocious maturation of perisomatic innervation by basket interneurons, enhanced inhibitory synaptic transmission, and earlier onset of ocular dominance plasticity. Intrigingly, PSA is also expressed in a subpopulation of adult cortical interneurons characterized by reduced structural features and connectivity (Gomez-Climent et al., 2011). Birth-dating analyses reveal that these interneurons are generated during embryonic development. They show a reduced density of perisomatic and peridendritic puncta expressing synaptic markers and receive less perisomatic synapses, when compared with interneurons lacking PSA-NCAM. Moreover, they have reduced dendritic arborization and spine density. Altogether these data indicate that PSA-NCAM expression is important for the connectivity of interneurons in the adult and that its regulation at the cellular level may play a role in the structural plasticity of inhibitory networks.

The environment also plays a role on the fate of PSA expressing cells. PSA^+ neuroblasts become mostly PSA^- glial cells after transplantation in a non-neurogenic environment (Seidenfaden et al., 2006). Some of the effects of PSA on cell fate could however be attributable to NCAM since the NCAM knock mouse displayed glial cells accumulating in the SVZ due to altered cell fate and/or disorganized RMS (Chazal et al., 2010), while the fate of neuroblasts remained mostly neuronal after PSA removal with endoN, despite some dispersion in the surrounding adult tissue (Battista & Rutishauser, 2010). PSA-NCAM has also been shown to be a prosurvival factor for immature neurons (Gascon et al., 2007). The mechanism underlying the effect of PSA on cell fate could thus involve the potentiating effect of PSA with regard to growth factors evidenced in other cell types or systems (Conchonaud et al., 2007; Muller et al., 2000; Vutskits et al., 2001; Zhang et al., 2004), the differentiation of neuroblasts induced by NCAM (Amoureux et al., 2000), or an upregulation of the p75 dependence receptor (Gascon et al., 2007).

Polysialylation of NCAM has also been shown to regulate neuritogenesis (Seidenfaden, 2006), and axonal growth (Doherty et al., 1990; Zhang et al., 1992) synaptic plasticity and activity-dependent cell remodeling, which has been extensively reviewed (Bonfanti & Theodosis, 2009; Muller et al., 2010; Kochlamazashvili et al., 2010). The effects of PSA-NCAM on axonal fasciculation and defasciculation in different systems have been reported and seem to depend on a combination of both intrinsic and context/environment dependent effects (reviewed by Durbec & Cremer, 2001).

At the cellular level, plastic events involve changes in physical/mechanical and chemical/molecular properties. Indeed, cell shape-dependent functions result from complex mechanical interactions between the cytoskeleton architecture and external conditions, be they cell-cell or cell-extracellular matrix adhesion contact-mediated, and their corresponding (mutually interactive) signaling machinery. It is noteworthy that polysialylation of NCAM influences signaling pathways such as ERK phosphorylation (Conchonaud et al., 2007) or FAK (Duveau & Fritschy, 2010), indicating that it is likely that these two types of signals cross-talk. It is important for future studies to elucidate how cross talk between these signals is coordinated to control neural cells structure and function. Ultimately, understanding how the highly interactive mechanical signaling can give rise to phenotypic changes is critical for targeting the underlying pathways that contribute to cell plasticity.

4. PSA controls plasticity at the tissue scale

4.1 PSA in a dynamic healthy environment: During development and in adult CNS areas of plasticity

High plasticity at the tissue level occurs during development when organogenesis takes place. At this scale, PSA has been known to play a major role in the formation of the nervous tissue and its organization as highly complex structures (Edelman, 1986). PSA critical role is reflected by PSA-NCAM spatiotemporal regulation throughout development and in adulthood. NCAM during development (embryonic and early postnatal) only exists in its polysialylated form, which then disappears in adulthood in most CNS areas (Figure 3).

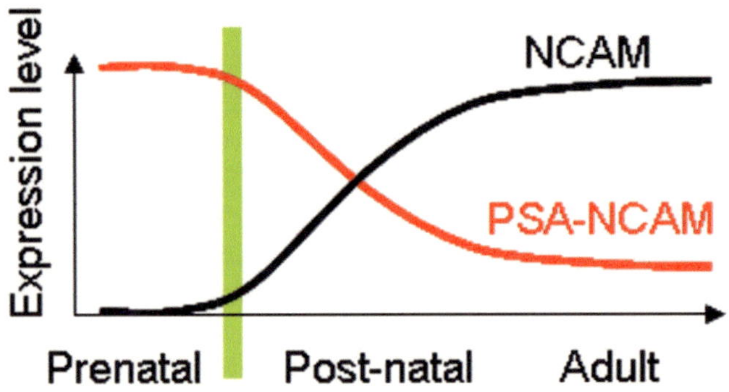

Fig. 3. Regulation of PSA-NCAM throughout development

Conversely to NCAM knock out mice which display a relatively mild phenotype (Chazal et al. 2000), mice lacking the polysialyltransferases ST8SiaII and ST8SiaIV, show a severe phenotype with specific brain wiring defects, progressive hydrocephalus, postnatal growth retardation, precocious death, and severe malformation of major brain axon tracts (corticospinal tract, anterior commissure, corpus callosum and internal capsule hypoplasia). Along with the growth retardation, polysialyltransferase double mutants appear increasingly cachectic and reveal severe deficits in simple tests of motor coordination, strength, and balance (Weinhold et al., 2005). Most of these phenotypes have been attributed to increased adhesion due to absence of PSA on NCAM, and illustrate the prominent aspect of the postranslational modification.

In the adult, following PSA removal by endoN, plasticity is affected in brain regions where PSA is normaly expressed. The first brain region is the SVZ, located at the start of the rostral migratory stream (RMS) pathway followed by PSA-NCAM expressing neuroblasts on their way to the olfactory bulb (OB), the later being constantly exposed to new odors and requiring permanent renewal of neurons for sensing the environment (Gheusi et al., 2000). NCAM deletion results in a 30% decrease in the size of the OB, and 10% in the overall brain size, associated with deficits in learning (Chazal et al., 2000; Cremer et al., 1994; Tomasiewicz et al., 1993).

The second main brain region is the subgranular zone (SGZ), from which newly born neurons emanate and integrate into the granular layer of the hippocampus, involved in memory and long term potentiation, a physiological strengthening of neural connections thought to underlie learning and memory. The activity-dependent expression of PSA at the synapse also suggests a role for this molecule in activity-induced synaptic plasticity and memory. Indeed in 2007, Lopez-Fernandez et al. demonstrated that up-regulation of PSA-NCAM in the dorsal hippocampus after contextual fear conditioning was involved in long-term memory formation. This article confirmed the previously published papers in which hippocampal upregulation of the polysialylated form of NCAM was shown to play a key role on spatial memory processes (Venero et al., 2006). The recall of acquired memories is initially dependent on the hippocampus for the process of cortical permanent memory formation, a dependency that decays with time in order to leave room for new memories to be acquired. Production of PSA-positive new neurons in the hippocampus increases the efficiency of the hippocampus, shortens the time where the recall is dependent on the hippocampus, allowing that physiological traces of old memories are promptly removed from the hippocampus to make room for new ones (Kitamura et al., 2009). New neurons enable the hippocampus to work more quickly. Many other studies also point out the functional significance of adult hippocampal neurogenesis in learning and memory (reviewed by Zhao et al., 2008). Therefore in this second brain system, PSA-NCAM is also connected to a brain function that requires plasticity, namely memory. The effect of a PSA-NCAM mimotope, PR21, developed in the laboratory and presented in 4.3. was investigated on spatial memory consolidation following injection of the peptide into the dorsal hippocampus of mice. The mimotope of PSA-NCAM improves memory consolidation requiring hippocampus, suggesting that it enhances the plasticity of this structure (Florian et al., 2006).

The existence of germinative zones associated with high PSA-NCAM labeling has initially been put to light in adult rodents. Apparently, this extends to other species such as the rabbit, the song bird, the primate, and up to the human CNS, although with species-specific organizations of germinative zones and pathways of migration (Bedard et al., 2006; Bernier et al., 2002; Luzzati et al., 2006; Marlatt et al., 2011; Pencea et al., 2001; Sawamoto et al., 2011). The existence of a human RMS, which is organized around a lateral ventricular extension reaching the OB, composed of PSA-NCAM$^+$ neuroblasts, that incorporate 5-bromo-2'-deoxyuridine and become mature neurons has been shown (Curtis et al., 2007; Kam et al., 2009; Sanai et al., 2004). Hipocampal adult human neurogenesis has also been demonstrated to take place (Ni Dhuill et al., 1999). The hypothalamo-neurohypophyseal axis is another system characterized by a dynamic PSA-NCAM expression, and plasticity due to its neuroendocrine physiology related to food intake, sleep-wake cycle and reproduction (Migaud et al., 2010). A strong immunolabeling for PSA-NCAM was initially observed in the rat and monkey hypothalamus (Bonfanti et al., 1992; Perrera et al., 1993; Theodosis et al., 1991) and more recently in the sheep (Franceschini et al., 2010). In this structure estrogen induced synaptic plasticity of gonadotropin releasing hormone (GnRH) neurons is concommittant with PSA-NCAM regulation by activity across the estrous cycle (Parkash & Kaur, 2007). PSA-NCAM$^+$ cells have also been described in the cortex of different species, like the adult rat (Gomez-Climent et al., 2011), rabbit, guinea pig and lizard, particularly in non-spatial learning and memory networks, where the labelled cells are not newly generated (Luzzati et al., 2009).

This indicates that polysialylation of NCAM plays a major role in CNS areas of physiological plasticity encompassing regenerative properties as well as refinement of neural networks. Long-standing questions include to what extent these regenerative properties recapitulate the events observed in development and to what extent are adult- or species-specific aspects deployed.

4.2 PSA in cancer

PSA-NCAM has the status of an oncofetal antigen. Many studies examined tumor versus control expression of NCAM and PSA epitopes in tissue specimens, as well as correlation between tumor expression and clinicopathological features. Generally, results showed a low constitutive expression of PSA-NCAM in control tissue, which reached a statistically significant increase in the tumor tissue. Likewise, the presence and number of metastases at surgery were correlated with PSA-NCAM expression. The data highlight the importance of taking into account PSA epitope when dealing with NCAM cell expression studies in tumor development and progression and the fact that PSA-NCAM could be an intersting biomarker to assess a better knowledge of brain tumors, help prognosis, design and evaluate therapies. We set up a specific and sensitive enzyme linked immunosorbent assay (ELISA) test for PSA-NCAM quantification, which correlated with PSA-NCAM semi quantitative analysis by immunohistochemistry, and thus provides an accurate quantitative measurement of PSA-NCAM content for the biopsies analyzed (Amoureux et al., 2010). Survival of patients with Glioblastoma Multiform (GBM) the most aggressive and frequent brain tumor, albeit without cure, is limited to one year on average, however significant variability in outcome is observed. We studied its expression in GBM and evaluated its prognosis value for overall survival (OS) and disease free survival (DFS). We showed that PSA-NCAM was expressed by approximately two thirds of the GBM at variable levels. On univariate analysis, PSA-NCAM content was an adverse prognosis factor for both OS and DFS. On multivariate analysis, PSA-NCAM expression was an independent negative predictor of OS and DFS. Furthermore, in glioma cell lines, PSA-NCAM level expression was correlated to the one of olig2, a transcription factor required for gliomagenesis (Amoureux et al., 2010). In addition to the prognosis value of PSA-NCAM in GBM, evidence from ours and others' studies have shown that PSA-NCAM is also a reliable marker in many types of cancers of the CNS and other organs, and that a correlation could be established between the level of PSA-NCAM and the severity of the disease and/or the response to treatment and relapse (Daniel et al., 2000, 2001; Figarella-Branger et al., 1990, 1996 ; Gluer et al., 1998; Trouillas et al., 2003). Another focus of intense investigation in the cancer field has been the existence and the role of cancer initiating cells in GBM (reviewed in Stiles & Rowitch, 2008). There is a strong possibility that the glial progenitors that populate the adult CNS are one source of gliomas with plasticity of immature glia (see part 3) underlying the phenotypic heterogeneity one sees in many infiltrating GBM. At present, however, there is no definitive way to distinguish a reactive/recruited progenitor or astrocyte from a glioma cell. There is accumulating evidence that $CD133^{+}$ cells are not the only elements in GBM with significant tumorigenic potential. Human gliomas also contain an abundance of cells that express markers normally expressed by adult glial progenitors, including PSA-NCAM providing another possible link between glial progenitor like cells and tumorigenesis. In addition, PSA-NCAM could be interesting to study and modulate GBM infiltration as it appears to recapitulate the migration of glial progenitor cells that occurs during brain development (Farin et al., 2006). In addition, treatment of GBM patients as previously suggested with chemicals that alter PSA expression with the goal to modulate polysialylation in tumors could be envisionned (Mahal et al., 2001).

4.3 PSA-mimicking agents in promoting tissue repair: Example of spinal cord injury

A striking feature of vertebrate regeneration is that it can take place only during development at embryonic and early postnatal stages, when the CNS is still in a phase characterized by high plasticity and remodelling. A specificity of this phase is the overall expression of PSA-NCAM as described above, which then disappears when the adult CNS is stable. From an evolutionary perspective, it is remarkable that the addition of this unique carbohydrate polymer to the membrane bound NCAM only appeared in vertebrates, indicating a role associated to a higher degree of complexity of the nervous system. The evolutionary perspective on PSA-NCAM appearance in vertebrates, placed in the context of repair, also strongly suggests that its disappearance in vertebrate adult nervous system may be one of the key factor that limit regeneration. Indeed, lower organisms have the ability to regenerate after adult CNS lesions and do not require expression of PSA-NCAM during development, whereas vertebrate axons can regenerate while PSA is still expressed but cannot during adulthood when PSA is absent. PSA-NCAM reexpression is common after lesions or induced demyelination of different fiber tracts and has been interpreted as the reactivation of a neurodevelopmental program (Bonfanti et al., 1996; Nait-Ousmenar et al., 1995).

Spinal cord injury (SCI) is a complex pathology, which does not lead to recovery in severe cases. Of specific interest to SCI, PSA-NCAM is expressed in the area lining the central canal of the spinal cord, much like the SVZ lines the brain lateral ventricules and in laminae I, II and X of adult rat spinal cord (Bonfanti et al., 1996). One of the main issues in SCI is the lack of ability of damaged and severed neurons to regenerate. It is undeniable that inhibitors, including chondroitin sulfate proteoglycans, along with non-proteoglycan inhibitors such as myelin-associated glycoprotein and NOGO, present in the disrupted myelin after axonal severing or the glial scar formed after injury constitute a major impediment to axonal re-growth (Silver & Miller, 2004). However, inhibiting one of them will not be sufficient to overcome the system that the vertebrate CNS has put in place to prevent regeneration. In addition, the limitation of such inhibitory strategies is related to the fact that all the inhibitors known to date, lead to the same downstream signaling pathways converging on the activation of Rho small GTPases and Rho-associated, coiled-coil containing protein kinase inhibiting axonal growth. Nevertheless, although not sufficient to overcome the inhibitory signals, spontaneous growth and plastic responses of CNS axons exist after SCI, especially in the first days after injury. Bareyre et al. (2004) and Courtine et al. (2008) unequivocally showed that spontaneous injury-induced structural and functional circuit rearrangements contribute to the spontaneous behavioral recovery after lesions in rodents. Such remodelling of spared system is probably crucial for rehabilitation in humans and offers some hope that a strategy stimulating these self-repair mechanisms may lead to better clinical results (Rosenzweig et al., 2010). It is therefore necessary to stimulate these spontaneous plastic changes. Properties of PSA-NCAM described in Part 3 such as its capacity to enhance cell mobility and promote axonal growth, both necessary to reconstruct functional cell connections after SCI strongly indicate that PSA-NCAM is a valuable, innovative and unique target to overcome the particular set of challenges of SCI. The use of

PSA-NCAM mimicking agents appear beneficial to recovery by shifting the balance between inhibitors and stimulators towards stimulators. A cyclic peptide, PSA mimetic peptide, PR-21, was selected from a phage library screening of 100 million dodecapeptides, by its ability to bind an antibody directed against PSA-NCAM (Torregrossa et al., 2004). This peptide was shown to be non immunogenic, and stable and to improve functional recovery after SCI in mice (Marino et al, 2009) as well as regeneration after periheral nerve crush (Mehanna et al., 2009). Earlier studies already had pointed out the importance of the embryonic form of NCAM after nerve injury (Daniloff et al., 1986, 1995). Another PSA mimetic peptide than PR21 was shown to improve SCI recovery in mice (Mehanna et al., 2010). Camand et al. (2004) also showed that PSA-NCAM is expressed by reactive astrocytes in the glial scar and that these molecular changes are correlated with the sprouting of axons, observed for cerebellar neurons (Dusart et al., 1999). Their observations suggest that PSA-NCAM re-expression is accompanied by the acquired permissiveness of axons to regenerate through the glial scar. El Maarouf & Rutishauser (2010) also suggested that PSA expression could be used as a strategy for promoting repair involving rebuilding of neural connections. Data showed that transfection of scar astrocytes with a construct that encodes polysialyltransferases improved the regeneration of corticospinal-tract axons through the glial scar (El Maarouf & Rutishauser, 2010) while overexpression of PSA by grafted Schwann cells, promoted functional recovery after SCI (Papastefanaki et al., 2007; Zhang et al., 2007). After having shown proof of efficacy in a mouse SCI model (Marino et al., 2009), we present herein data showing that the cyclic PSA mimotope PR21 also promotes SCI motor recovery in a clinically relevant rat model of SCI.

PR-21 efficacy was assessed using a chronic delivery paradigm over 14 days in a spinal cord impaction injury rat model. Spinal cord lesions were performed with the NYU impactor. Rats were anesthetized with an intraperitoneal injection of ketamine (75 mg/kg) and xylazine (10 mg/kg). A laminectomy was performed at T9-T10 level exposing the cord without disrupting the dura mater. The spinous processes of T8 and T11 were clamped to stabilize the spine, and the cord was subjected to weight drop impact using a 10 g rod dropped at the height of 12.5 mm. After the injury, the muscles and skin were closed in layers. An entry point for intrathecal cannulation was created immediately after spinal cord contusion injury. An incision of the ligamentum flavum between T12 and L1 was made using a pair of microscissors followed by a L1 laminectomy. The catheter attached to an Alzet minipump was then inserted into the subarachnoid space, and the tip of the catheter was carefully advanced up to 1.25 mm caudal to the injury epicenter. Two doses of PR21 (3 and 12 mg/kg) were tested.

We show, using this closer to human injury model, that PR-21 significantly enhanced motor recovery, measured using the standardised Basso-Bresnaham-Beattie motor BBB test (Basso et al., 1995), in rats when delivered intrathecally (Figure 4). The results show that PR-21 was also able to improve motor coordination.

A BBB score of 12 corresponds to a return to coordinated movements (Basso et al. 1995). This analysis showed that at day 70, 89% of 3mg/kg PR-21-treated animals could walk with occasional coordination versus 50% for vehicle-treated animals (Figure 5A). In addition, if a BBB score threshold of 14 corresponding to permanently coordinated animals (Basso et al. 1995) was applied, a significant higher proportion of 3mg/kg-treated animals could walk with a consistent coordination (Figure 5B) if compared to the vehicle-treated group (57% vs 30%).

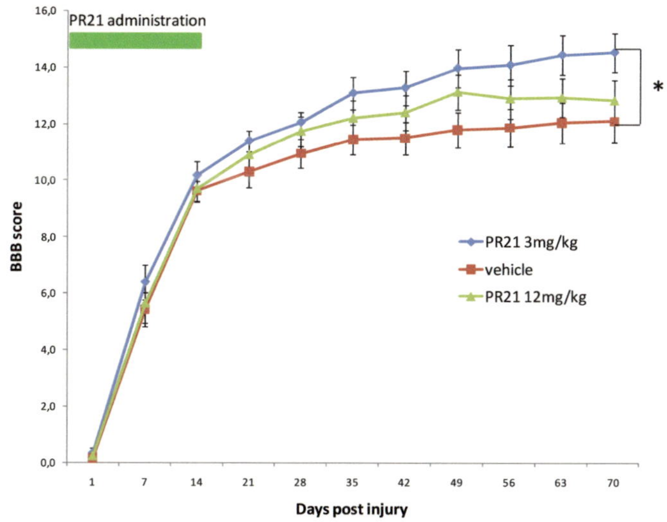

Fig. 4. Time course recovery of PR-21 treated (3mg/kg and 12mg/kg) and control rats (*$p<0.05$, one way ANOVA and Dunnet's post-hoc test).

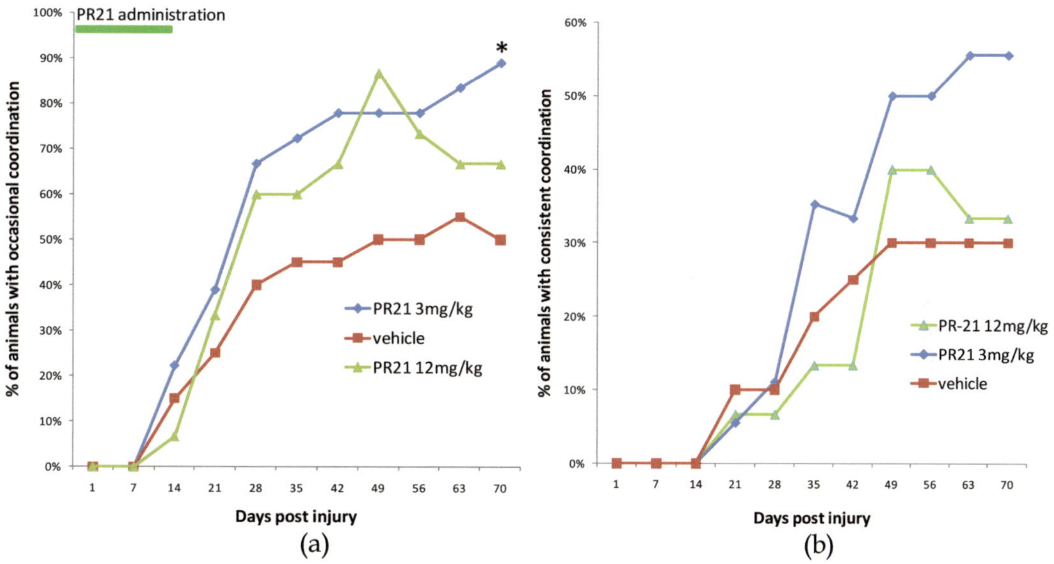

Fig. 5. Percentage of rats able to occasionally walk coordinating their fore- and hindlimbs (a) and with consistent coordination of their fore- and hindlimbs (b). (n=20); *$p<0.05$: Chi-Square/Fisher exact test.

Altogether, these results demonstrated that PR-21 at a dose of 3mg/kg clearly improves functional recovery of animals with contusion SCI. The dose of 12 mg/kg tended to improve motor performance but the results were not statistically significant. Further analysis by turbidimetry, a method used to assess compound solubility, revealed that PR21 solubilized for the dose of 12 mg/kg did not remain as soluble as for 3 mg/kg, which is a likely explanation for its lower efficiency.

In summary, in two independent animal SCI models (mice dorsal hemisection and rat contusion), and with two types of delivery (one-shot acute delivery in mice and 14 days continuous delivery in rats), a PSA-mimotope improved significantly motor recovery. The improved recovery was observed for global locomotion and hindlimb/forelimb coordination using the Basso Mouse Scale, motor coordination and balance using the Rotarod to test the ability of animals to remain on a rotating rod, and fine locomotor, sensory and proprioceptive performance using the grid walk test. Furthermore, PR-21 shortened the time of return to continence, a major issue in SCI, by 2-fold in the mouse model of SCI. The underlying mechanisms of improved recovery seems to be a decrease of the glial scar, containing regeneration inhibitors at the lesion site, and an increased density of serotonergic axons at and caudal to the lesion, these fibers being major players in the central pattern generators of locomotion (Marino et al., 2009). Other pathways such as the lesioned corticospinal tract seem to have also been positively affected by the treatment since the grid test revealed precise motor placement in which the corticospinal tract plays a significant role in the mouse model (Marino et al., 2009). Accordingly, providing the spinal cord with a mimotope of PSA-NCAM affects plasticity, supporting that PSA-NCAM is an important player in the pathological state of SCI.

5. Conclusion

How tissues regulate their size and morphology remains an incompletely answered question, particularly in the CNS, which shifts from a very plastic structure during embryonic and early postnatal stage to a static organ, isolated by the blood brain barrier in the adult. The presence of PSA-NCAM during developmental tissue morphogenesis or in immature cells throughout life in various species, as well as the use of enzymatic and chemical tools and genetic mice models, point to PSA-NCAM not only as a marker of plasticity, but also a structural and functional factor of plasticity conserved throughout vertebrate evolution. The transitory re-expression of PSA-NCAM after traumatic lesion or demyelination, hormonal or neurotransmitter induction, also suggests a role for PSA-NCAM in glial and neuronal plasticity, even in adulthood. Permanent re-expression such as in cancer illustrate a deregulation of plasticity leading to unwanted cell migration and metastasis. These effects seem mediated by the regulation of the mobility of its carrier, NCAM, within the membrane as well as the repulsive forces between cell membranes and their bound molecules. It is undeniable that these interactions are conditioning and conditioned by other molecular players, such as other ligand-receptor complexes, adhesion molecules, guidance cues, growth factors, locally, or within short or medium range, to coordinate cell and overall tissue homeostasis.

In the context of CNS trauma such as SCI, PSA appears to control the state of damaged neurons submitted to an intrinsic internal potential (repertoire of transcription factors, second messengers, cytoskeleton), exposed to an interactive external environment (extracellular matrix, other cells, growth factors etc.), via an interface identified by the membrane (3D organization of mobile receptors). Adaptive plasticity can be viewed as a physical and mechanical response, at the molecular and cellular level, that can be potentiated by endogenous regulation of PSA or PSA-like agents at the interface between the external and internal parts of the system, making possible plastic changes at the tissue scale.

6. References

Amoureux MC, Coulibaly B, Chinot O, Loundou A, Metellus P, Rougon G, Figarella-Branger D. (2010) Polysialic acid neural cell adhesion molecule (PSA-NCAM) is an adverse prognosis factor in glioblastoma, and regulates olig2 expression in glioma cell lines. BMC Cancer. Mar 10;10:91.

Amoureux MC, Cunningham BA, Edelman GM & Crossin KL. (2000) N-CAM binding inhibits the proliferation of hippocampal progenitor cells and promotes their differentiation to a neuronal phenotype. *J Neurosci.* May 15;20(10):3631-40.

Bachelin C, Zujovic V, Buchet D, Mallet J & Baron-Van Evercooren A. (2010) Ectopic expression of polysialylated neural cell adhesion molecule in adult macaque Schwann cells promotes their migration and remyelination potential in the central nervous system. *Brain.* Feb;133(Pt 2):406-20.

Bareyre FM, Kerschensteiner M, Raineteau O, Mettenleiter TC, Weinmann O & Schwab ME. (2004) The injured spinal cord spontaneously forms a new intraspinal circuit in adult rats, *Nat Neurosci.*; 7, 269–77.

Basso DM, Beattie MS, Bresnahan JC. (1995) A sensitive and reliable locomotor rating scale for open field testing in rats. J Neurotrauma. Feb;12(1):1-21.

Battista D & Rutishauser U. (2010) Removal of polysialic acid triggers dispersion of subventricularly derived neuroblasts into surrounding CNS tissues. *J Neurosci.* Mar 17;30(11):3995-4003.

Bédard A, Gravel C & Parent A. (2006) Chemical characterization of newly generated neurons in the striatum of adult primates. *Exp Brain Res.* Apr;170(4):501-12.

Bernier PJ, Bedard A, Vinet J, Levesque M & Parent A. (2002) Newly generated neurons in the amygdala and adjoining cortex of adult primates. *Proc Natl Acad Sci U S A.* Aug 20;99(17):11464-9.

Bonfanti L & Theodosis DT. (2009) Polysialic acid and activity-dependent synapse remodeling. *Cell Adh Migr.* Jan-Mar;3(1):43-50.

Bonfanti L, Merighi A & Theodosis DT. (1996) Dorsal rhizotomy induces transient expression of the highly sialylated isoform of the neural cell adhesion molecule in neurons and astrocytes of the adult rat spinal cord. *Neuroscience.* Oct;74(3):619-23.

Bonfanti L, Olive S, Poulain DA & Theodosis DT. (1992) Mapping of the distribution of polysialylated neural cell adhesion molecule throughout the central nervous system of the adult rat: an immunohistochemical study. *Neuroscience.* Jul;49(2):419-36.

Camand E, Morel MP, Faissner A, Sotelo C & Dusart I. (2004) Long-term changes in the molecular composition of the glial scar and progressive increase of serotonergic fibre sprouting after hemisection of the mouse spinal cord, *Eur J Neurosci*; 20:1161–76.

Charles P, Hernandez MP, Stankoff B, Aigrot MS, Colin C, Rougon G, Zalc B & Lubetzki C. (2000) Negative regulation of central nervous system myelination by polysialylated-neural cell adhesion molecule. *Proc Natl Acad Sci U S A.* Jun 20;97(13):7585-90.

Charles P, Reynolds R, Seilhean D, Rougon G, Aigrot MS, Niezgoda A, Zalc B & Lubetzki C. (2002) Re-expression of PSA-NCAM by demyelinated axons: an inhibitor of remyelination in multiple sclerosis? *Brain*. Sep;125(Pt 9):1972-9.

Chazal G, Durbec P, Jankovski A, Rougon G & Cremer H. (2000) Consequences of neural cell adhesion molecule deficiency on cell migration in the rostral migratory stream of the mouse. *J Neurosci*. Feb 15;20(4):1446-57.

Coman I, Barbin G, Charles P, Zalc B & Lubetzki C. (2005) Axonal signals in central nervous system myelination, demyelination and remyelination. *J Neurol Sci*. Jun 15;233(1-2):67-71.

Conchonaud F, Nicolas S, Amoureux MC, Ménager C, Marguet D, Lenne PF, Rougon G & Matarazzo V. (2007) Polysialylation increases lateral diffusion of neural cell adhesion molecule in the cell membrane. *J Biol Chem*. Sep 7;282(36):26266-74.

Courtès S, Vernerey J, Pujadas L, Magalon K, Cremer H, Soriano E, Durbec P & Cayre M. (2011) Reelin controls progenitor cell migration in the healthy and pathological adult mouse brain. *PLoS One*. May 6(5):e20430:1-14.

Courtine G, Song B, Roy RR, Zhong H, Herrmann JE, Ao Y, Qi J, Edgerton VR & Sofroniew MV. (2008) Recovery of supraspinal control of stepping via indirect propriospinal relay connections after spinal cord injury.*Nat Med*. Jan;14(1):69-74.

Cremer H, Lange R, Christoph A, Plomann M, Vopper G, Roes J, Brown R, Baldwin S, Kraemer P, Scheff S, Barthels D, Rajewsky K & Wille W. (1994). Inactivation of the N-CAM gene in mice results in size reduction of the olfactory bulb and deficits in spatial learning. *Nature* 367,455 -9.

Curtis MA, Kam M, Nannmark U, Anderson MF, Axell MZ, Wikkelso C, Holtås S, van Roon-Mom WM, Björk-Eriksson T, Nordborg C, Frisén J, Dragunow M, Faull RL & Eriksson PS. (2007) Human neuroblasts migrate to the olfactory bulb via a lateral ventricular extension. *Science*. Mar 2;315(5816):1243-9.

Daniel L, Durbec P, Gautherot E, Rouvier E, Rougon G, Figarella-Branger D. (2001) A nude mice model of human rhabdomyosarcoma lung metastases for evaluating the role of polysialic acids in the metastatic process. *Oncogene*.;20:997–1004.

Daniel L, Trouillas J, Renaud W, Chevallier P, Gouvernet J, Rougon G & Figarella-Branger D. (2000) Polysialylated-neural cell adhesion molecule expression in rat pituitary transplantable tumors (spontaneous mammotropic transplantable tumor in Wistar-Furth rats) is related to growth rate and malignancy. *Cancer Res*.;60:80–5.

Daniloff JK, Levi G, Grumet M, Rieger F, Edelman GM. (1986) Altered expression of neuronal cell adhesion molecules induced by nerve injury and repair. *J Cell Biol*. Sep;103(3):929-45.

Daniloff JK, Stuart Shoemaker R, Lee AF, Strain GM, Remsen LG. (1995) N-CAM promotes recovery in injured nerves. *Restor Neurol Neurosci*. Jan 1;7(3):137-44.

Decker L, Avellana-Adalid V, Nait-Oumesmar B, Durbec P & Baron-Van Evercooren A. (2000) Oligodendrocyte precursor migration and differentiation: combined effects of PSA residues, growth factors, and substrates. *Mol Cell Neurosci*. 2000 Oct;16(4):422-39.

Di Cristo G, Chattopadhyaya B, Kuhlman SJ, Fu Y, Bélanger MC, Wu CZ, Rutishauser U, Maffei & Huang ZJ. (2007) Activity-dependent PSA expression regulates inhibitory maturation and onset of critical period plasticity. *Nat Neurosci.* Dec;10(12):1569-77.

Doherty P, Cohen J, Walsh FS. (1990) Neurite outgrowth in response to transfected N-CAM changes during development and is modulated by polysialic acid. Neuron. Aug;5(2):209-19.

Durbec P & Cremer H. (2001) Revisiting the function of PSA-NCAM in the nervous system. *Mol Neurobiol.* Aug-Dec;24(1-3):53-64.

Durbec P & Rougon G. (2001) Neurite outgrowth in response to transfected N-CAM changes during development and is modulated by polysialic acid. *Mol Cell Neurosci.* Mar;17(3):561-76.

Dusart I, Morel MP, Wehrle R & Sotelo C. (1999) Late axonal sprouting of injured Purkinje cells and its temporal correlation with permissive changes in the glial scar. *J Comp Neurol.*; 408: 399-418.

Duveau V & Fritschy JM. (2010) PSA-NCAM-dependent GDNF signaling limits neurodegeneration and epileptogenesis in temporal lobe epilepsy. *Eur J Neurosci.* Jul;32(1):89-98.

Edelman GM. (1986). Cell adhesion molecules in neural histogenesis. *Annu. Rev. Physiol.* 48,417 -430.

El Maarouf A & Rutishauser U. (2010) Use of PSA-NCAM in repair of the central nervous system. *Adv Exp Med Biol.*;663:137-47.

Farin A, Suzuki SO, Weiker M, Goldman JE, Bruce JN & Canoll P. (2006) Transplanted glioma cells migrate and proliferate on host brain vasculature: a dynamic analysis. *Glia.*; 53:799–808.

Figarella-Branger D, Dubois C, Chauvin P, De Victor B, Gentet JC & Rougon G. (1996) Correlation between polysialic-neural cell adhesion molecule levels in CSF and medulloblastoma outcomes. *J Clin Oncol.*;14:2066–72.

Figarella-Branger DF, Durbec PL & Rougon G. (1990) Differential spectrum of expression of neural cell adhesion molecule isoforms and L1 adhesion molecules on human neuroectodermal tumors. *Cancer Res.*;50:6364–70.

Florian C, Foltz J, Norreel JC, Rougon G & Roullet P. (2006) Post-training intrahippocampal injection of synthetic poly-alpha-2,8-sialic acid-neural cell adhesion molecule mimetic peptide improves spatial long-term performance in mice. *Learn Mem.* May-Jun;13(3):335-41.

Franceschini I, Desroziers E, Caraty A & Duittoz A. (2010) The intimate relationship of gonadotropin-releasing hormone neurons with the polysialylated neural cell adhesion molecule revisited across development and adult plasticity. *Eur J Neurosci.* Dec;32(12):2031-41.

Fujimoto I, Bruses JL, Rutishauser U. (2001) Regulation of cell adhesion by polysialic acid. Effects on cadherin, immunoglobulin cell adhesion molecule, and integrin function and independence from neural cell adhesion molecule binding or signaling activity. J Biol Chem. Aug 24;276(34):31745-51.

Galuska SP, Geyer R, Gerardy-Schahn R, Mühlenhoff M & Geyer H. (2008) Enzyme-dependent variations in the polysialylation of the neural cell adhesion molecule (NCAM) in vivo. *J Biol Chem.* Jan 4;283(1):17-28.

Galuska SP, Oltmann-Norden I, Geyer H, Weinhold B, Kuchelmeister K, Hildebrandt H, Gerardy-Schahn R, Geyer R & Mühlenhoff M. (2006) Polysialic acid profiles of mice expressing variant allelic combinations of the polysialyltransferases ST8SiaII and ST8SiaIV. *J Biol Chem*. Oct 20;281(42):31605-15.

Gascon E, Vutskits L, Jenny B, Durbec P & Kiss JZ. (2007) PSA-NCAM in postnatally generated immature neurons of the olfactory bulb: a crucial role in regulating p75 expression and cell survival. *Development*. Mar;134(6):1181-90.

Gheusi G, Cremer H, McLean H, Chazal G, Vincent JD & Lledo PM. (2000) Importance of newly generated neurons in the adult olfactory bulb for odor discrimination. *Proc Natl Acad Sci U S A*. Feb 15;97(4):1823-8.

Glüer S, Schelp C, Madry N, von Schweinitz D, Eckhardt M & Gerardy-Schahn R. (1998) Serum polysialylated neural cell adhesion molecule in childhood neuroblastoma. *Br J Cancer*.;78:106–10.

Gómez-Climent MÁ, Guirado R, Castillo-Gómez E, Varea E, Gutierrez-Mecinas M, Gilabert-Juan J, García-Mompó C, Vidueira S, Sanchez-Mataredona D, Hernández S, Blasco-Ibáñez JM, Crespo C, Rutishauser U, Schachner M & Nacher J. (2011) The polysialylated form of the neural cell adhesion molecule (PSA-NCAM) is expressed in a subpopulation of mature cortical interneurons characterized by reduced structural features and connectivity. *Cereb Cortex*. May;21(5):1028-41.

Hack I, Bancila M, Loulier K, Carroll P & Cremer H. (2002) Reelin is a detachment signal in tangential chain-migration during postnatal neurogenesis. *Nat Neurosci*. Oct;5(10):939-45.

Hildebrandt H, Mühlenhoff M, Oltmann-Norden I, Röckle I, Burkhardt H, Weinhold B & Gerardy-Schahn R. (2009) Imbalance of neural cell adhesion molecule and polysialyltransferase alleles causes defective brain connectivity. *Brain*. Oct;132(Pt 10):2831-8.

Hu H, Tomasiewicz H, Magnuson T & Rutishauser U. (1996) The role of polysialic acid in migration of olfactory bulb interneuron precursors in the subventricular zone. Neuron. Apr;16(4):735-43.

Johnson CP, Fujimoto I, Rutishauser U & Leckband DE. (2005) Direct evidence that neural cell adhesion molecule (NCAM) polysialylation increases intermembrane repulsion and abrogates adhesion. *J Biol Chem*. Jan 7;280(1):137-45.

Kam M, Curtis MA, McGlashan SR, Connor B, Nannmark U & Faull RL. (2009) The cellular composition and morphological organization of the rostral migratory stream in the adult human brain. *J Chem Neuroanat*. May;37(3):196-205.

Kitamura T, Saitoh Y, Takashima N, Murayama A, Niibori Y, Ageta H, Sekiguchi M, Sugiyama H & Inokuchi K. (2009) Adult neurogenesis modulates the hippocampus-dependent period of associative fear memory. *Cell*. Nov 13;139(4):814-27

Kochlamazashvili G, Senkov O, Grebenyuk S, Robinson C, Xiao MF, Stummeyer K, Gerardy-Schahn R, Engel AK, Feig L, Semyanov A, Suppiramaniam V, Schachner M & Dityatev A.(2010) *J Neurosci*. Mar 17;30(11):4171-83.

Koutsoudaki PN, Hildebrandt H, Gudi V, Skripuletz T, Škuljec J & Stangel M. (2010) Remyelination after cuprizone induced demyelination is accelerated in mice deficient in the polysialic acid synthesizing enzyme St8siaIV. *Neuroscience.* Nov 24;171(1):235-44. Epub 2010 Sep 15.

Lopez-Fernandez MA, Montaron MF, Varea E, Rougon G, Venero C, Abrous DN & Sandi C. (2007) Upregulation of polysialylated neural cell adhesion molecule in the dorsal hippocampus after contextual fear conditioning is involved in long-term memory formation. *J Neurosci.* Apr 25;27(17):4552-61.

Luzzati F, Bonfanti L, Fasolo A & Peretto P. (2009) DCX and PSA-NCAM expression identifies a population of neurons preferentially distributed in associative areas of different pallial derivatives and vertebrate species. *Cereb Cortex.* May;19(5):1028-41.

Luzzati F, De Marchis S, Fasolo A & Peretto P. (2006) Neurogenesis in the caudate nucleus of the adult rabbit. *J Neurosci.* Jan 11;26(2):609-21.

Mahal LK, Charter NW, Angata K, Fukuda M, Koshland DE Jr & Bertozzi CR. (2001) A small-molecule modulator of poly-alpha 2,8-sialic acid expression on cultured neurons and tumor cells. *Science.;* 294:380–1.

Marino P, Norreel JC, Schachner M, Rougon G &Amoureux MC. (2009) A polysialic acid mimetic peptide promotes functional recovery in a mouse model of spinal cord injury. *Exp Neurol.* Sep;219(1):163-74.

Marlatt MW, Philippens I, Manders E, Czéh B, Joels M, Krugers H & Lucassen PJ. (2011) Distinct structural plasticity in the hippocampus and amygdala of the middle-aged common marmoset (Callithrix jacchus). *Exp Neurol.* Aug;230(2):291-301.

Mehanna A, Jakovcevski I, Acar A, Xiao M, Loers G, Rougon G, Irintchev A & Schachner M. (2010) Polysialic acid glycomimetic promotes functional recovery and plasticity after spinal cord injury in mice. *Mol Ther.* Jan;18(1):34-43.

Mehanna A, Mishra B, Kurschat N, Schulze C, Bian S, Loers G, Irintchev A & Schachner M. (2009) Polysialic acid glycomimetics promote myelination and functional recovery after peripheral nerve injury in mice. *Brain.* Jun;132(Pt 6):1449-62.

Migaud M, Batailler M, Segura S, Duittoz A, Franceschini I & Pillon D. (2010) Emerging new sites for adult neurogenesis in the mammalian brain: a comparative study between the hypothalamus and the classical neurogenic zones. *Eur J Neurosci.* Dec;32(12):2042-52.

Monnier PP, Beck SG, Bolz J & Henke-Fahle S. (2001) The polysialic acid moiety of the neural cell adhesion molecule is involved in intraretinal guidance of retinal ganglion cell axons. *Dev Biol.* Jan 1;229(1):1-14.

Muller D, Djebbara-Hannas Z, Jourdain P, Vutskits L, Durbec P, Rougon G & Kiss JZ. (2000). Brain-derived neurotrophic factor restores long-term potentiation in polysialic acid-neural cell adhesion molecule-deficient hippocampus. *Proc. Natl. Acad. Sci. USA* 97,4315 -20.

Muller D, Mendez P, Deroo M, Klauser P, Steen S & Poglia L.(2010) Role of NCAM in spine dynamics and synaptogenesis. *Adv Exp Med Biol.;*663:245-56.

Nait Oumesmar B, Vignais L, Duhamel-Clérin E, Avellana-Adalid V, Rougon G & Baron-Van Evercooren A. (1995) Expression of the highly polysialylated neural cell adhesion molecule during postnatal myelination and following chemically induced demyelination of the adult mouse spinal cord. *Eur J Neurosci*. Mar 1;7(3):480-91.

Ní Dhúill CM, Fox GB, Pittock SJ, O'Connell AW, Murphy KJ & Regan CM. (1999) Polysialylated neural cell adhesion molecule expression in the dentate gyrus of the human hippocampal formation from infancy to old age. *J Neurosci Res*. Jan 1;55(1):99-106.

Papastefanaki F, Chen J, Lavdas AA, Thomaidou D, Schachner M & Matsas R. (2007) Grafts of Schwann cells engineered to express PSA-NCAM promote functional recovery after spinal cord injury. *Brain*. Aug;130(Pt 8):2159-74.

Parkash J & Kaur G. (2007) Potential of PSA-NCAM in neuron-glial plasticity in the adult hypothalamus: role of noradrenergic and GABAergic neurotransmitters. *Brain Res Bull*. Oct 19;74(5):317-28.

Pencea V, Bingaman KD, Freedman LJ & Luskin MB. (2001) Neurogenesis in the subventricular zone and rostral migratory stream of the neonatal and adult primate forebrain. *Exp Neurol*. Nov;172(1):1-16.

Perera AD, Lagenaur CF & Plant TM. (1993) Postnatal expression of polysialic acid-neural cell adhesion molecule in the hypothalamus of the male rhesus monkey (Macaca mulatta). *Endocrinology*. Dec;133(6):2729-35.

Petridis AK, El-Maarouf A & Rutishauser U. (2004). Polysialic acid regulates cell contact-dependent neuronal differentiation of progenitor cells from the subventricular zone. *Dev. Dyn*. 230,675 -84.

Pincus P. (1991) Colloid stabilization with grafted polyelectrolytes. *Macromolecules*. 24:2912-19

Röckle I, Seidenfaden R, Weinhold B, Mühlenhoff M, Gerardy-Schahn R & Hildebrandt H. (2008) Polysialic acid controls NCAM-induced differentiation of neuronal precursors into calretinin-positive olfactory bulb interneurons. *Dev Neurobiol*. Aug;68(9):1170-84.

Rosenzweig ES, Courtine G, Jindrich DL, Brock JH, Ferguson AR, Strand SC, Nout YS, Roy RR, Miller DM, Beattie MS, Havton LA, Bresnahan JC, Edgerton VR & Tuszynski MH. (2010) Extensive spontaneous plasticity of corticospinal projections after primate spinal cord injury. *Nat Neurosci*. Dec;13(12):1505-10.

Rutishauser U. (1996) Polysialic acid and the regulation of cell interactions. *Curr Opin Cell Biol*. Oct;8(5):679-84.

Sanai N, Tramontin AD, Quiñones-Hinojosa A, Barbaro NM, Gupta N, Kunwar S, Lawton MT, McDermott MW, Parsa AT, Manuel-García Verdugo J, Berger MS & Alvarez-Buylla A. (2004) Unique astrocyte ribbon in adult human brain contains neural stem cells but lacks chain migration. *Nature*. Feb 19;427(6976):740-4

Sawamoto K, Hirota Y, Alfaro-Cervello C, Soriano-Navarro M, He X, Hayakawa-Yano Y, Yamada M, Hikishima K, Tabata H, Iwanami A, Nakajima K, Toyama Y, Itoh T, Alvarez-Buylla A, Garcia-Verdugo JM & Okano H. (2011) Cellular composition and organization of the subventricular zone and rostral migratory stream in the adult and neonatal common marmoset brain. *J Comp Neurol*. Mar 1;519(4):690-713.

Seidenfaden R, Desoeuvre A, Bosio A, Virard I & Cremer H. (2006) Glial conversion of SVZ-derived committed neuronal precursors after ectopic grafting into the adult brain. *Mol Cell Neurosci.* May-Jun;32(1-2):187-98.

Shi Q, Maruthamuthu V, Li F & Leckband D. (2010) Allosteric cross talk between cadherin extracellular domains. *Biophys J.* Jul 7;99(1):95-104.

Silver J & Miller JH. (2004) Regeneration beyond the glial scar, *Nat Rev Neurosci*;5 :146–516.

Stiles CD & Rowitch DH. (2008) Glioma stem cells: a midterm exam. *Neuron*; 58:832–46.

Theodosis DT, Rougon G & Poulain DA. (1991) Retention of embryonic features by an adult neuronal system capable of plasticity: polysialylated neural cell adhesion molecule in the hypothalamo-neurohypophysial system. *Proc Natl Acad Sci U S A.* Jul 1;88(13):5494-8.

Tomasiewicz H, Ono K, Yee D, Thompson C, Goridis C, Rutishauser U & Magnuson T. (1993). Genetic deletion of a neural cell adhesion molecule variant (N-CAM-180) produces distinct defects in the central nervous system. *Neuron* 11,1163 -74.

Torregrossa P, Buhl L, Bancila M, Durbec P, Schafer C, Schachner M & Rougon G. (2004) Selection of poly-alpha 2,8-sialic acid mimotopes from a random phage peptide library and analysis of their bioactivity. *J Biol Chem.* Jul 16;279(29):30707-14.

Trouillas J, Daniel L, Guigard MP, Tong S, Gouvernet J, Jouanneau E, Jan M, Perrin G, Fischer G, Tabarin A, Rougon G & Figarella-Branger D. (2003) Polysialylated neural cell adhesion molecules expressed in human pituitary tumors and related to extrasellar invasion. *J Neurosurg.*;98:1084–93.

Venero C, Herrero AI, Touyarot K, Cambon K, López-Fernández MA, Berezin V, Bock E & Sandi C. (2006) Hippocampal up-regulation of NCAM expression and polysialylation plays a key role on spatial memory. *Eur J Neurosci.* Mar;23(6):1585-95.

Vutskits L, Gascon E & Kiss JZ. (2003) Removal of PSA from NCAM affects the survival of magnocellular vasopressin- and oxytocin-producing neurons in organotypic cultures of the paraventricular nucleus. *Eur J Neurosci.* May;17(10):2119-26.

Weinhold B, Seidenfaden R, Röckle I, Mühlenhoff M, Schertzinger F, Conzelmann S, Marth JD, Gerardy-Schahn R & Hildebrandt H. (2005) Genetic ablation of polysialic acid causes severe neurodevelopmental defects rescued by deletion of the neural cell adhesion molecule. *J Biol Chem.* Dec 30;280(52):42971-7.

Yang P, Major D & Rutishauser U. (1994) Role of charge and hydration in effects of polysialic acid on molecular interactions on and between cell membranes. *J Biol Chem.* Sep 16;269(37):23039-44.

Yang P, Yin X & Rutishauser U. (1992) Intercellular space is affected by the polysialic acid content of NCAM. *J Cell Biol.* Mar;116(6):1487-96.

Zhang H, Miller RH, Rutishauser U. (1992) Polysialic acid is required for optimal growth of axons on a neuronal substrate. J Neurosci. Aug;12(8):3107-14.

Zhang H, Vutskits L, Calaora V, Durbec P & Kiss JZ. (2004) A role for the polysialic acid-neural cell adhesion molecule in PDGF-induced chemotaxis of oligodendrocyte precursor cells. *J Cell Sci.* 2004 Jan 1;117(Pt 1):93-103.

Zhang Y, Zhang X, Yeh J, Richardson P & Bo X. (2007) Engineered expression of polysialic acid enhances Purkinje cell axonal regeneration in L1/GAP-43 double transgenic mice. *Eur J Neurosci.*; 25: 351-61.

Zhao C, Deng W & Gage FH. (2008) Mechanisms and functional implications of adult neurogenesis. *Cell.* Feb 22;132(4):645-60.

17

What Does Maurocalcine Tell Us About the Process of Excitation-Contraction Coupling?

Michel Ronjat and Michel De Waard
Joseph Fourier University & INSERM U836, Grenoble Institute of Neuroscience France

1. Introduction

Cardiac and skeletal muscle contraction is triggered by the arrival of an action potential that locally and transiently depolarizes the plasma membrane. The entire chain of molecular events that link the arrival of the action potential and muscle contraction is called excitation-contraction coupling. This chapter will focus on the way membrane depolarization is sensed and how it triggers a massive increase in the cytosolic Ca^{2+} concentration. Plasma membrane depolarization is detected by L-type voltage-gated calcium channels whose activation allows Ca^{2+} entry into muscle cells. Cardiac and skeletal muscle voltage-gated calcium channels differ by their Ca^{2+} permeability. While cardiac channels lead to an important and rapid Ca^{2+} influx, skeletal muscle channel activation allows a moderate and rather slow Ca^{2+} entry (Bean, 1989). In fact, the molecular identity of the pore-forming channel subunit differs in both tissues, cardiac fibers expressing the $Ca_v1.2$ isoform while skeletal muscles express the $Ca_v1.1$ isoform. Nevertheless, the overall subunit composition of both channels, illustrated in Figure 1, bears interesting similarities.

The opening of these ion channels in response to plasma membrane depolarization results from a complex series of conformational changes occurring mainly in the pore-forming subunits. In these subunits, positively charged S4 segments act as voltage-sensing mechanical transducers by altering their position within the plasma membrane as a function of voltage value (Nakai, Adams, Imoto, & Beam, 1994). How movements of other structural elements are related to these S4 repositioning is still an open but particularly important question with regard to the excitation-contraction coupling process as we will see later. In both tissues, the activation of L-type voltage-gated calcium channels produces the opening of another calcium channel located in the membrane of the sarcoplasmic reticulum, the main intracellular Ca^{2+} store in muscles. This process leads to a transient and important increase in cytoplasmic Ca^{2+} concentration that is at the basis of muscle contraction. This particular ion channel presents several interesting features: i) it is sensitive to the plant alkaloid ryanodine (and is therefore named "ryanodine receptor"), ii) it is positioned in such a way that it faces the L-type channel in order to maximize activation efficiency, and iii) in both tissues, its activation and inactivation are controlled by cytoplasmic Ca^{2+} and Mg^{2+} concentrations (Figure 2).

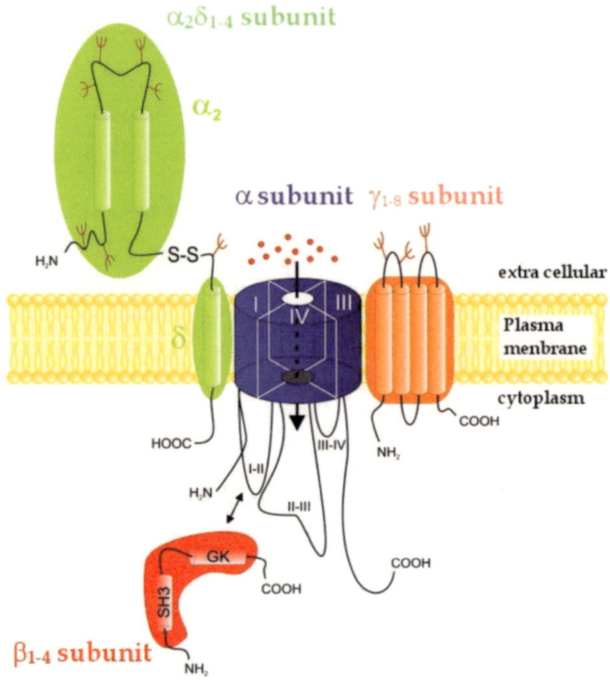

Fig. 1. Schematic representation of an L-type voltage-dependent calcium channel. $Ca_v\alpha$ is the voltage-sensing and pore-forming subunit, whereas $\alpha_2\delta$, β and γ are auxiliary subunits. $Ca_v\alpha$ is constituted of four homologous hydrophobic domains linked by cytoplasmic loops. The II-III loop, that links domains II and III, is the focus of this review. The β subunits is the only protein that is fully cytoplasmic.

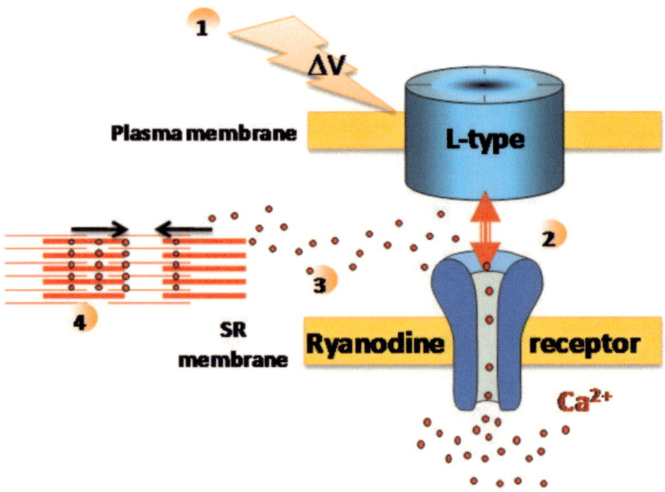

Fig. 2. Cartoon describing the general process of excitation-contraction coupling, starting by membrane depolarization (1), L-type channel activation and transfer of conformational change to the ryanodine receptor (2), ryanodine receptor opening and sarcoplasmic reticulum Ca^{2+} release (3), and muscle contraction (4).

As for the L-type calcium channels, two different types of ryanodine receptors are respectively expressed in skeletal (type 1 ryanodine receptor) and cardiac (type 2 ryanodine receptors) muscles. These two types of ryanodine receptors differ in their sensitivity to cytoplasmic Ca^{2+} concentration. Type 1 ryanodine receptor appears to be more sensitive to inhibition by high cytosolic Ca^{2+} concentration than type 2 ryanodine receptors. In contrast, there does not seem to be major differences with regard to activation by cytosolic Ca^{2+} (Chu, Fill, Stefani, & Entman, 1993; Fruen, Bardy, Byrem, Strasburg, & Louis, 2000; Michalak, Dupraz, & Shoshan-Barmatz, 1988). *In vivo,* cardiac type 2 ryanodine receptor appears to be less repressed by L-type calcium channel than type 1 ryanodine receptors as witnessed by the far greater likelihood to observe spontaneous channel openings (spark events) (H. Cheng, Lederer, & Cannell, 1993; Wier & Balke, 1999). The differences in biophysical properties between skeletal and cardiac L-type calcium channels on one hand, and skeletal and cardiac ryanodine receptors on the other hand, led to differences in excitation-contraction coupling mechanisms between these two tissue. In cardiac tissues, Ca^{2+} entering through activated L-type channels is sufficient to directly activate type 2 ryanodine receptors. This chain of events is called Ca^{2+}-induced Ca^{2+}-release. In this case, opening of the type 2 ryanodine receptor strictly depends on extracellular Ca^{2+} (Nabauer, Callewaert, Cleemann, & Morad, 1989). Indeed, the Ca^{2+} release process through type 2 ryanodine receptors is a U-shape function of voltage and follows the U-shape voltage-dependence of external Ca^{2+} entry. Moreover, blocking external Ca^{2+} entry precludes voltage-activated Ca^{2+} release through type 2 receptors. Type 1 ryanodine receptors in skeletal muscles differ from type 2 ryanodine receptors in their mode of activation. In spite of the capability of type 1 ryanodine receptors to perform Ca^{2+}-induced Ca^{2+}-release, the amplitude of Ca^{2+} entry through activated skeletal L-type channels is too low and its kinetic too slow to trigger a cardiac type Ca^{2+}-induced Ca^{2+}-release mechanism. Indeed, in these cells, the Ca^{2+} release process is no longer U-shaped in response to voltage increase and strictly follows the sigmoidal voltage-dependence of activation of L-type channels. Ca^{2+} permeability through skeletal L-type channels plays therefore a negligible role in the skeletal type excitation-contraction coupling. In skeletal muscles, the voltage-dependent trigger of type 1 ryanodine receptor activation is the changes in L-type channel conformation. This process has been termed voltage-induced Ca^{2+} release (Schneider & Chandler, 1973; Tanabe, Beam, Powell, & Numa, 1988). While it may seem comfortable for intellectual construction to oppose Ca^{2+}-induced Ca^{2+} release to voltage-induced Ca^{2+} release, one should keep in mind that such a drastic distinction between these two mechanisms also presents intrinsic restrictions. In skeletal muscles, the conformation transmission between L-type channel and type 1 ryanodine receptor requires a physical interaction between these channel types. While conceptually, the process of Ca^{2+}-induced Ca^{2+} release in cardiac tissues does not require a physical interaction between the two channels, it would be a mistake to eliminate the hypothesis that conformational changes in cardiac L-type channels can also be transmitted directly to type 2 ryanodine receptor. Conversely, while skeletal L-type conformational changes are essential to trigger Ca^{2+} release through type 1 ryanodine receptors in skeletal muscles, ruling out a contribution of Ca^{2+} (originating from internal sources) in the release of Ca^{2+} itself may also represent a shortcut attitude. In this review we will therefore not oppose the two modes of Ca^{2+}-release but simply take into account that in cardiac tissues external Ca^{2+} is an important factor, while in skeletal muscle conformational changes in L-type channels is a more important factor. We will assume that cardiac L-type channels might also be in interaction with the ryanodine type 2 receptor. The next paragraph will describe what is known in terms of molecular interactions between L-type channels and ryanodine receptors, focusing mainly on the most important determinants for excitation-contraction coupling.

2. Molecular determinants of the interaction between L-type channels and ryanodine receptors

2.1 Contribution of two separate domains of the II-III loop of $Ca_v1.1$ in voltage-induced Ca^{2+} release

Because voltage-induced Ca^{2+} release clearly relies on a direct interaction between skeletal muscle L-type and type 1 ryanodine receptor, most of the researches have focused on the interaction between these two channels. In this context, the use of animal models has been decisive in identifying molecular determinants critical for the excitation-contraction coupling process. Three mice models have been widely used: i) the dysgenic mice *mdg* corresponding to a natural knockout of the pore-forming subunit of skeletal L-type channels (Tanabe et al., 1988), ii) the dyspedic mice, a knock-out of type 1 ryanodine receptor (Takeshima et al., 1995), and iii) the β-null mice, a knock-out of the β auxiliary subunit of the skeletal L-type channel. All these gene knock-outs produce lethality at birth by defective respiratory muscle function implying that studies are done with primary myotubes isolated from late embryos.

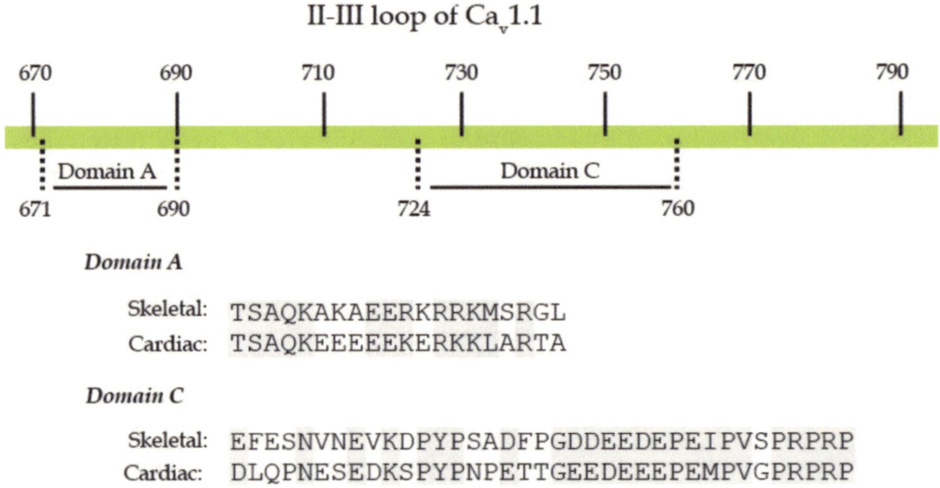

Fig. 3. Amino acid positions of skeletal muscle $Ca_v1.1$ *Domains A* and *C*. Amino acid sequences of these domains are compared to those of equivalent domains of the cardiac L-type $Ca_v1.2$ channel.

In a first series of precept experiments, *mdg* myotubes were used to express chimeras between the pore-forming $Ca_v\alpha$ subunits of skeletal and cardiac L-type channels in order to identify a skeletal muscle-specific L-type determinant responsible for voltage-induced Ca^{2+} release. This approach led to the identification of one particularly important determinant, the large cytoplasmic loop between the second and third hydrophobic domains (called II-III loop) of $Ca_v1.1$ (Nakai, Tanabe, Konno, Adams, & Beam, 1998) (see

Figure 3). According to structural predictions, this loop would be of 137 ($Ca_v1.1$) to 146 ($Ca_v1.2$) amino acids long. Further, restriction of the skeletal L-type domain responsible for voltage-induced Ca^{2+} release through type 1 ryanodine receptors locates a sequence of 36 amino acids, 62 amino acids downstream of the start of I-II loop sequence. This domain is called *Domain C*. While this region is undoubtedly important in $Ca_v1.1$, the sequence homology with the equivalent *Domain C* of $Ca_v1.2$ remains nevertheless quite high. The replacement of Ala^{739}, Phe^{741}, Pro^{742} and Asp^{744} of $Ca_v1.1$ by, respectively, Pro, Thr, Thr and Glu of $Ca_v1.2$ reduces voltage-induced Ca^{2+} release (Kugler, Grabner, Platzer, Striessnig, & Flucher, 2004; Kugler, Weiss, Flucher, & Grabner, 2004).

While these studies clearly demonstrated that *Domain C* is critical for allowing voltage-induced Ca^{2+}-release by L-type channels, the exact mechanism whereby this domain contributes to the process has still not been resolved. There is no solid evidence for a direct interaction between *Domain C* and the ryanodine receptor, suggesting that its mode of action is indirect or that this interaction relies on a transient conformation of the $Ca_v\alpha$ subunit controlled by voltage. Following the identification of the II-III loop as an important determinant in voltage-induced Ca^{2+} release, synthetic peptides corresponding to fragments of this loop were screened for their ability to interact with purified ryanodine receptors and modify their gating activity. Curiously, this approach led to the identification of a ryanodine receptor-interacting domain of the II-III loop that is different of *Domain C*. This sequence has been termed *Domain A* and corresponds to the amino acid region Thr^{671} to Leu^{690} of $Ca_v1.1$ (Figure 2). One particular feature of this domain is the presence of a stretch of basic residues (8 amino acids out of 20 are either Arg or Lys residues). *Domain A* of $Ca_v1.2$ bears sequence homology (10 residues are identical out of 20 and 3 have homology). In spite of this high sequence homology, cardiac *domain A* is considerably enriched in acidic residues suggesting that its function may have evolved differently than its skeletal muscle counterpart. Proof of functional differences between cardiac and skeletal *domain A* comes from the fact that skeletal *domain A* interacts with type 1 ryanodine receptor but not type 2 (O'Reilly et al., 2002). Conversely, cardiac *domain A* does not interact with type 1 ryanodine receptor (el-Hayek, Antoniu, Wang, Hamilton, & Ikemoto, 1995; O'Reilly et al., 2002). Curiously, there are no reports on the potential interaction between cardiac *domain A* and type 2 ryanodine receptor thereby illustrating the little credit the scientific community is giving to the possibility that cardiac L-type and type 2 ryanodine receptors physically interact. While direct interaction of skeletal *domain A* with type 1 ryanodine receptor, and evidence for functional effects (reviewed in paragraph 3), would ensure that this domain plays a critical role in excitation-contraction coupling, this possibility is hotly debated. In 2000, a study reported that the partial scrambling of skeletal *domain A* had little impact on excitation-contraction coupling (Proenza, Wilkens, & Beam, 2000). Careful examination of the scrambled sequence reveals that scrambling occurs on the ten last residues of *domain A* and that many of the residues remain in fact at the same position. Moreover a characteristic feature of *domain A*, a rich content in basic amino acid, is conserved in the scrambled sequence. At the functional level, while excitation-contraction coupling is preserved in principle, no quantification of the extent of this conservation is performed. While these data are interesting, it would have been preferable to perform a more extensive scrambling of the sequence (using the entire domain and avoiding conservation of position for some residues). Other evidences question the role of *domain A* in excitation-contraction coupling. Replacement of the skeletal $Ca_v1.1$ II-III loop by the II-III loop of *Musca domestica* $Ca_v\alpha$

channel, which has a poor homology with $Ca_v1.1$, largely prevents excitation-contraction coupling, but not completely, while reintegration of skeletal type *domain C* in this construct is enough to restore voltage-induced Ca^{2+}-release (Wilkens, Kasielke, Flucher, Beam, & Grabner, 2001). It should be mentioned that the chimera construct in which the entire II-III loop is replaced displays very little Ca^{2+} permeability, possibly questioning the correct functioning of the chimera channel. The study lacks also a chimera construct with the integration of skeletal muscle *domain A* rather than *domain C*. Also, expression of two skeletal $Ca_v1.1$ hemi-channels, one lacking *domain A*, reconstitutes voltage-induced Ca^{2+} release in *mdg* myotubes (Ahern, Arikkath et al., 2001). Similarly, complete deletion of *domain A* in $Ca_v1.1$ does not affect the amplitude of voltage-induced Ca^{2+} release, although it seems to speed the process (Ahern, Bhattacharya, Mortenson, & Coronado, 2001). Very curiously, in the later study, the same authors demonstrate that deletion of both *domains A* and *C* results in the reappearance of some voltage-induced Ca^{2+} release. While these studies indicate that the presence of *domain A* is not strictly necessary to observe voltage-induced Ca^{2+} release in skeletal muscles, it is hard to conclude that it has no function at all. Also, while *domain C* appears to be important to observe the excitation-contraction coupling process, the data tend to indicate that there might be some interdependence between II-III loop domains to exert their function. Indeed, it was observed *in vitro* that *domain C* exerts an inhibitory effect on the activating function of *domain A* on Ca^{2+} release from type 1 ryanodine receptor or on [^{3}H]-ryanodine binding (el-Hayek, Antoniu et al., 1995; Ikemoto & el-Hayek, 1998).

2.2 The skeletal muscle L-type channel contains other important determinants for excitation-contraction coupling

While the II-III loop of $Ca_v1.1$ has focused much of the attention in the comprehension of the excitation-contraction coupling process, it is important to mention that other relevant L-type channel determinants have been proposed to play a role in the process. Two of these determinants are present on the $Ca_v1.1$ pore-forming subunit (the III-IV loop and the C-terminal domain), and the second is constituted by the auxiliary β_{1a} subunit. Their implication is summarized hereunder.

Suspicion of the involvement of the III-IV loop of $Ca_v1.1$ arose as a consequence of the identification of a point mutation ($Arg^{1086}His$) associated to malignant hyperthermia, a skeletal muscle pathology in which inhalation of volatile anesthetics provokes enhanced Ca^{2+} release from the sarcoplasmic reticulum, excessive ATP hydrolysis, heat production and therefore muscle damage (Monnier, Procaccio, Stieglitz, & Lunardi, 1997). A functional analysis of this mutation highlights a putative role of this loop in controlling the voltage-dependence of the Ca^{2+} release thru type 1 ryanodine receptor and its caffeine sensitivity (Weiss et al., 2004). Caffeine is a well characterized exogenous activator of all types of ryanodine receptors favors Ca^{2+}-induced Ca^{2+} release by sensitizing the ryanodine receptors to activating cytosolic Ca^{2+}. A potential interaction site of this loop on type 1 ryanodine receptor has been identified that encompasses amino acids 922 to 1112 (Figure 4) (Leong & MacLennan, 1998b).

It is noteworthy that this III-IV loop is extremely conserved among $Ca_v\alpha$ subunits (with 46 amino acids identical on 53 between $Ca_v1.1$ and $Ca_v1.2$). This suggests that the cardiac III-IV loop may also interact with type 2 ryanodine receptor.

The C-terminus of $Ca_v1.1$ has also been shown to be an interesting determinant. The amino acid region 1543 to 1647 of $Ca_v1.1$ plays a role in the transport and targeting of $Ca_v1.1$ to the triad (Flucher, Kasielke, & Grabner, 2000; Proenza, Wilkens, Lorenzon, & Beam, 2000). The 1393 to 1527 amino acid region of $Ca_v1.1$ binds Ca^{2+} and calmodulin. In the absence of calmodulin, this domain has been shown to bind type 1 ryanodine receptor at amino acid position 3609 to 3643. A synthetic peptide corresponding to the 1487 to 1506 region of $Ca_v1.1$ inhibits both ryanodine binding and ryanodine receptor channel gating in bilayers (Sencer et al., 2001; Slavik et al., 1997). The authors of these findings suggest that these domains could participate to Ca^{2+} and/or calmodulin regulation of voltage-induced Ca^{2+} release, but more studies are warranted to confirm their importance.

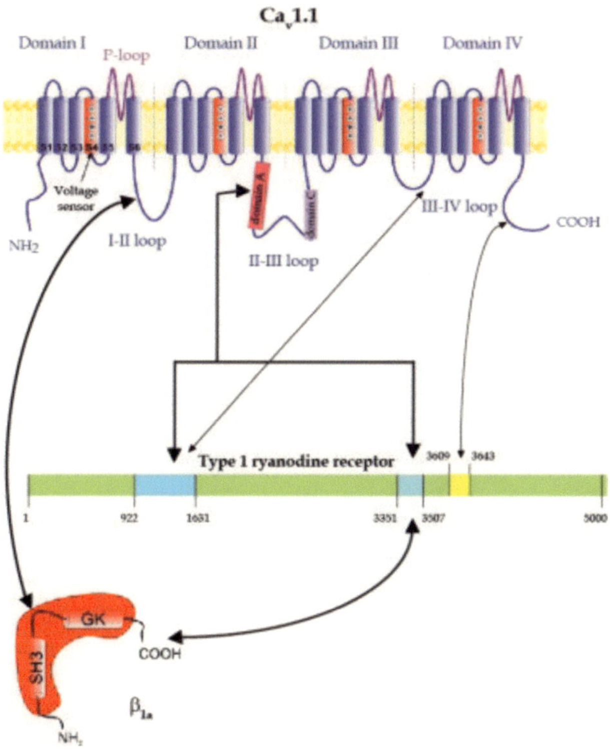

Fig. 4. Summary of known protein interactions between skeletal muscle L-type $Ca_v1.1$ and β subunits with type 1 ryanodine receptor. The amino acid positions of *Domains A* and *C* are defined in the text.

As mentioned above, absence of the β_{1a} subunit of the L-type calcium channel led to a complete loss of excitation-contraction coupling in skeletal muscle. Recently, this subunit has been shown to directly interact with the type 1 ryanodine receptor (W. Cheng, Altafaj, Ronjat, & Coronado, 2005). Binding experiments identified a cluster of positively charged residues of type 1 ryanodine receptor that control this interaction. Neutralization or deletion of these amino acids severely depress the amplitude of the depolarization induced Ca^{2+} release through type 1 ryanodine receptor, indicating that this interaction is important in the regulation of the excitation-contraction coupling in skeletal muscles.

3. Is *domain A* of relevance to excitation-contraction coupling?

Although the functional relevance of *domain A* is questioned by studies that have used chimera $Ca_v\alpha$ channels, we still believe that it is of interest to investigate the precise role of *domain A* in excitation-contraction coupling. Indeed, it is important to note the intrinsic limitations of several reports that rely on the use of chimera channels and *mdg* myotubes. For a chimera channel to integrate into the triad (the plasma membrane loci where skeletal muscle excitation-contraction takes place), it needs to preserve sufficient molecular determinants that guarantee normal cell trafficking and correct localization with regard to type 1 ryanodine receptor. The role of these determinants on excitation-contraction coupling, and specifically on voltage-induced Ca^{2+}-release, cannot be investigated. Also, these studies and conclusions are all based on the assumption (that is not yet proven) that Ca^{2+}-induced Ca^{2+}-release by cardiac L-type channels and type 2 ryanodine receptors can occur in the absence of conformational changes in $Ca_v1.2$ (invariably triggered by voltage changes). Testing this hypothesis would require to block voltage-dependent $Ca_v1.2$ charge movements without blocking Ca^{2+} permeability. While this is a provocative opinion, one could conclude that the chimera studies have demonstrated that skeletal muscle *domain C* governs the requirement on external Ca^{2+} to produce Ca^{2+} release: when skeletal *domain C* is present, external Ca^{2+} would not be required, while when absent, external Ca^{2+} is mandatory. *Domain C* is undoubtedly important in participating to voltage-induced Ca^{2+} release. However, how is *domain C* voltage-dependent and how does it transmit voltage-dependent conformational changes to type 1 ryanodine receptor without directly interacting with this channel?

In voltage-induced Ca^{2+}-release, which occurs in skeletal muscles, the initial trigger for Ca^{2+} release through type 1 ryanodine receptor is brought by the voltage change. However, voltage might be considered as the initial trigger for initiating the Ca^{2+} release process, and the released Ca^{2+}, once it has reached a critical concentration, may contribute to the process by promoting Ca^{2+}-induced Ca^{2+} release. This can be expected to occur as type 1 ryanodine receptor is perfectly capable to support Ca^{2+}-induced Ca^{2+} release instead of voltage-induced Ca^{2+} release such as when $Ca_v1.2$ replaces $Ca_v1.1$. While there is no doubt that type 1 ryanodine receptor can support both voltage- and Ca^{2+}-induced Ca^{2+} release, very little is known on the contribution of the different $Ca_v1.1$ *domains* (in particular *A* and *C*) to the balance between these two triggering processes. Among other arguments in favor of a contribution of *domain A* to excitation-contraction coupling is the fact that *domain A* is heavily charged. Obviously, voltage-induced Ca^{2+} release requires both the presence of $Ca_v1.1$ *domain C* in the context of an operational channel and a change in voltage, but it is expected that one type 1 ryanodine receptor-interacting $Ca_v1.1$ determinant is altering its conformation upon depolarization to influence excitation-contraction coupling. The basic nature of *domain A* is of interest in that respect since (i) it binds type 1 ryanodine receptor, (ii) influences the Ca^{2+} sensitivity of type 1 ryanodine receptor, and (iii) is susceptible to sense voltage changes owing to its close proximity with the plasma membrane. This latter point has never been investigated of course, but needless to say, *domain C* is not the most convincing candidate as voltage-sensor owing to its location within the channel and the lower net charge of the sequence. Earlier evidences that *domains A* and *C* may interact are also interesting since it could be envisioned that voltage-changes alter the interaction between these two domains thereby dynamically regulating excitation-contraction coupling. Finally, most of the studies on excitation-contraction coupling and on voltage-induced Ca^{2+} release in particular have focused their attention on the release process itself but not on the termination of the signal upon membrane repolarisation. In that respect, it would be interesting to have a careful examination of the role of *domain A* on the ending of voltage-induced Ca^{2+} release.

4. Animal toxins presenting an intriguing sequence homology with *domain A*

In the course of ongoing studies on ryanodine receptor channel pharmacology, the group of Coronado identified imperatoxin A as an effector of type 1 ryanodine receptor (Valdivia, Kirby, Lederer, & Coronado, 1992). The toxin has high affinity for the channel (close to 10 nM), produces an increase in channel opening probability by reducing the closure times, induces Ca^{2+} release from purified sarcoplasmic reticulum and increases [^{3}H]-ryanodine binding by conversion of the ryanodine binding site from a low to a high affinity state (el-Hayek, Lokuta, Arevalo, & Valdivia, 1995). Imperatoxin A was isolated from the venom of the African scorpion *Pandinus imperator*. It is a 33 amino acid peptide containing six cysteine residues and therefore three internal disulfide bridges. The pattern of connectivity results in a fold of the "inhibitor cysteine knot" type with three β-strands (Green et al., 2003). The peptide was first synthesized in 2007 (Zamudio et al., 1997). Since this initial discovery, other toxins have been identified, coming from the venom of other scorpions, that all share high sequence homology with imperatoxin A (Figure 5). These include maurocalcine in 2000 from *Scorpio maurus palmatus* (Mosbah et al., 2000), immediately synthesized in 2000 (Fajloun et al., 2000), two opicalcine variants in 2003 (S. Zhu, Darbon, Dyason, Verdonck, & Tytgat, 2003), hemicalcin in 2007 (Shahbazzadeh et al., 2007), and hadrucalcin in 2009 (Schwartz et al., 2009). All of these peptides have been tested on type 1 ryanodine receptor and shown to possess similar pharmacological properties as imperatoxin A. These toxins therefore define a family of functional homologous peptides.

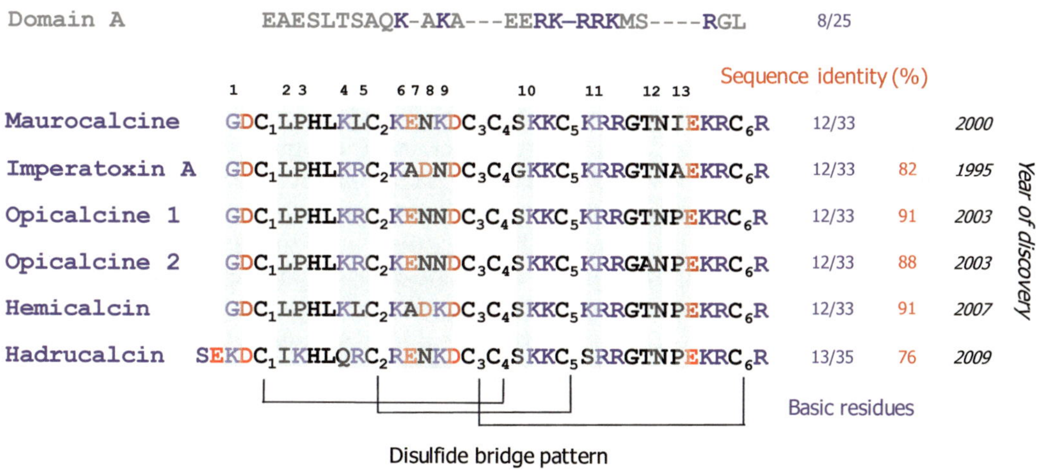

Fig. 5. Amino acid sequences and alignment of maurocalcine-like toxins acting on ryanodine receptors. Sequence homology with skeletal muscle *domain A* is also shown. Positively charged residues are shown in blue, sequence identities are given in red and the year of discovery of each toxin is also provided.

A close comparison of the peptide sequences reveals an important conservation. All peptides are heavily charged, containing mostly Lys residues and some Arg residues, and the net charge of these peptides is high (+7 if one excludes the potential basic charge of the His residue that is present in all toxins). Most of these basic residues are located on one face of the molecule, creating an important dipole moment in these toxins (Boisseau et al., 2006). A closer examination of these sequences also reveals that a limited stretch of the primary structure of the toxins is highly homologous to the primary structure of skeletal muscle *domain A* (Figure 4). Moreover, modification of *domain A*, meant to further improve the structural similarity between *domain A* and imperatoxin A and maurocalcine, led to a mutant *domain A* with increased functional effects on type 1 ryanodine receptors (Green et al., 2003). These effects include stimulation of Ca^{2+} release from sarcoplasmic reticulum vesicles and increased channel opening probability of type 1 ryanodine receptor incorporated in lipid bilayers. Of note, following the action of this optimized *domain A* peptide, these toxins have no cumulative effect on channel activation, strongly suggesting that *domain A* and the toxins share the same binding site on the ryanodine receptors. These experiments highlight in any case the functional relevance of this limited structural homology between *domain A* and the toxins. Further evidence for the importance of this cluster of residues comes from an elegant study of the group of Valdivia. The cardiac *domain A*, which is poorly activator of type 1 ryanodine receptor, can be modified to become a strong type 1 ryanodine receptor activator by selected mutations within this cluster of residues (X. Zhu, Gurrola, Jiang, Walker, & Valdivia, 1999). The mutations were chosen in such a way that cardiac *domain A* integrates the cluster of basic residues of *domain A* that is homologous to the toxin sequences.

5. *In vitro* functional similarities between maurocalcine and *domain A*

Among the various toxin activators of ryanodine receptors, maurocalcine is one of the best characterized. Besides sequence similarities with *domain A*, and the emerging evidence that both skeletal muscle *domain A* and maurocalcine share the binding sites on type 1 ryanodine receptor, there is also cumulative indication for functional similarities. We rapidly summarize these similarities hereunder. First, both *domain A* and maurocalcine enhance channel activity by promoting an increase in channel opening probability and the appearance of a subconductance state (Chen et al., 2003; Fajloun et al., 2000; Lukacs et al., 2008; O'Reilly et al., 2002). Both peptides differ however with regard to the level of subconductance state and the duration of channel opening in this subconductance state. Toxins promote long-lasting openings in the subconductance state that are barely observed with *domain A*. Interestingly, the various toxins also differ among each other with regard to the level of subconductance states (Chen et al., 2003; el-Hayek, Lokuta et al., 1995; Tripathy, Resch, Xu, Valdivia, & Meissner, 1998), suggesting that sequence divergences may control this subconductance level. Indeed, we observed that single point mutations in maurocalcine modify this subconductance level (Lukacs et al., 2008). The absence of long-lasting events with *domain A* might also reflect its significantly lower affinity for type 1 ryanodine receptor compared to the toxin. Second, in agreement with their potentiating effects on channel activity, both maurocalcine and *domain A* induce Ca^{2+} release from purified sarcoplasmic reticulum vesicles (Dulhunty et al., 1999; el-Hayek, Antoniu et al., 1995; Esteve et al., 2003). These stimulating effects of *domain A* on Ca^{2+} release are observed at high concentrations (above 10 μM in general) and are not consistently reported. In one study, application of peptide A was reported to decrease the caffeine- and Ca^{2+}-induced Ca^{2+} release (Chen et al., 2003), indicating in any case a relationship between the binding site of *domain A* and the

Ca^{2+}-sensitivity of the type 1 ryanodine receptors. In the same study, high concentrations of peptide A actually block the effect of maurocalcine, again suggesting that both peptides bind onto the same site on the type 1 ryanodine receptor. The variable response of type 1 ryanodine receptor to *domain A*, as can be concluded from the various reports, also suggest that peptide conformation or cofactors largely influence the response of type 1 ryanodine receptor to *domain A*. Third, both *domain A* and maurocalcine increase [^{3}H]-ryanodine binding (Chen et al., 2003; el-Hayek, Antoniu et al., 1995; Gurrola et al., 1999; Lu, Xu, & Meissner, 1994; X. Zhu et al., 1999). This stimulation is expected since conditions that trigger ryanodine channel opening are reported to produce a conversion in the binding site for ryanodine from a low affinity state to a higher one. The amplitude of the increase in [^{3}H]-ryanodine binding, triggered by *domain A*, maurocalcine or other toxins, appears quite variable. This stimulation depends on other factor, such as the concentrations of cytosolic Ca^{2+} (Esteve et al., 2003; Tripathy et al., 1998) and Mg^{2+} (unpublished observations). This dependence on cytosolic Ca^{2+} concentration deserves some important comments that may be of interest for the understanding of excitation-contraction coupling. Cytosolic Ca^{2+} concentration has a biphasic effect of [^{3}H]-ryanodine binding and channel gating, being an activator at low concentrations (above 100 nM) and inhibitor at higher concentrations (between 10 µM and 1 mM). Interestingly, maurocalcine and imperatoxin A were reported to alter this Ca^{2+}-dependence by sensitizing type 1 ryanodine receptors to both the stimulation and inhibitory effects of Ca^{2+} (Esteve et al., 2003; Tripathy et al., 1998). For instance, while 100 nM Ca^{2+} does not trigger Ca^{2+} release from type 1 ryanodine receptor, this same concentration will produce Ca^{2+}-induced Ca^{2+}-release in the presence of maurocalcine. This is an important observation because it means that binding of maurocalcine on type 1 ryanodine receptor makes it prone to perform Ca^{2+}-induced Ca^{2+} release. The analogy between maurocalcine and peptide A questions the possibility that *domain A* controls the ability of type 1 ryanodine receptors to perform Ca^{2+}-induced Ca^{2+} release. The possibility that this control is revealed during membrane depolarization opens interesting questions.

An important finding has been the identification of peptide A and toxin binding site on type 1 ryanodine receptors. A first report describes the location of biotinylated imperatoxin A binding sites on tetramers of type 1 ryanodine receptors (Samso, Trujillo, Gurrola, Valdivia, & Wagenknecht, 1999). This study illustrates that four imperatoxin A can bind onto a cytoplasmic domain of the tetramer of type 1 ryanodine receptors between the clam and handle domains, 11 nm away from the transmembrane pore. The far distance of binding of imperatoxin A indicates that its effect on channel gating occurs through an allosteric mechanism. In a 3D reconstruction of type 1 ryanodine receptor, the imperatoxin A binding site is in close physical proximity to the Ca^{2+}-calmodulin binding site (Wagenknecht & Samso, 2002). More precise location of the binding sites of the toxins has been investigated using recombinant fusion protein of fragments of the type 1 ryanodine receptor. Using biotinylated maurocalcine and peptide A, Altafaj and coll. identified two type 1 ryanodine receptor sequences that bind either *domain A* or maurocalcine (Altafaj et al., 2005). Weak interaction was observed with a fragment encompassing amino acid region 1021 to 1631 that contains the 37 amino acid sequence that was previously shown to bind to the skeletal muscle $Ca_v1.1$ II-III loop (Leong & MacLennan, 1998a). Stronger interaction was shown to occur with amino acid region 3351 to 3507. These findings are in agreement with the predicted localization of imperatoxin A binding site on the 3D structure of type 1 ryanodine receptor. They are also in agreement with the suggestion that these two regions are in close proximity in space, thereby suggesting that they constitute a single binding site. In any case, these data demonstrate for the first time that maurocalcine and *domain A* share the same binding site. Of importance, it has been observed that deletion of the 1272-1455 amino acid region within type 1 ryanodine receptor provokes the loss of depolarization induced Ca^{2+} release (Perez, Mukherjee, & Allen, 2003). Combined these observations reinforce the idea that *domain A* and its binding site are somehow involved in the control of voltage-induced Ca^{2+} release. While maurocalcine has limited functional effects on type 2 ryanodine receptors, it is interesting to note that it binds to this receptor through an homologous binding site (regions 1033 to 1622 and 3558 to 3609 of type 2 ryanodine receptors) (Altafaj et al., 2007). These results suggest that binding of the toxin can be dissociated from its functional effect. This possibility has lately been reinforced by unpublished observations showing that the effect of maurocalcine depends on the redox state of type 1 ryanodine receptor. Differences in redox states between type 1 and 2 ryanodine receptors may thus explain differences in maurocalcine effects on these two channels. Of note, the redox state of type 2 ryanodine receptor has been shown to control its activity under cardiomyocytes stretch conditions (Prosser, Ward, & Lederer, 2011).

6. Use of maurocalcine to understand excitation-contraction coupling in skeletal muscle cells

During the course of maurocalcine characterization, we have shown that external application of maurocalcine on cultured skeletal muscle myotubes produces a small, but detectable, transient cytoplasmic Ca^{2+} elevation (Esteve et al., 2003). This effect is also observed in the absence of external Ca^{2+} demonstrating the mobilization of intracellular stores. Finally, maurocalcine interferes with the action of 4-chloro-*m*-cresol indicating the involvement of type 1 ryanodine receptors. This was the first demonstration of a pharmacological effect of a maurocalcine-type toxin on intact muscle cells. Since then, similar observations have been made with imperatoxin A and hadrucalcin on cardiomycytes (Gurrola, Capes, Zamudio, Possani, & Valdivia, 2010; Schwartz et al., 2009). These observations have prompted us to investigate the cell penetration properties of maurocalcine. A series of studies have demonstrated since 2005 that maurocalcine belongs to the cell penetrating peptide family with vector properties that can be derived for various diagnostic, imaging and therapeutic applications (Aroui et al., 2009; Boisseau et al., 2006; Esteve et al., 2005; Jayagopal et al., 2009; Mabrouk et al., 2007; Poillot et al., 2010). A more detailed investigation of the mode of action of maurocalcine in developing skeletal muscle cells and adult fibers reveals that the peptide acts preferentially on type 1 ryanodine receptors that are uncoupled from the L-type calcium channels (Szappanos et al., 2005). Indeed, maurocalcine increases the frequency of spontaneous Ca^{2+} release events followed by the appearance of ember-like long lasting Ca^{2+} release events in permeabilized adult muscle fibers. Muscle permeabilization is expected to favor the uncoupling between type 1 ryanodine receptors and L-type calcium channels. These observations tend to indicate that maurocalcine has difficulties to act on type 1 ryanodine receptors when *domain A* of L-type channels already occupies the binding site on type 1 ryanodine receptor. Besides this effect of maurocalcine on uncoupled type 1 ryanodine receptors, the effect of the peptide was also investigated on L-type coupled type 1 ryanodine receptors by using high concentrations of the peptide to gain accessibility to its binding site on the ryanodine receptor. Maurocalcine was injected into adult skeletal muscle fibers and Ca^{2+} transients induced by membrane depolarization. In these experiments, incubation with maurocalcine does not induce *per se* a change in resting cytoplasmic Ca^{2+} concentrations or amplitude of the depolarization-induced Ca^{2+} release. Only minor effects of maurocalcine were observed on the onset kinetics and voltage-dependence of voltage-induced Ca^{2+} release. Although these effects are small, they tend to indicate that maurocalcine is able to occupy its binding site during depolarization. Interestingly, much stronger effects of maurocalcine were observed on the kinetics of the termination of Ca^{2+} release upon ending membrane depolarization (Pouvreau et al., 2006). These effects correlate with the duration and amplitude of membrane depolarization and confirm that depolarization progressively permits maurocalcine binding onto L-type coupled ryanodine receptors. Taken together, these experiments would suggest that *domain A* acts as a negative clamp on type 1 ryanodine receptor for Ca^{2+} release at resting membrane potential. In this configuration, its affinity for *domain A* is too high or the accessibility too restricted to allow maurocalcine binding to type 1 ryanodine receptor. Depolarization produces a change in conformation of *domain A* that reduces its affinity for the ryanodine receptor, thereby relieving the negative clamp for Ca^{2+} release. In this situation, maurocalcine gains access to the ryanodine receptor by competing with *domain A* for its binding site. Since maurocalcine does not trigger a gain of function with regard to Ca^{2+} release under membrane depolarization, this observation suggest that i) maurocalcine and *domain A* are functionally equivalent under depolarization, and ii) since maurocalcine is an activator of Ca^{2+} release, it is likely that *domain A* acquires activating functions under membrane depolarization. Upon membrane repolarization, *domain A* regains access to its binding site on the ryanodine receptor by increased affinity and competitive displacement of maurocalcine. The slow kinetics of maurocalcine displacement by *domain A* explains the slower termination of Ca^{2+} release. This sequence of events is depicted in Figure 6.

In the scheme of events presented in Figure 6, *domain A* acts as a functionally versatile ligand: a negative clamp under resting membrane potential and an activator under membrane depolarization. By analogy with maurocalcine action, we hypothesize that the activating function of *domain A* stems from an increase in sensitivity of type 1 ryanodine receptor to cytoplasmic Ca^{2+}. Two types of changes in Ca^{2+} sensitivity have been observed with maurocalcine: i) a shift in Ca^{2+} concentration required to activate Ca^{2+}-induced Ca^{2+}-release towards lowers concentrations (from a threshold of 100 to 10 nm Ca^{2+}), and ii) an enhanced amplitude of Ca^{2+}-induced Ca^{2+} release. With such a mechanism in hand, the initial trigger of Ca^{2+} elevation that sparks Ca^{2+}-induced Ca^{2+} release under these activating functions of *domain A* can be *domain C*. The Ca^{2+} sensitizing function of *domain A* under membrane depolarization highlights a new concept for excitation-contraction in skeletal muscles. Rather than opposing voltage-induced Ca^{2+} release and Ca^{2+}-induced Ca^{2+} release, the two phenomena might be closely linked. Voltage-induced Ca^{2+} release could be termed voltage-induced Ca^{2+}-induced Ca^{2+} release. The Ca^{2+} responsible for the induction of Ca^{2+}-induced Ca^{2+} release would originate from the initial opening of type 1 ryanodine receptor. In skeletal muscles, the presence of a negative clamp might be required to avoid spontaneous Ca^{2+} release events that otherwise would spark on Ca^{2+} release and contraction. In this scheme, the absence of *domain A* mainly alters the Ca^{2+} sensitivity of type 1 ryanodine receptors and does not prevent any activating function of *domain C*. When *domain A* is modified, change in Ca^{2+} dependence should be observed. However, this is extremely difficult to address in classical experimental conditions where cytosolic Ca^{2+} concentration is not under control.

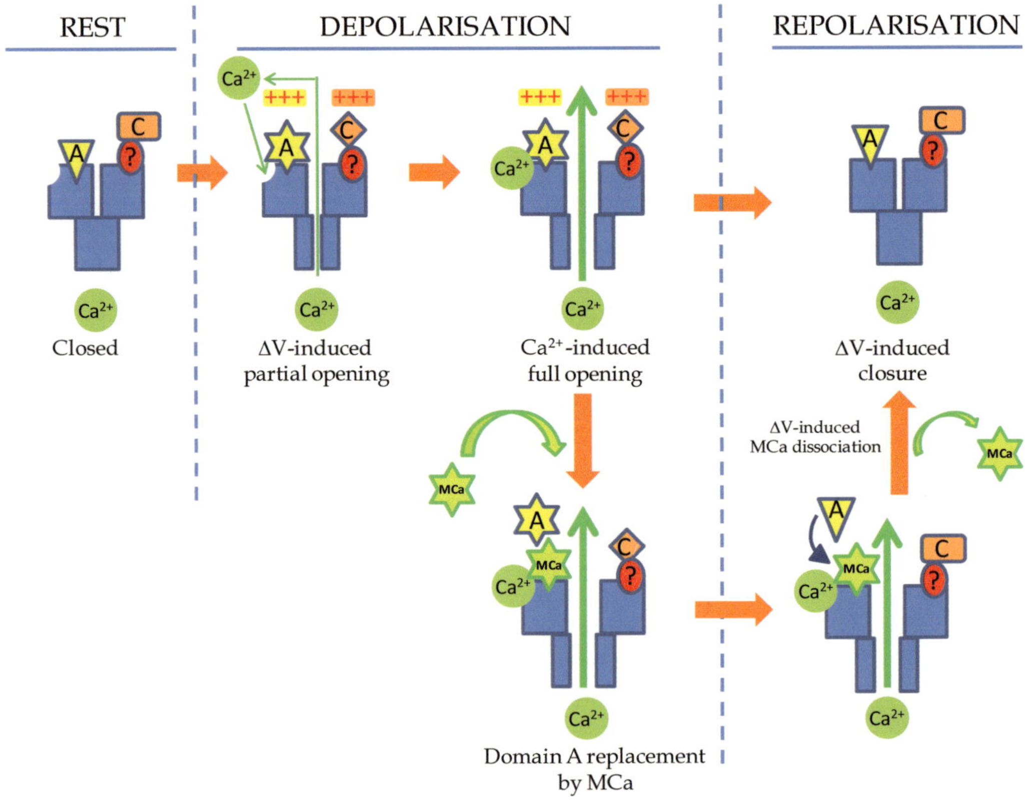

Fig. 6. Putative model of type 1 ryanodine receptor activation in skeletal muscles and role of *domain A*. A: *domain A*; C: *domain C*; ?: unknown putative *domain C* protein partner or receptor binding site. Membrane depolarization triggers *domain C*-mediated ryanodine receptor opening, while activating the Ca^{2+}-sensitizing function of *domain A*. Ca^{2+}-induced Ca^{2+}-release becomes the major sources of cytosolic Ca^{2+} elevation. *Domain A* interaction with the ryanodine receptor is weakened and maurocalcine can replace *domain A* with equivalent function. Upon membrane repolarization, the activating function of *domain C* is lost, *domain A* regains its affinity and the Ca^{2+} sensitivity is lost. If maurocalcine was occupying *domain A* binding site on the receptor, it is slowly displaced by the inhibitory conformation of *domain A* explaining the continuing Ca^{2+} release process in the interval, assuming that the activating function of maurocalcine is voltage-independent or at least less voltage-sensitive.

7. Conclusion

One common feature between *domain A* and maurocalcine is that both are positively charged. Interestingly, maurocalcine has been shown to possess a strong affinity for negatively charged membrane lipids (Boisseau et al., 2006; Ram et al., 2008), namely phosphatidylserine and phosphoinositides, but also gangliosides. Also, cell penetration of maurocalcine presents a voltage-sensitive component (Boisseau et al., 2006). These properties introduce three interesting questions. First, does *domain A* interact with the plasma membrane, and in particular with negatively charged lipids? Second, owing to its close proximity with the plasma membrane and its charged nature, does *domain A* sense changes in membrane potential? Third, if yes, does *domain A* possess some of the plasma membrane translocation properties of maurocalcine by physically invading the plasma membrane during depolarization? While these questions remain unanswered, they may be of interest because these considerations could impact the possible voltage-dependent relationship existing between *domain A* and type 1 ryanodine receptor. Owing to its privileged interaction with type 1 ryanodine receptor through a discrete binding site, *domain A* may also confer voltage-sensitivity to the ryanodine receptor. Maurocalcine, as a tool, clearly demonstrated that the domain of type 1 ryanodine receptor on which maurocalcine or skeletal *domain A* binds plays an important role in the control of channel opening through pore-distant allosteric modifications. Therefore, it is conceivable that depolarization-induced movement of *domain A* within this site could impact the allosteric change that control type 1 ryanodine channel opening. Of note, while skeletal muscle *domain A* has a net positive charge of +6, cardiac *domain A* has a net charge of 0 making it far less suitable to sense changes in membrane potential. By extension, these considerations of voltage-dependence could apply to other domains of the II-III loop of voltage-gated calcium channels. One striking feature of *domain C* is that, contrary to *domain A*, it is negatively charged (12 residues out of 36 are negatively charged in skeletal and cardiac *domain C* – net negative charge -9). On the basis of these observations, one can therefore question the existence of a voltage-sensitivity in the conformation of *domain C*. Also, taking over an idea that has been postulated in the past, it wouldn't be unlikely, considering the electrostatic complementarities of *domains A* and *C*, that both domains interact with each other and that this interaction is under the control of voltage changes. This questions the specific relationship that might exist between *domain A* and *domain C* for the control of voltage-induced Ca^{2+} release.

Most studies that have been performed up to now in order to understand the roles of channel cytoplasmic domains in excitation-contraction coupling have suffered from a lack of structural knowledge. The identification of a restricted binding site of maurocalcine and *domain A* on ryanodine receptors opens the door to a better definition of the critical interacting amino acid residues through RMN studies. These structural studies are currently underway and indicate that some specific type 1 ryanodine receptor amino acids are involved in the interaction with both *domain A* and maurocalcine (unpublished results). At last, the identification of the essential amino acids will allow refined functional studies based on more precise $Ca_v1.1$ or ryanodine receptor channel mutagenesis. In the process of comparing the functional homologies between *domain A* and maurocalcine, it would be of interest to substitute *domain A* by maurocalcine sequence in $Ca_v1.1$.

The importance of maurocalcine as a tool to investigate excitation-contraction coupling is demonstrated by the fact that, for the first time, this peptide allows a perturbation of the excitation-contraction coupling process in native cells without requiring the modification of any of the proteins involved, by mutagenesis or chimeras. Swapping domains between various ion channel isoforms is indeed a particularly tricky approach considering that there must be a delicate voltage-dependent conformational relationship between numerous important structural determinants. The use of maurocalcine has highlighted two important matters in the study of excitation-contraction coupling. First, it illustrates that the nature of the relationship between *domain A* and type 1 ryanodine receptor changes during membrane depolarization. Second, it uncovers the unexpected importance of *domain A* in the termination of Ca^{2+} release during membrane repolarization. The precise description of this voltage-induced type 1 ryanodine closure is lacking so far. Nevertheless, this result is the first evidence for the role of $Ca_v1.1$ *domain A* in this silencing process. The importance of this step in the control of Ca^{2+} homeostasis and contraction would justify that more studies be devoted to this issue.

The properties of maurocalcine to cross the plasma membrane suggest that it would be an interesting tool to modify excitation-contraction coupling in skeletal muscle *in vivo*. The fact that maurocalcine shares the same binding site than *domain A* on type 1 ryanodine receptor restricts its accessibility to L-type coupled ryanodine receptors. Obviously, it would thus be very interesting to investigate its effect on ryanodine receptors under conditions where its accessibility to the channel is not restricted, i.e. in the absence of $Ca_v1.1$. Effects of maurocalcine on *mdg* myotubes, lacking $Ca_v1.1$, are currently under investigation in our research team. These studies will provide important information on the effects of maurocalcine, and by inference on *domain A*, in a cellular context independent of the voltage-dependent channel environment and therefore also of the contribution of *domain C*.

The evidence for a voltage-modulated contribution of *domain A* to excitation-contraction coupling prompted us to emphasize the limits existing in the opposition between voltage-induced and Ca^{2+}-induced Ca^{2+} release mechanisms. Investigations focused on Ca^{2+}-induced Ca^{2+} release tend to take into account only the cell entry of Ca^{2+} from the extracellular space as trigger for Ca^{2+} release. The possible contribution of cytoplasmic or sarcoplasmic Ca^{2+} calcium in this process is not taken into account because of technical limitations. Indeed, it is quite difficult to maintain cytoplasmic Ca^{2+} at a certain concentration while at the same time measuring Ca^{2+} release from sarcoplasmic reticulum. Conversely, the possibility that type 2 ryanodine receptor Ca^{2+} release undergoes voltage-dependence might be masked by the voltage-dependence of external Ca^{2+} entry. If one takes into account these limitations, it is possible to draw a number of working hypotheses that do not necessarily oppose cardiac and skeletal muscle excitation-contraction coupling. Indeed, in light of the maurocalcine effects, we propose that the voltage-dependent modification of type 1 ryanodine receptor by *domain A* induces an increase of its sensitivity to cytoplasmic Ca^{2+}. Therefore, we propose to introduce the concept of voltage-induced-Ca^{2+} induced-Ca^{2+} release by which depolarization concomitantly promotes a change in Ca^{2+} sensitivity, through *domain A* of $Ca_v1.1$, and the opening, through *domain C*, of type 1 ryanodine receptor. In this model, the difference between cardiac and skeletal muscle excitation contraction coupling mechanism only resides on the absence of this change of Ca^{2+} sensitivity of the type 2 ryanodine receptor due to the absence of interaction of cardiac *domain A* with type 2 ryanodine receptor.

8. References

Ahern, C. A., Arikkath, J., Vallejo, P., Gurnett, C. A., Powers, P. A., Campbell, K. P., et al. (2001). Intramembrane charge movements and excitation- contraction coupling expressed by two-domain fragments of the Ca^{2+} channel. *Proc Natl Acad Sci U S A, 98*(12), 6935-6940.

Ahern, C. A., Bhattacharya, D., Mortenson, L., & Coronado, R. (2001). A component of excitation-contraction coupling triggered in the absence of the T671-L690 and L720-

Q765 regions of the II-III loop of the dihydropyridine receptor alpha(1s) pore subunit. *Biophys J, 81*(6), 3294-3307.

Altafaj, X., Cheng, W., Esteve, E., Urbani, J., Grunwald, D., Sabatier, J. M., et al. (2005). Maurocalcine and *domain A* of the II-III loop of the dihydropyridine receptor Cav 1.1 subunit share common binding sites on the skeletal ryanodine receptor. *J Biol Chem, 280*(6), 4013-4016.

Altafaj, X., France, J., Almassy, J., Jona, I., Rossi, D., Sorrentino, V., et al. (2007). Maurocalcine interacts with the cardiac ryanodine receptor without inducing channel modification. *Biochem J, 406*(2), 309-315.

Aroui, S., Ram, N., Appaix, F., Ronjat, M., Kenani, A., Pirollet, F., et al. (2009). Maurocalcine as a Non Toxic Drug Carrier Overcomes Doxorubicin Resistance in the Cancer Cell Line MDA-MB 231. *Pharm Res, 26*(4), 836-845.

Bean, B. P. (1989). Classes of calcium channels in vertebrate cells. *Annu Rev Physiol, 51*, 367-384.

Boisseau, S., Mabrouk, K., Ram, N., Garmy, N., Collin, V., Tadmouri, A., et al. (2006). Cell penetration properties of maurocalcine, a natural venom peptide active on the intracellular ryanodine receptor. *Biochim Biophys Acta, 1758*(3), 308-319.

Chen, L., Esteve, E., Sabatier, J. M., Ronjat, M., De Waard, M., Allen, P. D., et al. (2003). Maurocalcine and peptide A stabilize distinct subconductance states of ryanodine receptor type 1, revealing a proportional gating mechanism. *J Biol Chem, 278*(18), 16095-16106.

Cheng, H., Lederer, W. J., & Cannell, M. B. (1993). Calcium sparks: elementary events underlying excitation-contraction coupling in heart muscle. *Science, 262*(5134), 740-744.

Cheng, W., Altafaj, X., Ronjat, M., & Coronado, R. (2005). Interaction between the dihydropyridine receptor Ca^{2+} channel beta-subunit and ryanodine receptor type 1 strengthens excitation-contraction coupling. *Proc Natl Acad Sci U S A, 102*(52), 19225-19230.

Chu, A., Fill, M., Stefani, E., & Entman, M. L. (1993). Cytoplasmic Ca^{2+} does not inhibit the cardiac muscle sarcoplasmic reticulum ryanodine receptor Ca^{2+} channel, although Ca(2+)-induced Ca^{2+} inactivation of Ca^{2+} release is observed in native vesicles. *J Membr Biol, 135*(1), 49-59.

Dulhunty, A. F., Laver, D. R., Gallant, E. M., Casarotto, M. G., Pace, S. M., & Curtis, S. (1999). Activation and inhibition of skeletal RyR channels by a part of the skeletal DHPR II-III loop: effects of DHPR Ser687 and FKBP12. *Biophys J, 77*(1), 189-203.

el-Hayek, R., Antoniu, B., Wang, J., Hamilton, S. L., & Ikemoto, N. (1995). Identification of calcium release-triggering and blocking regions of the II-III loop of the skeletal muscle dihydropyridine receptor. *J Biol Chem, 270*(38), 22116-22118.

el-Hayek, R., Lokuta, A. J., Arevalo, C., & Valdivia, H. H. (1995). Peptide probe of ryanodine receptor function. Imperatoxin A, a peptide from the venom of the scorpion Pandinus imperator, selectively activates skeletal-type ryanodine receptor isoforms. *J Biol Chem, 270*(48), 28696-28704.

Esteve, E., Mabrouk, K., Dupuis, A., Smida-Rezgui, S., Altafaj, X., Grunwald, D., et al. (2005). Transduction of the scorpion toxin maurocalcine into cells. Evidence that the toxin crosses the plasma membrane. *J Biol Chem, 280*(13), 12833-12839.

Esteve, E., Smida-Rezgui, S., Sarkozi, S., Szegedi, C., Regaya, I., Chen, L., et al. (2003). Critical amino acid residues determine the binding affinity and the Ca^{2+} release efficacy of maurocalcine in skeletal muscle cells. *J Biol Chem, 278*(39), 37822-37831.

Fajloun, Z., Kharrat, R., Chen, L., Lecomte, C., Di Luccio, E., Bichet, D., et al. (2000). Chemical synthesis and characterization of maurocalcine, a scorpion toxin that activates Ca(2+) release channel/ryanodine receptors. *FEBS Lett, 469*(2-3), 179-185.

Flucher, B. E., Kasielke, N., & Grabner, M. (2000). The triad targeting signal of the skeletal muscle calcium channel is localized in the COOH terminus of the alpha(1S) subunit. *J Cell Biol, 151*(2), 467-478.

Fruen, B. R., Bardy, J. M., Byrem, T. M., Strasburg, G. M., & Louis, C. F. (2000). Differential Ca(2+) sensitivity of skeletal and cardiac muscle ryanodine receptors in the presence of calmodulin. *Am J Physiol Cell Physiol, 279*(3), C724-733.

Green, D., Pace, S., Curtis, S. M., Sakowska, M., Lamb, G. D., Dulhunty, A. F., et al. (2003). The three-dimensional structural surface of two beta-sheet scorpion toxins mimics that of an alpha-helical dihydropyridine receptor segment. *Biochem J, 370*(Pt 2), 517-527.

Gurrola, G. B., Arevalo, C., Sreekumar, R., Lokuta, A. J., Walker, J. W., & Valdivia, H. H. (1999). Activation of ryanodine receptors by imperatoxin A and a peptide segment of the II-III loop of the dihydropyridine receptor. *J Biol Chem, 274*(12), 7879-7886.

Gurrola, G. B., Capes, E. M., Zamudio, F. Z., Possani, L. D., & Valdivia, H. H. (2010). Imperatoxin A, a Cell-Penetrating Peptide from Scorpion Venom, as a Probe of Ca-Release Channels/Ryanodine Receptors. *Pharmaceuticals (Basel), 3*(4), 1093-1107.

Ikemoto, N., & el-Hayek, R. (1998). Signal transmission and transduction in excitation-contraction coupling. *Adv Exp Med Biol, 453*, 199-207.

Jayagopal, A., Su, Y. R., Blakemore, J. L., Linton, M. F., Fazio, S., & Haselton, F. R. (2009). Quantum dot mediated imaging of atherosclerosis. *Nanotechnology, 20*(16), 165102.

Kugler, G., Grabner, M., Platzer, J., Striessnig, J., & Flucher, B. E. (2004). The monoclonal antibody mAB 1A binds to the excitation--contraction coupling domain in the II-III loop of the skeletal muscle calcium channel alpha(1S) subunit. *Arch Biochem Biophys, 427*(1), 91-100.

Kugler, G., Weiss, R. G., Flucher, B. E., & Grabner, M. (2004). Structural requirements of the dihydropyridine receptor alpha1S II-III loop for skeletal-type excitation-contraction coupling. *J Biol Chem, 279*(6), 4721-4728.

Leong, P., & MacLennan, D. H. (1998a). A 37-amino acid sequence in the skeletal muscle ryanodine receptor interacts with the cytoplasmic loop between domains II and III in the skeletal muscle dihydropyridine receptor. *J Biol Chem, 273*(14), 7791-7794.

Leong, P., & MacLennan, D. H. (1998b). The cytoplasmic loops between domains II and III and domains III and IV in the skeletal muscle dihydropyridine receptor bind to a contiguous site in the skeletal muscle ryanodine receptor. *J Biol Chem, 273*(45), 29958-29964.

Lu, X., Xu, L., & Meissner, G. (1994). Activation of the skeletal muscle calcium release channel by a cytoplasmic loop of the dihydropyridine receptor. *J Biol Chem, 269*(9), 6511-6516.

Lukacs, B., Sztretye, M., Almassy, J., Sarkozi, S., Dienes, B., Mabrouk, K., et al. (2008). Charged surface area of maurocalcine determines its interaction with the skeletal ryanodine receptor. *Biophys J, 95*(7), 3497-3509.

Mabrouk, K., Ram, N., Boisseau, S., Strappazzon, F., Rehaim, A., Sadoul, R., et al. (2007). Critical amino acid residues of maurocalcine involved in pharmacology, lipid interaction and cell penetration. *Biochim Biophys Acta, 1768*(10), 2528-2540.

Michalak, M., Dupraz, P., & Shoshan-Barmatz, V. (1988). Ryanodine binding to sarcoplasmic reticulum membrane; comparison between cardiac and skeletal muscle. *Biochim Biophys Acta, 939*(3), 587-594.

Monnier, N., Procaccio, V., Stieglitz, P., & Lunardi, J. (1997). Malignant-hyperthermia susceptibility is associated with a mutation of the alpha 1-subunit of the human dihydropyridine-sensitive L-type voltage-dependent calcium-channel receptor in skeletal muscle. *Am J Hum Genet, 60*(6), 1316-1325.

Mosbah, A., Kharrat, R., Fajloun, Z., Renisio, J. G., Blanc, E., Sabatier, J. M., et al. (2000). A new fold in the scorpion toxin family, associated with an activity on a ryanodine-sensitive calcium channel. *Proteins, 40*(3), 436-442.

Nabauer, M., Callewaert, G., Cleemann, L., & Morad, M. (1989). Regulation of calcium release is gated by calcium current, not gating charge, in cardiac myocytes. *Science, 244*(4906), 800-803.

Nakai, J., Adams, B. A., Imoto, K., & Beam, K. G. (1994). Critical roles of the S3 segment and S3-S4 linker of repeat I in activation of L-type calcium channels. *Proc Natl Acad Sci U S A, 91*(3), 1014-1018.

Nakai, J., Tanabe, T., Konno, T., Adams, B., & Beam, K. G. (1998). Localization in the II-III loop of the dihydropyridine receptor of a sequence critical for excitation-contraction coupling. *J Biol Chem, 273*(39), 24983-24986.

O'Reilly, F. M., Robert, M., Jona, I., Szegedi, C., Albrieux, M., Geib, S., et al. (2002). FKBP12 modulation of the binding of the skeletal ryanodine receptor onto the II-III loop of the dihydropyridine receptor. *Biophys J, 82*(1 Pt 1), 145-155.

Perez, C. F., Mukherjee, S., & Allen, P. D. (2003). Amino acids 1-1,680 of ryanodine receptor type 1 hold critical determinants of skeletal type for excitation-contraction coupling. Role of divergence domain D2. *J Biol Chem, 278*(41), 39644-39652.

Poillot, C., Dridi, K., Bichraoui, H., Pecher, J., Alphonse, S., Douzi, B., et al. (2010). D-Maurocalcine, a pharmacologically inert efficient cell-penetrating peptide analogue. *J Biol Chem, 285*(44), 34168-34180.

Pouvreau, S., Csernoch, L., Allard, B., Sabatier, J. M., De Waard, M., Ronjat, M., et al. (2006). Transient loss of voltage control of Ca^{2+} release in the presence of maurocalcine in skeletal muscle. *Biophys J, 91*(6), 2206-2215.

Proenza, C., Wilkens, C., Lorenzon, N. M., & Beam, K. G. (2000). A carboxyl-terminal region important for the expression and targeting of the skeletal muscle dihydropyridine receptor. *J Biol Chem, 275*(30), 23169-23174.

Proenza, C., Wilkens, C. M., & Beam, K. G. (2000). Excitation-contraction coupling is not affected by scrambled sequence in residues 681-690 of the dihydropyridine receptor II-III loop. *J Biol Chem, 275*(39), 29935-29937.

Prosser, B. L., Ward, C. W., & Lederer, W. J. (2011). X-ROS signaling: rapid mechano-chemo transduction in heart. *Science, 333*(6048), 1440-1445.

Ram, N., Aroui, S., Jaumain, E., Bichraoui, H., Mabrouk, K., Ronjat, M., et al. (2008). Direct Peptide Interaction with Surface Glycosaminoglycans Contributes to the Cell Penetration of Maurocalcine. *J Biol Chem, 283*(35), 24274-24284.

Samso, M., Trujillo, R., Gurrola, G. B., Valdivia, H. H., & Wagenknecht, T. (1999). Three-dimensional location of the imperatoxin A binding site on the ryanodine receptor. *J Cell Biol, 146*(2), 493-499.

Schneider, M. F., & Chandler, W. K. (1973). Voltage dependent charge movement of skeletal muscle: a possible step in excitation-contraction coupling. *Nature, 242*(5395), 244-246.

Schwartz, E. F., Capes, E. M., Diego-Garcia, E., Zamudio, F. Z., Fuentes, O., Possani, L. D., et al. (2009). Characterization of hadrucalcin, a peptide from Hadrurus gertschi scorpion venom with pharmacological activity on ryanodine receptors. *Br J Pharmacol, 157*(3), 392-403.

Sencer, S., Papineni, R. V., Halling, D. B., Pate, P., Krol, J., Zhang, J. Z., et al. (2001). Coupling of RYR1 and L-type calcium channels via calmodulin binding domains. *J Biol Chem, 276*(41), 38237-38241.

Shahbazzadeh, D., Srairi-Abid, N., Feng, W., Ram, N., Borchani, L., Ronjat, M., et al. (2007). Hemicalcin, a new toxin from the Iranian scorpion Hemiscorpius lepturus which is active on ryanodine-sensitive Ca^{2+} channels. *Biochem J, 404*(1), 89-96.

Slavik, K. J., Wang, J. P., Aghdasi, B., Zhang, J. Z., Mandel, F., Malouf, N., et al. (1997). A carboxy-terminal peptide of the alpha 1-subunit of the dihydropyridine receptor inhibits Ca(2+)-release channels. *Am J Physiol, 272*(5 Pt 1), C1475-1481.

Szappanos, H., Smida-Rezgui, S., Cseri, J., Simut, C., Sabatier, J. M., De Waard, M., et al. (2005). Differential effects of maurocalcine on Ca^{2+} release events and depolarization-induced Ca^{2+} release in rat skeletal muscle. *J Physiol, 565*(Pt 3), 843-853.

Takeshima, H., Yamazawa, T., Ikemoto, T., Takekura, H., Nishi, M., Noda, T., et al. (1995). Ca(2+)-induced Ca^{2+} release in myocytes from dyspedic mice lacking the type-1 ryanodine receptor. *Embo J, 14*(13), 2999-3006.

Tanabe, T., Beam, K. G., Powell, J. A., & Numa, S. (1988). Restoration of excitation-contraction coupling and slow calcium current in dysgenic muscle by dihydropyridine receptor complementary DNA. *Nature, 336*(6195), 134-139.

Tripathy, A., Resch, W., Xu, L., Valdivia, H. H., & Meissner, G. (1998). Imperatoxin A induces subconductance states in Ca^{2+} release channels (ryanodine receptors) of cardiac and skeletal muscle. *J Gen Physiol, 111*(5), 679-690.

Valdivia, H. H., Kirby, M. S., Lederer, W. J., & Coronado, R. (1992). Scorpion toxins targeted against the sarcoplasmic reticulum Ca(2+)-release channel of skeletal and cardiac muscle. *Proc Natl Acad Sci U S A, 89*(24), 12185-12189.

Wagenknecht, T., & Samso, M. (2002). Three-dimensional reconstruction of ryanodine receptors. *Front Biosci, 7*, d1464-1474.

Weiss, R. G., O'Connell, K. M., Flucher, B. E., Allen, P. D., Grabner, M., & Dirksen, R. T. (2004). Functional analysis of the R1086H malignant hyperthermia mutation in the DHPR reveals an unexpected influence of the III-IV loop on skeletal muscle EC coupling. *Am J Physiol Cell Physiol, 287*(4), C1094-1102.

Wier, W. G., & Balke, C. W. (1999). Ca(2+) release mechanisms, Ca(2+) sparks, and local control of excitation-contraction coupling in normal heart muscle. *Circ Res, 85*(9), 770-776.

Wilkens, C. M., Kasielke, N., Flucher, B. E., Beam, K. G., & Grabner, M. (2001). Excitation-contraction coupling is unaffected by drastic alteration of the sequence surrounding residues L720-L764 of the alpha 1S II-III loop. *Proc Natl Acad Sci U S A, 98*(10), 5892-5897.

Zamudio, F. Z., Gurrola, G. B., Arevalo, C., Sreekumar, R., Walker, J. W., Valdivia, H. H., et al. (1997). Primary structure and synthesis of Imperatoxin A (IpTx(a)), a peptide activator of Ca^{2+} release channels/ryanodine receptors. *FEBS Lett, 405*(3), 385-389.

Zhu, S., Darbon, H., Dyason, K., Verdonck, F., & Tytgat, J. (2003). Evolutionary origin of inhibitor cystine knot peptides. *Faseb J, 17*(12), 1765-1767.

Zhu, X., Gurrola, G., Jiang, M. T., Walker, J. W., & Valdivia, H. H. (1999). Conversion of an inactive cardiac dihydropyridine receptor II-III loop segment into forms that activate skeletal ryanodine receptors. *FEBS Lett, 450*(3), 221-226.

18

^{13}C-Metabolic Flux Analysis and Metabolic Regulation

Yu Matsuoka[1] and Kazuyuki Shimizu[1,2]
[1]*Depatment of Bioscience and Bioinformatics, Kyushu Institute of Technology*
[2]*Institute of Advanced Bioscience, Keio University*
Japan

1. Introduction

Among the different levels of information in a cell, the most important information in understanding the complex metabolic control mechanism of the whole cell may be the metabolic flux distribution in the central metabolism (Sauer, 2006), as this is the manifestation of gene and protein expressions and the concentrations of intracellular metabolites (Matsuoka & Shimizu, 2010a) (Fig.1), where a set of metabolic fluxes (or enzyme activities) describes the cell physiology (Bailey, 1981). The information of the metabolic flux distribution is quite useful for metabolic engineering (Stephanopoulos, 1999).

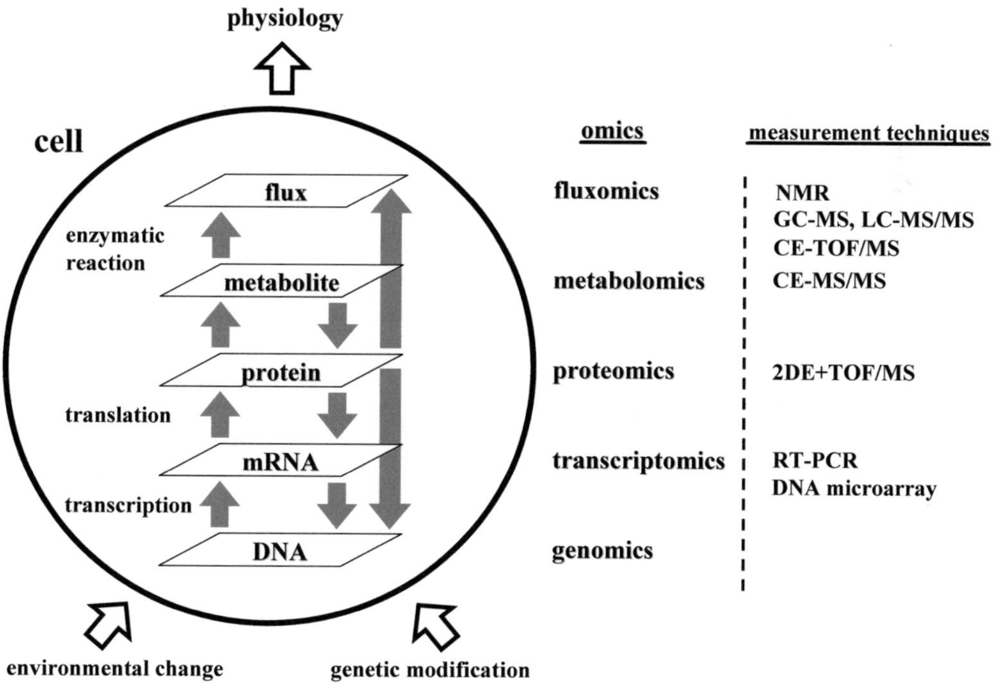

Fig. 1. Illustration of different levels of information in the cell.

In principle, metabolic flux analysis is based on mass conservation of key metabolites. The intracellular fluxes can be calculated from the measured specific rates by applying mass balances to these intracellular metabolites (Stephanopoulos et al., 1998) together with stoichiometric equations. The number of measureable extracellular fluxes is limited in practice, and the stoichiometric constraints often lead to an underdetermined algebraic system. Therefore, cofactor balances are sometimes required to be introduced into the stoichiometric model, or appropriate objective functions has to be introduced for optimization to determine the fluxes. The central metabolic pathway has both anabolic and catabolic functions, as it provides cofactors and building blocks for macromolecular system (anabolism) as well as energy (ATP) production (catabolism). The optimization may be made in terms of catabolism and anabolism, or both, under the constraints of the stoichiometric equations. Flux balance analysis (FBA) has been extensively used to predict the steady-state metabolic fluxes in order to maximize the cell growth rate (Edwards & Palsson, 2000; Schilling & Palsson, 1998; Price et al., 2003). However, the accuracy of the flux calculation depends on the validity of the cofactor assumptions and an appropriate choice of the objective function(s) employed (Schuetz et al., 2007). The presence of unknown reactions that generate or consume the cofactors may invalidate the assumption that their concentrations remain in balance, and the selected objective functions may not be appropriate, or their validity may be limited to certain states of the cell. Namely, this approach cannot essentially compute such fluxes as (1) recycled fluxes, (2) bidirectional fluxes, and (3) parallel fluxes due to the singularity of the stoichiometric matrix. Note, however, that the conventional FBA may be reasonably applied for the case without recycles such as anaerobic or micro-aerobic cultivation (Zhu & Shimizu, 2004, 2005). ^{13}C-Metabolic flux analysis has been developed to overcome such problems as stated above and to estimate *in vivo* fluxes in more accurately. In this chapter, ^{13}C-metabolic flux analysis (^{13}C-MFA) is explained with some appications. The analytical approach for ^{13}C-MFA is given to understand the feature of this method, and the basic principle of isotopomer dynamics is also explained. Finally, the metabolic regulation mechanism is also mentioned in view of global regulators.

2. ^{13}C-Metabolic flux analysis (^{13}C-MFA) and its application

For ^{13}C-metabolic flux analysis, isotopically labeled substrates are introduced to the cell, and the labeled carbon atoms are distributed throughout its metabolic network. The final isotopic enrichment in the intracellular metabolite pools can then be measured (Matsuoka & Shimizu, 2010a) (Fig. 2). The amino acids in biomass hydrolysate are much more abundant than their precursors in the cell metabolism, and it has been often used to deduce the labeling patterns of the intracellular metabolites from that of the proteinogenic amino acids, based on precursor-amino acids relationships (Szyperski, 1995). Either NMR spectroscopy or GC-MS has been extensively used for the isotopic tracer experiments. NMR is used to measure the positional ^{13}C enrichment (Szyperski, 1995; Marx et al., 1996). Although ^{13}C NMR is powerful and attractive for metabolic flux analysis, it requires relatively large amount of sample. On the other hand, GC-MS can easily analyze metabolites with much less amount of sample. As a result, it may be suitable for flux analysis of the culture processes that have low concentrations of biomass (Fischer & Sauer, 2003; Wittmann, 2007). Currently, these tracer techniques, in combination with direct extracellular flux measurements, are considered to be a powerful method for obtaining intracellular metabolic flux distribution (MFD) using only a few modeling assumptions (Stephanopoulos, 1999; Marx et al., 1996;

Schmidt et al., 1999; Wiechert & Graaf, 1997; Sauer, 2004). Alternatively, flux ratio analysis can be used to constrain the fluxes at the important branch point (Szyperski et al., 1999; Fischer & Sauer, 2003; Sauer et al., 1999; Hua et al., 2003).

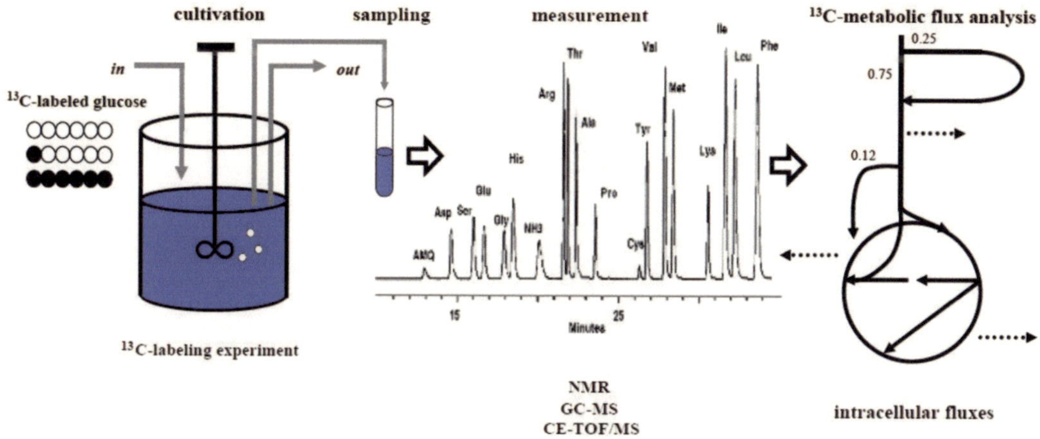

Fig. 2. ^{13}C-Metabolic flux analysis by ^{13}C-labeling experiment, the measurement of isotopomer distribution, and the flux determination.

^{13}C-Metaolic flux analysis has been applied to such microorganisms as *Escherichia coli*, *Corynebacterium glutamicum*, *Saccharomyces cerevisiae* (Blank & Sauer, 2004; Raghevendran et al., 2004), *Bacillus subtilis* (Sauer et al., 1997), acetic acid bacterium *Gluconacetobacter oboediens* (Sarkar et al., 2010), and others (Fuhrer et al., 2005), and its application has been extended to other organisms such as Cyanobacteria (Yang et al., 2002a, b), plant cells (Schwender et al., 2004), and mammarian cells (Sherry et al., 2004; Kelleher, 2004; Hellerstein, 2004; Sidorenko et al., 2008), including mouse (McCabe & Previs, 2004), brain (Rothman et al., 2003), neural cells (Selivanov et al., 2006), and cancer cells (Selivanov et al., 2005) and so on.

Let us consider the several single-gene knockout mutants such as *zwf*, *gnd* and *pgi* gene knockout. Figure 3 shows how the NMR spectra patterns change for *pgi* gene knockout mutant as compared to wild type, where it shows the relative intensities of ^{13}C-^{13}C scalar coupling multiplet spectra patterns of amino acid measured by NMR, where the samples were taken from the labeling experiments using a mixture of [U-^{13}C] and unlabeled [U-^{12}C] glucose conducted under aerobic steady state condition (Hua et al., 2003). The differences between NMR spectra patterns of wild type and *pgi* gene knockout mutant result in different flux distribution. Based on NMR data such as shown in Fig. 3, Fig. 4 shows the result of ^{13}C-metabolic flux analysis for wild type and *pgi* gene knockout mutant cultivated in the continuous culture at the dilution rate of 0.1 h^{-1} (Hua et al., 2003). It can be seen that the knockout of *pgi* gene, which codes for the first enzyme of the EMP (Embden-Meyerhoff-Parnas) pathway after the branch point at G6P, resulted in exclusive use of the oxidative pentose phosphate (PP) pathway for glucose catabolism. The overproduced NADPH inhibits the activity of G6PDH thus reduces the glucose consumption rate, resulting in the low growth rate. The glyoxylate pathway was activated, and the Entner-Doudoroff (ED) pathway was activated in this mutant. The activation of ED pathway may be due to reducing NADPH production as compared to employing 6PGDH pathway. The activation of the glyoxylate pathway is considered to be due to the feedback regulation to compensate for the lowered

OAA concentration caused by the lowered flux of EMP pathway and the lowered anaplerotic flux through Ppc to supply OAA.

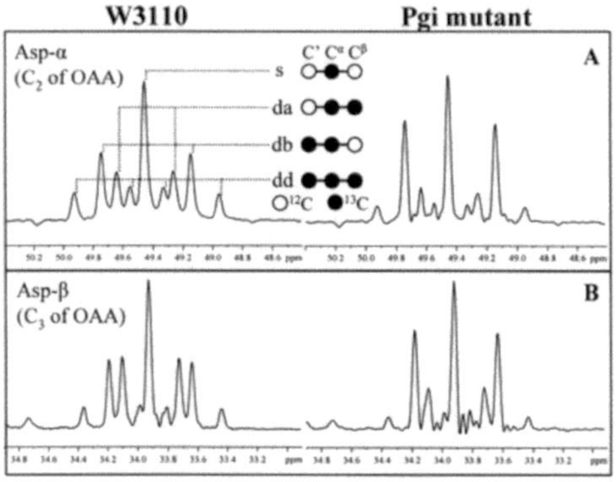

Fig. 3. ^{13}C-^{13}C scalar coupling multiplets observed for aspartate from glucose-limited chemostat cultures of *E. coli* W3110 (left) and the *pgi* mutant (right). The signals were extracted from the $\omega_1(^{13}C)$ cross sections in the [^{13}C,^{1}H]-COSY spectra. (A) Asp-α; (B) Asp-β. As indicated in panel A, the multiplets consist of a singlet (s), a doublet with a small coupling constant (da), a doublet split by a larger coupling constant (db), and a doublet of doublets (dd). Aspartate corresponds directly to its metabolic precursor, OAA.

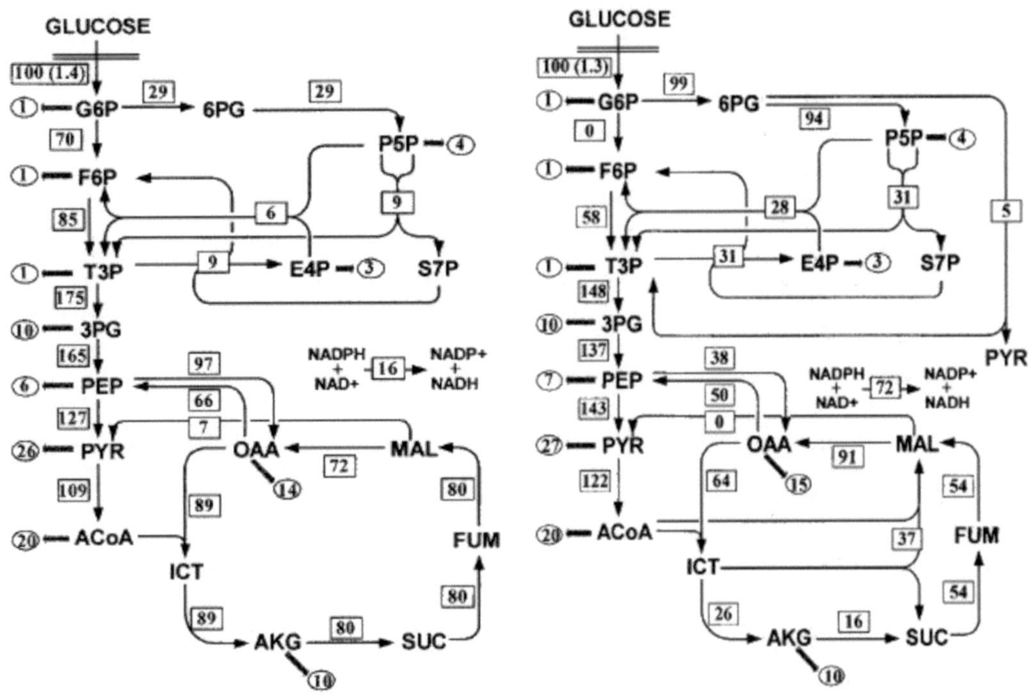

(a) (b)

Fig. 4. Metabolic flux distributions of (a) *E. coli* wild type and (b) its *pgi* gene knockout mutant under glucose-limited continuous cultivation at the dilution rate of 0.1 h^{-1}.

Let us consider next the *zwf* or *gnd* gene knockout mutants, where the ^{13}C-metabolic flux distributions are given in Fig. 5 in the cases of using glucose (Fig. 5a) and pyruvate (Fig. 5b) (Zhao et al., 2004). Since the flux through oxidative PP pathway was significantly decreased or totally blocked in the mutants grown on glucose, the metabolic network has to give a higher flux through EMP or ED pathway to the TCA cycle. The opposite relations are seen in mutants grown on pyruvate, where the deletion of genes caused a reduction in the fluxes through the TCA cycle.

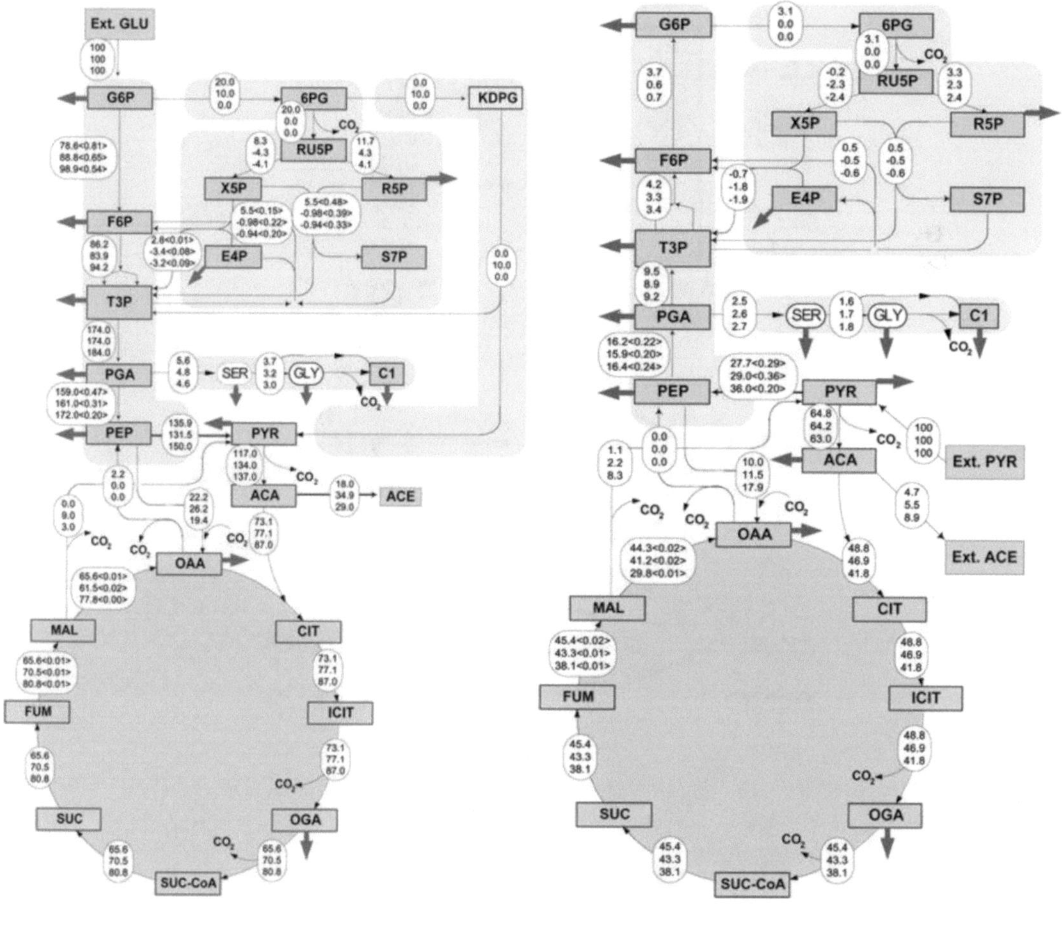

(a) (b)

Fig. 5. Metabolic flux distributions in chemostat culture of (a) glucose-grown and (b) pyruvate-grown *E. coli* parent strain (*upper values*), *gnd* (*middle values*), and *zwf* (*lower values*) mutants at D=0.2 h^{-1}.The exchange coefficients are shown in brackets for the reactions that were considered reversible. *Negative values* indicate the reversed pathway direction.

3. Analytical approach to ^{13}C-metabolic flux analysis

Let us consider how the fluxes are computed based on ^{13}C-labeled metabolites. For this, consider how the isotopomer pattern changes with respect to fluxes. Let us cosider TCA cycle for the case of using pyruvate or acetate as a carbon source, where fate of carbons is shown in Fig. 6 (Matsuoka & Shimizu, 2010b).

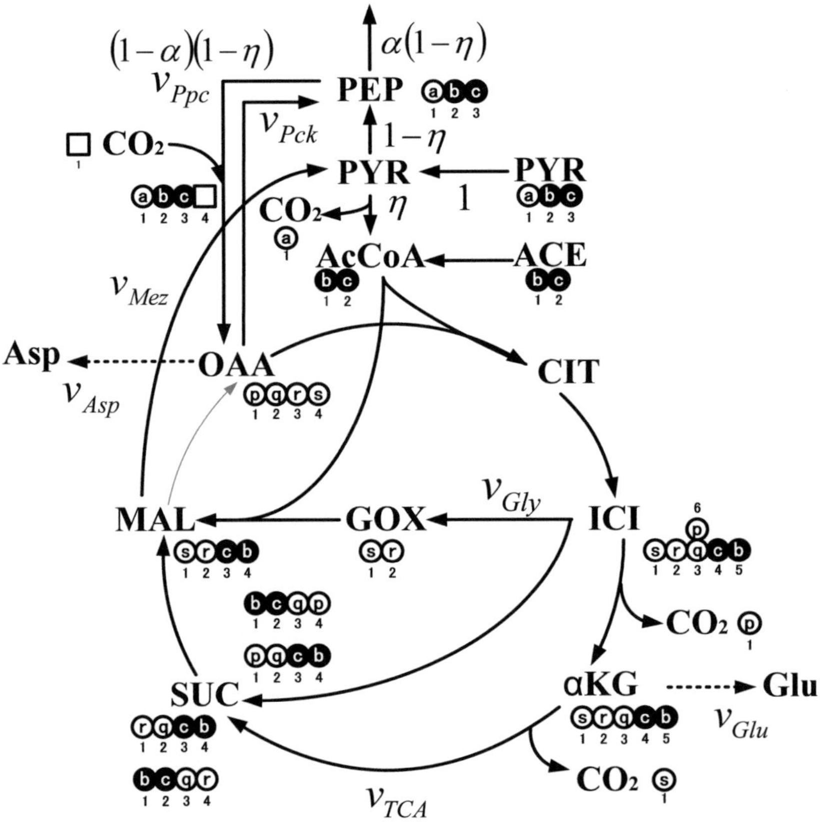

Fig. 6. Metabolic network for the change in isotopomer distribution: pyruvate or acetate as a carbon source.

Referring to Fig. 6, let η be the flux toward AcCoA from PYR, and let $(1-\eta)$ be the other flux toward PEP. Moreover, let α be the fraction toward gluconeogenic pathway from PEP, and let $(1-\alpha)$ be the other fraction toward anaplelotic pathway through Ppc to OAA. From the mass balance for OAA, the normalized molar formation rate of OAA is expressed as $v_{OAA}^{+} = v_{TCA} + 2v_{Gly} - v_{Mez} + v_{Ppc}$, and the molar consumption rate of OAA is expressed as $v_{OAA}^{-} = v_{TCA} + v_{Gly} + v_{Glu} + v_{Asp} + v_{Pck}$. Since $v_{OAA}^{+} = v_{OAA}^{-}$ holds at steady state, the following expression can be derived:

$$v_{Gly} + v_{Ppc} = v_{Glu} + v_{Asp} + v_{Mez} + v_{Pck} \tag{1}$$

Since $v_{Ppc} = (1-\alpha)(1-\eta) + v_{Pck}$ holds, the following expression can be derived:

$$(1-\eta)\alpha = 1-\eta + v_{Gly} - \left(v_{Glu} + v_{Asp} + v_{Mez}\right) \tag{2}$$

Then the isotopomer balance for O_1 may be expressed as

$$\begin{aligned}
&\frac{1}{2}\left(1-\frac{w}{1+2z}\right)\left[\left(O_3+O_{13}+O_{34}+O_{134}\right)A_0+\left(O_0+O_1+O_4+O_{14}\right)A_1\right] \\
&+\frac{z}{2}\left(1-\frac{w}{1+2z}\right)\begin{bmatrix}2\left(O_0+O_1+O_2+O_{12}\right)A_1+\left(O_1+O_{13}+O_{14}+O_{134}\right)A_0+ \\ \left(O_0+O_3+O_4+O_{34}\right)A_1\end{bmatrix} \\
&+yP_1C_0=\left(1+2z-w+y\right)O_1
\end{aligned} \tag{3}$$

where, $w \equiv v_{Mez}/v_{TCA}$, $y \equiv v_{Ppc}/v_{TCA}$, and $z \equiv v_{Gly}/v_{TCA}$. Note that the LHS of Eq. (3) is the formation rate of O_1, and the RHS is the consumption rate of O_1. In Eq. (3), A_i and P_j are the mole fractions of acetate and PYR isotopomers labeled at i-th and j-th carbons, respectively, and C_k $(k=0,1)$ is the mole fraction of labeled CO_2 through Ppc reaction. The similar equations can be derived for all the other isotopomers, and the set of equations can be expressed as

$$\Lambda_{AP} I_{OAA} = b_{AP} \tag{4}$$

where Λ_{AP} ($\in R^{16} \times R^{16}$) is the coefficient matrix, and b_{AP} ($\in R^{16}$) is the residual vector, where the detailed matrix and vector are given elsewhere (Matsuoka & Shimizu, 2010b). Eq. (4) may be solved as

$$I_{OAA} = \Lambda_{AP}^{-1} b_{AP} \tag{5}$$

if Λ_{AP} is nonsingular. In this case, Eq. (4) cannot be solved directly. Recall that the isotopomer fractions must be summed up to 1, namely the following equation must be satisfied:

$$\sum_i O_i = 1 \tag{6}$$

Therefore, Eq. (4) can be solved by replacing one of the equations using Eq. (6).

Let us consider the special case where 100 % [1-^{13}C] pyruvate is used as a carbon source, and it is set $v_{Mez} = 0$ and $v_{Pck} = 0$ for simplicity. Then Eq. (5) gives the following equation (Matsuoka & Shimizu, 2010b):

$$\begin{bmatrix} O_0 \\ O_1 \\ O_4 \end{bmatrix} = \begin{bmatrix} \frac{1}{1+y} \\ \frac{y}{1+y} \\ 0 \end{bmatrix} \tag{7}$$

In the case where 100 % [2-13C] pyruvate is used, the following equations can be derived (Matsuoka & Shimizu, 2010b):

$$O_1 = O_4 = \frac{1}{2(1+y)^2(1+2y)} \tag{8a}$$

$$O_2 = \frac{y}{1+y} \tag{8b}$$

$$O_{13} = O_{24} = \frac{y}{(1+y)(1+2y)} \tag{8c}$$

$$O_{14} = \frac{y}{(1+y)^2(1+2y)} \tag{8d}$$

In the case where 100 % [3-13C] pyruvate is used, the following expressions can be derived (Matsuoka & Shimizu, 2010b):

$$O_2 = 0 \tag{9a}$$

$$O_3 = \frac{y}{1+y} \tag{9b}$$

$$O_{13} = O_{24} = \frac{y}{(1+y)(1+2y)} \tag{9c}$$

$$\begin{bmatrix} O_{23} \\ O_{123} \\ O_{234} \end{bmatrix} = \begin{bmatrix} \dfrac{y}{(1+y)^2(1+2y)} \\ \dfrac{1}{2(1+y)^2(1+2y)} \\ \dfrac{1}{2(1+y)^2(1+2y)} \end{bmatrix} \tag{9d}$$

Let us consider the case of using acetate as a carbon source by letting $v_{Ppc} = 0$ in Fig. 6. If the mixture of [1-13C] acetate (A_1) and the unlabeled acetate is used, only 4 isotopomers (O_0, O_1, O_4, O_{14}) are present, and if $A_1 = 1$, those can be expressed as (Matsuoka & Shimizu, 2010b)

$$\begin{bmatrix} O_1 \\ O_4 \\ O_{14} \end{bmatrix} = \begin{bmatrix} \dfrac{1+2z}{(1+z)(2+3z)} \\ \dfrac{1}{2+3z} \\ \dfrac{z}{1+z} \end{bmatrix} \tag{10}$$

If [2-^{13}C] acetate (A_2) and the unlabeled acetate are used as a carbon source, 12 isotopomers are present, and those reduce to 3 isotopomers when $A_2 = 1$ as follows (Matsuoka & Shimizu, 2010b):

$$\begin{bmatrix} O_{23} \\ O_{123} \\ O_{234} \end{bmatrix} = \begin{bmatrix} \dfrac{z}{1+z} \\ \dfrac{1}{2+3z} \\ \dfrac{1+2z}{(1+z)(2+3z)} \end{bmatrix} \tag{11}$$

Figure 7 shows how the isotopomer distribution changes with respect to η for the case of using 100 % of [3-^{13}C] pyruvate as a carbon source (Matsuoka & Shimizu, 2010a). It indicates that OAA isotopomers as well as mass isotopomers change with respect to η in somewhat strange way, where Fig. 7c explains for the cases of $\eta = 0$ (left), $0 < \eta < 1$ (middle), and $\eta = 1$ (right). In Fig. 7a, O_3 increases as η becomes low. This is due to Ppc reaction (left figure in Fig. 7c). The case where η is low, most of the [3-^{13}C] pyruvate goes to OAA via Ppc and generates O_3. As proceeding TCA cycle, O_3 and [2-^{13}C] AcCoA (originated from [3-^{13}C] PYR since the 1st carbon of PYR is lost as CO_2 at PDHc reaction) generate O_{13} and O_{24} by the 1st turn of the TCA cycle, where O_{24} further generates O_{23} in the 2nd turn, and in turn O_{23} generates O_{123} and O_{234} at the 3rd turn of the TCA cycle (middle figure in Fig. 7c). Thus O_3, O_{13}, O_{24}, O_{23}, O_{123}, and O_{234} are eventually generated, where the amount of O_3 is the highest at low values of η, but the fractions of O_{13} and O_{24} become higher as TCA cycle flux increases. When η is high, where most flux goes through PDHc reaction, A_2 (P_3) generates O_2 and O_3 at the 1st turn of TCA cycle, and in turn O_2 and A_2 generate O_{23}, while O_3 and A_2 generate O_{13} and O_{24} in the 2nd turn. Moreover, O_{23} and A_2 generate O_{123} and O_{234} at the 3rd turn and so on. Namely, O_{123} and O_{234} become higher as cycling the TCA cycle, and thus those become higher at the steady state (right figure in Fig. 7c). The above observation implies that the fraction of O_3 becomes higher as η is low, whereas the fractions of O_{13} and O_{24} become higher as η increases, and then the fractions of O_{123} and O_{234} become higher as η becomes further higher. This is the reason why O_{13}, O_{24}, and O_{23} give maximum with respect to η as seen in Fig. 7a. Figure 7b shows how the mass isotopomers change with respect to η. It indicates that m_1 decreases while m_3 increases as η increases, whereas m_2 becomes maximum at certain value of η. This is due to the changing patterns of $O_{13} = O_{24}$ and O_{23} as seen in Fig. 7a. These illustrate the complexity of the ^{13}C-flux analysis in relation to TCA cycle. The transitions of the isotopomers with respect to turn of TCA cycle may be expressed in more systematic way by introducing the transition matrix.

Consider next the case of using glucose as a carbon source. Referring to Fig. 8 (Matsuoka & Shimizu, 2010a), let v_1 be the normalized flux toward glycolysis at glucose 6-phosphate (G6P) and let $1 - v_1$ be the flux toward oxidative pentose phosphate (PP) pathway, where the input flux to the system is normalized to be 1 without loss of generality. Then the mass balance equations can be derived for each metabolite. Moreover, isotopomer balances may be considered for each metabolite. Since glucose has 6 carbons, the number of its isotopomers is $2^6 = 64$ and let these be expressed as $G_0, G_1, \cdots, G_{123456}$, where G_0 denotes unlabeled glucose, G_1 denotes [1-^{13}C] glucose, and G_{123456} denotes [U-^{13}C] glucose and so on. The number of

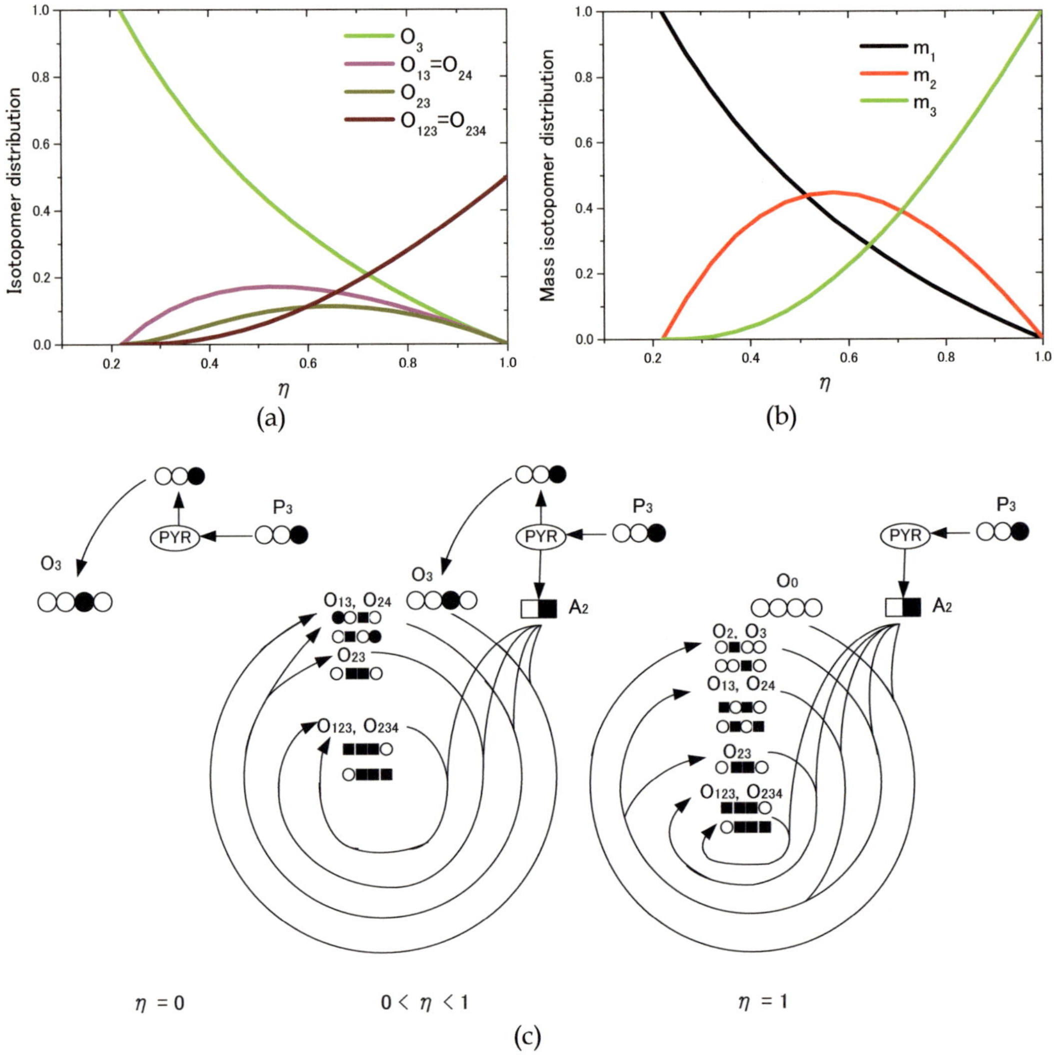

Fig. 7. Effect of η on (a) isotopomer and (b) mass isotopomer distributions for the case of using 100 % [3-13C] pyruvate ($\alpha = 0.4$). (c) schematic illustration of how the specific isotopomers are generated.

isotopomer balance equations is $2^6 = 64$ for G6P, $2^5 = 32$ for P5P (Ru5P, X5P, R5P), $2^7 = 128$ for S7P, $2^4 = 16$ for E4P, $2^6 = 64$ for F6P, and $2^3 = 8$ for GAP. The same number of isotopomers is present, and therefore, the isotopomer distribution for each metabolite can be obtained uniquely as a function of the labeling patterns of glucose on condition that the fluxes are given. Consider the case of using 100 % of [1-13C] glucose as a carbon source for simplicity. It may be easily shown that GAP_3 decreases, while GAP_0 increases as v_1 decreases, because the labeled 1st carbon is lost at 6PGDH reaction in the PP pathway, and thus the flux v_1 may be identified by the mass isotopomer patterns of GAP (or Ser).

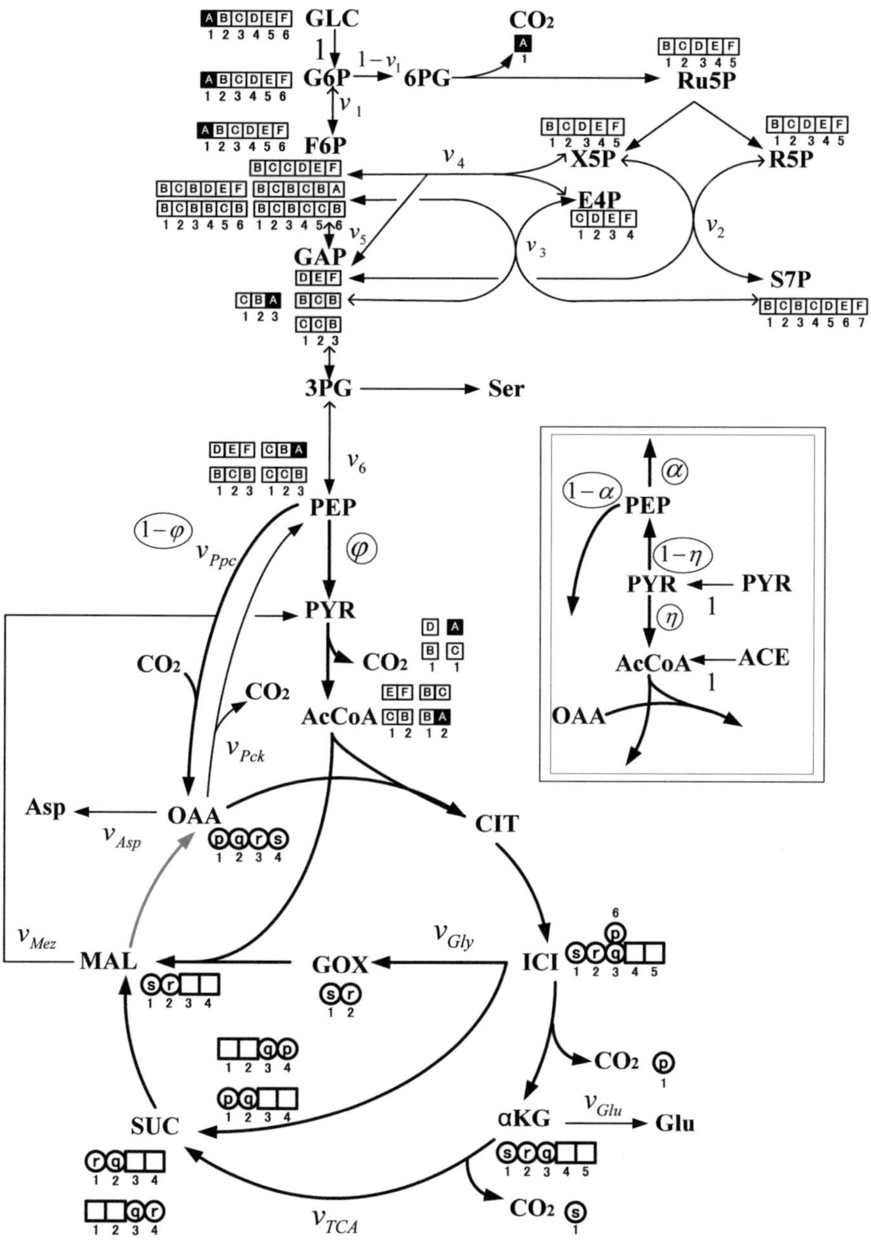

Fig. 8. Main metabolic pathways in the central metabolism together with the fate of [1-^{13}C] glucose.

Consider next how the isotopomers of GAP affect those of TCA cycle. Referring to Fig. 8, Let ϕ be the fraction from PEP to PYR through Pyk, and let $1-\phi$ be the rest of the fraction which flows from PEP to OAA by the anaplerotic pathway. Figure 9a shows the effect of ϕ on the isotopomer distribution of OAA, and Fig. 9b shows its effect on the mass isotopomer distribution using 100 % of [1-^{13}C] glucose for the case where $v_1 = 1$ for simplicity

(Matsuoka & Shimizu, 2010a). Note that O_i is the isotopomer of OAA labeled at i -th carbon(s) and m_j is the mass isotopomer, where it is the linear combination of the isotopomers of OAA. For example, $m_0 = O_0$, $m_1 = \sum_{i=1}^{4} O_i$, and so on. Unlike glycolysis and PP pathway, TCA cycle provides a variety of isotopomers as given in Fig. 9a, and Fig. 9b indicates that ϕ may be identified by MS measurement of $m_0 \sim m_3$. The reason why m_1 is less sensitive with respect to ϕ as compared to the other mass isotopomers is that O_2 increases while O_3 decreases as ϕ increases. Note that GAP_3 and GAP_0 are generated at Fba reaction from [1-^{13}C] glucose. Those bring O_3 and O_0 via Ppc reaction. Another isotopomers such as O_2 and O_{23} appear as proceeding TCA cycle (Matsuoka & Shimizu, 2010a). Although [1-^{13}C] glucose and/or [U-^{13}C] glucose are exclusively used in practice, [1, 2-^{13}C] glucose may be also used for the flux estimation.

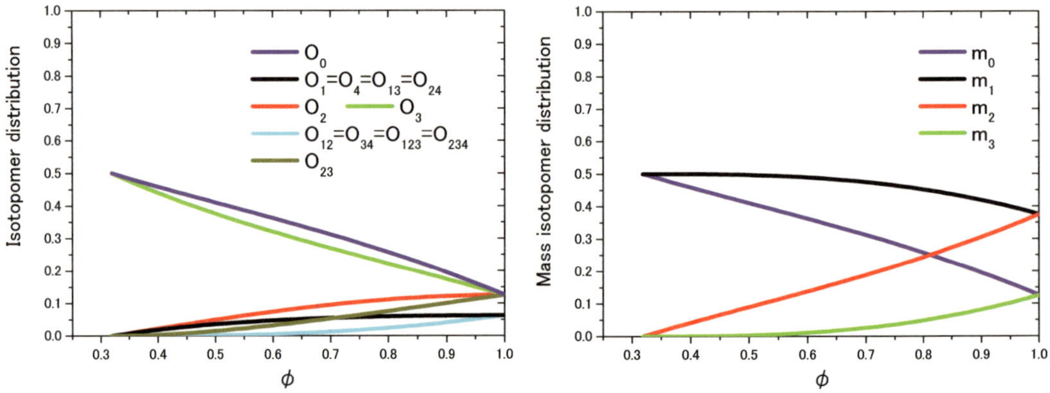

Fig. 9. Simulation result of (a) isotopomer and (b) mass isotopomer distributions of OAA with respect to ϕ for the case of using 100 % [1-^{13}C] glucose ($v_1 = 1.0$).

So far, we considered how the isotopomer distribution is expressed with respect to the flux. In practice, the fluxes must be found from the measured signals for the isotopomer distribution. Let these signals be obtained by either NMR or MS, and let these be $s_i\,(i = 1,2,\cdots)$. Then the fluxes may be obtained by minimizing the following objective function:

$$J = (\hat{s} - s)^T \sum_S^{-1} (\hat{s} - s) \tag{12}$$

where s represents the measured signal vector containing s_i, $\hat{s}$ is its estimated vector, and $\sum_S$ is the variance-covariance matrix associated with the measured errors.

4. Dynamic behavior of isotopomer distribution

The metabolic flux analysis with ^{13}C-labeled substrate has been made based on the steady state condition. However, the sampling is usually made during instationary phase due to economical reason, and unsteady correction is made by assuming the 1st order washout behavior (van Winden et al., 2001). In many practical applications, this may be justified from the viewpoint of the replacement of the broth by the labeled substrate. However, another dynamic behavior must be taken into account for the analysis of the TCA cycle, since the isotopomer distribution changes with respect to each turn of the TCA cycle. Although several analysis methods have been developed in the past for the special case of 100 % [2-^{13}C] acetate (Chance et al., 1983; Merle et al., 1996; Tran-Dinh et al., 1996a; Tran-Dinh et al., 1996b), here more general approach is considered.

Let Φ be the transition matrix which converts the isotopomer distribution at the i th turn to that at the $(i+1)$ th turn in the TCA cycle such that

$$x^{(i+1)} = \Phi \cdot x^{(i)} \tag{13}$$

where $x^{(i)}$ is the isotopomer vector at the i th turn of the TCA cycle. By considering the special cases of A_1=1, A_2=1, A_{12}=1, and A_0=1, respectively, and superimposing the results into one matrix, the general expression for Φ may be expressed as the matrix (Appendix A), letting $T \equiv v_{TCA}, G \equiv v_{Gly}$, $\beta \equiv 2^{-1}(T+G)$, and $\gamma \equiv T+2G$. Note that $T+2G=1$ holds. This expression for Φ may be useful in analyzing the transient behavior of the isotopomer distribution with respect to turn of the TCA cycle together with glyoxylate pathway. If we repeat the operation of Eq. (13), the following equation is obtained:

$$x^{(n)} = \Phi^n x^{(0)} \tag{14}$$

where $x^{(0)}$ is the isotopomer distribution vector before labeling experiment. This equation implies that the isotopomer distribution at the n th turn can be expressed once the initial state or at any known state was given, and this expression may be more useful than Eq. (13). Moreover, we can derive the same equations as Eq. (10) and (11) already derived in the steady state analysis if we set $n \to \infty$ (Appendix B). Thus, Eq. (14) can show not only steady state but also dynamic behavior of isotopomer. Appendix C shows the details of Φ^n .

As stated before, most of the flux analysis has been made based on the steady state assumption. However, this assumption must be carefully considered for the TCA cycle, since the ^{13}C-labeled isotopomer distribution changes for each turn of the TCA cycle. Figure 10a shows how the isotopomer distribution changes with respect to turn of the TCA cycle, where 100 % [2-^{13}C] acetate was used as a carbon source, and only TCA cycle was assumed to be active. Figure 10a indicates that O_2 and O_3 appear at the 1st turn, and O_{13} and O_{24} as well as O_{23} appear at the 2nd turn, but those reduced to zero as further cycling in the TCA cycle. On the other hand, O_{123} and O_{234} appear after the 3rd turn and converged to the steady state values. The converged values coincide with the values obtained by setting $z=0$ in the analytical expression (11). This result may be explained as follows: The C_2 position of AcCoA becomes C_4 position of CIT and αKG, and C_3 position of SUC, while C_5 position of CIT and αKG are never labeled if C_1 position of AcCoA is not labeled (Fig. 6). Randomization of carbons in SUC is made by allowing ^{13}C in the C_3 of SUC to be distributed equally to either C_2 or C_3 position of MAL in the first turn of the cycle. Thus at the end of the first turn of the cycle, O_2 and O_3 are produced with equal portions (Fig. 10a), which in the 2nd turn of the cycle generate C_{34} and C_{24} of CIT, C_{34} and C_{24} of αKG, and C_{23} and C_{13} of SUC. C_{23} of SUC only produces O_{23} since O_{23} is equivalent to O_{32}, while C_{13} of SUC produces equal amount of O_{24} and O_{13}. Condensation of these with C_2 of AcCoA (A_2) produces C_{234},C_{134}, and C_{246} of CIT, which eventually produce O_{123} and O_{234}.

Figure 10b shows the case where the glyoxylate pathway is active as well as TCA cycle. It indicates that at the first turn, O_2 and O_3 are generated, while O_{23},O_{123} and O_{234} appear as each turn is repeated, resulting in the eventual convergence to the steady state, where O_{23} newly appears as compared with the case of only TCA cycle. The converged values are the same as those computed using Eq. (11).

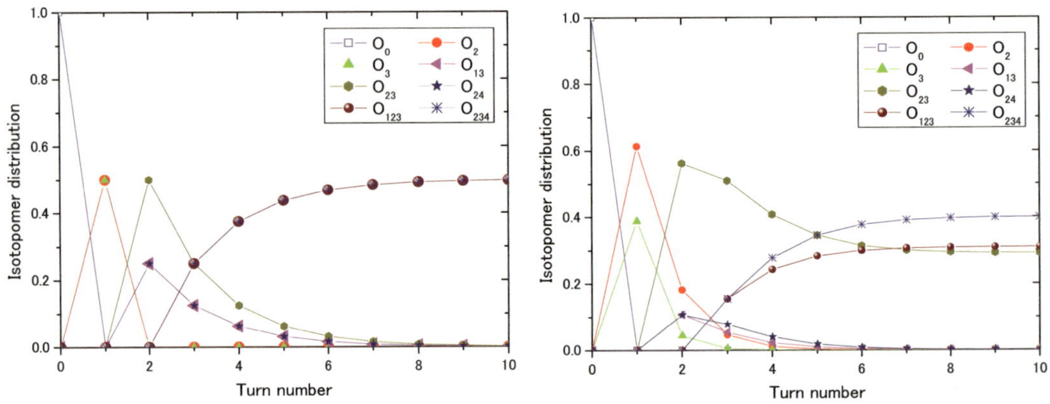

(a) (b)

Fig. 10. Isotopomer distribution with respect to turn of the TCA cycle with glyoxylate pathway: Case of using the 100 % [2-^{13}C] acetate where (a) only TCA cycle is active, and (b) glyoxylate pathway as well as TCA cycle is active.

5. Metabolic regulation by global regulators in response to culture environment

The main goal of metabolic engineering is to improve the metabolic phenotype through genetic modifications (Bailey, 1991; Stephanopoulos & Vallino, 1991). Most of the past metabolic engineering approach has been considered to improve a particular biosynthetic capacity through engineering the target pathway. The resulting phenotypes are, however, often suboptimal and not satisfactory due to distant effects of genetic modifications or regulatory influences. It is, therefore, strongly desirable to take into account the metabolic regulation mechanism for metabolic engineering. Roughly speaking, metabolic regulation is made by recognition of culture environment (or cell's state) and the adjustment of the metabolism, where global regulators and/or sigma factors play important roles in regulating the metabolic pathway genes. Note that the high-throughput techniques for transcriptomics, proteomics, metabolomics, and fluxomics have the potential to disclose the metabolic regulation mechanism, but most of them provide a snapshot from one stage, and the methodology in interpreting different levels of information is not established yet (Vemuri & Aristidou, 2005). In order to put forward advanced fermentation toward biofuel production etc., it is quite important to understand the effect of culture environment on the metabolism based on gene level regulation, and modify the cell metabolism based on such information. In the followings, a brief explanation on the carbon catabolite regulation is given.

Global regulator	Metabolic pathway gene
Cra	+ : *aceBAK, acnA, cydB, fbp, icdA, pckA, pgk, ppsA* – : *acnB, adhE, eda, edd, pfkA, ptsHI, pykF, zwf*
Crp	+ : *acnAB, aceEF, acs, focA, fumA, fur, gltA, malT, manXYZ, mdh, mlc, pckA, pdhR, pflB, pgk, ptsG, sdhCDAB, sucABCD, ugpABCEQ* – : *cyaA, lpdA, rpoS*
ArcA/B	+ : *cydAB, focA, pflB* – : *aceBAK, aceEF, acnAB, cyoABCDE, fumAC, gltA, icdA, lpdA, mdh, nuoABCDEFGHIJKLMN, pdhR, sdhCDAB, sodA, sucABCD*
Fnr	+ : *acs, focA, frdABCD, pflB, yfiD* – : *acnAB, cyoABCDE, cydAB, fnr, fumA, icdA, ndh, nuoABCDEFGHIJKLMN, sdhCDAB, sucABCD*
IclR	– : *aceBAK, acs*
FadR	+ : *iclR*
SoxR/S	+ : *acnA, fumC, fur, sodA, zwf*
Mlc	– : *crr, manXYZ, malT, ptsG, ptsHI*
RpoS	+ : *acnA, acs, adhE, fumC, gadAB, osmC, poxB, talA, tktB* – : *ompF*
Fur	– : *entABCDE, sodA*
IclR	– : *aceBAK, acs*
PdhR	– : *aceEF, lpdA*
PhoB	+ : *phoABERU, pstSCAB*

Table 1. Effect of global regulators on the metabolic pathway gene expressions.

Among the culture environment, carbon sources are by far important for the cell from the point of view of energy generation (catabolism) and biosynthesis (anabolism). Catabolite regulation is made when different carbohydrates are present in a medium, where glucose is preferentially consumed, and the uptake of other carbon sources is repressed. The center of the regulatory network for catabolite repression in *Escherichia coli* is the phosphoenol pyruvate (PEP): carbohydrate phosphotransferase systems (PTSs). These systems are involved in both transport and phosphorylation of carbohydrates. The PTS in *E. coli* consists of two common cytoplasmic proteins, EI (enzyme I) encoded by *ptsI* and HPr (histidine-phosphorylatable protein) encoded by *ptsH*, as well as carbohydrate-specific EII (enzyme II) complexes. The glucose-specific PTS in *E. coli* consists of the cytoplasmic protein $EIIA^{Glc}$ encoded by *crr* and the membrane-bound protein $EIICB^{Glc}$ encoded by *ptsG*, which transport and concomitantly phosphorylate glucose (Fig. 11). The phosphoryl groups are transferred from PEP via successive phosphorelay reactions in turn by EI, HPr, $EIIA^{Glc}$ and $EIICB^{Glc}$ to glucose. The cAMP-Crp complex and the repressor Mlc are involved in the regulation of *ptsG* gene and *pts* operon expressions. It has been demonstrated that unphosphorylated $EIICB^{Glc}$ can relieve the *ptsG* gene expression by sequestering Mlc from its binding sites through protein-protein interaction depending on the glucose concentration. In contrast to Mlc, which represses the expressions of *ptsG*, *ptsHI* and *crr* (Plumbridge, 1998), cAMP-Crp complex activates *ptsG* gene expression (De Reuse & Danchin, 1988). These two antagonistic regulatory mechanisms guarantee a precise adjustments of *ptsG* expression levels under various conditions (Bettenbrock et al., 2006). It should be noted that unphosphorylated $EIIA^{Glc}$ inhibits the uptake of other non-PTS carbohydrates by the so-called inducer exclusion (Aiba, 1995), while phosphorylated $EIIA^{Glc}$ ($EIIA^{Glc}$-P) activates adenylate cyclase (Cya), which generates cAMP from ATP and leads to an increase in the intracellular cAMP level (Park et al., 2006). In the absence of glucose, Mlc binds to the upstream of *ptsG* gene and prevents its transcription. If glucose is present in the medium, the amount of unphosphorylated $EIICB^{Glc}$ increases due to the phosphate transfer to glucose. In this situation, Mlc binds to $EIICB^{Glc}$, and thus it does not bind to the operator of *pts* genes (Bettenbrock et al., 2006; Tanaka et al., 2000; Lee et al., 2000). The overall catabolite regulation mechanism is briefly shown in Fig. 12 (Matsuoka & Shimizu, 2011). The catabolic regulation phenomenon has been modeled by several researchers (Bettenbrock et al., 2006; Kremling et al.; 2004; Kremling & Gilles, 2001).

In addition to cAMP-Crp, which acts depending on the level of glucose concentration, the catabolite repressor/activator protein (Cra) originally characterized as the fructose repressor (FruR) controls the carbon flow in *E. coli* (Moat et al., 2002; Saier, 1996; Saier & Ramseier, 1996). The carbon uptake and glycolysis genes such as *ptsHI*, *pfkA*, *pykF*, *zwf* and *edd-eda* are repressed, while gluconeogenic pathway genes such as *ppsA*, *fbp*, *pckA*, *icdA* and *aceA*, *B* are activated by Cra (Moat et al., 2002; Saier & Ramseier, 1996) (Table 1, Fig. 11). It has been known that the mutant defective in *cra* gene is unable to grow on gluconeogenic substrates such as pyruvate, acetate and lactate (Saier et al., 1996). Since gluconeogenic pathway genes are deactivated, and the glycolysis genes are activated by *cra* gene knockout, the glucose uptake rate may be enhanced, but it must be careful since *icdA*, *aceA*, *B*, and *cydB* genes are repressed, while *zwf* and *edd* gene expressions are activated and thus ED pathway is activated by *cra* gene knockout (Sarkar et al., 2008).

The effects of other culture environments than carbon source are given elsewhere (Matsuoka & Shimizu, 2011).

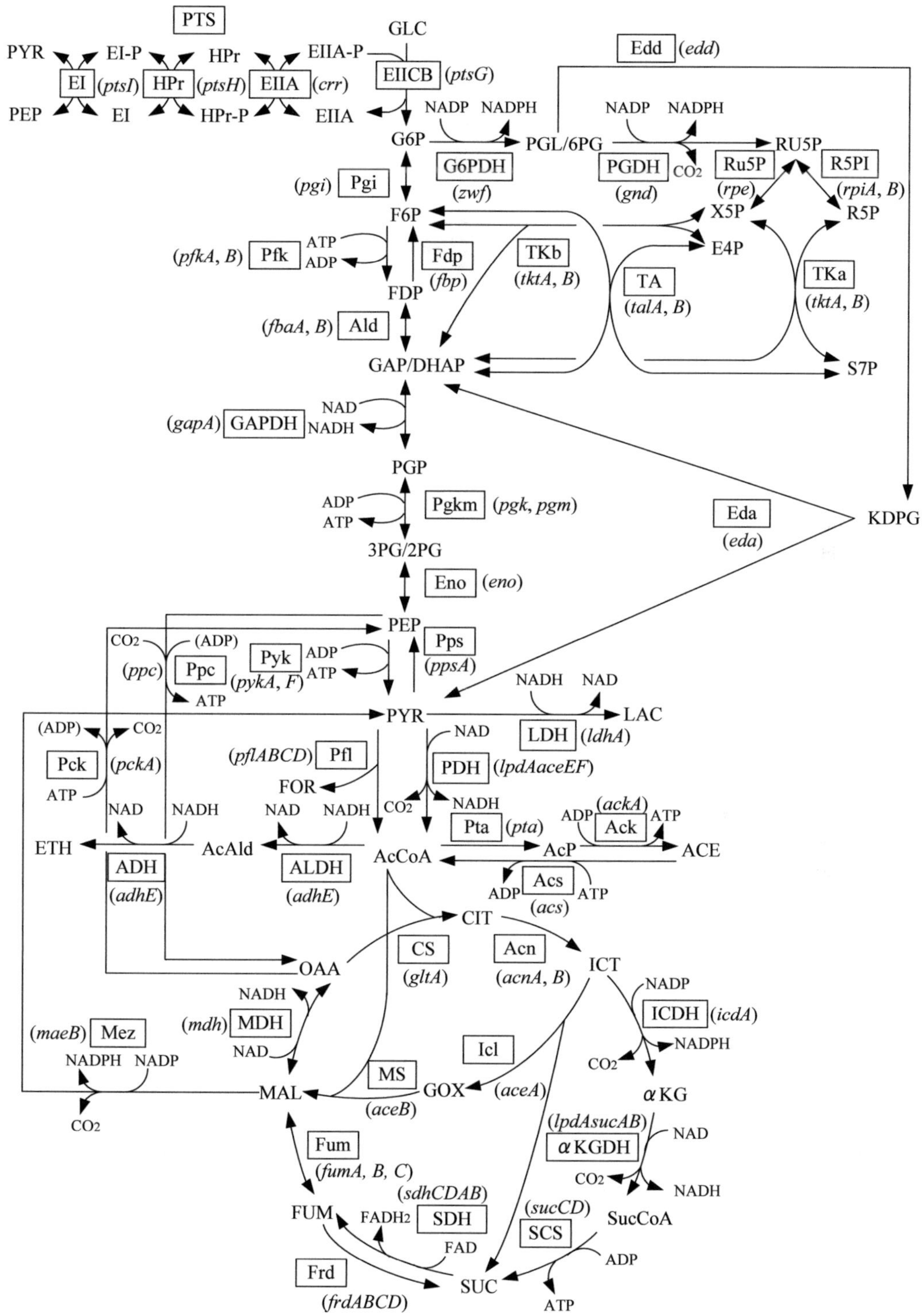

Fig. 11. Main metabolic pathways of *Escherichia coli*.

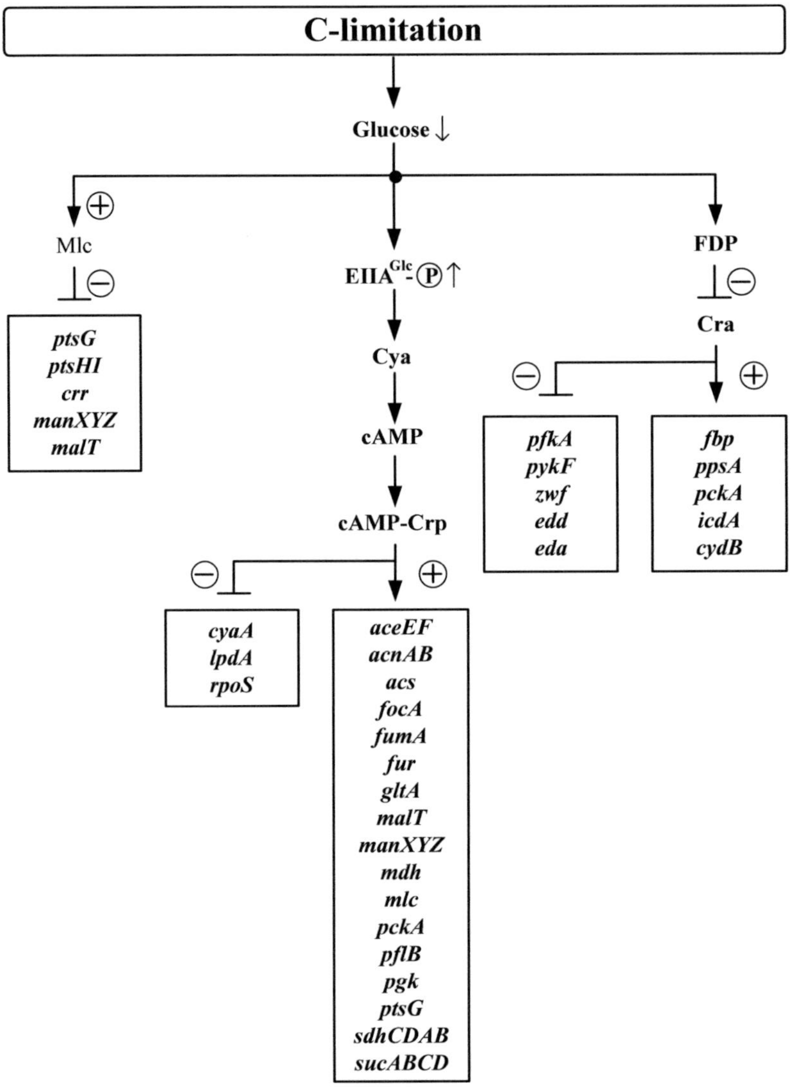

Fig. 12. Transcriptional regulation in response to carbon catabolite regulation.

6. Conclusion

^{13}C-Metabolic flux analysis is quite useful to understand the cell metabolism. In this chapter, analytical approach for ^{13}C-metabolic flux analysis is also explained to make the relationship between fluxes and isotopomer patterns to be transparent. Metabolic regulation occurs in response to culture environment. The global regulators sense such a change, and those regulate metabolic pathway genes. To predict cell metabolism in response to environmental change, the relationships between culture environment and global regulators and the relationships between global regulators and metabolic pathway genes have to be figured out. For this, it is desired to integrate different levels of information as well as ^{13}C-metabolic flux distribution.

7. Appendix A: The general expression for Φ.

$\Phi \equiv$

γA_0	$(T+G)A_0$	GA_0	GA_0	$(T+G)A_0$	GA_0	0	TA_0	0	0	GA_0	0	0	0	0	0
$(\beta+G)A_1$	$(2^{-1}T+G)A_1 +2^{-1}GA_0$	GA_1	$2^{-1}GA_1 +2^{-1}TA_0$	βA_1	GA_1	βA_0	$2^{-1}TA_1 +2^{-1}GA_0$	0	0	$2^{-1}GA_1 +2^{-1}TA_0$	0	0	βA_0	0	0
$(\beta+G)A_2$	$(2^{-1}T+G)A_2$	$GA_2 +\beta A_0$	$2^{-1}GA_2$	βA_2	$GA_2 +2^{-1}TA_0$	0	$2^{-1}TA_2$	$2^{-1}GA_0$	βA_0	$2^{-1}GA_2$	0	$2^{-1}TA_0$	0	$2^{-1}GA_0$	0
βA_2	$2^{-1}TA_2$	βA_0	$2^{-1}GA_2 +GA_0$	βA_2	$2^{-1}TA_0$	GA_0	$2^{-1}TA_2$	$(2^{-1}G+G)A_0$	βA_0	$2^{-1}GA_2$	GA_0	$2^{-1}TA_0$	0	$2^{-1}GA_0$	0
βA_1	$2^{-1}TA_1 +2^{-1}GA_0$	0	$2^{-1}GA_1 +2^{-1}TA_0$	$\beta A_1 +GA_0$	0	βA_0	$2^{-1}TA_1 + (2^{-1}G+G)A_0$	0	GA_0	$2^{-1}GA_1 +2^{-1}TA_0$	0	GA_0	βA_0	0	0
$(\beta+G)A_{12}$	$(2^{-1}T+G)A_{12}$	GA_{12}	$2^{-1}GA_{12}$	βA_{12}	$2^{-1}GA_0 +GA_{12}$	0	$2^{-1}TA_{12}$	$2^{-1}TA_0$	0	βA_{12}	βA_0	$2^{-1}GA_0$	0	$2^{-1}TA_0$	βA_0
0	$2^{-1}GA_2$	βA_1	$GA_1 +2^{-1}TA_2$	0	$2^{-1}TA_1$	$GA_1 + \beta A_2$	$2^{-1}GA_2$	$(2^{-1}G+G)A_1$	βA_1	$2^{-1}TA_2$	GA_1	$2^{-1}TA_1$	βA_2	$2^{-1}GA_1$	0
0	GA_1	0	TA_1	GA_1	0	$(T+G)A_1$	$2GA_1$	0	GA_1	TA_1	0	GA_1	$(T+G)A_1$	0	0
0	0	$(T+G)A_2$	GA_2	0	TA_2	GA_2	0	$2GA_2$	$(T+G)A_2$	0	GA_2	TA_2	0	GA_2	0
0	$2^{-1}GA_2$	βA_1	$2^{-1}TA_2$	GA_2	$2^{-1}TA_1$	βA_2	$(2^{-1}G+G)A_2$	$2^{-1}GA_1$	$\beta A_1 +GA_2$	$2^{-1}TA_2$	0	$2^{-1}TA_1 +GA_2$	βA_2	$2^{-1}GA_1$	0
βA_{12}	$2^{-1}TA_{12}$	0	$2^{-1}GA_{12}$	βA_{12}	$2^{-1}GA_0$	0	$2^{-1}TA_{12}$	$2^{-1}TA_0$	0	$GA_0 + \beta A_{12}$	βA_0	$2^{-1}GA_0$	GA_0	$(2^{-1}T+G)A_0$	$(\beta+G)A_0$
0	0	βA_{12}	GA_{12}	0	$2^{-1}GA_2 +2^{-1}TA_{12}$	GA_{12}	0	$2^{-1}TA_2 + (2^{-1}G+G)A_{12}$	βA_{12}	0	$\beta A_2 +GA_{12}$	$2^{-1}GA_2 +2^{-1}TA_{12}$	0	$2^{-1}TA_2 +2^{-1}GA_{12}$	βA_2
0	$2^{-1}GA_{12}$	0	$2^{-1}TA_{12}$	GA_{12}	$2^{-1}GA_1$	βA_{12}	$(2^{-1}G+G)A_{12}$	$2^{-1}TA_1$	GA_{12}	0	βA_1	$2^{-1}GA_1 +GA_{12}$	βA_{12}	$2^{-1}TA_1$	$(T+G)A_1$
0	$2^{-1}GA_{12}$	0	$2^{-1}TA_{12}$	0	$2^{-1}GA_1$	βA_{12}	$2^{-1}GA_{12}$	$2^{-1}TA_1$	0	GA_1	βA_1	$2^{-1}GA_1$	$GA_1 + \beta A_{12}$	$(2^{-1}T+G)A_1$	γA_1
0	0	βA_{12}	0	0	$2^{-1}GA_2 +2^{-1}TA_{12}$	0	0	$2^{-1}TA_2 +2^{-1}GA_{12}$	βA_{12}	GA_2	βA_2	$2^{-1}GA_2 +2^{-1}TA_{12}$	GA_2	$(2^{-1}T+G)A_2 +2^{-1}GA_{12}$	$(\beta+G)A_2$
0	0	0	0	0	GA_{12}	0	0	TA_{12}	0	GA_{12}	$(T+G)A_{12}$	GA_{12}	GA_{12}	$(T+G)A_{12}$	γA_{12}

8. Appendix B: Derivation of eq. (10) and (11) using Φ^n as $n \to \infty$

1. The case where $A_2 = 1$:

By inspection, if we assume n to be large enough, Φ^n can be reduced to

$$\Phi^n = \begin{bmatrix} 2G & G & G \\ 0.5T & 0.5(T+G) & 0.5T \\ 0.5T & 0.5(T+G) & 0.5T+G \end{bmatrix}^n \quad \text{(B1)}$$

where only the terms associated with O_{23}, O_{123} and O_{234} were left. Note that Φ can be diagonalized by multiplying the eigen matrix P and its inverse such that

$$P^{-1}\Phi P = \begin{bmatrix} \lambda_1 & 0 & 0 \\ 0 & \lambda_2 & 0 \\ 0 & 0 & \lambda_3 \end{bmatrix} \quad \text{(B2)}$$

where $\lambda_i \ (i=1,2,3)$ are the distinct eigen values of Φ. Since we have the relationship

$$\left(P^{-1}\Phi P\right)^n = P^{-1}\Phi^n P \quad \text{(B3)}$$

Φ^n can be expressed as

$$\Phi^n = P\left(P^{-1}\Phi P\right)^n P^{-1} = P\begin{bmatrix} \lambda_1{}^n & 0 & 0 \\ 0 & \lambda_2{}^n & 0 \\ 0 & 0 & \lambda_3{}^n \end{bmatrix} P^{-1} \tag{B4}$$

Here λ_i $(i=1,2,3)$ and P are obtained for Φ as follows:

$$\lambda_1 = G, \quad \lambda_2 = 2G+T, \quad \lambda_3 = \frac{G}{2}$$

$$P = \begin{bmatrix} 1 & 1 & 0 \\ 0 & \dfrac{1}{T+G}\left\{\dfrac{T(2G+T)^2}{G(3G+2T)} - T\right\} & 1 \\ -1 & \dfrac{T(2G+T)}{G(3G+2T)} & -1 \end{bmatrix} \tag{B5}$$

Then Φ^n could be calculated as

$$\Phi^n = \begin{bmatrix} G^n & (2G+T)^n & 0 \\ 0 & (2G+T)^n \dfrac{1}{T+G}\left\{\dfrac{T(2G+T)^2}{G(3G+2T)} - T\right\} & \left(\dfrac{G}{2}\right)^n \\ -G^n & (2G+T)^n \dfrac{T(2G+T)}{G(3G+2T)} & -\left(\dfrac{G}{2}\right)^n \end{bmatrix} \begin{bmatrix} \dfrac{T}{G+T} & -\dfrac{G}{G+T} & -\dfrac{G}{G+T} \\ \dfrac{G}{G+T} & \dfrac{G}{G+T} & \dfrac{G}{G+T} \\ -\dfrac{T}{3G+2T} & \dfrac{3G+T}{3G+2T} & -\dfrac{T}{3G+2T} \end{bmatrix}$$

$$= \begin{bmatrix} \dfrac{T}{G+T}G^n + \dfrac{G}{G+T}(2G+T)^n & \dfrac{G}{G+T}\left\{-G^n + (2G+T)^n\right\} & \dfrac{G}{G+T}\left\{-G^n + (2G+T)^n\right\} \\ \dfrac{T}{3G+2T}\left\{(2G+T)^n - \left(\dfrac{G}{2}\right)^n\right\} & \dfrac{T}{3G+2T}(2G+T)^n + \dfrac{3G+T}{3G+2T}\left(\dfrac{G}{2}\right)^n & \dfrac{T}{3G+2T}\left\{(2G+T)^n - \left(\dfrac{G}{2}\right)^n\right\} \\ -\dfrac{T}{G+T}G^n + \dfrac{T(2G+T)}{(3G+2T)(G+T)}(2G+T)^n + \dfrac{T}{3G+2T}\left(\dfrac{G}{2}\right)^n & \dfrac{G}{G+T}G^n + \dfrac{T(2G+T)}{(3G+2T)(G+T)}(2G+T)^n - \dfrac{3G+T}{3G+2T}\left(\dfrac{G}{2}\right)^n & \dfrac{G}{G+T}G^n + \dfrac{T(2G+T)}{(3G+2T)(G+T)}(2G+T)^n + \dfrac{T}{3G+2T}\left(\dfrac{G}{2}\right)^n \end{bmatrix} \tag{B6}$$

Noting that G is less than 1 and $2G+T=1$, we have for $n \to \infty$

$$\Phi^\infty = \begin{bmatrix} \dfrac{G}{G+T} & \dfrac{G}{G+T} & \dfrac{G}{G+T} \\ \dfrac{T}{3G+2T} & \dfrac{T}{3G+2T} & \dfrac{T}{3G+2T} \\ \dfrac{T(2G+T)}{(3G+2T)(G+T)} & \dfrac{T(2G+T)}{(3G+2T)(G+T)} & \dfrac{T(2G+T)}{(3G+2T)(G+T)} \end{bmatrix} \tag{B7}$$

Then, we can derive the following relationships:

$$\begin{bmatrix} O_{23} \\ O_{123} \\ O_{234} \end{bmatrix} = \begin{bmatrix} \dfrac{G}{G+T} \\ \dfrac{T}{3G+2T} \\ \dfrac{T(2G+T)}{(3G+2T)(G+T)} \end{bmatrix} \tag{B8}$$

which is equivalent to Eq. (11) already derived previously for the steady state.

2. The case where $A_1 = 1$:

In the similar was as above, Φ^n is expressed for n to be large enough as follows:

$$\Phi^n = \begin{bmatrix} 2G & G & G \\ 0.5T & 0.5(T+G) & 0.5T \\ 0.5T & 0.5(T+G) & 0.5T+G \end{bmatrix}^n \tag{B9}$$

where only the terms associated with O_{14}, O_4 and O_1 are left. Then the steady state values could be obtained as follows:

$$\begin{bmatrix} O_{14} \\ O_4 \\ O_1 \end{bmatrix} = \begin{bmatrix} \dfrac{G}{G+T} \\ \dfrac{T}{3G+2T} \\ \dfrac{T(2G+T)}{(3G+2T)(G+T)} \end{bmatrix} \tag{B10}$$

where this equation is equivalent to Eq. (10) as previously derived for the steady state.

9. Appendix C: Isotopomer pattern of OAA at the n th turn of the TCA cycle together with glyoxylate pathway

Let us consider deriving the general case for $A_2 = 1$, where Φ can be reduced from 16×16 to 8×8 due to the limited appearance of the isotopomer as

$$\Phi = \begin{bmatrix} 0 & 0 & 0 & 0 & 0 & 0 & 0 & 0 \\ 2^{-1}(T+G)+G & G & 2^{-1}G & 0 & 0 & 0 & 0 & 0 \\ 2^{-1}(T+G) & 0 & 2^{-1}G & 0 & 0 & 0 & 0 & 0 \\ 0 & 0 & 2^{-1}T & 2^{-1}(T+G) & 0 & 0 & 0 & 0 \\ 0 & T+G & G & G & 2G & T+G & G & G \\ 0 & 0 & 2^{-1}T & 2^{-1}(T+G) & 0 & G & 0 & 0 \\ 0 & 0 & 0 & 0 & 2^{-1}T & 0 & 2^{-1}(T+G) & 2^{-1}T \\ 0 & 0 & 0 & 0 & 2^{-1}T & 0 & 2^{-1}(T+G) & 2^{-1}T+G \end{bmatrix} \tag{C1}$$

Let this matrix be subdivided into 4 submatrices as

$$\Phi = \begin{bmatrix} A(4\times 4) & 0(4\times 4) \\ B(4\times 4) & D(4\times 4) \end{bmatrix} \tag{C2}$$

Then it can be shown that the following equation holds

$$\Phi^n = \begin{bmatrix} A^n & 0 \\ B^n = \sum_{r=0}^{n-1} D^r B A^{n-r-1} & D^n \end{bmatrix} \tag{C3}$$

where A^n and D^n can be expressed as

$$A^n = \begin{bmatrix} 0 & 0 & 0 & 0 \\ -\frac{T+G}{G}2^{-n}G^n + \frac{T+2G}{G}G^n & G^n & -2^{-n}G^n + G^n & 0 \\ \frac{T+G}{G}2^{-n}G^n & 0 & 2^{-n}G^n & 0 \\ -\frac{T+G}{G}2^{-n}G^n + 2^{-n}(T+G)^n & 0 & -2^{-n}G^n + 2^{-n}(T+G)^n & 2^{-n}(T+G)^n \end{bmatrix} \tag{C4a}$$

$D^n =$ (C4b)

$\frac{T}{T+G}G^n + \frac{G}{T+G}(T+2G)^n$	$-\frac{T^2+TG+G}{T+G}G^n + T\left(nG^{n-1}+G^n\right) + \frac{G}{T+G}(T+2G)^n$	$-\frac{G}{T+G}G^n + \frac{G}{T+G}(T+2G)^n$	$-\frac{G}{T+G}G^n + \frac{G}{T+G}(T+2G)^n$
0	G^n	0	0
$-\frac{T}{2T+3G}2^{-n}G^n + \frac{T}{2T+3G}(T+2G)^n$	$-TG^{n-1} + \frac{2T(G+T)}{G(2T+3G)}2^{-n}G^n + \frac{T}{2T+3G}(T+2G)^n$	$\frac{T+3G}{2T+3G}2^{-n}G^n + \frac{T}{2T+3G}(T+2G)^n$	$-\frac{T}{2T+3G}2^{-n}G^n + \frac{T}{2T+3G}(T+2G)^n$
$\frac{T}{T+G}G^n + \frac{T}{2T+3G}2^{-n}G^n + \frac{T(T+2G)}{(T+G)(2T+3G)}(T+2G)^n$	$\frac{T^2+TG+G}{T+G}G^n + T\left(-nG^{n-1} + \frac{-TG+T-G}{TG}G^n\right) \frac{2T(G+T)}{G(2T+3G)}2^{-n}G^n + \frac{T(T+2G)}{(T+G)(2T+3G)}(T+2G)^n$	$\frac{G}{T+G}G^n \frac{T+3G}{2T+3G}2^{-n}G^n + \frac{T(T+2G)}{(T+G)(2T+3G)}(T+2G)^n$	$\frac{G}{T+G}G^n + \frac{T}{2T+3G}2^{-n}G^n + \frac{T(T+2G)}{(T+G)(2T+3G)}(T+2G)^n$

10. Acknowledgement

This research was supported in part by Strategic International Cooperative Program, Japan Science and Technology Agency (JST).

11. References

Aiba, H. (1995). The *lac* and *gal* operons today. In "Regulation of Gene expression in *Escherichia coli.*", In: Lin, E. C. C. & Lynch, A. S. (Ed.), Chap 9, pp. 182-200, Landes

Bailey, J. E. (1991). Toward a science of metabolic engineering. *Science*, vol.252, pp. 1668-1675

Bettenbrock, K.; Fischer, S.; Klemling, A.; Sauter, F. T.; Gilles, E. D. (2006) A quantitative approach to catabolite repression in *Escherichia coli. J. Biol. Chem.* Vol.281, pp. 2578-2584

Blank, L. M. & Sauer, U. (2004). TCA cycle activity in *Saccharomyces cerevisiae* is a function of the environmentally determined specific growth and glucose uptake rate. *Microbiology*, vol.150, pp. 1085-1093

Chance, E. M.; Seeholzer, S. H.; Kobayashi, K.; Williamson, J. R. (1983). Mathematical analysis of isotope labeling in the citric acid cycle with applications to ^{13}C NMR studies in perfused rat hearts. *J. Biol. Chem.*, vol.258, No.22, pp. 13785-13794

Edwards, J. S. & Palsson, B. O. (2000). The *Escherichia coli* MG1655 in silico metabolic genotype: its definition, characteristics, and capabilities. *PNAS*, vol.97, pp. 5528-5533

Fischer, E. & Sauer, U. (2003). Metabolic flux profiling of *Escherichia coli* mutants in central carbon metabolism by GC-MS. *Eur. J. Biochem.*, vol.270, pp. 880-891

Fuhrer, T.; Fischer, E.; Sauer, U. (2005). Experimental identification and quantification of glucose metabolism in seven bacterial species. *J. Bacteriol.*, vol.187, pp. 1581-1590

Hellerstein, M. K. (2004). New stable isotope-mass spectrometric techniques for measuring fluxes through intact metabolic pathways in mammarian systems: introduction of moving pictures into functional genomics and biochemical phenotyping. *Metab. Eng.*, vol.6, pp. 85-100

Hua, Q.; Yang, C.; Baba, T.; Mori, H.; Shimizu, K. (2003). Responses of the central carbon metabolism in *Escherichia coli* to phosphoglucose isomerase and glucose-6-phosphate dehydrogenase knockouts. *J. Bacteriol.*, vol.185, pp. 7053-7067

Kelleher, J. K. (2004). Probing metabolic pathways with isotopic tracers: insights from mammarian metabolic pathways. *Metab. Eng.*, vol.6, pp. 1-5

Kremling, A.; Fischer, S.; Sauter, T.; Bettenbrock, K.; Gilles, E. D. (2004). Time hierarchies is the *Escherichia coli* carbohydrate uptake and metabolism. *BioSystems*, vol.73, pp. 57-71

Kremling, A. & Gilles, E. D. (2001). The organization of metabolic reaction networks. II. Signal processing in hierarchical structured functional units. *Metab. Eng.*, vol. 3, pp. 138-150

Lee, S. J.; Boos, W.; Bouché, J. P.; Plumbridge, J. (2000). Signal transduction between a membrane-bound transporter, PtsG, and a soluble transcription factor, Mlc, of *Escherichia coli. EMBO J.*, vol.19, pp. 5353-5361

Marx, A.; Graaf, A. A.; Wiechert, W.; Eggeling, L.; Sahm, H. (1996). Determination of the fluxes in the central metabolism of *Corynebacterium glutamicum* by nuclear magnetic resonance spectroscopy combined with metabolite balancing. *Biotechnol. Bioeng.*, vol.49, pp. 111-129

Matsuoka, Y. & Shimizu, K. (2010a). Current status of ^{13}C-metabolic flux analysis and future perspectives. *Process Biochemistry*, vol.45, pp. 1873-1881

Matsuoka, Y. & Shimizu, K. (2010b). The relationships between the metabolic fluxes and ^{13}C-labeled isotopomer distribution for the flux analysis of the main metabolic pathways. *Biochem. Eng. J.*, vol. 49, pp. 326-336

Matsuoka, Y. & Shimizu, K. (2011) Metabolic regulation in *Escherichia coli* in response to culture environments via global regulators. *Biotechnol. J.*, DOI: 10.1002/biot.201000447

McCabe, B. J.& Previs, S. F. (2004). Using isotope tracers to study metabolism: Application in mouse models. *Metabolic Eng.*, vol.6, pp. 25-35

Merle, M.; Martin, M.; Villegier, A.; Canioni, P. (1996). Mathematical modelling of the citric acid cycle for the analysis of glutamate isotopomers from cerebellar astrocytes incubated with [1-^{13}C] glucose. *Eur. J. Biochem.*, vol.239, pp. 742-751

Moat, A. G.; Foster, J. W.; Spector, M. P. (2002). (Ed.), In Microbial Physiology, John Wiley & Sons

Park, Y. H.; Lee, B. R.; Seok, Y. J.; Peterkofsky, A. (2006). *In vitro* reconstruction of catabolite repression in *Escherichia coli*. *J. Biol. Chem.*, vol.281, pp. 6448-6454

Plumbridge, J. (1998). Expression of *ptsG*, the gene for the major glucose PTS transporter in *Escherichia coli*, is repressed by Mlc and induced by growth on glucose. *Mol. Microbiol.*, vol.29, pp. 1053-1063

Price, N. D; Papin, J. A.; Schilling, C. H.; Palsson, B. O. (2003). Genome-scale microbial *in silico* models: the constraints-based approach. *Trends Biotechnol.*, vol.21, pp. 162-169

Raghevendran, V.; Gombert, A. K.; Christensen, B.; Kotter, P.; Nielsen, J. (2004). Phenotypic characterization of glucose repression mutants of *Saccharomyces cerevisiae* using experiments with 13C-labelled glucose. *Yeast*, vol. 21, pp. 769-779

De Reuse, H.; Danchin, A. (1988). The *ptsH*, *ptsI* and *crr* genes of the *Escherichia coli* phosphoenolpyruvate-dependent phosphotransferase system: a complex operon with several modes of transcription. *J. Bacteriol.*, vol.179, pp. 3827-3837

Rothman, D. L.; Behar, K. L.; Hyder, F.; Shulman, R. G.. (2003). *In vivo* NMR studies on the glutamate neurotransmitter flux and neuroenergetics: implications for brain function. *Ann. Rev. Physiol.*, vol.65, pp. 401-427

Saier, M. H., Jr. (1996). Cyclic AMP-independent catabolite repression in bacteria. *FEMS Microbiol. Lett.*, vol.138, pp. 97-103

Saier, M. H., Jr. & Ramseier, T. M. (1996). The catabolite repressor/activator (Cra) protein of enteric bacteria. *J. Bacteriol.*, vol.178, pp. 3411-3417

Saier, M. H., Jr.; Ramseier, T. M.; Reizer, J. (1996). Regulation of carbon utilization, in: Neidhardt, F. C., Curtiss III, R., Ingraham, J. L., Lin, E. C. C., Low, K. B., Magasanik, B., Reznikoff, W. S., Riley, M., Schaechter, M., and Umbarger, H. E. (Ed.), *Escherichia coli* and *Salmonella*: cellular and molecular biology, 2nd ed., vol. 1. ASM Press, pp.1325-1343, Washington, D.C.

Sarkar, D.; Siddiquee, K. A.; Araúzo-Bravo, M. J.; Oba, T.; Shimizu, K. (2008). Effect of *cra* gene knockout together with *edd* and *iclR* genes knockout on the metabolism in *Escherichia coli*. *Arch. Microbiol.*, vol. 190, pp. 559-571

Sarkar, D.; Yabusaki, M.; Hasebe, Y.; Ho, P. Y.; Kohmoto, S.; Kaga, T; Shimizu, K. (2010). Fermentation and metabolic characteristics of *Gluconacetobacter oboediens* for different carbon sources. *Appl. Microbiol. Biotechnol.*, vol. 87, pp. 127-136

Sauer, U. (2004). High-throughput phenomics: experimental methods for mapping fluxomes. *Curr. Opin. Biotechnol.*, vol.15, pp. 58-63

Sauer, U. (2006). Metabolic networks in motion: 13C-based flux analysis. *Molecular systems biology*, 2:62.

Sauer, U.; Hatzimanikatis, V.; Bailey, J. E.; Hochuli, M.; Szyperski, T.; Wüthrich, K. (1997). Metabolic fluxes in riboflavin-producing *Bacillus subtilis*. *Nat. Biotechnol.*, vol.15, pp. 448-452

Sauer, U.; Lasko, D. R.; Fiaux, J.; Hochuli, M.; Glaser, R.; Szyperski, T.; Wüthrich, K.; Bailey, J. E. (1999). Metabolic flux ratio analysis of genetic and environmental modulations of *Escherichia coli* central carbon metabolism. *J. Bacteriol.*, vol.181, pp. 6679-6688

Schilling, C. H. & Palsson, B. O. (1998). The underlying pathway structure of biochemical reaction netwoks. *PNAS*, vol.95, pp. 4193-4198

Schmidt, K.; Nielsen, J.; Villadsen, J. (1999). Quantitative analysis of metabolic fluxes in *Escherichia coli*, using two-dimensional NMR spectroscopy and complete isotopomer models. *J. Biotechnol.*, vol.71, pp. 175-190

Schuetz, R.; Kuepfer, L.; Sauer, U. (2007). Systematic evaluation of objective functions for predicting intracellular fluxes in *Escherichia coli*. *Mol. Syst. Biol.*, vol.3, pp. 1-14

Schwender, J.; Goffman, F.; Ohlrogge, J.; Shachar-Hill, Y. (2004). Rubisco without the Calvin cycle improves the carbon efficiency of developing green seeds. *Nature*, vol.432, pp. 779-782

Selivanov, V. A.; Meshalkina, L. E.; Solovjeva, O. N.; Kuche, P. W.; Ramos-Montoya, A.; Kochetov, G. A.; Lee, P. W. N.; Cascante, M. (2005). Rapid simulation and analysis of isotopomer distributions using constraints based on enzyme mechanisms: an example from HT29 cancer cells. *Bioinformatics*, vol.21, pp. 3558-3564

Selivanov, V. A.; Sukhomlin, T.; Centelles, J. J.; Lee, P. W. N.; Cascante, M. (2006). Integration of enzyme kinetic models and isotopomer distribution analysis for studies of in situ cell operation. *BMC Neuroscience*, vol.7, pp. 1-15

Sherry, A. D.; Jeffrey, F. M.; Malloy, C. R. (2004). Analytical solutions for 13C isotopomer analysis of complex metabolic conditions: substrate oxidation, multiple pyruvate cycles, and gluconeogenesis. *Metab. Eng.*, vol.6, pp. 12-24

Sidorenko, Y.; Wahl, A.; Dauner, M.; Genzel, Y.; Reichl, U. (2008). Comparison of metabolic flux distributions for MDCK cell growth in glutamine- and pyruvate-containing media. *Biotech. Prog.*, vol.24, pp. 311-320

Stephanopoulos, G. N. (1999). Metabolic fluxes and metabolic engineering. *Metab. Eng.*, vol.1, pp. 1-11

Stephanopoulos, G. N.; Aristidou, A. A.; Nielsen, J. (1998). Metabolic Engineering, Acad. Press, SanDiego

Stephanopoulos, G.. & Vallino, J. J. (1991). Network rigidity and metabolic engineering in metabolite overproduction. *Science*, vol. 252, pp. 1675-1681

Szyperski, T. (1995). Biosynthetically directed fractional ^{13}C-labeling of proteinogenic amino acids —An efficient analytical tool to investigate intermediary metabolism. *Eur. J. Biochem.*, vol.232, pp. 433-448

Szyperski, T.; Glaser, R. W.; Hochuli, M.; Fiaux, J. (1999). Bioreaction network topology and metabolic flux ratio analysis by biosynthetic fractional ^{13}C labeling and two-dimensional NMR spectroscopy. *Metab. Eng.*, vol.1, pp. 189-197

Tanaka, Y.; Kimata, K.; Aiba, H. (2000). A novel regulatory role of glucose transporter of *Escherichia coli*: membrane sequestration of a global repressor Mlc. *EMBO J.*, vol.19, pp. 5344-5352

Tran-Dinh, S.; Beganton, F.; Nguyen, T.; Bouet, F.; Herve, M. (1996b). Mathematical model for evaluating the Krebs cycle flux with non-constant glutamate-pool size by ^{13}C-NMR spectroscopy. Evidence for the existence of two types of Krebs cycles in cells. *Eur. J. Biochem.*, vol.242, pp. 220-227

Tran-Dinh, S.; Bouet, F.; Huynh, Q.; Herve, M. (1996a). Mathematical models for determining metabolic fluxes through the citric acid and the glyoxylate cycles in *Saccharomyces cerevisiae* by ^{13}C-NMR spectroscopy. *Eur. J. Biochem.*, vol.242, pp. 770-778

Vemuri, G. N. & Aristidou, A. A. (2005). Metabolic engineering in the –omics era: elucidating and modulating regulatory networks. *Microbiol. Mol. Biol. Rev.*, vol.69, pp. 197-216

Wiechert, W. & de Graaf A. A. (1997). Bidirectional reaction steps in metabolic networks: I. Modeling and simulation of carbon isotope labeling experiments. *Biotech. Bioeng.*, vol.55, pp. 101-117

van Winden, W.; Schipper, D.; Verheijen, P.; Heijen, J. (2001). Innovations in generation and analysis of 2D [^{13}C, ^{1}H] COSY NMR spectra for metabolic flux analysis purpose. *Metab. Eng.*, vol.3, pp. 322-343

Wittmann, C. (2007). Fluxome analysis using GC-MS. *Microbiol. Cell Factories*, vol.6, pp. 1-17

Yang, C.; Hua, Q.; Shimizu, K. (2002). Metabolic flux analysis in Synechocystis using isotope distribution from ^{13}C-labeled glucose. *Metab. Eng.*, vol.4, pp. 202-216

Yang, C.; Hua, Q.; Shimizu, K. (2002). Integration of the information from gene expression and metabolic fluxes for the analysis of the regulatory mechanisms in *Synechocystis*. *Appl. Microbiol. Biotech.*, vol.58, pp. 813-822

Zhao, J.; Baba, T.; Mori, H.; Shimizu, K. (2004). Global metabolic response of *Escherichia coli* to *gnd* or *zwf* gene-knockout, based on ^{13}C-labeling experiments and the measurement of enzyme activities. *Appl. Microbiol. Biotech.*, vol.64, pp. 91-98

Zhu, J. & Shimizu, K. (2004). The effect of *pfl* genes knockout on the metabolism for optically pure D-lactate production by *Escherichia coli*. *Appl. Microbiol. Biotechnol.*, vol.64, pp. 367-375

Zhu, J. & Shimizu, K. (2005). Effect of a single-gene knockout on the metabolic regulation in *E. coli* for D-lactate production under microaerobic condition. *Metab. Eng.*, vol.7, pp. 104-115

19

QCM as Cell-Based Biosensor

Tsong-Rong Yan, Chao-Fa Lee and Hung-Che Chou
Institute of Bioengineering, Tatung University
Taiwan, R.O.C.

1. Introduction

1.1 History of the quartz crystal microbalance

Piezoelectricity is defined as electric polarization produced by mechanical strain in certain crystals, the polarization being proportional to the strain (Cady, 1946). The Curies first observed piezoelectricity in 1880 as a potential difference generated across two surfaces of a quartz crystal under strain. The converse piezoelectric effect, the deformation of a piezoelectric material by an applied electric field, was predicted by Lippmann (Lippmann, 1881). In 1959 Sauerbrey provided a description and experimental proof of the mass-frequency relation for foreign layers deposited on thickness-shear mode crystals that are still widely used today for determination of mass-sensing format is commonly referred to as the QCM. The mass increases when a film is deposited onto the crystal surface and monolayer sensitivity is easily reached (Lu & Lewis, 1972). However, when the film thickness increases, viscoelastic effects come into play. In the late 80's, it was recognized that the QCM can also be operated in liquids (Konash & Bastiaans, 1980), if proper measures are taken to overcome the consequences of the large damping and viscoelastic effects contribute strongly to the resonance properties (Bruckenstein & Shay, 1985; Ward & Buttery, 1990). Today, microweighing is one of several uses of the QCM (Johannsmann, 2008). Measurements of viscosity and more general, viscoelastic properties, are of much importance as well. The "non-gravimetric" QCM is by no means an alternative to the conventional QCM. Many researchers, who use quartz resonators for purposes other than gravimetry, have continued to call the quartz crystal resonator "QCM". Actually, the term "balance" makes sense even for non-gravimetric applications if it is understood in the sense of a force balance. At resonance, the force exerted upon the crystal by the sample is balanced by a force originating from the shear gradient inside the crystal. This is the essence of the small-load approximation.

1.2 Biological applications of the QCM

The use of biological coatings, such as enzymes and antibodies, was a natural progression from the initial piezoelectric (Pz) sensor development. The first reported gas phase biosensor was in fact an enzyme-based Pz sensor (Guilbault, 1983). The first Pz immunosensor for microbial pathogens was developed by Muramatsu et al. in 1986 (Muramatsu et al., 1986). Their Pz crystals were coated with antibodies against *Candida albicans*. The antigen was detected from 1×10^6 to 5×10^8 cells mL^{-1}. The sensor proved to be specific with no detectable response observed with the other species tested. Conventional bacterial detection methods generally require laborious procedures and many hours or even days for complete analysis. The relatively large mass of bacterial cells, combined with the availability of antibodies to most species means that Pz immunosensor detection offers a very attractive alternative to microbiological methods. Generally giving results in minutes, with adequate sensitivity and selectivity. Su & Li design the automatic QCM immunosensor system was fabricated by immobilizing anti-Salmonella polyclonal antibodies on 8 MHz AT-cut quartz crystal with Protein A method. The oscillation frequency of the QCM sensor was monitored in real time (Su & Li, 2004). The direct detection of nucleic acid interactions based on the use of acoustic wave devices was first provided by Fawcett et al. (Fawcett et al., 1988). They described a quartz crystal microbalance (QCM)-based biosensor for DNA detection by immobilizing single-stranded DNA onto quartz crystals and detecting the mass changes after hybridization. Since this early work, a number of articles have appeared employing similar procedures, resulting in microgravimetric measurements based on nucleic acids (Itamar et al., 2002). A QCM detection system consisted of a sensitive element (QCM wafer), an inlet subsystem, and a frequency acquisition unit. In Hao et al. work, a DNA probe functionalized quartz crystal microbalance (QCM) biosensor was developed to detect *B. anthracis* based on the recognition of its specific DNA sequences (Hao et al., 2011).

1.3 The QCM in cell biological

The QCM was already well-known and established as an analytical tool for studying adsorption phenomena at the solid-liquid interface (Martin et al., 1991; Buttry & Ward, 1992) when its potential for studying cell-substrate adhesion was recognized. It was then recognized that the adhesion of cells to the quartz surface also induced a shift in resonance frequency that was shown to be linearly correlated with the fractional surface coverage (Gryte et al., 1993; Redepenning et al., 1993). Time-resolved measurements of the resonance frequency were then used to follow the attachment and spreading of cells to the quartz surface, with extraordinary time resolution. Comparison with established cytological techniques has proven that the QCM readout reports reliably on the number of cells on the surface and the time course of adhesion (Wegener et al., 1998). Specifically, they have demonstrated that QCM is a convenient method for measuring in real time the attachment of African Green Monkey kidney cells as well as their detachment that accompanies cell death caused by either NaOH addition or virus infection. Additionally, the observed QCM responses are consistent with attached cell layers that behave as viscous films rather than rigid masses, illustrating that care must be exercised in the interpretation of QCM measurements involving cell adhesion or binding events.

Impedance analysis of cell-covered gold-film electrodes was first introduced by Giaever and Keese (Giaever & Keese, 1991, 1993), who showed that the frequency dependent electrical impedance for such a system is dependent on the impedance of the cell-free electrode, the resistance between adjacent cells, Rb, the capacitance of the cell membranes, Cm, and an additional impedance contribution that is associated with the current flow between the basal plasma membrane and the substrate. The resistance, Rb, is dependent on the width of the intercellular cleft between adjacent cells and the establishment of tight junctions, and is therefore a sensitive parameter for cell–cell contacts. Thus, impedance analysis of the shear displacement (from now on referred to as "quartz mode") and impedance analysis of the cell-covered surface electrode (referred to as "cell mode") may provide independent information about the establishment of cell–substrate and cell–cell contacts. A computer-controlled relay allows switching between both modes.

Pioneering works using the QCM showed that viscoelastic properties of films deposited on the sensor surface can be examined using frequency changes (Kanazawa & Gordon, 1985; Okahata et al., 1989). However, measurement of alone is not easily and specifically related to structural or conformational changes occurring at the sensor surface, especially for cytoskeletal changes that imply cellular volume and mass redistribution over the adhesion surface. Additionally, the energy dissipation needs to be monitored to detect cytoskeletal changes because cell responses can be interpreted (or modeled) as a purely liquid effect as well as a purely elastic mass effect (Marx et al., 2003). Most real scenarios are actually in between these two limits. Rodahl et al. extended the QCM technique by performing the direct monitoring of energy dissipation (D) considered a useful indicator of the viscoelastic properties of the material deposited at the sensor surface (Rodahl et al., 1996a, 1996b). Among the different QCM-D investigations of cells, it was showed that the interaction between cells and a surface produced a unique and reproducible dissipation against frequency plot, also referred to as a QCM-D signature (Fredriksson, 1998a, 1998b).

These new tools were called as the whole-cell QCM biosensor. Initial use of the QCM was for measuring chemicals in the gas phase binding the QCM surface and more recently as a tool in analytical electrochemistry.

In the past decades the QCM has been used to create biosensors. Infrequently, whole cells have been studied at the QCM surface (Gryte et al., 1993; Muratsugu et al., 1997; Wegener et al., 1998). Marx suggests the potential of this cellular biosensor for the real-time identification or screening of all classes of biologically active drugs or biological macromolecules that affect cellular attachment and cellular spreading, regardless of their molecular mechanism of action (Marx et al., 2003, 2007). The interconnected cytoskeleton structural elements depicted provide targets for drugs or macromolecules that bind and alter their structures or dynamical properties. Biosensor output is proportional to a particular detected concentration of the drug or macromolecule effector.

1.4 Scope of the present study

Applications for the QCM have been invested in our laboratory for a long-term. We combine some surface modify methods to immobilize the oligonucleotide probe on the gold electrode surface of QCM that could use to detect the PCR product of *Vibrio parahaemolyticus* and breast cancer. We study a modified antibody-coated QCM biosensor will be constructed for detecting food pathogenic bacteria *Vibrio parahaemolyticus* and to find the most suitable surface modification method. We have recently started to observe and research the cell growth process on QCM. Utilize a silver electrode QCM mass sensor to detect the physiology of cells by the plasma surface modification as a protection film (Chou et al., 2009). Develop a new method that utilizes the QCM technique to measure the tight junction of the Caco-2 cell (Chou & Yan, 2010). Using QCM to design a cell detection platform is an important application of QCM. In the process of system design and application, there still have some points, which are easily overlooked or are not understood by researchers. Most of the problems can be separated into three types. One is the cell culture environment conditions that affect the system signals, another is the system unification or lack of communication over regulations, and the other is the over-analysis of the experimental data (Chou et al., 2009). A QCM has been constructed and used for chemical sensing and biochemical detecting. Frequency shift has a relation with mass changed in the gaseous condition. However, the velocity, density, and viscosity also cause frequency changed in liquid solution. We develop a new method and system to investigate the variable value of amplitude and frequency from the oscillator of quartz (Lee et al., 2009). A biosensor consists of the immobilized biological recognition element (e.g. antibody, DNA, enzyme, receptor, microorganism, or cell) in intimate contact with a signal transducer (electrochemical, optical, thermal, or acoustic signal) that together permit analysis of chemical properties or quantities. We show that the QCM technique is a sensitive method not only to measure quasi-static cell adhesion processes but also dynamic changes of mechanical cell properties. A new cell-based QCM biosensor system which can real-time monitor the adhesion and growth of animal cells will be developed. A QCM chip sensor can measure the frequency, amplitude, trans-epithelial resistance, and A/C impedance. These materials were all fabricated under the cell unit to monitor cell growth conditions (Lee et al., 2010).

2. Electronic behavior of quartz resonators

QCM (Quartz crystal microbalance) refers to a sensor based on a piezo-electric. A QCM can be used in fixed and immobility area, and the mass can be found while changing the surface of the QCM with the relation between mass and frequency in Sauerbrey formula (Sauerbrey, 1959). The change of frequency (Δf) is proportional to the added mass (ΔM) Eq. (1). The characteristic can merely be adapted to special application. QCM can be used as gas detectors when coated with various materials (King, 1964).

$$\Delta f = \frac{2 f_0^2}{\sqrt{\mu_q \rho_q}} \frac{\Delta M}{A} \tag{1}$$

f_0, μ_q, ρ_q, and A denote the specific resonant frequency, density, shear modulus and the piezoelectrically active area of quartz, respectively.

The use of biological coatings, such as enzymes and antibodies, was a natural progression from the initial QCM development. Guilbault reported an enzyme-based QCM sensor in gas phase (Guilbault, 1983). A quartz immunosensor for microbial pathogens was developed by Muramatsu et al. (Muramatsu et al., 1986). Since then, liquid QCM began to be applied to cell biology, biotechnology, immunology and electrochemistry. Kanazawaand and Gordan prospected that Δf as a linear function of a square root of $\rho_q \times \eta_q$ and $\rho_w \times \eta_w$ (density, ρ and viscosity, η), as shown in Eq. (2) (Kanazawa & Gordon, 1985). It was discovered that changes in the physical-chemical property of the contacting materials/media will result in frequency changes of the quartz crystal resonator when used in liquid phase (Muramatsu et al., 1988; Shumacher, 1990).

$$\Delta f_w \propto \sqrt{\rho_q \times \eta_q} - \sqrt{\rho_w \times \eta_w} \tag{2}$$

ρ_w, η_w represent the change of dissolvent in liquid phase.

However, these sealed apparatuses open up the surface functional analysis. A kind of variability situation involves in a liquid media, there are some and important significant phenomena due to liquid contact. For the quartz crystal in contact with liquid, the resonant resistance change was first derived by Muramatsu et al., as shown in Eq. (3).

$$\Delta R = (2\pi 2\rho_1 \eta)^{1/2} A/k^2 \quad (3)$$

where k is the electromechanical coupling factor.

References are also made to other types of electrochemical detection techniques, such as impedance analysis that measures both resistance and reactance (Mirsky et al., 1997). The oscillation was damped and the Q factor was significantly reduced (Auge et al., 1994). Q factor formula can be described in liquid applications as shown in Eq. (4).

$$Q \approx \frac{\pi \rho_q v_q^2}{\sqrt{2\omega \rho_w \eta_w}} \quad (4)$$

where v_q is the velocity of the liquid at the surface and ω is the angular frequency.

Impedance analysis of cell-covered gold-film electrodes and the frequency dependent electrical impedance for such a system is dependent on the impedance associated with the current flow between the basal plasma membrane and the substrate (Giaever & Keese, 1991, 1993). More information about changes in dissipation (D) can be obtained via either impedance spectroscopy by recording the oscillation decay or the freely oscillating crystal after rapid excitation at resonance (Rodahl et al., 1995), the $f = f_0 - f_r$ and D are obtained directly via the relation as shown in Eq. (5):

$$D = \frac{1}{\pi f t} \quad (5)$$

where f_0 is the resonance frequency of the crystal, and f_r is a constant reference frequency.

Another variety of the reaction from electrochemical and quartz crystal microbalance measurement provides a powerful utility in studying solution phase transformations (Bard & Faulkner, 2001). When solution phase reactions are considered, there is also a capacitive component of the current that decays exponentially in time according to the following equation, as shown in Eq. (6).

$$I_c(t) = \frac{\Delta E}{R} e^{-t/RC} \quad (6)$$

where R is resistance (ohms) and C is the capacitance (F).

The above mentioned equation is useful for analysis on QCM sensor. In situ localizes a non-rigid coating, the damping behavior of the crystal is also required and can be evaluated by QCM-D (Edvardsson et al., 2006), electrical impedance (Wudy et al., 2008) and trans-epithelial resistance (TEER). To establish QCM cell-culture instrument, it is a useful method to monitor the interactions between the cells with the surface of a non-invasive type. The attitude about the entire cells investigated with the process of spreading, separation, movement or ingestion (Tan et al., 2008).

3. A construction of QCM measurement

3.1 Overview of the system Integration

A quartz crystal microbalance cell-culture incubator system includes environmental control system, QCM detecting system and data collection system. An environmental control system is divided into two major parts, incubator and cell-container. A QCM detecting system is a real-time measurement of the shift in resonance frequency and energy dissipation due to changes in mass and viscoelastic properties. A trans-epithelial resistance is also monitored in this system. This system was designed two channels of QCM to monitor the grow cell. Use 8-bits processor (Pic18F45J10, Microchip) and a complex programmable logic device (CPLD, LCMXO640, Lattice) to construct measurement and control circuit. All signal materials can make a hand-shaking, recording and controlled via computer synchronically (Fig. 1).

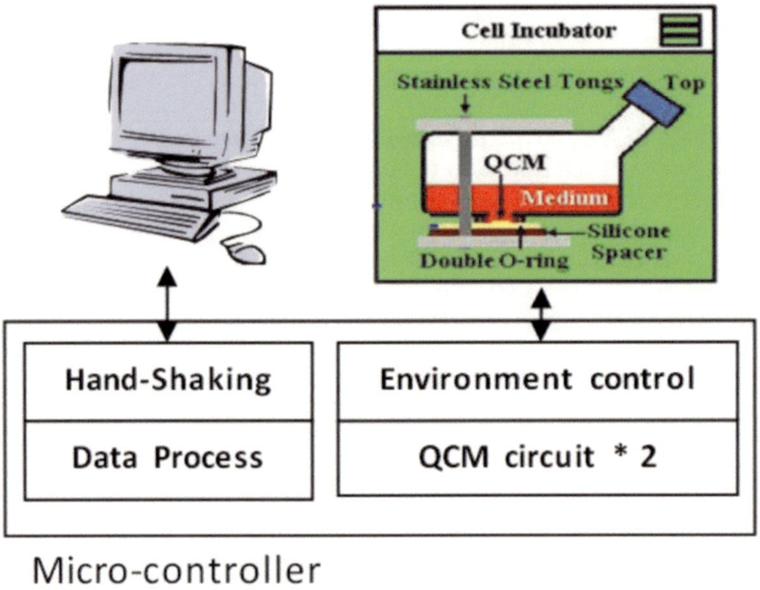

Fig. 1. Schematic diagram of the proposed QCM cell detection system.

Cells can be isolated from tissues for *ex vivo* culture in several ways. Mononuclear cells can be released from soft tissues by enzymatic digestion with enzymes such as collagenase, trypsin, or pronase, which break down the extracellular matrix. Alternatively, pieces of tissue can be placed in growth media, and the cells that grow out are available for culture. There are numerous well established cell lines representative of particular cell types. Fig. 2A shows the schematic system design of this study. The system includes three parts: the cell incubator chamber, the QCM sensor chip, and the circuit system. Signals are detected by a sensor chip, stored on a personal computer (PC), and are processed by custom made software. The cell incubator chamber is made of Pyrex1 glass, together with stainless steel tongs and other accessories to sterilize the autoclave directly. The QCM biosensor used in this study was purchased from a local source (Mercury Electronics Ind. Co., Ltd., Taiwan) and only the bare crystal was fixed on a ceramic base.

Fig. 2B shows that the QCM is inlaid on the ceramic substrate so that the twist force does not directly contact the chip surface. This design stabilizes the chip effectively during oscillation when the chip is connected to the incubator chamber. The gold electrode connected to the chip is behind the ceramic substrate to avoid any biotoxicity caused by line materials in the connect circuit contacting the cell culture solution. By having the ceramic substrate on the double o ring, it not only prevents solution leakage from the cell incubator chamber, but also reduces the possibility of contamination. All chip materials that contact the cells have improved biocompatibility so they do not influence the growth of cells. A Glass tank used to culture the cell and has a detection portion in the underside. Glass tank is made of Pyrex® glass and chip material choice has a good biocompatibility. All materials do not affect cell growth.

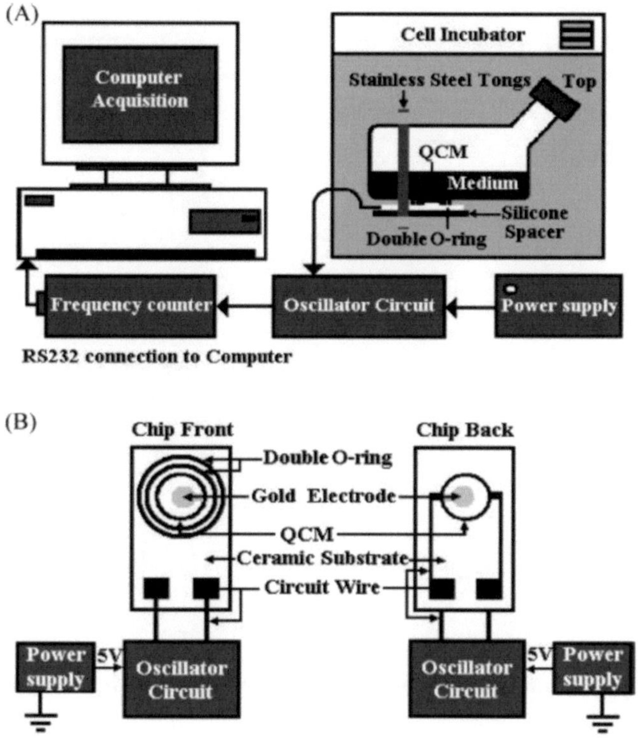

Fig. 2. Schematic diagram of the proposed QCM cell detection system. (A) Schematic diagram of the measurement system hardware setup used the QCM; (B) Schematic diagram of the QCM sensor chip.

3.2 Circuit description

3.2.1 Self-resonant circuit and amplitude capture

A Pierce oscillator uses an inverting amplifier so that its phase of 180° must be shifted to 360° through the feedback network, in order to fulfill the phase condition (Pierce, 1923). In our system, self-resonant circuit was designed with a 50% duty cycle crystal oscillator, based on the Pierce oscillator. A peak-detector can be applied as a sample and to be held as the amplitude of oscillator. We used two op-amps to amplify the differential input signals and a diode to hold the peak of amplitude. Fig. 3. Shows the circuit using two comparators (AD8564, Analog Devices) and one op-amp (MC33204, On Semiconductor). We designed a 50% duty cycle crystal oscillator. One comparator produces the resonance with quartz. The other just generates a reverse clock. The AD8564 is a quad 7 ns comparator. The fast 7 ns propagation delay makes a good choice for timing circuit. The MC33204 is a general operational amplifier. This op-amp offers a balance reference voltage. This circuit can generate an equal width square clock. The output frequency of square clock is up to 10 MHz and be examined in this experiment.

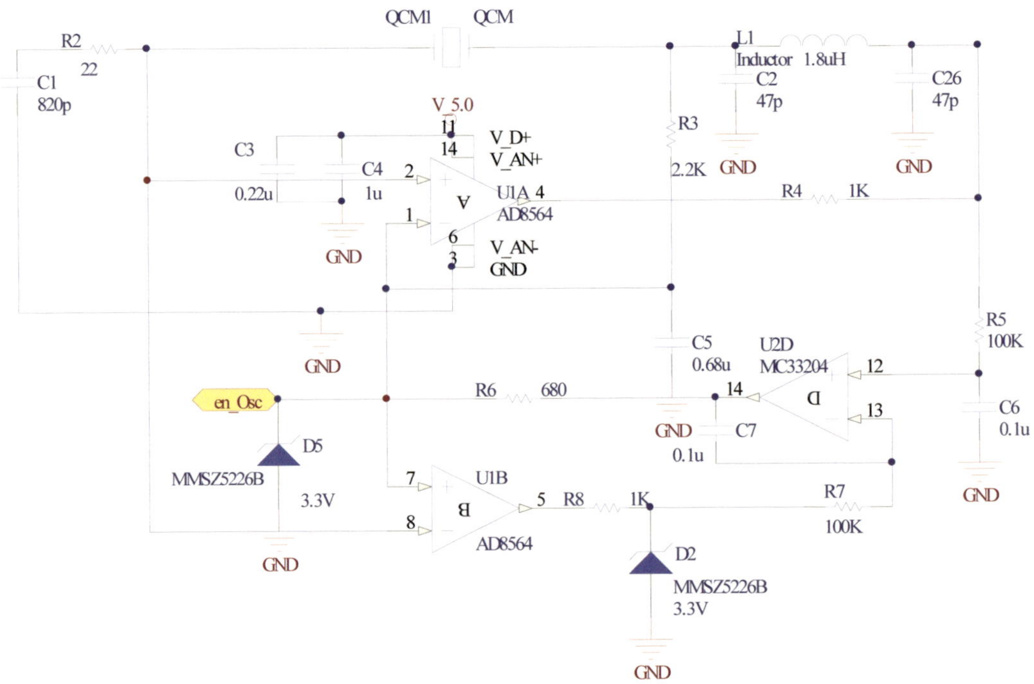

Fig. 3. A 50% duty cycle crystal oscillator design used two comparators and one op-amp.

3.2.2 Dual mode measurement system with frequency and impedance analysis

Accordingly, this topic directed to integrate these measurement materials of frequency, amplitude, and an impedance measurement system into a same measurement system. In an embodiment of this circuit, a voltage control oscillator (VCO) and a signal generator for the frequency scanning were included. The measurement circuit may generate output of the frequency scanning signal through a phase lock loop (PLL). Thus, the QCM system can simultaneously measure these properties of the object on QCM. Simple electrical impedance was described with measuring amplitudes of opposition to alternating current. A frequency generator (4-16 MHz) was built in this system, based on a phase-locked loop (PLL) circuit (TLC2933, TI). The TLC2933 was designed for PLL systems and was composed of a voltage-controlled oscillator (VCO) and an edge-triggered-type phase frequency detector. The circuit structure in this embodiment of this topic is further illustrated below in detail. Fig. 4 is a circuit diagram of a measurement system. The detailed circuits of the oscillation circuit (OC), the clock generation circuit (CGC), the enable circuit (EC), the bias circuit (BC), and the sample and hold circuit (SHC) can be obtained. The switch unit (SW) can be implemented as a two-to-one changeover switch or a multiplexer. When the SW switches to the frequency scanning signal (FSS), FSS output by the CGC is transmitted to the quartz (Q) and to be scanned through the SW, so as to scan the Q, followed by generating an output of the impedance sensing signal IMS at the other end of the Q. In addition, it should be noted that the clock output signal (COS) was a generated clock of QCM resonant by the OC and the amplitude output signal (AOS) was an amplified signal of IMS by the SHC.

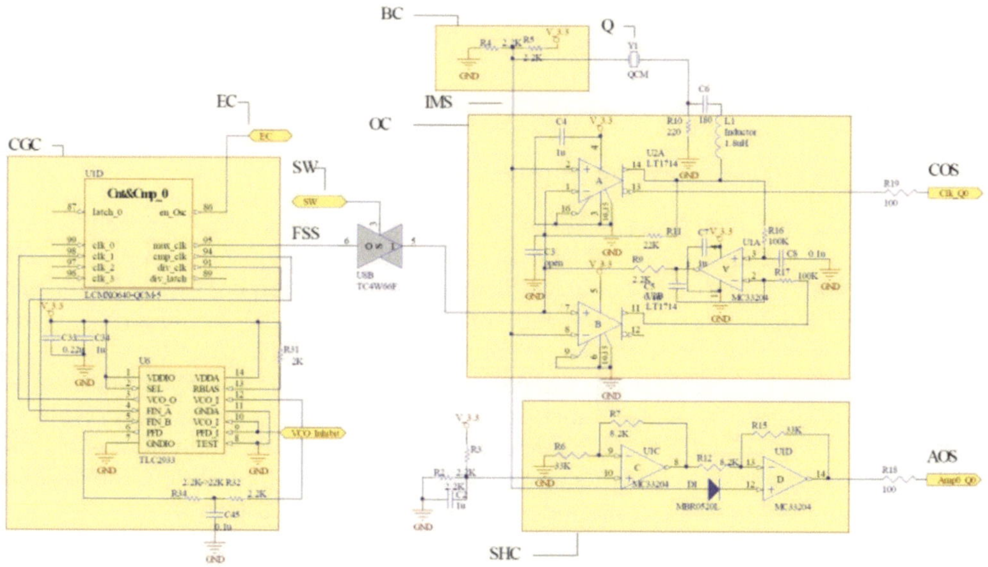

Fig. 4. Dual mode measurement system with frequency and impedance Analysis

3.2.3 Trans-epithelial resistance circuit

A cell measurement system, particularly a cell measurement system integrated with a QCM and a technique measuring TEER, is suitable to measure the changes of the frequency and the TEER. Currently, as the techniques are improved, QCMs are gradually used as a sensor in fields such as biological and medical science. Not only are the parameters of QCMs measured, the TEER can also be measured by ion flows between a QCM electrode and an external electrode in our system. A circuit diagram of the cell measurement system is shown in Fig. 5. The cell measurement includes: a power unit (PU), an oscillation module (OM), a frequency-monitoring module (FMM), a level-shift unit (LSU), a quartz crystal sensing module (QSM), a periodic wave-generation module (PWGM), a low-pass filtration module (LPFM), and a control module (CM).The PWGM includes a resistor R1 and a third electrode (E3), and the E3 contacts with a cell lines to be tested as TEER.

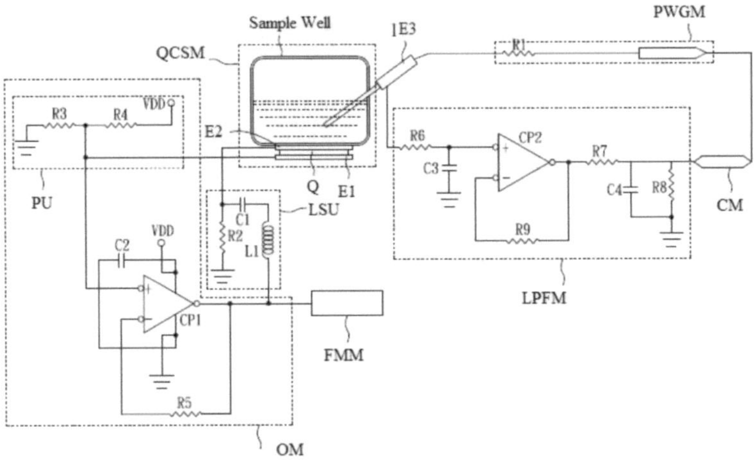

Fig. 5. A circuit of trnas-epithelia electric resistance Measurement

4. Characters of QCMs from cell monitoring

Early reports indicated that QCM is based on the absorption of target molecules on the crystal surface causing concomitant change of its resonant frequency (Fawcett et al., 1988). Therefore, QCM can be applied not only to chemical gas adsorption but also to biomolecular detection. When applied to biomolecular reactions, the QCM technique can detect pathogenic bacteria in food (Park & Kim, 1998) and plant viruses in plant tissue (Eun et al., 2002). Despite the many potential advantages of QCM, relatively few attempts have been made to extend QCM to the measurement of a living cell. For example, researchers have used QCM to examine cell adhesion (Khraiche et al., 2005), apoptosis (Braunhut et al., 2005), and shear-induced senescence (Jenkins et al., 2004) in the past. In addition, the QCM is a versatile and non-invasive in vitro method for real-time characterization of cells.

4.1 A silver electrode QCM using plasma treatment

4.1.1 Reason of using a silver electrode QCM

The QCM sensors in previous bio-experiments were typically ready-made gold electrode QCMs sold by specific manufacturers. These QCMs are not only expensive, but also lack standards. These is advantages make it hard for researchers to analyze the relationships between cell and chip parameters. However, normal industrial QCMs in the electronics industry market are significantly cheaper and come in more varieties. Moreover, the only difference among QCM chips manufactured by specific manufacturer is that some use a silver material electrode. This change in QCM electrode material causes three major flaws when using silver electrode QCMs in cell detection. First, in the cell culture process, the silver electrode peels off the sensor (Fig. 6A). Second, the general method of helping cells adhere to the electrode by a hydrophilic surface reaction with H2SO4 or NH3 solution product does not work with silver QCMs because silver can react with H2SO4 and NH3, causing the silver electrode to dissolve (Fig. 6B). Third, silver electrode raises potential biotoxicity issues. This study designs a protective film using a plasma-polymerized and photo-induced polymerization reaction process to achieve an optimum design. The resulting film not only prevents the silver electrode from peeling but also can wrap silver toxicity in a protective film. Using optimum photo-induced polymerization to increase the hydrophilic level of the surface after plasma-polymerized production film makes it possible to use a silver electrode QCM in cell detection.

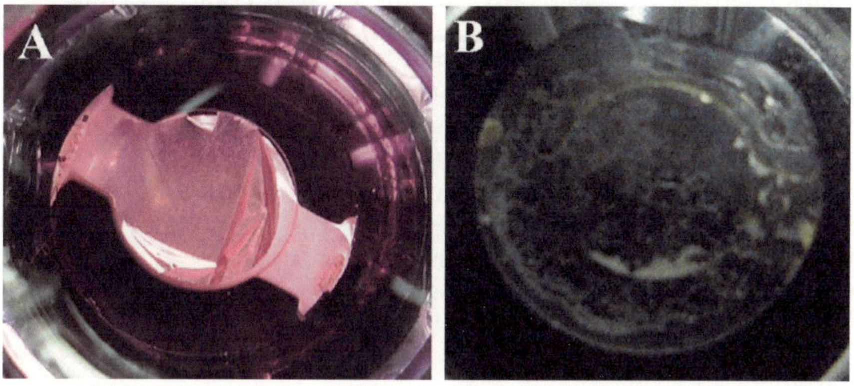

Fig. 6. (A) The silver electrode of the QCM was peeled off by Dulbecco's Modified Eagle's Medium. (B) The silver electrode was dissolved by NH3 solution.

4.1.2 Preparation of plasma-polymerized film and photo-induced polymerization

To devise a layer of protective film and solve the problem listed above, selecting a plasma-polymerized film. This is because plasma-polymerized organic layers have many advantageous properties, such as good adhesion to most substrates, excellent uniformity and thickness control, and no pin-hole formation. The elemental composition and physical and chemical properties of organic layers can be varied over a wide range because of their highly branched and cross-linked structures. Due to their biocompatible characteristics, organic films can accept a large amount of biological components, such as cells and tissues that can be loaded onto the surface of the organic layer (Jenkins et al., 2004).

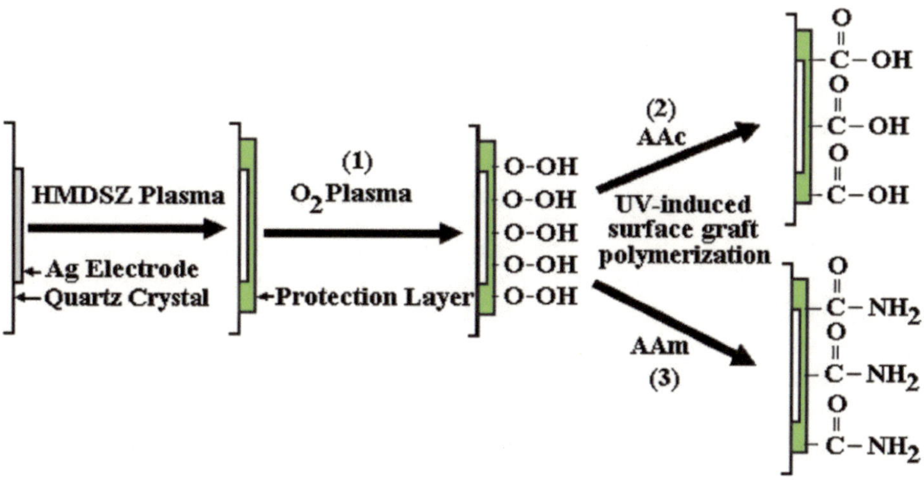

Fig. 7. The scheme of the experiment reaction process.

Fig. 7 depicts the modification method for binding a protective film and photo-induced polymerization on the QCM. The plasma system used in this study is the same as the one previously shown in (Chen et al., 2007). First, the QCM chip was placed in a RF plasma chamber after cleaning. Silver electrodes on the QCM surface were deposited in organic interlayer films as hexamethyldisilazane (HMDSZ) monomers using the plasma technique. For optimal reaction parameters of deposition, the RF plasma instrument used 100 mTorr of monomer with a plasma power of 30 W. This study tested the change on the QCM chip surface, using HMDSZ reaction times of 1, 3, 5, 7, and 10 min to form a protective film. After plasma deposition, the silver electrode QCM surface was made hydrophilic by oxygen plasma photo-induced polymerization using the traditional process (Chen et al., 2007; Elsom et al., 2008). This study presents three methods to increase surface hydrophilicity. The first method uses oxygen plasma 40 mTorr oxygen with a power of 100 Wand time of 1 min. After oxygen plasma deposition, two other types of hydrophilic surface methods were designed using surface-grafted acrylamide (AAm) and acrylic acid (AAc) 10% solution using 1000 W of UV light irradiation for 30 min (Fig. 8). Then AAm and AAc homopolymerwere removed from the grafted film by washing the QCM in distilled water three times and drying it in the air.

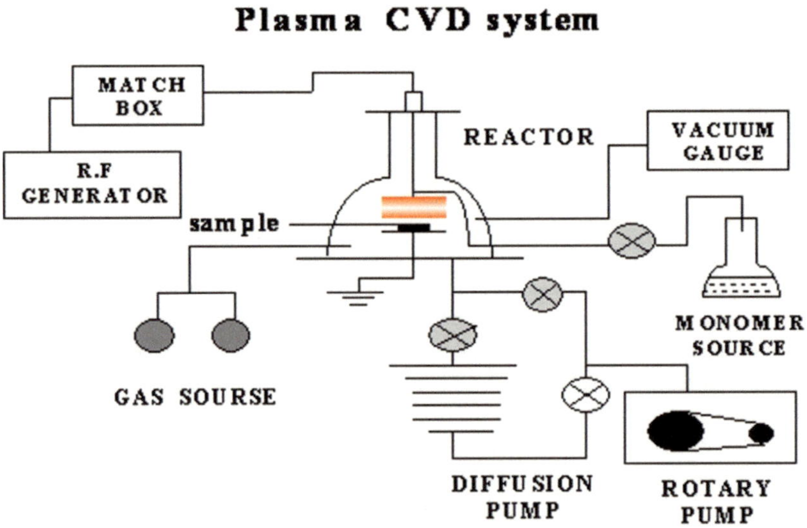

Fig. 8. The plasma treatment system with radio frequency

4.1.3 Measurement of cell adhesion by cell QCM platform

The silver QCM surface was first covered with a plasma polymerized film and photo-induced polymerization to render it resistant to solution and hydrophilic. The QCM surface was then cleaned by rinsing with distilled water and air drying. Next, the chip was sterilized by spraying it with ethanol and exposing it to UV light for 30 min. After combining the cell incubator chamber and the QCM sensor (Fig. 9), 500 ul medium was added to the incubator chamber to complete the connection of chip circuit line and oscillator circuit. When turning on the system, testing cells 1×10^5 were added until the system became stable. After achieving a stable signal, the cell incubator chamber was placed in a cell incubator to process the culture.

4.1.4 Results

Fig. 10 shows the results of applying a silver electrode QCM to cell detection. For untreated silver electrode QCMs, the frequency decreases quickly (Fig. 10A), and becomes worse after seeding the cells (Fig. 10B).The process of falling off decreases the chip's shaking ability, which causes the frequency to drop. Experiments with the HMDSZ protective film and photo-induced polymerization of AAm show that frequency remains steady in silver QCMs without adding cells (Fig. 10A). Adding cells simply caused the cells to grow on the chip surface (Fig. 10B). The frequency declines slowly at first, and then increases as the number of cells increases. After cell growth reaches a steady state, Δf also remains steady. As the medium nutrients run out, the quantity of dead cells increases, which cause the frequency to rise slowly. The same result appears in other papers utilizing gold QCMs (Zhou et al., 2000; Nimeri et al., 1998).

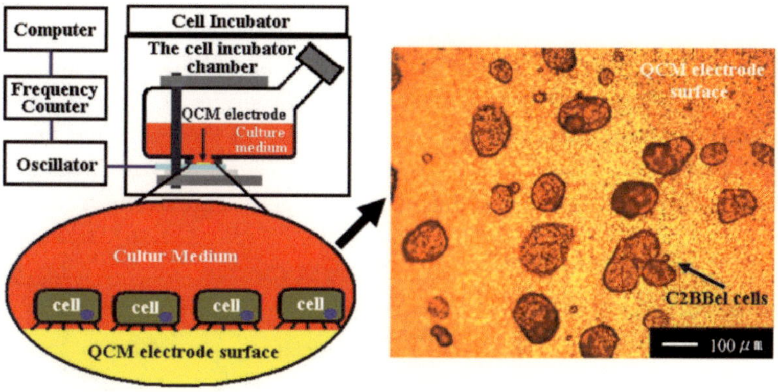

Fig. 9. Schematic diagram of the cell detection system hardware setup used for the QCM, and the condition of C2BBel cells growth on the QCM surface.

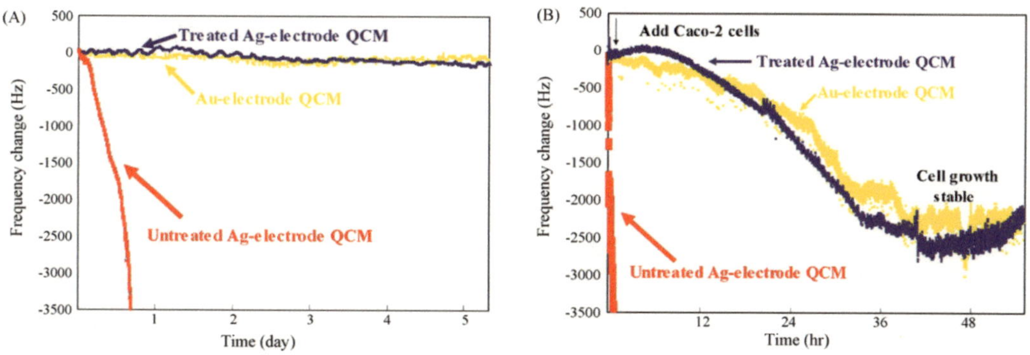

Fig. 10. (A) Frequency change with time of QCM in response to the gold electrode (positive control), the sliver electrode and the treated-sliver electrode QCM. (B) Added 1×10^6 Caco-2 cells on the gold electrode, sliver electrode and the treatment silver electrode QCM surface, respectively. The Au electrode QCM was gold electrode QCM for positive control group with no any others method to treatment, and the treated Ag electrodes were HMDSZ plasma-polymerization times at 10 min, oxygen plasma at 1 min, photo-induced polymerization of AAm. The data points for every 10 s and made a shift line.

4.2 Frequency response of QCM to detect the epithelial cell tight junction

4.2.1 QCM frequency characterization

We selected 9MHz AT-cut QCM (7mm diameter) sensors to be laid on a ceramic substrate. The fractional frequency change caused by the adsorption of the target molecules is estimated by the Sauerbrey (1959) equation that was modified by Jenkins et al. (2004) as shown in Eq. (7):

$$\frac{\Delta m}{A} = -(\mu\rho)^{0.5}\frac{\Delta f}{2\pi f_0^2} \tag{7}$$

where Δf is the change in frequency, f_0 is the resonance frequency of the crystal (9 MHz at 25 °C), Δm is the change of weight on the quartz crystal surface (g), and q is the density of quartz (2.648 g cm^{-3}). The m is the shear modulus of quartz (2.947x10^{11} dyn cm^{-2}); A is the electrode area (9.08 mm^2). The aforementioned formula shows that the electrode plays an important role in QCM detection. The QCM detection range is determined by the electrode surface area (Li et al. 2005). Nimeri et al. (1998) also mentioned that the highest and the lowest sensitivities occur at the center and the perimeter of the electrode, respectively. This implies that QCM detection should be focused on the center of the electrode. If the weight of electrode surface is constant, but liquid viscosity-density change in the solution above the

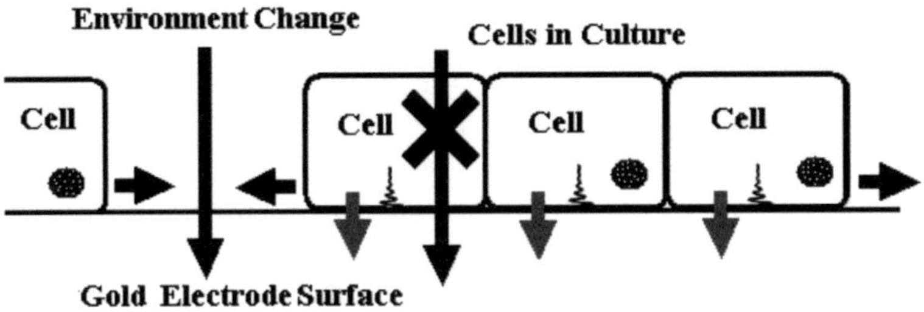

Fig. 11. Principle diagram of a QCM chip detecting Caco-2 cell tight junction permeability.

QCM crystal surface will also produce a frequency change. Talib et al. (2006) reported that different QCM solutions produce different signal frequencies. But there is only a constant distance at the electrode surface in which QCM can detect the change of solution. The QCM sensor in depth detection range decayed quickly at liquid because the QCM sensor shear wave was weakened in the solution. The QCM detection range from electrode surface up to the solution in the water is around 0.25 um, which represents only 1/10 of the thickness of a cell (Nimeri et al., 1998). This is why QCM cannot detect environmental changes for the top layer of a cell. Based on the points in the previous paragraph, Fig. 11 shows that when the cell grows on the QCM surface, the electrode can only interact with cells that adhere to the surface. Therefore, QCM cannot detect solution changes beyond the cells. The more cells growing on the electrode surface, the smaller the surface of the electrode that is exposed to the solution. Therefore, when if we change the cell solution in samples where there is no impact to the physiology of cells, at least in the short-term, only the parts of the electrode surface without cell adhesion can be detected by changing of solution. The consequence is that the changing of QCM surface solution reduces the total frequency change. When the growing cells develop a completely tight junction membrane, replacing the solution produces only minimal frequency changes. The aforementioned concept described is the foundation used to test the completeness of the tight junction membrane in this study.

4.2.2 Result

In our study, we measured Δf changes every three days during cell cultivation. After testing, the chip was removed and analyzed for cell growth by a metallographic microscope. The population of cells (black dots or area) increased as the time elapsed on the surface of the sensor. There are a few black spots shown after 3 d (Fig. 12A); but grows into colonies in 9 d (Fig. 12B), and finally covers the entire sensor surface in 15 d (Fig. 12C). These facts prove that

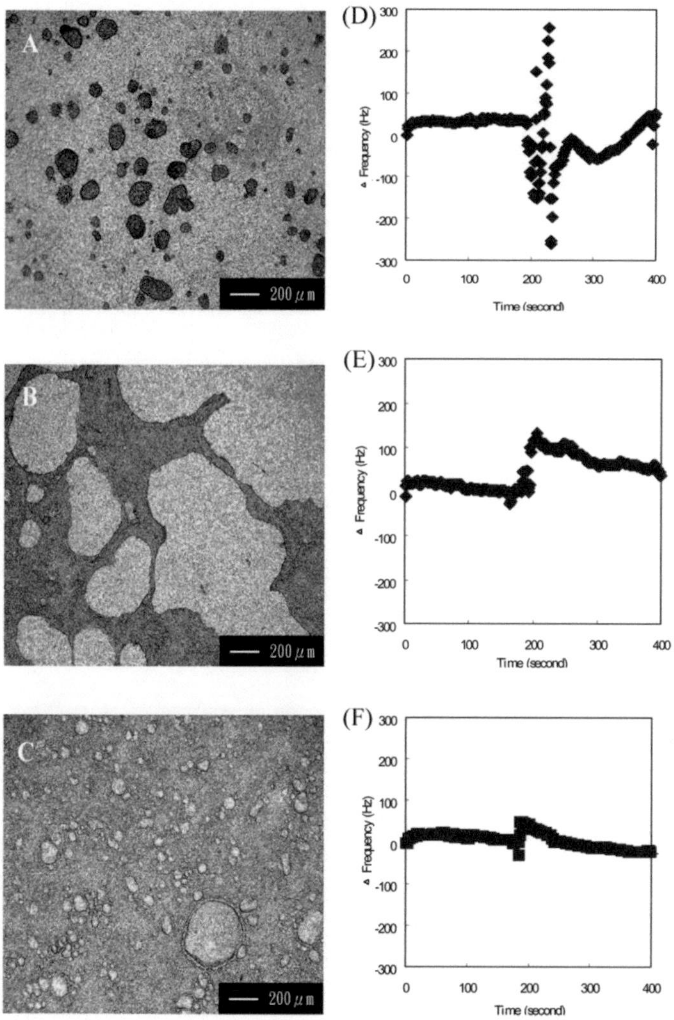

Fig. 12. Microscopic images of Caco-2 cell growth on the QCM cell sensor at (A) 3 days, (B) 9 days, (C) 15 days and measured frequency (Δf) change of Caco-2 cell growth on the QCM cell sensor at (D) 3 days, (E) 9 days, (F) 15 days.

cells can grow normally on the chip. The frequency profiles for cells cultured at 3, 9, and 15 d are shown in Fig. 12D, E, and F, respectively. The PBS was added to the chamber to allow the chip to react at 200 sec. As the cell growth quantity increased on the QCM surface, Df decreased when adding fixed quantities of PBS. Table 1 shows the amount of change for all reactions. These results demonstrate that the amount of reaction change for blank chips was about 258 Hz. As cell growth continued, the amount of frequency change decreased. The changing amount of frequency was less than 100 Hz after 12 d of incubation. After incubating for over 15 d, the amount of frequency change did not decrease further. Theoretically, when Caco-2 cell forms a tight junction structure, close cell linkage reduces the degree of frequency change to zero. In addition to system noise, frequency did not reach zero due to possible physical shock and cell response of adding PBS. In addition, the quantity of frequency change was not stable at 0 d. We consider that the chip was not stable at the beginning. Fig. 13 shows that cell growth increased over time while the frequency change and the TEER value decreased. The black line (Δ) represents the change of frequency results, and the gray line (○) represents the control group TEER values. The result shows that the QCM and TEER detection system gave similar results with respect to cell over time, and that the QCM method successfully detected cell tight junction integrality.

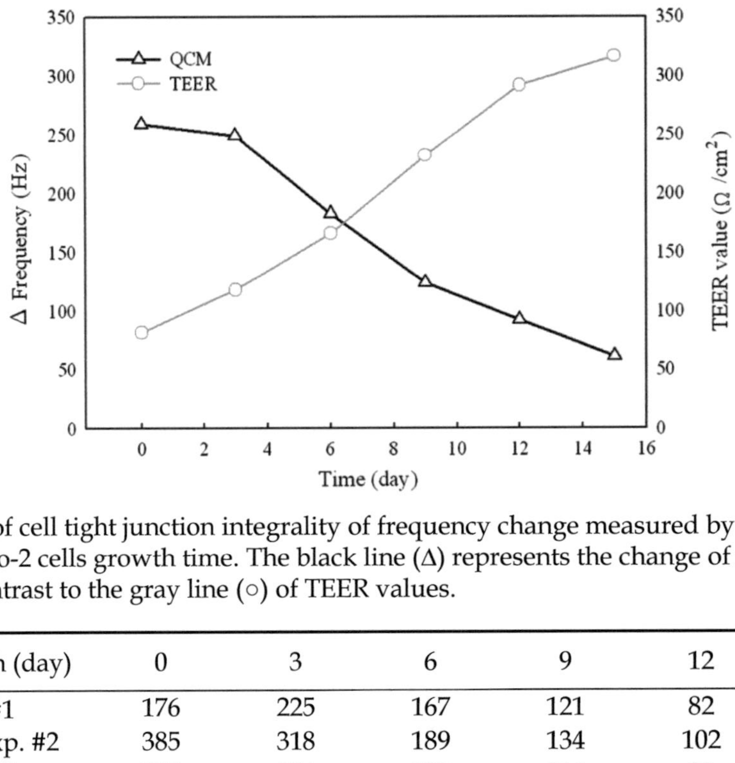

Fig. 13. Plot of cell tight junction integrality of frequency change measured by QCM and TEER vs.Caco-2 cells growth time. The black line (Δ) represents the change of frequency results in contrast to the gray line (○) of TEER values.

Cell growth (day)	0	3	6	9	12	15
Exp. #1	176	225	167	121	82	97
Δf (Hz) Exp. #2	385	318	189	134	102	42
Exp. #3	215	204	192	116	92	45
Average	258.7	249.0	182.7	123.7	92.0	61.3
S.D.*	111.1	60.7	13.7	9.3	10.0	30.9

*S.D: standard deviation.

Table 1. The frequency change ($\Delta f, Hz$) at different cell growth time (days)

4.3 Composite response of long-term cell monitoring from QCMs

4.3.1 Caco-2 cells adhesion state

In the long-term monitoring of the Caco-2 cells growth on the surface of the fabricated QCM-cell system, the frequency decreased down rapidly during the period from 24 to 48 hours and then turned into a stable state. The situation was observed by Khraiche et al. (2005). There were similar results in our system (Fig. 14). All data revealed that cells attached and grown well on the surface of the QCM. The frequency turned into a stable state

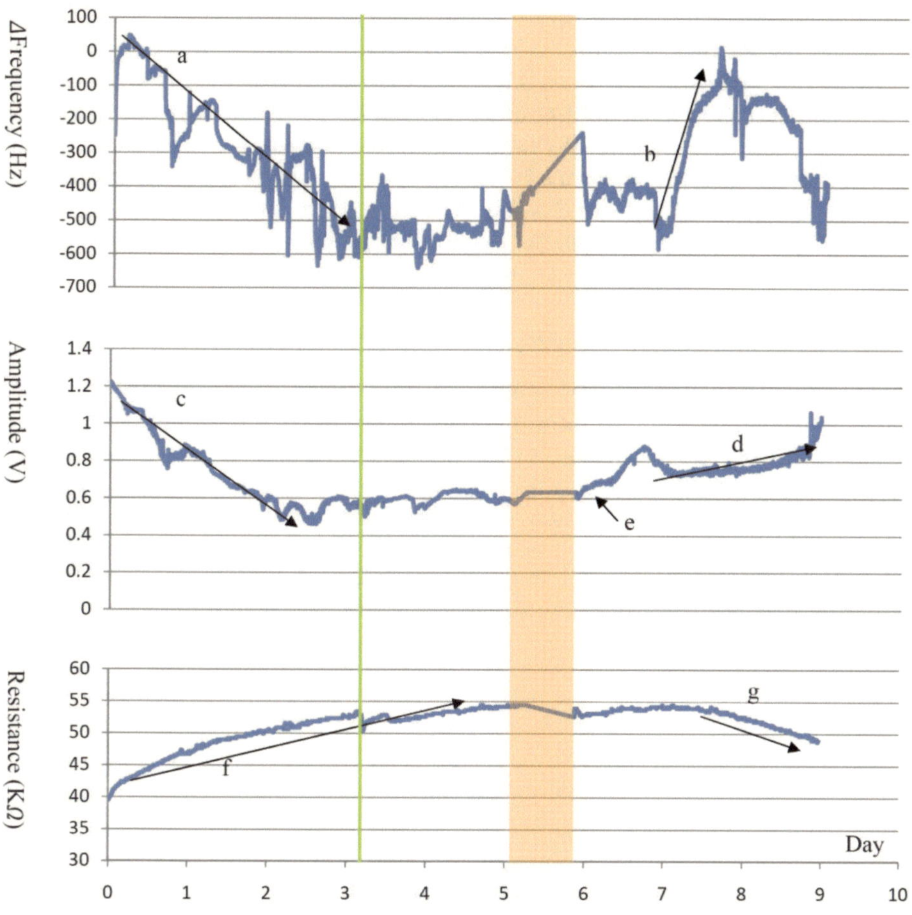

Fig. 14. Caco-2 cells growth condition was investigated using QCM.

after 3 days (arrow a: Δf=600 Hz) and the amplitude showed the leading trend for about 1 day (arrow c: amplitude was from 1.2 to 0.5 V). The TEER value continued to rise until the 5th day (arrow f: resistance was from 40 to 55 KΩ). The change of weight affecting the quartz resonant frequency and can be detected in the range for about 0.25 μm above the quartz surface in a solution state (Talib et al., 2006). The affected depth was only one-tenth of cells in thickness, the change of frequency can be detected with the cells grown on the surface when cells attached on QCM, and the change of frequency cannot monitor the process of cell metabolism. The amplitude showed the leading trend possibly caused by the complex weight and ions effect, and the change of amplitude seemed to represent a response of QCM that was not more sensitive enough than the change of frequency. Nonetheless, we have a more interesting discussion about the change of amplitude which will be mentioned later. The TEER value does not continue to rise after the sixth day, which was likely caused by stopping the replacement of the medium. Thus, the TEER value reflected the integration degree of the cell membrane and the intact of tight junction which we could see from the rational curve.

4.3.2 Caco-2 cells death process

We found a dramatic change after stopping the replacement of medium. The frequency rose quickly to 600 Hz within 24 hours after the 7th day. However, the amplitude increased slowly (arrow d: amplitude was from 0.7 to 0.8 V). The change of TEER value was lagged behind the change of frequency for about 12 hours (Fig. 14, arrow g: resistance was from 55 to 48 KΩ). The frequency response can be explicated by the reduced mass on the QCM surface. All means that the cells are in the process of death. We can propose the cells membrane rupture caused the TEER value decreased. We cannot explain the curve of the singular point. The sudden change in amplitude may be caused by changes in environmental change. The appearance may be a precursor of cell death.

4.3.3 Caco-2 cells impedance analysis

The impedance spectrum of a QCM sensor contains massive information in details that is particularly valuable in sensor applications (Sun, et al., 1996). Measurement of the change in the resonant frequency (8.9~9.1 MHz) of the quartz crystal and the damping of its vibration can be extracted from the electrical impedance spectrum. Fig. 15 shows the sample of Caco-2 cells impedance analysis. From day 0 to day 5, we found the peak drop ping down because of the energy dissipation of QCM, the result was reasonable. The Caco-2 cells grew and attached on the surface of quartz, causing the energy loss of oscillation. From day 7 to day 9, the dissipation was reduced implying the loading attached was reduced on the surface of QCM. This could possibly be explained as the evidence of the Caco-2 cells gradually dying.

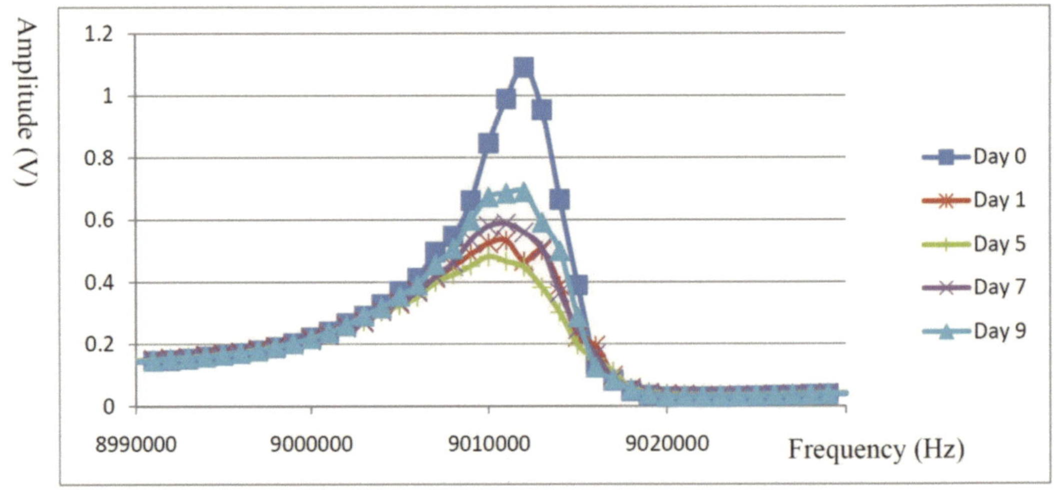

Fig. 15. Caco-2 cells impedance analysis.

4.3.4 Caco-2 cells treated with trypsin-EDTA

One of our objectives was to design a cell-based QCM as a drug screening tool. Traditional biosensors are important and powerful tools in the drug discovery process. However, biosensors are limited to using a purified target molecule immobilized to the sensor surface. Cell-based QCM can simplify the process of biological mechanism in vivo situation. A Trypsin-EDTA (0.05% Trypsin, 0.53 mM EDTA-4Na) solution was used widely for dissociation of tissues and cell monolayer. Therefore, we used trypsin-EDTA as a poison and performed this simple test after cells grew on the QCM on day 6. There were three formulations as shown in the follows:

Formula 1: Trypsin-EDTA, 3 ml. (100% v/v of Trypsin-EDTA)

Formula 2: Trypsin-EDTA, 2 ml. + DMEM, 1 ml (66.7% v/v of Trypsin-EDTA)

Formula 3: Trypsin-EDTA, 1.5 ml + DMEM, 1.5ml (50% v/v of Trypsin-EDTA)

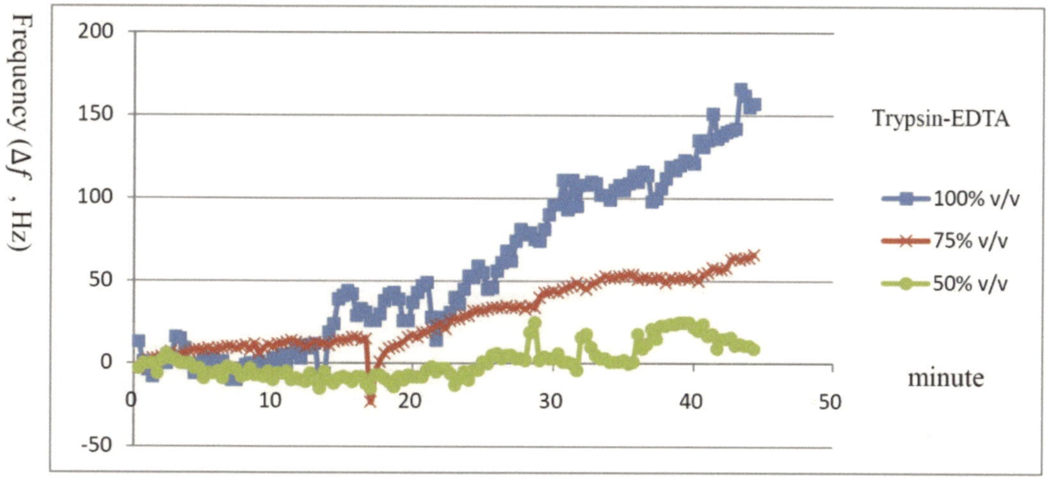

Fig. 16. Shows the change of frequency of QCM after the Caco-2 cells were treated with trypsin-EDTA.

After adding trypsin-EDTA, the Caco-2 cells began to detach from the golden-film of QCM, which made the frequency rapid increase as we predicted (Fig. 16). When Caco-2 cell medium was replaced by formula 1, a maximum frequency increasing variation ($\Delta f \cong 120 Hz$) took place and second one was shown in formula 2 ($\Delta f \cong 120 Hz$) within the period of 40 minutes. We speculated the increase of frequency influenced by the weight loss of cells on the electrode surface. Of course, the density and viscosity decreasing also can influence the resonance of quartz. Formula 3 shows only a little impact exigency ($\Delta f \cong 25 Hz$), which could be neutralized by DMEM. Nonetheless, the similar result also appeared in the change of amplitude (Fig. 17). We can observed some difference change in Δf and ΔA. There exist a vigorously curve ($\Delta A \cong -0.2$ to $0.62 V$) at the beginning 20 minutes after application of formula 1. And there were similar upward trend lines after treated with formula 2 and 3. Formula 1, the Caco-2 cells were separated and isolated, causing cell-tight junction associated fracture. Formula 2 & 3, the Caco-2 cells still remained intact cell tight junction. This may be the result of the complexity indifferent ion concentration and cells junction permeability as we expected. Table 2 shows the Caco-2 cells survival testing by MTT assay (570 nm wavelength). We assume the Caco-2 cells survival rate is 100% which is to be treated with Formula 3 (OD: 1.172). Therefore the Caco-2 cells survival rate treated with formula 1 (OD: 0.889) and 2 (OD: 0.938) were 76% and 80%. We believe these results were dissimilar to our experiment and we could not understand these Caco-2 cells likely to represent a good health condition or not. However we can illustrate the system by measuring amplitude and frequency of QCM which provide different biological detection in vivo.

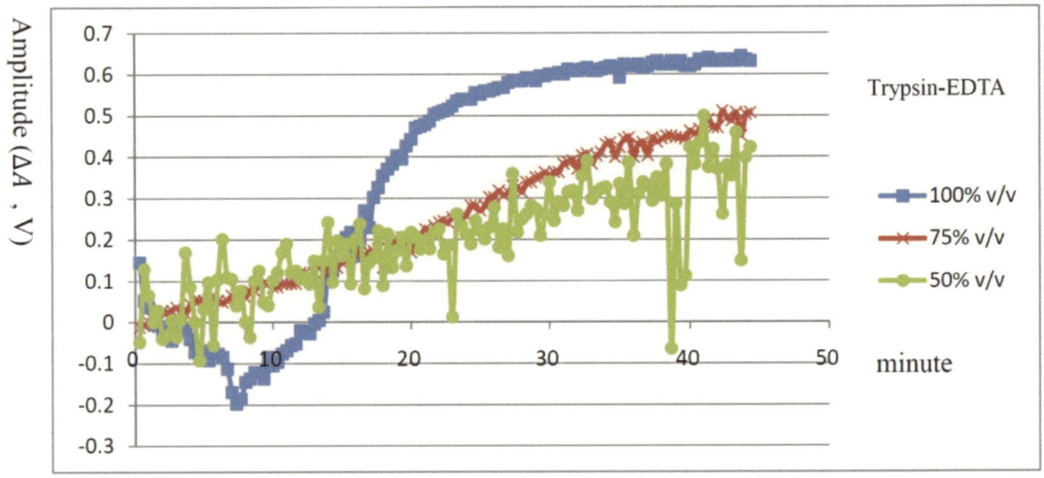

Fig. 17. Shows the change of amplitude of QCM after the Caco-2 cells were treated with trypsin-EDTA.

Trypsin-EDTA	1	2	3	Mean	Survive (%)
100% v/v	0.917	0.860	0.889	0.889 ± 0.028	76%
66.7% v/v	0.971	0.924	0.919	0.938 ± 0.029	80%
50% v/v	1.160	1.174	1.182	1.172 ± 0.011	100%
Blank	0.133	0.128	0.125	0.129 ± 0.004	---

Table 2. Survival rate of the Caco-2 cells measured by MTT assay.

5. Conclusion

The QCM cell detection system was successfully designed in our laboratory. This system proves the model of Caco-2 cell growth. Using the HMDSZ of plasma-polymerized film and photo-induced polymerization of AAm is a feasible way to improve the hydrophilicity of silver electrodes. The proposed modification method involves binding a protective film. The longer the plasma-polymerized reaction time, the greater the film deposition and thickness have. An increased film thickness extends the film protection effect of sliver electrode, the HMDSZ film extends the unpeeling time of silver electrodes from 1 to 7 days. The cell growth experiments in the current study show no difference between amine groups or hydroxyl groups. However, the AAm-treated electrode offered better protection than the

AAc-treated electrode. In summary, the experiment results of this study prove the feasibility of using a treated silver electrode QCM to detect cells.

During the process of cell growth and tight junction formation, this system detected frequency changes based on QCM reactions with different solutions. Results suggest that cells stay close together when the amount of frequency change is below 100 Hz. The method is comparable to, but more versatile and convenient than, the TEER. The result showed that the QCM and TEER detection system gave similar results with respect to cells over time. This detection system makes it possible to detect the tight junction of Caco-2 cells and can be used for other cell detection in the future.

We developed a cell-based QCM measurement system for long-term monitoring cell growth in this study. This system provides an investigation of cell adhesion, proliferation, and death process. The cell growth conditions, medium composition and other factors will influence the QCM oscillation frequency, amplitude, and TEER value. To observe cell growth, prior studies showed that the crystal oscillation frequency can reflect the surface quality changes only for the cell membrane level. We investigate a trend of the changes in medium composition, cells metabolic, or intact of cell tight junction. The QCM resonance frequency and amplitude, the trans-epithelial electric resistance, and the electric impedance analysis, all illustrate the state of the cells in this study. Although these mechanisms are still uncertain, we believe that they could provide a method to develop a cell measurement technology for long-term observation of Caco-2 cell growth. This QCM-cell system opens QCM technology application fields and facilitates the researchers to follow-up screening and conducive for drugs development of related fields in vivo.

6. Acknowledgements

We gratefully acknowlede the finicial support of the National Science Council, Taiwan (NSC 98-2221-E-036-023-MY2 and NSC 100-2221-E-036-035).

7. References

Auge, J.; Hauptmanna, P.; Eichelbaumb, F. & Rosler, S. (1994). Quartz crystal microbalance sensor in liquids, *Sens. Actuators B: Chem.*, 18, pp. 518-522.

Bard, A.J. & Faulkner, L.R. (2001). Electrochemical Method. Fundamentals and Applications, second ed. Wiley, New York, 2001. *Biosens. Bioelectron,* 23, 1259-68.

Bruckenstein, S. & Shay, M. (1985). Experimental aspects of use of the quartz crystal microbalance in solution. *Electrochim. Acta,* 30, pp. 1295-1300

Buttry, D.A. & Ward, M.D. (1992). Measurement of interfacial processes at electrode surfaces with the electrochemical quartz crystal microbalance. *Chem. Rev.,* 92, pp. 1355-1379.

Braunhut, S.J.; McIntosh, D.; Vorotnikova, E.; Zhou, T. & Marx, K.A. (2005). Detection of apoptosis and drug resistance of human breast cancer cells to taxane treatments using quartz crystal microbalance biosensor technology. *Assay Drug Dev. Technol.,* 3, pp. 77-89

Cady, W.G. (1946). Piezoelectricity, *McGraw-Hill,* New York and London.

Chou, H.C.; Yan, T.R. & Chen, K.S. (2009). Detecting cells on the surface of a silver electrode quartz crystal microbalance using plasma treatment. *Colloids and Surfaces B: Biointerfaces,* 73, pp.244–249

Chou, H.C.; Yan, T.R., & Lee, C.F. (2009). Matters needing attention for applying the quartz crystal microbalance technique to detect the cell morphology. *BIOMED. ENG-APP. BAS. C.*, 21, pp. 415-420.

Chou, H.C.; Yan, T.R. (2010). Applying the quartz crystal microbalance technique to detect the epithelial cell tight junction integrality of Caco-2 cells. *Anal. Lett.*, 43, pp. 2009~2018

Curie, P. & Curie, J. (1880). Développement, par pression, de l'électricité polaire dans les cristaux hémièdres à faces inclines. *Acad. Sci.*, 91, pp. 294-295.

Chen, K.S.; Chen, S.C.; Lin, H.R.; Yan T.R. & Tseng, C.C. (2007). A novel technique to immobilize DNA on surface of a quartz crystal microbalance by plasma treatment and graft polymerization. *Mater. Sci. Eng. C: Biomim. Supramol. Syst.*, 27, pp. 716.

Elsom, J.; Lethem, M.I.; Rees, G.D. & Hunter A.C. (2008). Novel quartz crystal microbalance based biosensor for detection of oral epithelial cell–microparticle interaction in real-time, *Biosens. Bioelectron.*, 23, pp. 1259.

Edvardsson, M.; Rodahl, M. & Hook, F. (2006). Investigation of binding event perturbations caused by elevated QCM-D oscillation amplitude, *Analyst.*, 131, pp. 822-828.

Eun, A.J.C.; Huang, L.; Chew, F.T.; Li, S.F.Y. & Wong, S.M. (2002). Detection of two orchid viruses using quartz crystal microbalance (QCM) immunosensors. *J. Virol. Methods*, 99, pp. 71-79

Fawcett, N.C.; Evans J.A.; Chien, L.C. & Flowers, N. (1988). Nucleic acid hybridization detected by piezoelectric resonance. *Anal. Lett.*, 21, pp. 1099-1114

Fredriksson, C.; Khilman, S.; Kasemo, B. & Steel, D.M. (1998a). In vitro real-time characterization of cell attachment and spreading. *J. Mater. Sci.: Mater. Med.*, 9 (12), pp. 785–788.

Fredriksson, C.; Kihlman, S.; Rodahl, M. & Kasemo, B. (1998b). The piezoelectric quartz crystal mass and dissipation sensor: a means of studying cell adhesion. *Langmuir*, 14, pp. 248–251.

Giaever, I. & Keese, C.R. (1991). Micromotion of mammalian cells measured electrically. *Proc. Natl. Acad. Sci. USA*, 88, pp. 7896–7900.

Giaever, I. & Keese, C.R. (1993). A morphological biosensor for mammalian cells. *Nature*, 366, pp. 591–592.

Gryte, D.M.; Ward, M.D. & Hu, W.S. (1993). Real-time measurement of anchorage-dependent cell adhesion using a quartz crystal microbalance. *Biotechnol. Prog.*, 9, pp. 105-108

Guilbault, G.G. (1983). Determination of formaldehyde with an enzyme coated piezoelectric crystal. *Anal. Chem.*, 55, pp. 1682-1684.

Hao, R.Z.; Song, H.B & Zuo, G.M. (2011). DNA probe functionalized QCM biosensor based on gold nanoparticle amplification for *Bacillus anthracis* detection. *Biosens. Bioelectron.*, 26, pp. 3398–3404

Itamar, W.; Patolsky, F. Weizmann, Y. & Willner, B. (2002). Amplified detection of single-base mismatches in DNA using microgravimetric quartz-crystal-microbalance transduction. *Talanta*, 56, pp. 847–856.

Jenkins, M.; Horsfall, M.; Mathew, D.; Scanlon, M.; Jayasekara, R. & Lonergan, G. T. (2004). Application of a quartz crystal microbalance to evaluate biodegradability of starch by *Bacillus subtilis*. *Biotechnol. Lett.*, 26, pp. 1095–1099.

Johannsmann D. (2008). "Viscoelastic, mechanical, and dielectric measurements on complex samples with the quartz crystal microbalance". *Physical Chemistry, Chemical Physics*, 10, pp. 4516–4534.

Kanazawa, K.K. & Gordon, J.G. (1985). The oscillation frequency of a quartz resonator in contact with a liquid. *Anal. Chim. Acta*, 175, pp. 99–105

Khraiche,M.L.; Zhou, A. & Muthuswamy, J. (2005). Acoustic sensor for monitoring adhesion of Neuro-2A cells in real-time, J. *Neurosci. Methods*, 144, pp. 1–10.

King, W.H. (1964). Piezoelectric sorption detector, *Anal. Chem.*, 36, pp. 1735-1739.

Konash, P.L. & Bastiaans, G.J. (1980). Piezoelectric crystals as detectors in liquid chromatography. *Anal. Chem.*, 52, pp. 1929-1931

Lee, C.F; Yan, T. R. & Chou, H.C. (2009). Improve the data acquisition system of a qcm sensor by increasing the sampling rate of frequency and amplitude. *BIOMED. ENG-APP. BAS. C.*, 21 6, pp. 405–410

Lee, C.F.; Yan, T.R.; Chou, H.C. & Lin, Z.Y. (2010). A quartz crystal microbalance cell-culture incubator system. *2010 INTERNATIONAL INNOVATION AND INVENTION CONFERENCE*, pp.42~48, Taipei, Taiwan, September 26~27, 2010

Li, J.; Thielemann, C.; Reuning, U. & Johannsmann, D. (2005). Monitoring of integrin-mediated adhesion of human ovarian cancer cells to model protein surfaces by quartz crystal resonators: evaluation in the impedance analysis mode. *Biosens Bioelectron.*, 20, pp. 1333-1340

Lippmann, G. (1881). Principe de conservation de l'électricité. *Anal. Chim. Phys.*, 24, pp. 145-178.

Lu, C.S. & Lewis, O. (1972). Investigation of film-thickness determination by oscillating quartz resonators with large mass load. *J. Appl. Phys.*, 43, pp. 4385-4390.

Martin, S.J.; Granstaff, V.E.& Frye, F.C. (1991). Characterization of a quartz crystal microbalance with simultaneous mass and liquid loading. *Anal. Chem.*, 63, pp. 2272-2281.

Mirsky, V.M.; Riepl, M. & Wolfbeis, O.S. (1997). Capacitive monitoring of protein immobilization and antigen-antibody reactions on monomolecular alkylthiol films on gold electrodes, *Bios. Bioelec.*, 12, pp. 977–989.

Muramatsu, H.; Kijiwara K., Tamiya E., & Karube I. (1986). Piezoelectric immuno sensor for the detection of *Candida albicans* microbes. *Anal. Chim. Acta*, 188, pp. 257-261

Muramatsu, H.; Tamiya, E. & Karube I. (1988). Computation of equivalent circuit parameters of quartz crystals in contact with liquids and study of liquid properties, *Anal. Chem.*, 60, pp. 2142-2146.

Muratsugu, M.; Romanschin, A.D. & Thompson, M. (1997). Adhesion of human platelets to collagen detected by Cr51 labelling and acoustic wave sensor, *Anal. Chim. Acta*, 342, pp. 23–29.

Marx, K.A.; Zhou, T.; Warren M. & Braunhut, S.J., (2003). Quartz crystal microbalance study of endothelial cell number dependent differences in initial adhesion and steady-state behavior: evidence for cell-cell cooperatively in initial adhesion and spreading. *Biotechnol. Prog.* 19 3, pp. 987–999.

Marx, K.A.; Zhou, T.; Montrone, A.; McIntosh, D. & Braunhut, S.J. (2007). A comparative study of the cytoskeleton binding drugs nocodazole and taxol with a mammalian cell quartz crystal microbalance biosensor: Different dynamic responses and energy dissipation effects. *Anal. Biochem.*, 361, pp. 77–92

Nimeri, G.; Fredriksson, C.; Elwing, H.; Liu, H.; Rodahl, M. & Kasemo, B. (1998). Quartz crystal microbalance-based measurements of shear-induced senescence in human embryonic kidney cells. *Surf. B: Biointerf.*, 11, pp. 225.

Okahata, Y.; Kimura, K. & Ariga, K. (1989). Detection of the phase transition of Langmuir-Blodgett films on a quartz-crystal microbalance in an aqueous phase. *J. Am. Chem. Soc.*, 111 26, pp. 9190–9194.

Park, I.S. & Kim, N. (1998). Thiolated salmonella antibody immobilization onto the gold surface of piezoelectric quartz crystal. *Biosens Bioelectron.*, 13, pp. 1091-1097

Pierce, G.W. (1923). Piezoelectric crystal resonators and crystal oscillators applied to the precision calibration of wavemeters. *Proc. Am. Acad. Arts Sci.*, 59, pp. 81-106.

Redepenning, J.; Schlesinger T.K.; Mechalke E.J.; Puleo D.A. & Bizios R. (1993). Osteoblast attachment monitored with a quartz crystal microbalance. *Anal Chem.*, 65, pp. 3378-3381

Rodahl, M.; Hook, F.; Krozer, A.; Brzezinski, P. & Kasemo, B. (1995). Quartz crystal microbalance setup for frequency and Q–factor measurements in gaseous and liquid environments. *Rev. Sci. Instr.*, 66, pp. 3924-3930.

Rodahl, M., B. Kasemo. (1996a). A simple setup to simultaneously measure the resonant frequency and the absolute dissipation factor of a quartz crystal microbalance. *Rev Sci Instr*, 67, pp. 3238-3241

Rodahl, M. & Kasemo, B. (1996b). Frequency and Dissipation-factor response to localized liquid deposits on a QCM electrode. *Sensors and Actuator B*, 37, pp. 111-116.

Sauerbrey, G. (1959). The use of quartz oscillators for weighing thin layers and for microweighing. *Z. Physik*, 155, pp. 206-222.

Shumacher, R. (1990). The quartz crystal microbalance: a novel approach to the in situ investigation of interfacial phenomena at the solid/liquid junction, *Angew. Chem. Int. Ed.*, 29, pp. 329–338.

Su, X.L. & Li, Y. (2004). An automatic quartz crystal microbalance immunosensor system for salmonella detection. *ASAE Annual Meeting:* 047043.

Sun, H.T.; Faccio, M.; Cantalini, C. & Pelino, M. (1996). Impedance analysis and circuit simulation of quartz resonator in water at different temperatures. *Sens. Actuators B*, 32, pp. 169-173.

Tan, L.; Jia, X.; Jiang, X.F.; Zhang, Y.Y.; Tan, H.; Yao, S.Z. & Xie, Q.G. (2008). Real-time monitoring of the cell agglutination process with a quartz crystal microbalance. *Anal. Biochem.*, 383, pp. 130–136.

Talib, Z.A.; Baba, Z.; Kurosawa, S. & Sidek, H.A.A. (2006). A. Kassimb and W.M.M. Yunus, Frequency behavior of a quartz crystal microbalance (QCM) in contact with selected solutions. *Am. J. Appl. Sci.*, 3, pp. 1853-1858.

Ward, M.D. & Buttery, D.A. (1990). In situ interfacial mass detection with piezoelectric transducers. *Science 249*, pp. 1000-1007.

Wegener, J.; Janshoff A. & Galla, H.J. (1998). Cell adhesion monitoring using a quartz crystal microbalance: comparative analysis of different mammalian cells lines. *Eur. Biophys. J.*, 28, pp. 26–37.

Wudy, F.; Multerer, M.; Stock, C.; Schmeer, G. & Gores, H.J. (2008). Rapid impedance scanning QCM for electrochemical applications based on miniaturized hardware and high-performance curve fitting. *Electrochim. Acta.*, 53, pp. 6568–6574.

Zhou, T.; Marx, K.A.; Warren, M.; Schulze, H. & Braunhut, S.J. (2000). The quartz crystal microbalance as a continuous monitoring tool for the study of endothelial cell surface attachment and growth. *Biotechnol. Prog.*, 16, pp. 268.